Contemporary Clinical Neuroscience

Series Editor
Mario Manto, Division of Neurosciences, Department of Neurology
CHU-Charleroi, Charleroi, Belgium, University of Mons, Mons, Belgium
Charleroi, Belgium

Contemporary Clinical Neurosciences bridges the gap between bench research in the neurosciences and clinical neurology work by offering translational research on all aspects of the human brain and behavior with a special emphasis on the understanding, treatment, and eradication of diseases of the human nervous system. These novel, state-of-the-art research volumes present a wide array of preclinical and clinical research programs to a wide spectrum of readers representing the diversity of neuroscience as a discipline. The book series considers proposals from leading scientists and clinicians. The main audiences are basic neuroscientists (neurobiologists, neurochemists, geneticians, experts in behavioral studies, neurophysiologists, neuroanatomists), clinicians (including neurologists, psychiatrists and specialists in neuroimaging) and trainees, graduate students, and PhD students.

Volumes in the series provide in-depth books that focus on neuroimaging, ADHD (attention deficit hyperactivity disorder and other neuropsychiatric disorders, neurodegenerative diseases, G protein receptors, sleep disorders, addiction issues, cerebellar disorders, and neuroimmune diseases. The series aims to expand the topics at the frontiers between basic research and clinical applications. Each volume is available in both print and electronic form.

Elizabeth A. Thomas • Georgia M. Parkin
Editors

Biomarkers for Huntington's Disease

Improving Clinical Outcomes

Springer

Editors
Elizabeth A. Thomas
Department of Neurobiology and Behavior
University of California
Irvine, CA, USA

Georgia M. Parkin
Department of Neurosciences
University of California
San Diego, CA, USA

ISSN 2627-535X ISSN 2627-5341 (electronic)
Contemporary Clinical Neuroscience
ISBN 978-3-031-32814-5 ISBN 978-3-031-32815-2 (eBook)
https://doi.org/10.1007/978-3-031-32815-2

This Springer imprint is published by the registered company Springer Nature Switzerland AG
The registered company address is: Gewerbestrasse 11, 6330 Cham, Switzerland

Paper in this product is recyclable.

Preface

Huntington's disease (HD) is a fatal, inherited, neurodegenerative disorder, characterized by chorea, other motor instabilities, psychiatric manifestations, and cognitive decline. George Huntington, an American physician, was the first person to provide a comprehensive description of adult-onset HD in 1872, including its autosomal-dominant pattern of inheritance. The HD gene, Huntingtin (*HTT*), was the first disease-associated gene to be mapped to a human chromosome, in 1983. Ten years later, researchers identified the disease mutation: an expanded triplet nucleotide repeat within exon 1 of the *HTT* gene. Although genetic testing for the *HTT* mutation is now straightforward, enormous variability can exist over the timing of disease onset, severity of symptoms, and course of disease progression; hence, biomarkers to track symptom onset and progression are greatly needed in the field. Early genetic testing provides an opportunity for clinical interventions aimed at delaying onset and/or slowing progression of disease; however, current treatments for HD are limited, with only two FDA-approved drugs available to manage chorea. Encouragingly, however, several disease-modifying treatment approaches are in the therapeutic pipeline, with more than 200 clinical studies, and many more preclinical studies, in the works. As the number and complexity of clinical trials increases, so must accompanying biomarker options, especially in the context of identifying premanifest individuals who may be near or within optimal treatment windows. This book represents the first volume focused solely on biomarkers for HD and offers a distinct resource that is informative, not only for clinicians and those involved in clinical trial design, but also for researchers, physicians, and healthcare professionals. This edited volume is written by top leaders in the field and takes an interdisciplinary approach to cover a broad spectrum of biomarker types, in order to provide the latest advances in the development of biochemical, molecular, imaging, and digital biomarkers that have been investigated for HD. With the ultimate goal of improving patient care and outcomes, the development of disease-associated biomarkers has never been more important for the field.

Department of Neurobiology and Behavior
University of California, Irvine, CA, USA
Elizabeth A. Thomas

Department of Neurosciences
University of California, San Diego, CA, USA
Georgia M. Parkin

Acknowledgments

This work would not be possible without the time, effort, and patience of all individuals with Huntington's disease, and their families. Clinical research studies take time, and the provision of biofluid samples – blood, saliva, and particularly cerebrospinal fluid – as well as quantitative neuroimaging can cause discomfort for the patients. Outside of biomarker collection, clinical research study visits may involve cognitive, motor, functional, and psychiatric assessments, which can be time-consuming and sometimes exhaustive. We are deeply grateful for this commitment.

We would also like to acknowledge and respect the animal lives sacrificed for the advancement of research, as well as the many hours of work dedicated by researchers toward all studies and findings included in this book.

It is our sincere hope that this book will assist in the collaborative advancement of Huntington's disease biomarker research, including the use of biomarkers as prognostic and therapeutic markers, thereby giving back to all families who have dedicated their time and samples to us.

Contents

Contributors

Jamie L. Adams Department of Neurology, University of Rochester, Rochester, NY, USA

Center for Health + Technology, University of Rochester, Rochester, NY, USA

Hanadi Ananbeh Laboratory of Applied Proteome Analyses and Research Center PIGMOD, Institute of Animal Physiology and Genetics of the Czech Academy of Sciences, Libechov, Czech Republic

Daniele Bertoglio Molecular Imaging Center Antwerp (MICA), University of Antwerp, Antwerp, Belgium

µNeuro Research Centre of Excellence, University of Antwerp, Antwerp, Belgium

Bio-Imaging Lab, University of Antwerp, Antwerp, Belgium

Maria Björkqvist Brain Disease Biomarker Unit, Wallenberg Neuroscience Center, Department of Experimental Medical Science, Lund University, Lund, Sweden

Kateřina Budková Laboratory of Applied Proteome Analyses, Institute of Animal Physiology and Genetics, Czech Academy of Sciences, Libechov, Czech Republic

Henrik Carlsson Department of Medical Sciences, Uppsala University, Uppsala, Sweden

Jakub Červenka Laboratory of Applied Proteome Analyses, Institute of Animal Physiology and Genetics, Czech Academy of Sciences, Libechov, Czech Republic

Natalia Chunga Department of Neurology, University of Rochester, Rochester, NY, USA

Jody Corey-Bloom Department of Neurosciences, University of California San Diego, San Diego, CA, USA

Wenzhen Duan Division of Neurobiology, Department of Psychiatry and Behavioral Sciences, Department of Neuroscience, Johns Hopkins University School of Medicine, Baltimore, MD, USA

Ida Erngren Department of Medical Sciences, Uppsala University, Uppsala, Sweden

Liesbeth Everix Molecular Imaging Center Antwerp (MICA), University of Antwerp, Antwerp, Belgium

μNeuro Research Centre of Excellence, University of Antwerp, Antwerp, Belgium

Erin Furr Stimming Department of Neurology, McGovern Medical School, The University of Texas Health Science Center, Houston, TX, USA

Steven W. Granger Salimetrics, LLC, Carlsbad, CA, USA

Carolina Gubert Florey Institute of Neuroscience and Mental Health, University of Melbourne, Parkville, VIC, Australia

Anthony J. Hannan Florey Institute of Neuroscience and Mental Health, University of Melbourne, Parkville, VIC, Australia

Department of Anatomy and Physiology, University of Melbourne, Parkville, VIC, Australia

Kateřina Vodičková Kepková Laboratory of Applied Proteome Analyses, Institute of Animal Physiology and Genetics, Czech Academy of Sciences, Libechov, Czech Republic

Jaime Kulisevsky Movement Disorders Unit, Neurology Department, Hospital Sant Pau, Barcelona, Spain

Institut d'Investigacions Biomèdiques-Sant Pau (IIB-Sant Pau), Barcelona, Spain

Universitat Autònoma de Barcelona (UAB), Medicine Department, Barcelona, Spain

Kim Kultima Department of Medical Sciences, Uppsala University, Uppsala, Sweden

Blair R. Leavitt Center for Brain Health, University of British Columbia Hospital, Vancouver, BC, Canada

Centre for Molecular Medicine and Therapeutics, Department of Medical Genetics and Division of Neurology, Department of Medicine, University of British Columbia, Vancouver, BC, Canada

Chloe J. Love School of Medicine, Faculty of Health, Geelong Warnum Ponds Campus, Deakin University, Geelong, VIC, Australia

Bethany A. Masson Florey Institute of Neuroscience and Mental Health, University of Melbourne, Parkville, VIC, Australia

Georgia M. Parkin Department of Neurosciences, University of California, San Diego, CA, USA

Zhong Pei Department of Neurology, The First Affiliated Hospital, Sun Yat-sen University, Guangdong Provincial Key Laboratory of Diagnosis and Treatment of Major Neurological Diseases, National Key Clinical Department and Key Discipline of Neurology, Guangzhou, China

Jesús Pérez-Pérez Movement Disorders Unit, Neurology Department, Hospital Sant Pau, Barcelona, Spain

Institut d'Investigacions Biomèdiques-Sant Pau (IIB-Sant Pau), Barcelona, Spain

Universitat Autònoma de Barcelona (UAB), Medicine Department, Barcelona, Spain

Fabricio Pio Center for Brain Health, University of British Columbia Hospital, Vancouver, BC, Canada

Wendy Qin Florey Institute of Neuroscience and Mental Health, University of Melbourne, Parkville, VIC, Australia

Lori Quinn Teachers College, Columbia University, New York, NY, USA

Natalia P. Rocha Department of Neurology, McGovern Medical School, The University of Texas Health Science Center, Houston, TX, USA

Jon Rodríguez-Antigüedad Movement Disorders Unit, Neurology Department, Hospital Sant Pau, Barcelona, Spain

Institut d'Investigacions Biomèdiques-Sant Pau (IIB-Sant Pau), Barcelona, Spain

Universitat Autònoma de Barcelona (UAB), Medicine Department, Barcelona, Spain

Abera Saeed Centre for Eye Research Australia, Royal Victorian Eye and Ear Hospital, Melbourne, VIC, Australia

Ophthalmology, Department of Surgery, University of Melbourne, Melbourne, VIC, Australia

Helena Kupcova Skalnikova Laboratory of Applied Proteome Analyses and Research Center PIGMOD, Institute of Animal Physiology and Genetics of the Czech Academy of Sciences, Libechov, Czech Republic

Institute of Biochemistry and Experimental Oncology, First Faculty of Medicine, Charles University, Prague, Czech Republic

Steven Staelens Molecular Imaging Center Antwerp (MICA), University of Antwerp, Antwerp, Belgium

μNeuro Research Centre of Excellence, University of Antwerp, Antwerp, Belgium

Rita Suchá Laboratory of Applied Proteome Analyses, Institute of Animal Physiology and Genetics, Czech Academy of Sciences, Libechov, Czech Republic

Antonio L. Teixeira Department of Psychiatry and Behavioral Sciences, McGovern Medical School, University of Texas, Houston, TX, USA

Elizabeth A. Thomas Department of Neurobiology and Behavior, Institute for Interdisciplinary Salivary Bioscience Research, University of California, Irvine, CA, USA

Department of Neuroscience, The Scripps Research Institute, La Jolla, CA, USA

Petr Vodička Laboratory of Applied Proteome Analyses, Institute of Animal Physiology and Genetics, Czech Academy of Sciences, Libechov, Czech Republic

Eleni Voukali Laboratory of Applied Proteome Analyses, Institute of Animal Physiology and Genetics, Czech Academy of Sciences, Libechov, Czech Republic

Emma M. Waddell Center for Health + Technology, University of Rochester, Rochester, NY, USA

Peter van Wijngaarden Centre for Eye Research Australia, Royal Victorian Eye and Ear Hospital, Melbourne, VIC, Australia

Ophthalmology, Department of Surgery, University of Melbourne, Melbourne, VIC, Australia

Huajing You Department of Neurology, The First Affiliated Hospital, Sun Yat-sen University, Guangdong Provincial Key Laboratory of Diagnosis and Treatment of Major Neurological Diseases, National Key Clinical Department and Key Discipline of Neurology, Guangzhou, China

About the Editors

Elizabeth A. Thomas received her Bachelor's degree in Biochemistry from the University of California (UC), Berkeley, in 1989, and her Ph.D. in Pharmacology from the UC Irvine, School of Medicine in 1994. Her international appointments include time as an Adjunct Research Officer at the Mental Health Research Institute of Victoria in Melbourne, Australia, followed by an Honorary Senior Research Fellow at The Florey Institute of Neuroscience and Mental Health, also in Melbourne, Australia. She is currently the Laboratory Director of the Institute for Interdisciplinary Salivary Bioscience Research at UC Irvine, and also holds positions of Research Professor in the Department of Neurobiology and Human Behavior at UC Irvine, and Adjunct Associate Professor in the Department of Neuroscience at the Scripps Research Institute. She has studied Huntington's disease for nearly 25 years, where her research has spanned basic science focusing on gene expression regulation in the striatum, novel therapeutic approaches in preclinical Huntington's mouse models, and, more currently, biomarker research studies on clinical populations. She serves on the Advisory board for the Huntington's Disease Society of America Center of Excellence at UC San Diego. Dr. Thomas has more than 100 peer-reviewed publications to date, and her work has appeared in top journals, including *The Lancet*, *Nature Neuroscience*, *Proceedings of the National Academy of Sciences*, *Cell Stem Cell*, *Aging Cell*, *Human Molecular Genetics*, *Biological Psychiatry*, and *Molecular Psychiatry*.

Georgia M. Parkin received her Bachelor's degree in Biochemistry from the University of Melbourne, Australia, in 2011, her Master's degree in Neuroscience from the University of Zurich, Switzerland, in 2014, and her Ph.D. in Neuroscience from The Florey Institute of Neuroscience and Mental Health, and the University of Melbourne, Australia, in 2019. Her work has spanned the fields of traumatic brain injury, psychiatric illness with a focus on schizophrenia and bipolar disorder, and Huntington's disease. Dr. Parkin is academically trained in laboratory-based research, with her past studies focusing on cell culture and mouse models, as well as human subjects and clinical patients, and involving biospecimens such as post-mortem brain tissue, blood, saliva, and cerebrospinal fluid. Concurrent to, and following, completion of her Ph.D., Dr. Parkin has also specialized in clinical biomarker research, through appointments at the Murdoch Children's Research Institute and Royal Children's Hospital, Australia, and the Institute for Interdisciplinary Salivary Bioscience Research at the University of California (UC), Irvine. She currently works in the Department of Neurosciences at UC San Diego and is part of the Huntington's Disease Society of America Center of Excellence at this site. Her published work has been featured in journals including *The Lancet*, *Nature Neuroscience*, *Neuroscience and Biobehavioral Reviews*, *Bipolar Disorders*, *International Journal of Molecular Sciences*, among others; her research has also been shared by multiple news sites.

Part I
Introduction

The Utility of Biomarkers for Huntington's Disease

Elizabeth A. Thomas

Abbreviations

AD	Alzheimer's disease
BEST	Biomarkers, EndpointS, and other Tools
CSF	Cerebral spinal fluid
FDA	Food and Drug Administration
HD	Huntington's disease
HTT	Huntington gene
HTT	Huntingtin protein
MRI	Magnetic resonance imaging
mHTT	Mutant huntingtin protein
NIH	National Institutes of Health
PET	Positron emission tomography
UHDRS	Unified Huntington's Disease Rating Scale

1 Introduction

Biomarkers – the measurement and quantification of factors that reflect biological change – play crucial roles in multiple aspects of neurodegenerative diseases, such as Huntington's disease (HD). These aspects include discovery and translational science, medical and drug development, as well as patient care and outcomes. The field of biomarker research has experienced significant and continuous growth over the past decade, leading to rapid advancements in biomarker development that go beyond the classical fluid biomarkers (Califf, 2018). The use of innovative

E. A. Thomas (✉)
Department of Neurobiology and Behavior, Institute for Interdisciplinary Salivary Bioscience Research, University of California, Irvine, CA, USA

Department of Neuroscience, The Scripps Research Institute, La Jolla, CA, USA
e-mail: eathoma1@uci.edu

E. A. Thomas, G. M. Parkin (eds.), *Biomarkers for Huntington's Disease*, Contemporary Clinical Neuroscience, https://doi.org/10.1007/978-3-031-32815-2_1

biospecimens, such as saliva and extracellular vesicles, has dramatically increased avenues in which to explore non-, or minimally-invasive options for biomarker research. Recent advances in various 'omics' technologies have enabled quantitative monitoring of various biological molecules in a high-throughput manner; these approaches have extended beyond genomics and proteomics, to include microbiomics, lipidomics and metabolomics, all of which provide a means for novel biomarker discovery. With the increase in telehealth applications, many clinical trials now include virtual visits in their study protocols, which has led to the increased utility of digital biomarkers. Digital biomarkers have the potential to transform clinical care by providing opportunities for remote symptom monitoring and disease tracking.

In this introductory chapter to the book, "Biomarkers for Huntington's Disease – Improving Clinical Outcomes", I provide an overview of biomarker classes, the types of biomarkers utilized to date in HD and the different areas of need for the HD field. Subsequent chapters in this book will cover these topics in detail. It is now becoming clear that biomarkers are critically needed, not only for the HD medical and research fields, but also for the patient and caregiver communities.

2 Huntington's Disease Characteristics

HD is a genetic neurodegenerative disorder, characterized by chorea, other motor system dysfunction, psychiatric manifestations and cognitive decline. In 1872, the American physician George Huntington provided the first comprehensive description of adult-onset HD, highlighting its autosomal-dominant inheritance pattern. Over a century later, the gene responsible for this disease was mapped to a chromosome in 1983, followed by the identification of the disease-causing mutation – an expanded triplet nucleotide (CAG) repeat within exon 1 of the Huntingtin (*HTT*) gene – in 1993. Today, this mutation can be easily detected through genetic testing, allowing confirmation of the genetic status of individuals at risk of developing HD. In the USA HD occurs at a rate of approximately 13–15 cases per 100,000, which is slightly higher than previously estimated (Exuzides et al., 2022). Additionally, there are another ~200,000 individuals at a genetic risk of developing the disease: individuals who have one biological parent with HD, but have not been genetically tested themselves. HD pathogenesis is largely caused by the expression of the mutant Huntingtin protein (mHTT), which leads to the formation of soluble protein oligomers and insoluble aggregates that cause disruption of many intracellular pathways, including transcriptional abnormalities, synaptic dysfunction, mitochondrial toxicity and immune systems disturbances, among others (Ross & Tabrizi, 2011; Reddy & Shirendeb, 2012; Johri et al., 2013; Thomas, 2019; Valadão et al., 2020). These events subsequently contribute to the development of motor system abnormalities (i.e. chorea), cognitive decline, and psychiatric manifestations, ultimately leading to death approximately 15–20 years after disease onset from complications such as respiratory or cardiac failure (Heemskerk & Roos, 2012).

In the clinic, the determination of disease onset typically relies on the identification of clear motor symptoms, such as chorea. Subtle cognitive impairments, affecting executive function and visuospatial ability, and psychiatric disturbances, including irritability, anxiety, depression, and obsessive-compulsive behaviors, can be detected up to 10 years before diagnosis. The number of CAG repeats in the *HTT* gene is the main determining factor of the age of disease onset, which is an important feature of this disease. However, there is much variability in the age of symptom onset in HD and it is thought that both genetic and environmental factors play important roles in this variation (Andrew et al., 1993; Wexler, 2004; Andresen et al., 2007; Arrasate & Finkbeiner, 2012), hence better markers are needed to more accurately predict and diagnose onset. The typical age-of-onset for HD is in the fourth or fifth decade of life, however 5% of cases (with CAG repeats of >60) show juvenile onset (van Dijk et al., 1986; Nance & Myers, 2001) with another ~10% of individuals showing late onset of the disease (>60 years of age). Diagnosis of HD in aged individuals can be missed due to the perceived low likelihood of developing HD at a later life stage and heterogeneity in symptom presentation (which may be associated with age-related decline). The use of prognostic biomarkers will be particularly important for individuals with late onset HD, as the association between CAG repeat number and age of onset decreases with increasing biological age (Chaganti et al., 2017).

3 Biomarker Categories

There are several definitions of biomarkers, but in its simplest term, a biomarker is defined as a characteristic that is objectively measured and evaluated as an indicator of a normal biological process, pathogenic process or a response to an exposure or therapeutic intervention (FDA-NIH Biomarker Working Group, 2016). However, there is significant confusion over the fundamental definitions and concepts of biomarkers and their use in research and clinical practice. In attempts to clarify the definitions of different biomarkers and to better understand their appropriate applications, the U.S. Food and Drug Administration and the National Institutes of Health Joint Leadership Council created the BEST (Biomarkers, EndpointS, and other Tools) resource in 2016 (FDA-NIH Biomarker Working Group, 2016). The BEST resource contains a biomarker glossary that outlines the relationships, connections and dependencies among biomarker terms and places them in the context of their respective uses in patient care, clinical research, or therapeutic development (FDA-NIH Biomarker Working Group, 2016). According to the FDA-NIH Biomarker Working Group, there are seven types of classes of biomarkers for neurodegenerative disorders based on their specific use or the clinical application: diagnostic, monitoring, pharmacodynamic, or response, predictive, prognostic, safety, and susceptibility/risk biomarkers (Fig. 1).

- Diagnostic Biomarkers: These biomarkers are used as a diagnostic tool to detect or confirm the presence of a neurodegenerative disorder. In HD, a genetic test will provide information about whether one will receive a diagnosis of HD in their lifetime, however, diagnosis of clinical disease onset is still needed.
- Monitoring Biomarkers: Monitoring biomarkers can be used to monitor the progression of disease trajectories, or a monitor responses to a disease intervention or treatment. These biomarkers are most widely utilized for clinical trials in HD, but could also be used to track the natural disease progression in HD patients.
- Pharmacodynamic or Response Biomarkers: The ability to determine the effectiveness of a given treatment can be assessed by pharmacodynamic biomarkers. These are essential tools in clinical trials and therapeutic drug development.
- Predictive Biomarkers: These biomarkers are used to predict the development of clinical manifestation, or outcomes of a given intervention or treatment.
- Prognostic Biomarkers: Although similar to predictive biomarkers, prognostic biomarkers are routinely used to set trial entry and exclusion criteria or to identify higher-risk populations. In addition, prognostic biomarkers are especially important for predicting the risk of an event or poor outcome in an individual.
- Safety Biomarkers: These biomarkers allow one to identify the possibility of developing toxicity or to detect the presence of toxic events after exposure to a medical intervention. They are also beneficial in identifying patients in which particular therapies should not be started due to significant safety risks.
- Susceptibility or Risk Biomarkers: These biomarkers are used to evaluate the risk of developing a specific neurodegenerative disease, or the transition from a healthy to a disease state. Susceptibility/risk biomarkers: are most often use in epidemiological studies.

All of these classes apply to HD, with the exception of susceptibility/risk biomarkers (Fig. 1), given that genetic testing is available for HD to determine if one has the

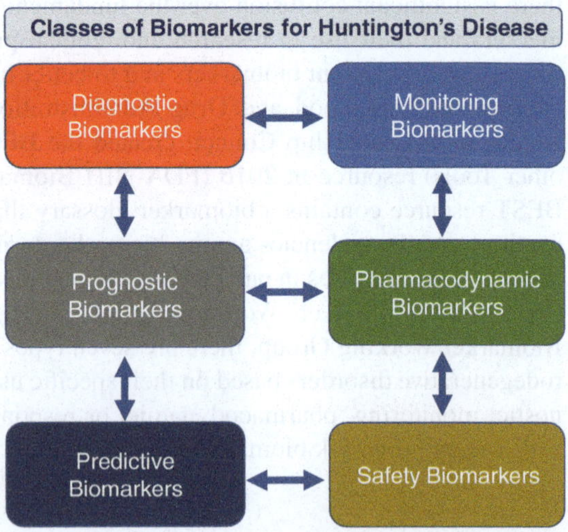

Fig. 1 Summary of biomarkers classes. According to the FDA-NIH Biomarker Working Group, there are seven types of biomarkers for neurodegenerative disorders based on their specific use or the clinical application: diagnostic, monitoring, pharmacodynamic, or response, predictive, prognostic, safety, and susceptibility/risk biomarkers

genetic mutation. Biomarkers are not limited to one class, as a single marker could have utility in multiple categories. One such example is the mHTT protein itself. mHTT measured in CSF has been shown to increase with disease progression (Rodrigues et al., 2020) and levels have been shown to correlate with clinical measures of disease severity (Southwell et al., 2015; Byrne et al., 2018; Rodrigues et al., 2020). Additionally, mHTT has also been used as a pharmacodynamic or response biomarker, with respect to clinical trials targeting the HTT gene. For example, a dose-dependent reduction of CSF mHTT was observed in a Phase I/IIa clinical trial evaluating an *HTT*-targeted antisense oligonucleotide (tominersen) (Tabrizi et al., 2019).

4 Biomarker Sources

The above-mentioned biomarker categories can consist of biomarkers derived from many sources. These include biofluid, imaging and digital biomarkers; these have been extensively explored in neurodegenerative diseases, including HD and are outlined briefly below.

Biofluid Biomarkers

Fluid biomarkers are those derived from biological fluids, such as cerebral spinal fluid (CSF), blood (serum, plasma or whole blood), urine, oral fluid/saliva, sweat and even tears (Fig. 2). All of these biofluids contain a wealth of proteins, glycoproteins, nucleic acids, hormones, lipids and metabolites, which comprise a variety of biomolecules that can serve as biomarkers for disease states. By far the most common biofluid for biomarker research is blood, either in the form of whole blood, serum or plasma, all of which are extensively used. However, CSF is the biofluid whose molecular composition most reflects structural and functional changes in the brain, hance, it has traditionally been thought to represent the most promising biofluid for biomarker discovery in HD and other neurodegenerative disorders.

Despite the enormous utility of CSF and blood fluids, their sample collection requires invasive techniques, which can limit their clinical utility as biomarker sources. Hence, researchers have turned to the use of minimally- or non-invasive biofluids, such as saliva, urine, sweat and tears (Fig. 2). These alternative biofluids have received increased attention over the past decade as novel matrices for biomarker discovery (Khamis et al., 2017; Farah et al., 2018; Thomas, 2020; Adigal et al., 2021; Saha et al., 2022). It has become increasingly appreciated that these non-invasive fluids can reflect changes in health status, making them attractive biofluids for many purposes.

In addition to the HTT protein mentioned above, neurofilament light (NfL) is probably the next more frequently studied fluid biomarker in HD and other

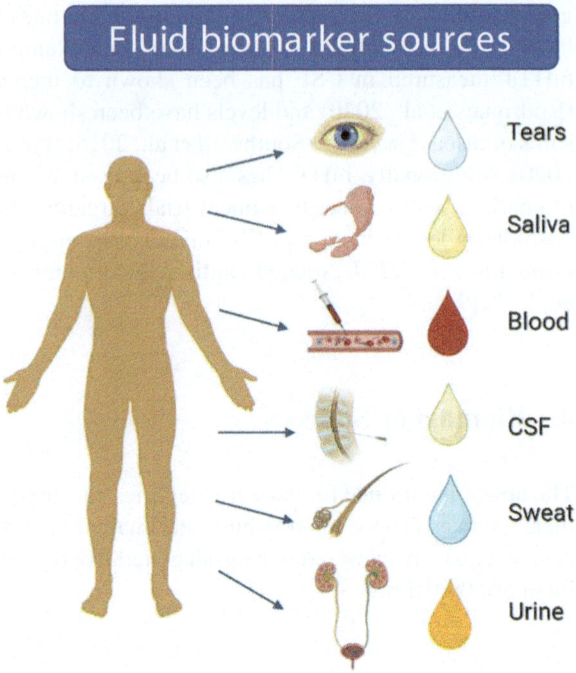

Fig. 2 Fluid biomarker sources. Fluid biomarkers come from diverse sources in the body. These include cerebral spinal fluid (CSF), blood (serum, plasma or whole blood), urine, oral fluid/saliva, sweat and tears. All of these biofluids contain a wealth of proteins, glycoproteins, nucleic acids, hormones, lipids and metabolites, which make up a variety of biomolecules that can serve as biomarkers

neurodegenerative disorders, including Alzheimer's disease (AD), fronto-temporal dementia and amyotrophic lateral sclerosis (Byrne et al., 2017, 2018; Cruickshank et al., 2020; Rodrigues et al., 2020; Scahill et al., 2020). NfL is exclusively expressed in neurons and is released into the extracellular space following axonal degeneration or neuronal damage (Petzold, 2005; Bridel et al., 2019). Importantly, significant associations have previously been reported between either plasma or serum NfL and those levels measured in the CSF, for a number of CNS disorders (Meeter et al., 2016; Steinacker et al., 2018; Hu et al., 2019; Mattsson et al., 2019). Notably, NfL is being used in HD clinical trials as an exploratory biomarker to monitor disease progression and to assess therapeutic efficacy. However, it remains unknown if biofluid measures of NfL will be validated in a manner that reflects clinical benefit.

Imaging Biomarkers

Imaging modalities, such as magnetic resonance imaging (MRI), computed tomography, X-ray, ultrasound, positron emission tomography (PET), single photon emission computed tomography, optical imaging and light microscopy, among others, have been employed for identification of quantitative imaging biomarkers in many neurodegenerative diseases. The processing of imaging data involves the extraction of information from an image for subsequent analysis. The image processing and analysis tasks include enhancement, registration, segmentation and classification steps; the determination of parameters for each of these steps play important roles in the consistent and accurate evaluation of quantitative imaging biomarkers.

Several features make imaging biomarkers prime candidates for biomarker research, such as their non-invasive nature, standardized data acquisition and processing, and quality control, as well as the ease of transferring data over long distances, which would benefit multi-site studies. Neuroimaging modalities have been extensively investigated in HD and have contributed significantly to our understanding of the disease's natural progression. Striatal atrophy, as determined by different types of MRI, has been the most consistent, sensitive, and robust finding in HD obtained through several observational studies, i.e. PREDICT and TRACK HD. These studies have confirmed striatal atrophy as an early biomarker that precedes clinical diagnosis of the disease (Paulsen et al., 2008; Tabrizi et al., 2011). Additionally, over the past 5 years, technological advances have allowed the development of PET ligand for the HTT protein which allow for the detection and quantification of misfolded HTT protein (Liu et al., 2020; Herrmann et al., 2021). These ligands show incredible potential to advance the field, much in the way that amyloid beta and tau PET tracers have advanced the field of AD research.

Digital Biomarkers

The field of digital biomarkers is rapidly advancing, offering new avenues for assessing various aspects of the overall health and behavior of individuals. With the availability of sensors and personal devices, continuous and real-time collection of data has become possible, providing valuable insights into complex measures, such as psychological state, exercise levels, cognitive abilities, eating patterns, motion, and tremors. These data primarily come from emerging sources like smartphones and wearable electronic devices, supported by innovative technologies that enable streaming and storage of complex information. These devices may enable remote at-home personalized health monitoring and substantially reduce the healthcare costs (Sempionatto et al., 2022). As a result, standards for evaluating these digital biomarkers are currently being developed.

It is expected that digital biomarkers will introduce entirely new measures for phenomena that are already assessed in clinical practice. It is even possible that total daily activity or a combination of peak and continuous activity could serve as better predictors of disease onset, prognosis, or treatment response compared to existing measures. This can be likened to advances in hypertension, whereby frequent blood pressure measurements taken during both activity and rest, are likely to provide superior indicators of hypertension, or even treatment responses, compared to the conventional measurement of resting blood pressure while seated.

5 Need for Biomarkers in HD

The past decade has seen substantial advances in the development of biomarkers that go beyond the classical fluid biomarkers, including the evolution of imaging and digital biomarkers all of which have been extensively explored in HD (Fig. 3) and are described throughout this book. Robust and reliable biomarkers are critically needed in the HD field to diagnose and predict disease onset, monitor disease progression and assess treatment responses. Such biomarkers are needed not only for clinical trial purposes, but also to inform patients and their caregivers.

One of the most important areas for biomarker impact is to predict and diagnosis the onset of manifest HD. The period prior to overt motor symptom presentation is the premanifest or presymptomatic stage, sometimes also called “pre-HD”. The premanifest stage can be further divided into prodromal, early premanifest and late premanifest stages. Throughout this book, different designations for the premanifest stage have been made, as there has not yet been a universally-adopted staging system to standardize staging. However, an exciting advance in this area is the recently described Huntington’s Disease Integrated Staging System (HD-ISS) which is a biological framework for categorizing HD patients into four groups based on

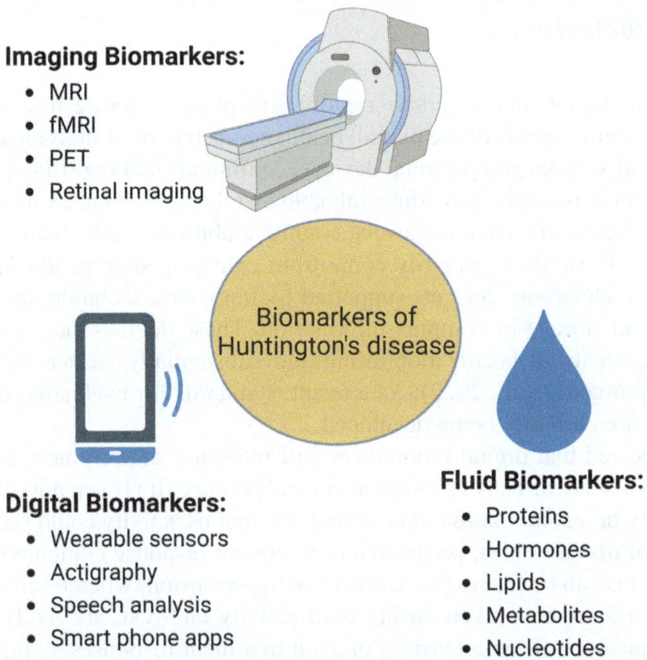

Fig. 3 Overview of biomarkers in Huntington’s disease. Summary of the main types of biomarkers use in HD, which include imaging, biofluid and digital biomarkers. magnetic resonance imaging (MRI); functional magnetic resonance imaging (fMRI); positron emission tomography (PET). Created with BioRender.com

biological, clinical, and functional assessments (Tabrizi et al., 2022). One problem for implementation is that it is not possible, or even likely, that all variables can be measured in every study. For example, Stage 1 requires information on striatal atrophy which can only be defined by MRI data. Because not all patients will have quantitative imaging data, the incorporation of fluid or digital biomarkers will help refine these categories (Parkin et al., 2023).

Biomarkers for Clinical Trials

The current options for treating Huntington's disease are limited, as there are only two drugs approved by the FDA, namely tetrabenazine (Xenazine) and deutetrabenazine (Austedo), which can help manage chorea. However, there are numerous ongoing clinical studies (as of 2023) exploring various treatment approaches aimed at targeting the HTT gene itself, or modifying the progression of the disease. A summary of past and current clinical trials conducted on HD patients are reviewed in Chap. 17. As the number and complexity of these trials increase, the need for suitable biomarkers also becomes crucial.

The ability to identify or precisely predict disease onset is of utmost importance in the future of clinical trials, and thus far determining the appropriate stage of the disease for a specific trial has been the subject of considerable debate. The primary objective would be to identify patients who are most likely to experience symptom progression during the trial period, while still being at a disease stage within which therapeutic treatment could have an effect. This will allow better comparisons between placebo and drug-treated groups; that is, if placebo-treated patients do not show any worsening of disease symptoms during the trial period, it will be difficult to determine the potential benefits of the investigational drug being tested. Biomarkers that would enable the precise prediction of disease onset would be of high value in these cases.

The recently proposed HD-ISS staging system (Tabrizi et al., 2022) was designed primarily to be implemented in clinical trials and has great potential to improve clinical trial success rates by optimizing patient inclusion and stratification. The HD-ISS already utilizes MRI quantitative imaging biomarkers to discern Stage 0 from Stage 1 (Tabrizi et al., 2022); however, to date, current clinical trials only recruit individuals who are in Stage 2 or 3, hence biomarkers have not yet fully been implemented in this context. Further consideration of fluid and digital biomarkers in this model could further improve staging of subjects for clinical trials and this could also lead to the much-needed standardization across ongoing and future clinical studies.

Although biomarkers are critically needed to stratify patients for clinical trials, pharmacodynamic and monitoring biomarkers are also essential to track drug efficacy and target engagement, as is the case with all clinical trials, not just those involving HD patients. Further, clinical trial endpoints will require robust biomarkers to determine if goals, such as drug target engagement, are achieved or if there are safety concerns associated with a particular drug. Digital biomarkers offer several potential benefits as monitoring biomarkers, as they allow for continuous monitoring, real-time data collection, and the ability to capture subtle changes that may go unnoticed in traditional assessments. These factors can provide a more dynamic and comprehensive view of a patient's health status during the trial period. Examples of digital biomarkers utilized in HD are described in Chap. 18. Overall, the addition of biomarkers to clinical trial design will certainly lead to improved clinical trial success and outcomes.

Biomarker Utility for Patients and Caregivers

In addition to utility in clinical trials, biomarkers are also needed for the HD patients and caregivers themselves to provide information for planning the individual's care needs accordingly. Subtle cognitive impairments, affecting executive function and visuospatial ability, as well as psychiatric disturbances, including irritability, anxiety, depression, and obsessive-compulsive behaviors, can be detected up to 10 years before expected disease onset. Having biomarkers of these early signs and symptoms would be of particular importance for family members and gene mutation carriers as it would allow them to easily recognize and document the early signs of the disease in a more definitive manner. The same biomarkers that could predict disease onset for the purpose of clinical trial inclusion would also help physicians and caregivers with many aspects of patient care including healthcare and lifestyle planning, as well as deciding whether the patient should seek investigational and potential early therapeutic intervention. Digital biomarkers may serve this purpose particularly well, as they could detect subtle changes in a person's condition that may be more prominent during particular day-to-day activities outside the scope of a clinical assessment, and at an early stage; this is crucial for HD, whereby it is now thought that early intervention might significantly impact disease progression.

While the new HD-ISS system was designed for use in clinical trials, an additional formal staging system to monitor symptom and disease progression could have additional utility for patient and caregivers. The Unified Huntington's Disease Rating Scale (UHDRS), which was developed as a clinical rating scale to assess clinical performance and capacity in manifest HD patients (Huntington Study Group, 1996), is still widely used today for clinical assessments and in clinical trials; however, it does not address the early features of the disease, such as behavioral changes mentioned above, which are often reported as being the most problematic for patients and their family members. Further, the UHDRS was not necessarily designed for home care use. A staging system targeted for families and caregivers could help determine how far along their loved one is in behavioral, cognitive or motor decline over time and possibly what to expect when the patient moves from one stage to the next. The inclusion of biomarkers into such a staging system, or even implemented on their own, would have significant added-value in this area, and could represent invaluable tools for home care.

6 Conclusions

Biomarkers are critical to the foundation of science, medicine and healthcare. This field has seen extensive growth over the past decade in the development of new tools, advances in the sensitivity of measurements, and innovation in computation and analysis. At the same time, continuous progress in our capacity to store, collate, and compute massive amounts of information is fundamentally changing our understanding of both biology and clinical outcomes. These areas are coming together to rapidly advance the field. Further, it is becoming clear that a single biomarker of any type or class will likely not be suitable to address the biomarker needs of the patients, physicians and those involved in clinical trials. A single biomarker is also unlikely to address and span all heterogeneous and multifaceted stages of HD. Interrelationships of multiple biomarkers will need to be compiled depending on the intended outcome. With the ultimate goal of treating patients, the development of disease-associated biomarkers has never been more important.

References

Adigal, S. S., Rizvi, A., Rayaroth, N. V., John, R. V., Barik, A., Bhandari, S., et al. (2021). Human tear fluid analysis for clinical applications: Progress and prospects. *Expert Review of Molecular Diagnostics, 21*, 767–787.

Andresen, J. M., Gayán, J., Djoussé, L., Roberts, S., Brocklebank, D., Cherny, S. S., et al. (2007). The relationship between CAG repeat length and age of onset differs for Huntington's disease patients with juvenile onset or adult onset. *Annals of Human Genetics, 71*, 295–301.

Andrew, S. E., Goldberg, Y. P., Kremer, B., Telenius, H., Theilmann, J., Adam, S., et al. (1993). The relationship between trinucleotide (CAG) repeat length and clinical features of Huntington's disease. *Nature Genetics, 4*, 398–403.

Arrasate, M., & Finkbeiner, S. (2012). Protein aggregates in Huntington's disease. *Experimental Neurology, 238*, 1–11.

Bridel, C., van Wieringen, W. N., Zetterberg, H., Tijms, B. M., Teunissen, C. E., Alvarez-Cermeño, J. C., et al. (2019). Diagnostic value of cerebrospinal fluid neurofilament light protein in neurology: A systematic review and meta-analysis. *JAMA Neurology, 76*, 1035–1048.

Byrne, L. M., Rodrigues, F. B., Blennow, K., Durr, A., Leavitt, B. R., Roos, R. A. C., et al. (2017). Neurofilament light protein in blood as a potential biomarker of neurodegeneration in Huntington's disease: A retrospective cohort analysis. *Lancet Neurology, 16*, 601–609.

Byrne, L. M., Rodrigues, F. B., Johnson, E. B., Wijeratne, P. A., De Vita, E., Alexander, D. C., et al. (2018). Evaluation of mutant huntingtin and neurofilament proteins as potential markers in Huntington's disease. *Science Translational Medicine, 10*, eaat7108.

Califf, R. M. (2018). Biomarker definitions and their applications. *Experimental Biology and Medicine (Maywood, N.J.), 243*, 213–221.

Chaganti, S. S., McCusker, E. A., & Loy, C. T. (2017). What do we know about Late Onset Huntington's Disease? *Journal of Huntington's Disease, 6*, 95–103.

Cruickshank, T., Bartlett, D., Govus, A., Hannan, A., Teo, W. P., Mason, S., et al. (2020). The relationship between lifestyle and serum neurofilament light protein in Huntington's disease. *Brain and Behavior: A Cognitive Neuroscience Perspective, 10*, e01578.

Exuzides, A., Reddy, S. R., Chang, E., Ta, J. T., Patel, A. M., Paydar, C., et al. (2022). Epidemiology of Huntington's disease in the United States medicare and medicaid populations. *Neuroepidemiology, 56*, 192–200.

Farah, R., Haraty, H., Salame, Z., Fares, Y., Ojcius, D. M., & Said Sadier, N. (2018). Salivary biomarkers for the diagnosis and monitoring of neurological diseases. *Biomedical Journal, 41*, 63–87.

FDA-NIH Biomarker Working Group. (2016). BEST (Biomarkers, EndpointS, and other Tools) resource. Silver Spring (MD): Food and Drug Administration (US); Bethesda (MD): National Institutes of Health (US). www.ncbi.nlm.nih.gov/books/NBK326791/

Heemskerk, A. W., & Roos, R. A. (2012). Aspiration pneumonia and death in Huntington's disease. *PLOS Currents, 4*, RRN1293.

Herrmann, F., Hessmann, M., Schaertl, S., Berg-Rosseburg, K., Brown, C. J., Bursow, G., et al. (2021). Pharmacological characterization of mutant huntingtin aggregate-directed PET imaging tracer candidates. *Scientific Reports, 11*, 17977.

Hu, H., Chen, K. L., Ou, Y. N., Cao, X. P., Chen, S. D., Cui, M., et al. (2019). Neurofilament light chain plasma concentration predicts neurodegeneration and clinical progression in nondemented elderly adults. *Aging (Albany NY), 11*, 6904–6914.

Huntington Study Group. (1996). Unified Huntington's Disease Rating Scale: Reliability and consistency. *Movement Disorders, 11*, 136–142.

Johri, A., Chandra, A., & Flint Beal, M. (2013). PGC-1α, mitochondrial dysfunction, and Huntington's disease. *Free Radical Biology & Medicine, 62*, 37–46.

Khamis, M. M., Adamko, D. J., & El-Aneed, A. (2017). Mass spectrometric based approaches in urine metabolomics and biomarker discovery. *Mass Spectrometry Reviews, 36*, 115–134.

Liu, L., Prime, M. E., Lee, M. R., Khetarpal, V., Brown, C. J., Johnson, P. D., et al. (2020). Imaging mutant huntingtin aggregates: Development of a potential PET ligand. *Journal of Medicinal Chemistry, 63*, 8608–8633.

Mattsson, N., Cullen, N. C., Andreasson, U., Zetterberg, H., & Blennow, K. (2019). Association between longitudinal plasma neurofilament light and neurodegeneration in patients with Alzheimer disease. *JAMA Neurology, 76*, 791–799.

Meeter, L. H., Dopper, E. G., Jiskoot, L. C., Sanchez-Valle, R., Graff, C., Benussi, L., et al. (2016). Neurofilament light chain: A biomarker for genetic frontotemporal dementia. *Annals of Clinical Translational Neurology, 3*, 623–636.

Nance, M. A., & Myers, R. H. (2001). Juvenile onset Huntington's disease – Clinical and research perspectives. *Mental Retardation and Developmental Disabilities Research Reviews, 7*, 153–157.

Parkin, G. M., Thomas, E. A., & Corey-Bloom, J. (2023). Plasma NfL as a prognostic biomarker for enriching HD-ISS stage 1 categorisation: A cross-sectional study. *eBioMedicine, 93*, 104646.

Paulsen, J. S., Langbehn, D. R., Stout, J. C., Aylward, E., Ross, C. A., Nance, M., et al. (2008). Detection of Huntington's disease decades before diagnosis: The predict-HD study. *Journal of Neurology, Neurosurgery, and Psychiatry, 79*, 874–880.

Petzold, A. (2005). Neurofilament phosphoforms: Surrogate markers for axonal injury, degeneration and loss. *Journal of the Neurological Sciences, 233*, 183–198.

Reddy, P. H., & Shirendeb, U. P. (2012). Mutant huntingtin, abnormal mitochondrial dynamics, defective axonal transport of mitochondria, and selective synaptic degeneration in Huntington's disease. *Biochimica et Biophysica Acta, 1822*, 101–110.

Rodrigues, F. B., Byrne, L. M., Tortelli, R., Johnson, E. B., Wijeratne, P. A., Arridge, M., et al. (2020). Mutant huntingtin and neurofilament light have distinct longitudinal dynamics in Huntington's disease. *Science Translational Medicine, 12*, eabc2888.

Ross, C. A., & Tabrizi, S. J. (2011). Huntington's disease: From molecular pathogenesis to clinical treatment. *Lancet Neurology, 10*, 83–98.

Saha, T., Del Caño, R., la De Paz, E., Sandhu, S. S., & Wang, J. (2022). Access and management of sweat for non-invasive biomarker monitoring: A comprehensive review. *Small*, e2206064.

Scahill, R. I., Zeun, P., Osborne-Crowley, K., Johnson, E. B., Gregory, S., Parker, C., et al. (2020). Biological and clinical characteristics of gene carriers far from predicted onset in the Huntington's disease Young Adult Study (HD-YAS): A cross-sectional analysis. *Lancet Neurology, 19*, 502–512.

Sempionatto, J. R., Lasalde-Ramírez, J. A., Mahato, K., Wang, J., & Gao, W. (2022). Wearable chemical sensors for biomarker discovery in the omics era. *Nature Reviews Chemistry, 6*, 899–915.

Southwell, A. L., Smith, S. E., Davis, T. R., Caron, N. S., Villanueva, E. B., Xie, Y., et al. (2015). Ultrasensitive measurement of huntingtin protein in cerebrospinal fluid demonstrates increase with Huntington disease stage and decrease following brain huntingtin suppression. *Scientific Reports, 5*, 12166.

Steinacker, P., Anderl-Straub, S., Diehl-Schmid, J., Semler, E., Uttner, I., von Arnim, C. A. F., et al. (2018). Serum neurofilament light chain in behavioral variant frontotemporal dementia. *Neurology, 91*, e1390–e1401.

Tabrizi, S. J., Leavitt, B. R., Landwehrmeyer, G. B., Wild, E. J., Saft, C., Barker, R. A., et al. (2019). Targeting huntingtin expression in patients with Huntington's disease. *The New England Journal of Medicine, 380*, 2307–2316.

Tabrizi, S. J., Scahill, R. I., Durr, A., Roos, R. A., Leavitt, B. R., Jones, R., et al. (2011). Biological and clinical changes in premanifest and early stage Huntington's disease in the TRACK-HD study: The 12-month longitudinal analysis. *Lancet Neurology, 10*, 31–42.

Tabrizi, S. J., Schobel, S., Gantman, E. C., Mansbach, A., Borowsky, B., Konstantinova, P., et al. (2022). A biological classification of Huntington's disease: The Integrated Staging System. *Lancet Neurology, 21*, 632–644.

Thomas, E. A. (2019). Epigenetic mechanisms in Huntington's disease. In O. Binda (Ed.), *Chromatin signaling and neurological disorders* (Vol. 7, pp. 73–95). Elsevier.

Thomas, E. A. (2020). *Salivary biomarkers and neurodegenerative conditions*. Springer.

Valadão, P. A. C., Santos, K. B. S., Ferreira, E. V. T. H., Macedo, E. C. T., Teixeira, A. L., Guatimosim, C., et al. (2020). Inflammation in Huntington's disease: A few new twists on an old tale. *Journal of Neuroimmunology, 348*, 577380.

van Dijk, J. G., van der Velde, E. A., Roos, R. A., & Bruyn, G. W. (1986). Juvenile Huntington disease. *Human Genetics, 73*, 235–239.

Wexler, N. S. (2004). Venezuelan kindreds reveal that genetic and environmental factors modulate Huntington's disease age of onset. *Proceedings of the National Academy of Sciences of the United States of America, 101*, 3498–3503.

Part II
Biofluid Sources for Huntington's Disease Biomarkers

Cerebrospinal Fluid Biomarkers in Huntington's Disease

Fabricio Pio and Blair R. Leavitt

Abbreviations

DOPAC	3,4-dihydroxyphenylacetic acid
BDNF	Brain derived neurotrophic factor
CSF	Cerebrospinal fluid
DA	Dopamine
GABA	Gamma-aminobutyric acid
HVA	Homovanilic acid
HTT	Huntingtin protein
HD	Huntington's disease
LP	Lumbar puncture
MSNs	Medium spiny projection neurons
mHTT	Mutant huntingtin protein
GGEL	NE-(omega-1-glutamyl)-1-LYSINE
NfL	Neurofilament light chain
NMDA	N-methyl D-aspartate
PDYN	Prodynorphin
PENK	Proenkephalin
TREM2	Triggering receptor expressed on myeloid cells-2
TFC	Total Functional Capacity
TMS	Total Motor Score
UHDRS	Unified Huntington's Disease Rating Scale

F. Pio
Center for Brain Health, University of British Columbia Hospital, Vancouver, BC, Canada

B. R. Leavitt (✉)
Center for Brain Health, University of British Columbia Hospital, Vancouver, BC, Canada

Centre for Molecular Medicine and Therapeutics, Department of Medical Genetics and Division of Neurology, Department of Medicine, University of British Columbia, Vancouver, BC, Canada
e-mail: bleavitt@cmmt.ubc.ca

E. A. Thomas, G. M. Parkin (eds.), *Biomarkers for Huntington's Disease*,
Contemporary Clinical Neuroscience, https://doi.org/10.1007/978-3-031-32815-2_2

1 Background

The cerebrospinal fluid (CSF) is an important and valuable source of information for many neurological disorders because of its close contact with the brain. The first known records that specifically mention the cerebrospinal fluid and its anatomical spaces date back to ancient Egypt, between 3000 and 2500 BC (Deisenhammer, 2015). A number of authors refer to the Smith papyrus as the first occurrence of the CSF in the medical literature (Clarke & O'Malley, 1971). Many famous names turn up in context with the history of the cerebrospinal fluid, including Hippocrates, who, referred as "water" surrounding the brain and, Galen of Pergamon, who called it "excremental liquid" (Hadju, 2003; Herbowski, 2013).

Despite these early reports, based on their descriptions of a "watery humour" and "spirituous lymph" filling the ventricles it is Andreas Vesalius (1514–1564, Belgium) and Emanuel Swedenborg (1688–1772, Sweden) to whom the discovery of CSF is usually attributed in the medical literature (Vecchio, 2017). Swedenborg summarized his observations on the brain, spinal cord, and blood circulation in a manuscript written between 1741 and 1744. However, he did not have medical credentials and was unable to find a publisher. An important contribution to our understanding of how the CSF changes in disease was made by Thomas Willis (1621–1675) an English physician who is remembered primarily for his discovery of the "circle of Willis" and the eleventh cranial nerve. In 1664 Willis described "a liquid" in the aqueduct of Sylvius that connects the ventricles. He reported that in cases of "epidemic fever" (i.e. meningitis) the consistency of the "liquid" was altered (Hadju, 2003).

Francois Magendie (1783–1855), a French physiologist, studied the properties of cerebrospinal fluid by experimenting on living animals. He discovered the foramen Magendie, the opening in the roof of the fourth ventricle. His description of the CSF as a fluid in which the brain and the spinal cord are suspended like the fetus in the uterus opened thinking about roles of the CSF in health and diseases (Tubbs et al., 2008). Four decades after Magendi's studies, rapid advances occurred in understanding the role of CSF. In 1891, W. Essex Wynter (1860–1945) treated tuberculous meningitis by tapping the spinal subarachnoid space (Hadju, 2003). In the same year, Heinrich Quincke (1842–1922) popularized lumbar puncture by presenting the technique at the German Congress of Medicine and advocating its use for diagnostic and therapeutic purposes (Quincke, 1891). Albrecht von Haller (1708–1777), a Swiss physician who is regarded as the first European physiologist and the father of modern physiology observed in 1747 that CSF was secreted into the ventricles (Herbowski, 2018) an observation that is considered a landmark in the development of physiology as a distinct branch of science.

The first full account of the cerebrospinal fluid was given by Domenico Cotugno (1736–1822), an Italian anatomist, physiologist, and professor of surgery at the University of Naples, in 1764, and for some time, it was referred as "liquor cotugnii". Contugno never referred to CSF in humans, but he described it in fishes and turtles (Peltier, 1988). There is also wide consensus that Heinrich Quincke performed the first diagnostic lumbar puncture (LP) in 1891 which paved the way for modern CSF diagnostic procedures. His studies in the cerebrospinal fluid included cytomorphological analysis, protein assay, spinal glucose levels and bacterial examination (Deisenhammer, 2015). William Mestrezat (1883–1928) and Harvey W. Cushing (1869–1939) helped to bring the scientific study of the CSF into the modern era. In 1912, Mestrezat gave the first accurate description of the chemical composition of the CSF (Oliveira et al., 2020). In 1914, Harvey W. Cushing demonstrated that the CSF was secreted by the choroid plexus (Lehtinen et al., 2013).

Interest in CSF increased rapidly, both in clinical practice and in research, when the LP was introduced to collect CSF samples in living human patients. Novel biochemical and analytical techniques developed in the twentieth century were rapidly applied to the analysis of CSF both in health and disease states. The cerebrospinal fluid was originally thought to be produced primarily the choroid plexus (Lehtinen et al., 2013), which is described as a cauliflower-like growth of blood vessels covered by a thin layer of epithelial cells. It projects into the temporal horns, of each lateral ventricle, posterior portion of the third and roof of the fourth ventricles. Because of controversies concerning the production of the cerebrospinal fluid, the recently formulated Bulat-Klarica-Orešković hypothesis describes the CSF production and absorption (CSF exchange) as constant and present everywhere in the CSF system, and although the CSF is partially produced by the choroid plexus, it is mainly formed as a consequence of water filtration between the capillaries and interstitial fluid (Orešković et al., 2017). The CSF has been described as the third circulation and more recently as part of the brain glymphatic system. It provides buoyancy, nutrients, and endogenous waste product removal for the brain by bulk flow into the venous and lymphatic systems and by carrier-mediated resorptive transport system in the choroid plexus.

CSF volume in the average adult is estimated to be 150 ml with a distribution of 125 ml within the subarachnoid spaces and 25 ml within the ventricles (Lehtinen et al., 2013). About 500 ml of CSF is produced each day and the net flow of CSF through the ventriculo-subarachnoid system is on average approximately 0.4 ml/min, with more CSF secreted at night than during the day. The CSF is mainly made up of Na^+, Cl^- and HCO_3^- with lesser amounts of K^+, Mg^{++}, Ca^{++}, certain vitamins (such as folate, Ascorbic acid, thiamine, and pyridoxine); and peptides and proteins actively transported into the CSF from blood or synthesized in the choroid plexus and then transferred into the CSF such as transthyretin, insulin like growth factor, and brain derived neurotrophic factor. The chemical composition of healthy CSF is similar to plasma, as is the osmotic pressure and Na++ concentration. However, its chloride ion content is 15% greater; K+ is approximately 40% less and glucose 30% less than that of the plasma. The primary source of absorption of the cerebrospinal fluid is through the arachnoid villi, microscopic finger-like inward projections of the arachnoid membrane that penetrate into the venous sinuses. The CSF presents many exogenous compounds to the choroid plexus for metabolism or removal indirectly cleaning the extracellular space of the brain.

The CSF contains certain essential substances including several micronutrients, ions, peptides and proteins (Lehtinen et al., 2013). These substances penetrate into the brain from the CSF and are essential for brain health. Although the vast majority of macronutrients such as glucose, amino acids, lactate, and many micronutrients, hormones, vitamins and minerals are transported directly from blood into the brain by specialized mechanisms in the brain capillaries. Although most macronutrients and certain micronutrients enter the brain mainly from blood directly through the blood-brain-barrier, a few substrates are transported solely or mainly from blood via the choroid plexus into the CSF and then diffuse slowly into the extra-cellular space of the brain for delivery into the brain cells. This choroid plexus - CSF- extra cellular brain space route applies to several micronutrients and ions, and it is now considered one of the main sources of folate and vitamin C for brain cells.

There has been clear recognition of free exchange of molecules mainly by diffusion, across ependymal and pia interfaces and flowing ventricular and subarachnoid CSF. The CSF flow is not unidirectional, but is bidirectional, pulsing back and forth through the aqueduct of Sylvius with each heartbeat. CSF flow is affected by both posture and inspiration. Body temperature, blood gases (ph, pO_2, pCO_2) and plasma osmolality may also affect choroid plexus blood flow and alter CSF production (Lehtinen et al., 2013). A major difference between the CNS and the rest of the body is normally there are very few immunoglobulins in the CSF and negligible white cells unlike the plasma and lymph.

2 Huntington's Disease (HD)

The discovery of the gene mutation that causes HD in 1993 by the Huntington Disease Collaborative Study Group was a major advance in the history of genetics (Reviewed in Chial, 2008). Almost three decades after the gene mutation was identified, this devastating disease remains incurable and no treatment has yet been proven to successfully prevent or delay the onset or progression of the disease (Roos, 2010). The disease-causing mutation in the *HTT* gene is an expanded CAG repeat in exon 1 that produces a mutant polyglutamine-expanded mutant form of the huntingtin protein (mHTT) (Fodale et al., 2020). Mutant huntingtin pathologically affects many cellular and molecular processes primarily through a toxic-gain-of-function (Labbadia & Morimoto, 2013). It aggregates and is toxic to the neuronal cells, particularly to vulnerable medium spiny neurons in the caudate/putamen (striatum). Mutant huntingtin causes selective neuronal dysfunction and death, ultimately resulting in the symptoms of Huntington's disease (Landles & Bates, 2004).

While many aspects of huntingtin's normal cellular functions remain poorly understood, there is a growing body of evidence that the protein plays an important role in endocytosis, intracellular vesicular trafficking, synaptic functioning, autophagy, and transcription. Mutant huntingtin has been linked to multiple pathologic cellular processes including impaired transcription, oxidative stress, dysregulated autophagy, aggregate formation, and disrupted homeostasis. This toxic gain-of-function acquired by the mutant huntingtin leads to many cellular abnormalities, but the critical insults and exact pathways that are the primary cause of neuronal death are still unknown (Ross et al., 2014).

Current clinical outcome measures currently used in clinical trials, such as the Unified Huntington's Disease Rating Scale (UHDRS) - although well-defined and accepted by regulatory agencies - lack the ability to distinguish between symptom relief and amelioration of the underlying pathological processes. They tend to be subjective in nature, susceptible to inter- and intra-rater variability, and generally lack the ability to detect pre-clinical changes that occur in carriers who are yet to develop overt symptoms (Weir et al., 2011). In more advanced stages of HD, knowledge is also limited about utility of these scales to track the course of clinical manifestations. There is a lack of sensitive disease outcome measures to track disease progression throughout the disease course in HD (Ross et al., 2014). Ceiling and floor effects of the UHDRS hamper the detection of changes in patients with both pre-manifest and advanced HD. These limitations make disease monitoring difficult and complicate efficacy measurements for therapeutic interventions in HD trials.

3 CSF Biomarkers in HD

The development of clinical biomarkers is crucial for trials of prospective therapies and also for assessment of disease progression (Ross et al., 2014; Disatnik et al., 2016). In Huntington's disease, biomarkers could be used for many purposes, including, diagnostic, by aiding in a more precise definition of onset; prognostic, indicating how the disease will develop in an individual, for natural history by accurately determining the disease stage and also, for pharmacokinetic and pharmacodynamic and efficacy parameters (Weir et al., 2011). The ideal biomarker would be quantitative, easy and cheap to process but mostly reliable. Biofluid markers that would reflect the neuropathology are highly desirable. Studies done with more accessible fluids such as blood and urine have shown mostly negative results, except in very defined situations.

The cerebrospinal fluid, although starting as a plasma ultrafiltrate, receives substantive contributions from the brain. Twenty per cent of its protein content is brain-derived, suggesting that relevant biomarkers for HD neuropathology can be identified (Reiber, 2001; Begcevic et al., 2016; González et al., 2018). The history of the search for Huntington's disease biomarkers in the cerebrospinal fluid dates back to the early 1970s (Manyam et al., 1978). However, low sample numbers, inconsistent procedures, unreliable quantification techniques and lack of targets to investigate that had a plausible link to the pathology of Huntington's disease, limited the utility of these initial studies. The earliest studies using the cerebrospinal fluid of HD mutant gene expansion carriers focused on neurotransmitters, as potential and quantifiable, surrogate for neuronal function and dysfunction. More recent efforts to elucidate CSF biomarkers in HD have focused on improved CSF sampling methods, longitudinal collections, and more comprehensive biochemical analyses.

HDClarity is a multisite and multinational initiative supported by the CHDI Foundation that aims to establish the largest highest-quality repository of HD CSF in the world. Samples of CSF and blood from clinically well-characterised HD mutation carriers and matched controls are being longitudinally collected and stored in a central repository to expedite research into biomarkers and to enable development of novel treatments for HD (Rodrigues et al., 2022). CSF and blood samples are collected using a standardised lumbar puncture protocol from participants in six cohorts: healthy controls; early Pre-manifest HD; late Pre-manifest HD; early Manifest HD; moderate Manifest HD, and advanced Manifest HD. A report on data from February 2016 to September 2019 confirmed that research LPs in HD are feasible, acceptable to the HD community, and that the HDClarity protocol has a manageable safety profile (Rodrigues et al., 2022).

4 Neurotransmitter Biomarkers

GABA (Gamma-Aminobutyric Acid)

In 1977, the first study quantifying GABA levels in Huntington's disease CSF, showed decreased levels of GABA in 19 Huntington's disease patients when compared to 26 normal controls (Enna et al., 1977). This study was done 5 years before a genetic marker and 15 years prior to the discovery of the causative gene in HD. Those results suggested that there could be a change in GABA before symptoms manifest. Bonnet et al. (1987) found that total GABA and homocarnosine to be significantly higher in the cerebrospinal fluid of Huntington's patients. A later study using Isoniazid (plus pyridoxine) demonstrated a three-fold increase in CSF GABA levels following treatment compared to placebo but failed to improve clinical measurements (Manyam et al., 1981). In this study, isoniazid plus pyridoxine also increased the CSF levels of aspartate, asparagine, homocarnosine, ornithine, histidine, alfa- amynobutiric acid, isoleucine and alanine, and the levels of beta alanine in the CSF and plasma, which contributed to the undesired clinic response observed with the treatment, according to Manyam et al. (1987). GABA has not been studied since 1987.

Choline and Other Amino Acids

Another neurotransmitter that caught attention was choline. Since the caudate is one of the most affected structures in Huntington's disease and it has the highest acetylcholinesterase activity in the brain, CSF levels of choline and acetylcholinesterase were quantified (Manyam et al., 1990). Choline was found to be decreased but acetylcholinesterase levels showed no difference in HD patients when compared to controls.

Another study, in which, 27 amino acids were quantified, lower levels of asparagine, isoleucine, leucine, phenylalanine, histidine, arginine, alfa aminoadipic amino acid and homocarnosine were found when compared to non-choreic controls (Byrne & Wild, 2016). Tyrosine was the only amino acid found to be increased. There has been no replication of these studies on amino acids in HD since the direct genetic testing for the mutation was established and it remains unclear to this date if quantification of amino acids in CSF could be informative of disease onset or progression (Byrne & Wild, 2016).

Dopamine

The striatum, one of the main structures affected by Huntington's disease pathology, shows prominent loss of post-synaptic dopamine receptors (Blumenstock & Dudanova, 2020). It is thought that the imbalance in dopamine signalling created by this loss, with the consequent reduction in dopamine reuptake, is a contributor to the choreic movements that are typical of the disorder (Chen et al., 2013; Cepeda et al., 2014; Reiner et al., 2011). Studies of dopamine metabolites in HD, mostly homovanilic acid (HVA), have been contradictory (reviewed in Byrne & Wild, 2016). Some have shown no difference in HD, while others found decreased levels of HVA. In one study, increased levels of dopamine (DA), homovanilic acid (HVA) and 3,4-dihydroxyphenylacetic acid (DOPAC) among others were found (García Ruiz et al., 1995). Garcia Ruiz quantified HVA, 5-HIAA and tryptophan in a diversity of neurological conditions and found that, although HVA was lower in Huntington's disease when compared to other neurological disorders, it is not a reliable marker of dopaminergic activity and that the potential role of the dopaminergic pathway metabolites as biomarkers in HD is still unclear (Byrne & Wild, 2016).

Proenkephalin and Prodynorphin

Proenkephalin (PENK) and prodynorphin (PDYN) are peptides produced by the striatal medium spiny projection neurons (MSNs) under dopaminergic signaling. Post-mortem brain of HD patients and rodent models have demonstrated to have reduced levels of PDYN transcripts and the neuropeptide dynophin is decreased in HD CSF (Al Shweiki et al., 2021). One study in 47 patients with HD, showed levels of PENK and PDYN to be significantly decreased compared to all other groups and were associated with increasing disease severity scores in HD (Barschke et al., 2022). The decline in these peptides appears to be associated with disease pathogenesis and these may represent severity markers in HD, reflecting ongoing striatal neurodegeneration of the MSNs (Niemela et al., 2021).

5 Tryptophan and the Kynurenine Pathway

Tryptophan is a neurotransmitter amino acid. It produces quinolinic acid and picolinic acid upon degradation through the kynurenine pathway. Quinolinic acid is an N-methyl-d-aspartate (NMDA) receptor agonist and has been proposed to contribute to the neuropathology of HD by increasing excitotoxicity in the striatum (Lugo-Huitrón et al., 2013). The direct intra-striatal administration of quinolinic acid showed to be selectively toxic to the medium spiny neurons producing deficits similar to some aspects of Huntington's disease (Harris et al., 1998). The enzyme kynurenine mono-oxygenase has been considered a high priority target for therapeutic development. One study using radioenzymatic assay showed no significant difference in the levels of quinolinic acid in the CFS of HD patients (Schwarcz et al., 1988). However, this study had only 10 HD participants and 7 controls among other issues such as, no information on the processing of the CSF, controls had schizophrenia, there was no standardization of dietary conditions. In 1992 Heyes and colleagues quantified several kynurenine pathway metabolites in several distinct neurological disorders. The number of HD sample was small and CSF quinolinic acid was found to be elevated in inflammatory disorders, but not in neurodegenerative diseases (Heyes et al., 1992). The most comprehensive evaluation to date used high-performance liquid-chromatography to measure tryptophan, kynurenine, kynurenic acid, 3- hydroxykynurenine, anthranilic acid and quinolinic acid in CSF from 20 healthy individuals compared with 20 pre-manifest HD and 40 manifest HD subjects. None of these metabolites were found to have significant group differences in CSF levels (Rodrigues et al., 2021).

6 Transglutaminase Activity Pathway

In Huntington's disease an expanded polyglutamine amino acid sequence confers a gain of toxicity to the huntingtin protein and transglutaminase enzymes have been implicated in the aggregation and toxicity of mutant huntingtin (Lesort et al., 1999). NE-(omega-1-glutamyl)-1-LYSINE (GGEL) is produced from transglutaminase reaction and can be used as a marker of its activity. Jeitner *et al*, found significant higher levels of GGEL in the CSF of 36 HD patients compared to 27 controls (Jeitner et al., 2001). Other glutamylpolyamines that could reflect alterations of transglutaminase activity and candidates for biomarkers including alfa-gltutamylspermidine, gama-glutamylputrescine and bis-y-glutamylputrescine were also increased in manifest HD compared to controls.

7 Cysteamine

Cysteamine and cystamine both inhibit transglutaminases but by different mechanisms. The involvement of increased transglutaminase activity in neurodegenerative diseases is supported by the observation that genetic inactivation or inhibition of various transglutaminases in animal models slows progression of these diseases (Van Raamsdonk et al., 2005). Cysteamine is approved for the treatment of cystinosis and has been evaluated for both Huntington's disease and non-alcoholic fatty liver disease. One study aimed to determine plasma, CSF, and tissue (liver, kidney, muscle) cysteamine levels following intraduodenal delivery of the drug in rats pre-treated and naïve to cysteamine and to estimate the hepatic first-pass effect on cysteamine (Dohil et al., 2014). There was no difference in CSF and tissue cysteamine levels between naïve and treated groups, although cysteamine was more rapidly cleared in the treated group.

8 Immune System Biomarkers

The activation of the immune system has been suggested to be a crucial element in the late-stage pathogenesis of Huntington's disease (Ellrichmann et al., 2013) and based on this hypothesis various potential CSF biomarkers related to inflammation have been studied. In addition, un-biased proteomic studies of CSF from HD subjects have identified a number of inflammatory proteins to be elevated compared to control CSF samples (reviewed in Silajdžić & Björkqvist, 2018). In a small study with 23 mutation carriers (20 manifest and 3 pre manifest HD) and 14 healthy controls, CSF levels of various inflammatory cytokines (IL6; IL8, YKL-40) were higher in HD mutation carriers than in controls (Rodrigues et al. 2016b). YKL-40 is a marker of microglial activation (Dichev et al., 2020; Muszyński et al., 2017; Llorens et al., 2017) that was found to be associated with clinical measurements of disease progression in HD CSF (Rodrigues et al. 2016b). YKL-40 levels in CSF had independent clinical predictive power in HD beyond the association with age and CAG repeats, predicting disease severity. The activation of the innate immune system may be a direct effect of mutant huntingtin on myeloid immune cells, suggested by the increased levels of the microglia-associated proteins chitotriosidase and YKL-40, which are considered markers of neuroinflammation, in the CSF of HD patients (Rodrigues et al. 2016b).

Neurogranin is a post-synaptic protein that regulates the availability of calmodulin. Calmodulin is a calcium-modulated protein that senses calcium levels and relays signals to various calcium sensitive enzymes, ion channels and other proteins. Triggering receptor expressed on Myeloid cells-2 (TREM2) is a surface cell receptor expressed by myeloid cells, monocytes, macrophages and microglia, its activation is inhibitory to the immune response (Dichev et al., 2020; Muszyński et al., 2017). Although, studies have shown TREM2 and neurogranin to be increased in HD CSF, they are not considered useful as biomarkers for disease progression in HD (Byrne et al. 2018a; Xiang et al., 2020). Neurosecretory protein VGF belongs to the granin family of neuropeptides. VGF and VGF-derived peptides have been repeatedly identified in well powered and well-designed multiomic studies as dysregulated in neurodegenerative diseases. Expression differences of VGF-derived peptides have also been associated with Huntington's disease (Quinn et al., 2021).

9 Brain-Derived Neurotrophic Factor (BDNF)

Brain derived neurotrophic factor is implicated in the survival of striatal neurons. BDNF is reduced in Huntington's disease possibly because mutant huntingtin impairs its cortical-striatal transport contributing to striatal neurodegeneration. The quantification of BDNF in CSF has been performed in the HD-CSF cohort using an ultrasensitive immunoassay. BDNF concentration was below the limit of detection of the conventional Elisa. Using the ultrasensitive methods, BDNF was quantifiable in all samples, but it did not differ between controls and HD mutation carriers in CSF (Ou et al., 2021). CSF BDNF levels were not associated with clinical scores or MRI brain volumes in HD, and had poor ability to discriminate controls from HD mutation carriers and premanifest from manifest HD. It was concluded that CSF BDNF was unlikely to be a useful biomarker for HD progression (Ou et al., 2021).

10 Neurofilaments

Neurofilament Light Chain Protein

Neuroaxonal damage is the pathological substrate of permanent disability in various neurological disorders (Khalil et al., 2018). Reliable quantification and longitudinal follow-up of this neuronal damage is important for assessing disease activity, monitoring treatment responses, facilitating treatment development and determining prognosis (Khalil et al., 2018). Neurofilament light chain (NfL) has the potential to be a global diagnostic biomarker for multiple neurodegenerative diseases (reviewed in Wang et al., 2019). NfL is a neuronal cytoplasmic protein highly expressed in large calibre myelinated axons (Gaetani et al., 2019). Its levels increase in cerebrospinal fluid and blood proportionally to the degree of axonal damage in a variety of neurological disorders, including inflammatory, neurodegenerative, traumatic and cerebrovascular diseases (Gaetani et al., 2019). New immunoassays able to detect biomarkers at ultralow levels have allowed for the measurement of NfL in blood and in CSF, thus making it possible to easily and repeatedly measure NfL for monitoring disease course in HD. Many studies have been done on neurofilament light chain as a possible biomarker in Huntington's disease.

NfL studies have shown to be elevated in many neurological conditions when compared with healthy controls and is a well-established biomarker for neuroaxonal damage, suggesting that NfL levels are a sensitive indicator of disease progression and therapeutic outcome (reviewed in Bridel et al., 2019). One longitudinal study over 8 weeks showed CSF mHTT, CSF NfL and plasma NfL accurately distinguished between controls and Huntington's disease mutant carriers (Byrne et al. 2018b). In addition, this study showed that the NfL concentration in the CSF and plasma was also able to segregate pre-manifest and manifest HD subjects.

Rodrigues and colleagues showed that mHTT and NfL have distinct longitudinal dynamics in Huntington's disease. In a cohort study with 80 participants, over a period of 24 months, mHTT and NfL in the CSF and NfL in blood were measured (Rodrigues et al., 2020). The results demonstrated that baseline CSF NfL values were better predictors of clinical disease status, subsequent clinical progression, and brain atrophy than the rate of change in NfL. Another conclusion from this study was that CSF levels of NfL were more useful than mHTT levels for monitoring progression and as a prognostic biomarker for HD disease state.

In an additional study, the CSF NfL levels of HD patients participating in a clinical trial (SAT-HD) and pre-manifest carriers were compared to CSF levels from matched healthy controls (Szejko et al., 2018). CSF levels of NfL were significantly higher in all HD subjects and pre-manifest carriers compared to controls. NfL levels correlated with age and CAG repeat number in HD patients and pre-manifest carriers but did not correlate with age in the control group. One of the limitations of this study was the small number of pre-manifest carriers and a lack of individuals with advanced HD. Despite these caveats, these findings support the previous studies' findings that CSF NfL levels are a highly sensitive, but non-specific, marker of neuronal (likely axonal) damage in HD mice (Soylu-Kucharz et al., 2017).

CSF NfL has shown the strongest effect size of all measurements and was the only measure associated with predicted years to onset in one study (Scahill et al., 2020). Higher NfL values were found in individuals closer to the predicted onset. 53% of individuals with premanifest-HD had CSF NfL in the normal range suggesting that this biomarker becomes abnormal about 24 years from predicted onset, which may be a potential appropriate time to initiate disease modifying therapies. CSF NfL in this early stage would be the most suitable biomarker to monitor progression and eventually efficacy showing superior sensitivity from plasma NfL, contrary to individuals close to disease onset when CSF and plasma NfL have near equivalence (Scahill et al., 2020).

Byrne and co-workers have presented convincing data on NfL concentrations in CSF and plasma with striking prognostic power for progression in clinical HD (Byrne et al., 2017). Monitoring NfL levels in CSF and plasma may also be a marker of relevance in pre-clinical studies. In this study, CSF NfL concentrations were associated with body weight loss and motor dysfunction, both robust and progressive features of HD (Byrne et al., 2017). It correlates with the rate of worsening of cognition, functional disability, and striatal atrophy beyond the effects age and CAG repeat size in HD (Byrne et al., 2017).

Tau Protein

Tau is another axonal protein which has often been shown to be altered in CSF samples from neurodegeneration diseases (Rodrigues et al. 2016a). CSF tau levels can be quantified using enzyme linked immunosorbent assays. In a study using healthy controls and HD pre-symptomatic and symptomatic subjects, 76 participants were enrolled in a cross-sectional multicenter international pilot study. After age adjustments, CSF tau levels were significantly elevated in HD expansion carriers when compared to healthy controls. It also predicted motor, cognitive and functional disability in patients. A validation cohort study with 11 pre-manifest and 12 manifest HD participants comparing tau and NfL as biomarkers for HD (Niemelä et al., 2017) and concluded that NfL had strong correlations with all items of UHDRS including total functional capacity (TFC), total motor score (TMS), and 5-year probability of disease onset. In this study CSF tau had slightly weaker correlations with TMS and TFC. These studies support the suggestion that CSF tau is a promising biomarker of disease progression and possibly of therapeutic response in HD and other neurodegenerative diseases (Schraen-Maschke et al., 2008).

11 Mutant Huntingtin

Quantification of disease associated proteins in the CSF has been critical for the study and treatment of several neurodegenerative disorders (Wild et al., 2015). However, mHTT is present at very low levels in the cerebrospinal fluid of HD patients. Evidence suggests that mutant huntingtin is toxic and at the DNA level somatic CAG repeat expansion in vulnerable cells influences disease course (Tabrizi et al., 2020). Viral expression of a mutant huntingtin fragment restricted to the striatum resulted in detectable mutant huntingtin in the CSF and targeted lowering of mutant huntingtin in the striatum of HD mice using an antisense oligonucleotide caused a significant reduction of mutant huntingtin in the CSF (Caron et al. 2022a). Genetic inactivation of mutant huntingtin in the forebrain also resulted in a significant reduction of mutant huntingtin in the CSF (Caron et al. 2022a). It required the development of an ultrasensitive single molecule counting mHTT immunoassay, with high specificity for mHTT to be feasible (Fodale et al., 2017). The analysis of the cerebrospinal fluid of HD mutation carriers and controls showed quantifiable levels of mHTT in nearly all mutation carriers while it was undetectable in control CSF samples (Wild et al., 2015). This study also showed that the level of mHTT is associated with proximity to predicted disease onset and corelated with cognitive and motor function decline (Wild et al., 2015). Another study to validate the ultrasensitive mHTT detection in the cerebrospinal fluid using the single molecule counting immunoassay concluded that the concentration of mutant huntingtin in the CSF may serve both as a disease progression biomarker and, as a readout for mHTT lowering therapeutic approaches (Southwell et al., 2015).

A highly sensitive mutant huntingtin detection assay utilising micro-bead-based immunoprecipitation and flow cytometry also found similar correlations between CSF mHTT and clinical measures in HD patients, and in addition demonstrated for the first time that mutant huntingtin levels in CSF reflect brain levels and were decreased following huntingtin lowering interventions (Southwell et al., 2015). These results supported the use of CSF mHTT analysis as a clinical biomarker in the first successful huntingtin lowering clinical trial (Tabrizi et al., 2019). An important safety consideration with huntingtin-lowering therapies is the degree to which any treatment targets mHTT selectively or targets both normal and mHTT, but this cannot be determined with assays that are specific for only polyglutamine expanded mHTT (Leavitt et al., 2020).

A recently developed ultrasensitive bead based single molecule counting immunoassay to quantify HTT protein in a polyglutamine length-independent manner (total mutant huntingtin and non-expanded wild type huntingtin) was assessed in control and HD participant CSF sample (Fodale et al., 2022). The selectivity and specificity of the assay was assessed, and a preliminary analytical qualification undertaken to enable its clinical use. A previously developed mutant huntingtin assay was also used to analyze CSF from the same control and HD participants (Fodale et al., 2017). The results suggest that there is a correlation between mutant huntingtin and total (polyglutamine length-independent) huntingtin levels in human CSF as assessed by this assay.

12 Metals

The importance of metal biology in neurodegenerative diseases such as Huntington disease is well- documented with evidence of interaction between metals such as copper, zinc, iron and manganese and mutant huntingtin pathology. Levels of copper, zinc, iron and manganese were measured in a sub-set of HDClarity CSF samples from control, premanifest, manifest and late-manifest HD participants (Pfalzer et al., 2022). Elevation of copper, manganese and zinc in the CSF were found early in the disease, prior to alterations in canonical biomarkers of HD. CSF iron was found to be elevated in manifest HD. These changes were not seen in the plasma and demonstrate that there are alterations in metal biology selectively in the CSF that occur prior to elevations in neurofilaments and mutant huntingtin.

13 Oxytocin

The hypothalamic neuropeptide oxytocin has been measured in a cohort of HD gene expansion carriers compared to controls and correlated to disease progression and social cognition (Hellem et al., 2022). Levels of oxytocin were found to be significantly lowered by 33.5% in HD gene expansion carriers compared to controls. Among the HD gene carriers divided by groups with or without cognitive impairment, oxytocin levels were decreased in patients with cognitive symptoms. The level of CSF oxytocin may represent an objective and comparable measure that could be used as a state biomarker for cognitive impairment.

14 Metabolic Changes

It hasn't been established yet how metabolic changes in the Huntington's disease brain correlate to progression across the full spectrum of early to late-stage disease. In one study, the metabolic profile was studied across manifest HD from early to advanced stages through metabolomic analysis by mass spectrometry in plasma and CSF (McGarry et al., 2020). In CSF, worsening disease was associated with nominally significant increase in NAD^+, arginine, saturated long chain free fatty acids, diacylglycerides, triacylglycerides and sphingomyelins. Diacylglycerides and triacyglycerides species associated with clinical progression were different in plasma and CSF suggesting different metabolic preferences for these compartments. NAD^+ levels were increased, correlating with disease progression, this was an expected finding. Data suggested that defects in the urea cycle, glycine, and serine metabolism may be under-recognized in the progression of HD pathology.

In another study using liquid chromatography mass spectrometry, the metabolome of CSF was analysed from premanifest and manifest HD subjects as well as controls (Herman et al., 2019). Inter-group differences showed that tyrosine metabolism, including tyrosine, thyroxine, L-dopa and dopamine was significantly altered in manifest compared to premanifest HD. These metabolites demonstrated moderate to strong associations with measures of disease severity and symptoms. Thyroxine and dopamine also correlated with 5-year risk of onset in premanifest HD subjects. Phenylalanine and the purine metabolism was also significantly altered, but was not strongly associated with disease severity (Herman et al., 2019). Decreased levels of lumichrome were commonly found in mutant huntingtin carriers and levels correlated with the 5-year risk of disease onset in premanifest carriers. CSF metabolome studies can be used to characterize molecular pathogenesis which may be an essential contributor to the future development of novel therapies (Herman et al., 2019).

15 Endolysosomal/Autophagy System

The endolysosomal/autophagy system is dysfunctional in HD, potentially contributing to disease pathogenesis and representing a potential target for therapeutic intervention. Using parallel reaction monitoring mass spectrometry analysis of multiple endolysosomal proteins in the CSF of 60 HD mutation carriers and 20 healthy controls none of 18 proteins measured displayed significant differences in concentration between HD patients and controls with generalized linear models controlling for age and CAG (Lowe et al., 2020). No significant differences across disease stage were found in any of three components: representing lysosomal hydrolases, binding/transfer proteins and innate immune system/peripheral proteins. Nevertheless, several proteins were associated with measures of disease severity and cognition, most notably amyloid precursor protein which displayed strong correlations with UHDRS total motor score, symbol digits modalities and stroop word reading (Lowe et al., 2020). Although unlikely to have a role as disease state CSF biomarkers for HD; several proteins in this pathway have demonstrated association with clinical severity in HD and warrant further study.

16 Micro-RNAs

Micro-RNAs are small noncoding RNA molecules with a bonded nucleotide base that negatively regulate mRNA levels in a sequence specific manner. miRNAs are abundant in the CNS and play important roles in various neuronal processes such as synaptic development, maturation, and plasticity. miRNAs are resistant to degradation by ribonuclease. Studies of human HD prefrontal cortex identified 75 significantly altered miRNAs, including several that were associated with age at HD motor onset or the level of neuropathologic involvement in the striatum (Reed et al., 2018). In one study cross-sectional CSF miRNA levels in prodromal and symptomatic participants with Huntington's disease were assessed. Participants were recruited from the larger cohort of the observational Predict-HD study which followed the clinical progression of participants with premanifest-HD over time. 6 miRNAs were found to be significantly increased in CSF (Wang & Zhang, 2020; Claassen & Torres-Russotto, 2018). These micro-RNAs (miR 135b-3; miR 140-5p; miR-520f-3p; miR-3958-5p, miR-4317 and miR-8082) could be potential biomarkers for the early diagnosis of HD (Wang & Zhang, 2020). A pilot study evaluating miR-323b-3p showed that it is upregulated in individuals carrying an expanded *HTT* allele (Ferradelschi et al., 2021).

17 Novel Exploratory CSF Proteins

In one recent study, 26 pre-specified proteins were analysed in CSF from premanifest-HD, manifest-HD and controls using parallel reaction monitoring mass spectrometry. Strong association of altered select CSF proteins with clinical measures of disease severity were reported. In manifest-HD C1QB, CNR1, PPP1R1B, TTR, Nfl and PDYN and with years to predict disease onset in the premanifest-HD, C4B, CTSD, IGHG1, TTR, ALB. These CSF proteins panels that were identified could improve discrimination of premanifest from controls; early/mid-stage HD from premanifest and late-stage from early-mid stage, demonstrating that combinations of CSF proteins can outperform individual marker for stratifying individuals based on HD mutation status and disease severity (Caron et al. 2022b).

18 Conclusions

With several clinical trials testing novel therapeutic approaches for treating Huntington's disease in progress and more expected soon - improved biomarkers are needed to monitor disease progression, target engagement, and efficacy. Cerebrospinal fluid is currently the lead candidate source of such biomarkers. Due to the inaccessibility of brain tissue for analysis, CSF concentrations of multiple potential biomarkers have been investigated for their ability to correlate with clinical features in HD and/or reflect changes in mHTT expression and resultant neuropathology in the brain. Robust biomarker assays have now been developed to detect changes in CSF levels of mHTT, total HTT, and neurofilament light chain. These and other potential biomarkers are being qualified for clinical use in the multiple HD clinical trials that are currently underway or planned for the near future (Table 1).

Table 1 Summary table of published biochemical biomarkers in HD CSF in the last 5 years

z	Year	Article	Authors	Publisher
1	2017	Tau or neurofilament light-Which is the more suitable biomarker for Huntington's disease?	Niemelä V, Landtblom AM, Blennow K, Sundblom	PLoS One. 2017 Feb 27;12(2):e0172762
2	2017	Neurofilament light protein in CSF and blood is associated with neurodegeneration and disease severity in Huntington's disease R6/2 mice	Soylu-Kucharz R, Sandelius Å, Sjögren M, Blennow K, Wild EJ, Zetterberg H, Björkqvist M	Sci Rep. 2017 Oct 26;7(1):14114
3	2017	Neurofilament light protein in blood as a potential biomarker of neurodegeneration in Huntington's disease: a retrospective cohort analysis	Byrne LM, Rodrigues FB, Blennow K, Durr A, Leavitt BR, Roos RAC, Scahill RI, Tabrizi SJ, Zetterberg H, Langbehn D, Wild EJ	Lancet Neurol. 2017 Aug;16(8):601–609
4	2017	YKL-40 as a Potential Biomarker and a Possible Target in Therapeutic Strategies of Alzheimer's Disease	Muszyński P, Groblewska M, Kulczyńska-Przybik A, Kułakowska A, Mroczko B	Curr Neuropharmacol. 2017;15(6):906–917
5	2017	YKL-40 in the brain and cerebrospinal fluid of neurodegenerative dementias	Llorens, F., Thüne, K., Tahir, W. *et al.*	. *Mol Neurodegeneration* **12**, 83 (2017)
6	2017	Validation of Ultrasensitive Mutant Huntingtin Detection in Human Cerebrospinal Fluid by Single Molecule Counting Immunoassay	Fodale V, Boggio R, Daldin M, Cariulo C, Spiezia MC, Byrne LM, Leavitt BR, Wild EJ, Macdonald D, Weiss A, Bresciani A	J Huntingtons Dis. 2017;6(4):349–361
7	2018	Cerebrospinal fluid total protein reference intervals derived from 20 years of patient data	. González, Quevedo; VEDO, A., Sánchez, M., González García, S	Journal of Laboratory and Precision Medicine, North America, 3,4, apr. 2018

(continued)

Table 1 (continued)

z	Year	Article	Authors	Publisher
8	2018	Neurofilaments as biomarkers in neurological disorders	Khalil M, Teunissen CE, Otto M, Piehl F, Sormani MP, Gattringer T, Barro C, Kappos L, Comabella M, Fazekas F, Petzold A, Blennow K, Zetterberg H, Kuhle J	Nat Rev Neurol. 2018 Oct; 14(10):577–589
9	2018	A Critical Evaluation of Wet Biomarkers for Huntington's Disease: Current Status and Ways Forward	Silajdžić, Edina and Björkqvist, Maria	J Huntingtons Dis. 2018;7(2):109–135
10	2018	MicroRNAs in CSF as prodromal biomarkers for Huntington disease in the PREDICT-HD study	Reed ER, Latourelle JC, Bockholt JH,Bregu J, Smock J, Paulsen JS, Myers RH	Neurology. 2018 Jan 23;90(4):e264–e272
11	2018	Evaluation of mutant huntingtin and neurofilament proteins as potential markers in Huntington's disease	Byrne LM, Rodrigues FB, Johnson EB, Wijeratne PA, De Vita E, Alexander DC, Palermo G, Czech C, Schobel S, Scahill RI, Heslegrave A, Zetterberg H, Wild EJ	Sci Transl Med. 2018 Sep 12;10(458)
12	2018	CSF microRNA in patients with Huntington disease	Claassen DO, Torres-Russotto D	Neurology. 2018 Jan 23;90(4):151–152
13	2018	Quantification of the Light Subunit of Neurofilament Protein in Cerebrospinal Fluid of Huntington's Disease Patients	Szejko N, Picón C, García-Caldentey J, de Yebenes JG, Alvarez-Cermeño JC, Villar LM, López-Sendón Moreno JL	PLoS Curr. 2018 Aug 31;10
14	2018	Cerebrospinal fluid neurogranin and TREM2 in Huntington's disease	Byrne LM, Rodrigues FB, Johnson EB, De Vita E, Blennow K, Scahill R, Zetterberg H, Heslegrave A, Wild EJ	Sci Rep. 2018 Mar 9;8(1):4260
15	2019	Alterations in the tyrosine and phenylalanine pathways revealed by biochemical profiling in cerebrospinal fluid of Huntington's disease subjects	Herman S, Niemelä V, Emami Khoonsari P, Sundblom J,Burman J, Landtblom AM, Spjuth O, Nyholm D, Kultima K	Sci Rep. 2019 Mar 11;9(1):4129
16	2019	Neurofilament Light Chainin Cerebrospinal Fluid and Blood as a Biomarker for Neurodegenerative Diseases: A Systematic Review and Meta-Analysis	Wang SY, Chen W, Xu W, Li JQ,Hou XH, Ou YN, Yu JT, Tan L	Medicine (Baltimore). 2022 Mar 4;101(9):e28932

(continued)

Table 1 (continued)

z	Year	Article	Authors	Publisher
17	2019	Phase 1–2a IONIS-HTTRx Study Site Teams. Targeting Huntingtin Expression in Patients with Huntington's Disease	Tabrizi SJ, Leavitt BR, Landwehrmeyer GB, Wild EJ, Saft C, Barker RA, Blair NF, Craufurd D, Priller J, Rickards H, Rosser A, Kordasiewicz HB, Czech C, Swayze EE, Norris DA, Baumann T, Gerlach I, Schobel SA, Paz E, Smith AV, Bennett CF, Lane RM	N Engl J Med. 2019 Jun 13;380(24): 2307–2316
18	2019	Neurofilament light chain as a biomarker in neurological disorders	Gaetani L, Blennow K, Calabresi P, *et al*	Journal of Neurology, Neurosurgery & Psychiatry 2019;**90:**870–881
19	2019	Diagnostic Value of Cerebrospinal Fluid Neurofilament Light Protein in Neurology: A Systematic Review and Meta-analysis	. Bridel C, van Wieringen WN, Zetterberg H, Tijms BM, Teunissen CE; and the NFL Group; Alvarez-Cermeño JC, Andreasson U, Axelsson M, Bäckström DC, Bartos A, Bjerke M, Blennow K, Boxer A, Brundin L, Burman J, Christensen T, Fialová L, Forsgren L, Frederiksen JL, Gisslén M, Gray E, Gunnarsson M, Hall S, Hansson O, Herbert MK, Jakobsson J, Jessen-Krut J, Janelidze S, Johannsson G, Jonsson M, Kappos L, Khademi M, Khalil M, Kuhle J, Landén M, Leinonen V, Logroscino G, Lu CH, Lycke J, Magdalinou NK, Malaspina A, Mattsson N, Meeter LH, Mehta SR, Modvig S, Olsson T, Paterson RW, Pérez-Santiago J, Piehl F, Pijnenburg YAL, Pyykkö OT, Ragnarsson O, Rojas JC, Romme Christensen J, Sandberg L, Scherling CS, Schott JM, Sellebjerg FT, Simone IL, Skillbäck T, Stilund M, Sundström P, Svenningsson A, Tortelli R, Tortorella C, Trentini A, Troiano M, Turner MR, van Swieten JC, Vågberg M, Verbeek MM, Villar LM, Visser PJ, Wallin A, Weiss A, Wikkelsø C, Wild EJ	JAMA Neurol. 2019 Sep 1;76(9):1035–1048

(continued)

Table 1 (continued)

z	Year	Article	Authors	Publisher
20	2020	Huntingtin-Lowering Therapies for Huntington Disease: A Review of the Evidence of Potential Benefits and Risks	Al Shweiki MR, Oeckl P, Pachollek A, Steinacker P, Barschke P, Halbgebauer S, Anderl-Straub S, Lewerenz J, Ludolph AC, Bernhard Landwehrmeyer G, Otto M	JAMA Neurol. 2020 Jun 1;77(6):764–772
21	2020	Cerebrospinal fluid endo-lysosomal proteins as potential biomarkers for Huntington's disease	Lowe AJ, Sjödin S, Rodrigues FB, Byrne LM, Blennow K, Tortelli R, Zetterberg H, Wild EJ	PLoS One. 2020 Aug 17;15(8): e0233820
22	2020	Cross-sectional analysis of plasma and CSF metabolomic markers in Huntington's disease for participants of varying functional disability: a pilot study	McGarry A, Gaughan J, Hackmyer C, Lovett J, Khadeer M,Shaikh H, Pradhan B, Ferraro TN, Wainer IW, Moaddel R	Sci Rep. 2020 Nov 24;10(1):20490
23	2020	Mutant huntingtin and neurofilament light have distinct longitudinal dynamics in Huntington's disease	Rodrigues FB, Byrne LM, Tortelli R, Johnson EB, Wijeratne PA,Arridge M, De Vita E, Ghazaleh N, Houghton R, Furby H, Alexander DC, Tabrizi SJ, Schobel S, Scahill RI, Heslegrave A, Zetterberg H, Wild EJ	Sci Transl Med. 2020 Dec 16; 12 (574): eabc2888
24	2020	Cortical and Striatal Circuits in Huntington's Disease	Blumenstock S, Dudanova I	Front Neurosci. 2020 Feb 6;14:82
25	2020	Analysis of mutant and total huntingtin expression in Huntington's disease murine models	Fodale, V., Pintauro, R., Daldin, M. *et al*	*Sci Rep* **10**, 22137 (2020)
26	2020	YKL-40 and neuron-specific enolase in neurodegeneration and neuroinflammation	Dichev V, Kazakova M, Sarafian V	Rev Neurosci. 2020 Jul 28;31(5):539–553
27	2020	Neurogranin: A Potential Biomarker of Neurological and Mental Diseases	Xiang Y, Xin J, Le W, Yang Y	Front Aging Neurosci. 2020 Oct 6;12:584743
28	2020	Biological and clinical characteristics of gene carriers far from predicted onset in the Huntington's disease Young Adult Study (HD-YAS): a cross-sectional analysis	Scahill RI, Zeun P, Osborne-Crowley K, Johnson EB, Gregory S, Parker C,Lowe J, Nair A, O'Callaghan C, Langley C, Papoutsi M, McColgan P, Estevez-Fraga C, Fayer K, Wellington H, Rodrigues FB, Byrne LM, Heselgrave A, Hyare H, Sampaio C, Zetterberg H, Zhang H, Wild EJ, Rees G, Robbins TW, Sahakian BJ, Langbehn D, Tabrizi SJ	Lancet Neurol. 2020 Jun;19(6):502–512

(continued)

Table 1 (continued)

z	Year	Article	Authors	Publisher
29	2020	Circulating Exosomal miRNA as Diagnostic Biomarkers of Neurodegenerative Diseases	Wang L, Zhang L	Front Mol Neurosci. 2020 Apr 15;13:53
30	2020	Huntington disease: new insights into molecular pathogenesis and therapeutic opportunities	Tabrizi SJ, Flower MD, Ross CA, Wild EJ	Nat Rev Neurol.2020 Oct; 16(10) :529–546
31	2021	VGF as a biomarker and therapeutic target in neurodegenerative and psychiatric diseases	Quinn JP, Kandigian SE, Trombetta BA, Arnold SE, Carlyle BC	Brain Commun. 2021Oct 27; 3(4) :fcab261
32	2021	Kynurenine pathway metabolites in cerebrospinal fluid and blood as potential biomarkers in Huntington's disease	Rodrigues FB, Byrne LM, Lowe AJ, Tortelli R, Heins M,Flik G, Johnson EB, De Vita E, Scahill RI, Giorgini F, Wild EJ	J Neurochem. 2021 Jul;158(2):539–553
33	2021	Brain-derived neurotrophic factor in cerebrospinal fluid and plasma is not a biomarker for Huntington's disease	Ou ZA, Byrne LM, Rodrigues FB, Tortelli R, Johnson EB, Foiani MS, Arridge M, De Vita E, Scahill RI, Heslegrave A, Zetterberg H, Wild EJ	Sci Rep. 2021 Feb 10;11(1):3481
34	2021	Cerebrospinal Fluid Levels of Prodynorphin-Derived Peptides are Decreased in Huntington's Disease	Al Shweiki MR, Oeckl P, Pachollek A, Steinacker P, Barschke P, Halbgebauer S, Anderl-Straub S, Lewerenz J, Ludolph AC, Bernhard Landwehrmeyer G, Otto M	Mov Disord. 2021 Feb;36(2):492–497
35	2021	Proenkephalin Decreases in Cerebrospinal Fluid with Symptom Progression of Huntington's Disease	Niemela V, Landtblom AM, Nyholm D, Kneider M, Constantinescu R, Paucar M, Svenningsson P, Abujrais S, Burman J, Shevchenko G, Bergquist J, Sundblom J	Mov Disord. 2021 Feb;36(2):481–491
36	2021	Circulating hsa-miR-323b-3p in Huntington's Disease: A Pilot Study	Ferradelschi, M; Romano, S; Giglio, S; Romano, C; Morena, E; Mechelli, R; Annibali, V; Ubaldi, M; Buscarinu, MC; Umeton, R; Sani, G; Vecchione, A; Salvetti, M; Ristori, G	Front. Neurol., 05 May 2021 Sec. Neurogenetics Volume 12 – 2021
37	2022	Cerebrospinal fluid biomarkers for assessing Huntington disease onset and severity	Caron NS, Haqqani AS, Sandhu A, Aly AE, Findlay Black H,Bone JN, McBride JL, Abulrob A, Stanimirovic D, Leavitt BR, Hayden MR	Brain Commun. 2022 Nov 25; 4(6) :fcac309

(continued)

Table 1 (continued)

z	Year	Article	Authors	Publisher
38	2022	Cerebrospinal fluid mutant huntingtin is a biomarker for huntingtin lowering in the striatum of Huntington disease mice	Caron NS, Banos R, Aly AE, Xie Y, Ko S, Potluri N, Anderson C,Black HF, Anderson LM, Gordon B, Southwell AL, Hayden MR	Neurobiol Dis. 2022 May;166:105652
39	2022	Quantifying Huntingtin Protein in Human Cerebrospinal Fluid Using a Novel Polyglutamine Length-Independent Assay	. Fodale V, Pintauro R, Daldin M, Spiezia MC, Macdonald D, Bresciani	. J Huntingtons Dis. 2022;11(3):291–305
40	2022	Cerebrospinal fluid levels of proenkephalin and prodynorphin are differentially altered in Huntington's and Parkinson's disease	Barschke P, Abu-Rumeileh S, Al Shweiki MHDR, Barba L, Paolini Paoletti F,Oeckl P, Steinacker P, Halbgebauer S, Gaetani L, Lewerenz J, Ludolph AC, Landwehrmeyer GB, Parnetti L, Otto M	J Neurol. 2022 Sep;269(9):5136–5143
41	2022	Decreased CSF oxytocin relates to measures of social cognitive impairment in Huntington's disease patients	Hellem MNN, Cheong RY, Tonetto S, Vinther-Jensen T, Hendel RK, Larsen IU, Nielsen TT, Hjermind LE, Vogel A, Budtz-Jørgensen E, Petersén Å, Nielsen JE	Parkinsonism Relat Disord. 2022 Jun; 99:23–29
42	2022	Alterations in metal homeostasis occur prior to canonical markers in Huntington disease	Pfalzer AC, Yan Y, Kang H, Totten M, Silverman J, Bowman AB, Erikson K, Claassen DO	Sci Rep. 2022 Jun 20;12(1):10373

References

Al Shweiki, M. R., Oeckl, P., Pachollek, A., Steinacker, P., Barschke, P., Halbgebauer, S., Anderl-Straub, S., Lewerenz, J., Ludolph, A. C., Bernhard Landwehrmeyer, G., & Otto, M. (2021, February). Cerebrospinal fluid levels of prodynorphin-derived peptides are decreased in Huntington's disease. *Movement Disorders, 36*(2), 492–497.

Barschke, P., Abu-Rumeileh, S., Al Shweiki, M. H. D. R., Barba, L., Paolini Paoletti, F., Oeckl, P., Steinacker, P., Halbgebauer, S., Gaetani, L., Lewerenz, J., Ludolph, A. C., Landwehrmeyer, G. B., Parnetti, L., & Otto, M. (2022, September). Cerebrospinal fluid levels of proenkephalin and prodynorphin are differentially altered in Huntington's and Parkinson's disease. *Journal of Neurology, 269*(9), 5136–5143.

Begcevic, I., Brinc, D., Drabovich, A. P., Batruch, I., & Diamandis, E. P. (2016, May). Identification of brain-enriched proteins in the cerebrospinal fluid proteome by LC-MS/MS profiling and mining of the Human Protein Atlas. *Clinical Proteomics, 15*(13), 11.

Blumenstock, S., & Dudanova, I. (2020, February). Cortical and striatal circuits in Huntington's disease. *Frontiers in Neuroscience, 6*(14), 82.

Bonnet, A. M., Tell, G., Schechter, P. J., Grove, J., Saint-Hilaire, M. H., De Smet, Y., & Agid, Y. (1987). Cerebrospinal fluid GABA and homocarnosine concentrations in patients with Friedreich's ataxia, Parkinson's disease, and Huntington's chorea. *Movement Disorders, 2*(2), 117–123.

Bridel, C., van Wieringen, W. N., Zetterberg, H., Tijms, B. M., Teunissen, C. E., and the NFL Group, Alvarez-Cermeño, J. C., Andreasson, U., Axelsson, M., Bäckström, D. C., Bartos, A., Bjerke, M., Blennow, K., Boxer, A., Brundin, L., Burman, J., Christensen, T., Fialová, L., Forsgren, L., Frederiksen, J. L., Gisslén, M., Gray, E., Gunnarsson, M., Hall, S., Hansson, O., Herbert, M. K., Jakobsson, J., Jessen-Krut, J., Janelidze, S., Johannsson, G., Jonsson, M., Kappos, L., Khademi, M., Khalil, M., Kuhle, J., Landén, M., Leinonen, V., Logroscino, G., Lu, C. H., Lycke, J., Magdalinou, N. K., Malaspina, A., Mattsson, N., Meeter, L. H., Mehta, S. R., Modvig, S., Olsson, T., Paterson, R. W., Pérez-Santiago, J., Piehl, F., Pijnenburg, Y. A. L., Pyykkö, O. T., Ragnarsson, O., Rojas, J. C., Romme Christensen, J., Sandberg, L., Scherling, C. S., Schott, J. M., Sellebjerg, F. T., Simone, I. L., Skillbäck, T., Stilund, M., Sundström, P., Svenningsson, A., Tortelli, R., Tortorella, C., Trentini, A., Troiano, M., Turner, M. R., van Swieten, J. C., Vågberg, M., Verbeek, M. M., Villar, L. M., Visser, P. J., Wallin, A., Weiss, A., Wikkelsø, C., & Wild, E. J. (2019, September 1). Diagnostic value of cerebrospinal fluid neurofilament light protein in neurology: A systematic review and meta-analysis. *JAMA Neurology, 76*(9), 1035–1048.

Byrne, L. M., Rodrigues, F. B., Blennow, K., Durr, A., Leavitt, B. R., Roos, R. A. C., Scahill, R. I., Tabrizi, S. J., Zetterberg, H., Langbehn, D., & Wild, E. J. (2017, August). Neurofilament light protein in blood as a potential biomarker of neurodegeneration in Huntington's disease: A retrospective cohort analysis. *Lancet Neurology, 16*(8), 601–609. Erratum in: *Lancet Neurology*, 2017 Sept, *16*(9), 683.

Byrne, L. M., Rodrigues, F. B., Johnson, E. B., De Vita, E., Blennow, K., Scahill, R., Zetterberg, H., Heslegrave, A., & Wild, E. J. (2018a, March 9). Cerebrospinal fluid neurogranin and TREM2 in Huntington's disease. *Scientific Reports, 8*(1), 4260.

Byrne, L. M., Rodrigues, F. B., Johnson, E. B., Wijeratne, P. A., De Vita, E., Alexander, D. C., Palermo, G., Czech, C., Schobel, S., Scahill, R. I., Heslegrave, A., Zetterberg, H., & Wild, E. J. (2018b, September 12). Evaluation of mutant huntingtin and neurofilament proteins as potential markers in Huntington's disease. *Science Translational Medicine, 10*(458), eaat7108.

Byrne, L. M., & Wild, E. J. (2016, January). Cerebrospinal fluid biomarkers for Huntington's disease. *Journal of Huntington's Disease*, 1–13.

Caron, N. S., Banos, R., Aly, A. E., Xie, Y., Ko, S., Potluri, N., Anderson, C., Black, H. F., Anderson, L. M., Gordon, B., Southwell, A. L., & Hayden, M. R. (2022a, May). Cerebrospinal fluid mutant huntingtin is a biomarker for huntingtin lowering in the striatum of Huntington disease mice. *Neurobiology of Disease, 166*, 105652.

Caron, N. S., Haqqani, A. S., Sandhu, A., Aly, A. E., Findlay Black, H., Bone, J. N., McBride, J. L., Abulrob, A., Stanimirovic, D., Leavitt, B. R., & Hayden, M. R. (2022b, November 25). Cerebrospinal fluid biomarkers for assessing Huntington disease onset and severity. *Brain Communications, 4*(6), fcac309.

Cepeda, C., Murphy, K. P., Parent, M., & Levine, M. S. (2014). The role of dopamine in Huntington's disease. *Progress in Brain Research, 211*, 235–254.

Chen, J. Y., Wang, E. A., Cepeda, C., & Levine, M. S. (2013, July). Dopamine imbalance in Huntington's disease: A mechanism for the lack of behavioral flexibility. *Frontiers in Neuroscience, 4*(7), 114.

Chial, H. (2008). Huntington's disease: The discovery of the Huntingtin gene. *Nature Education, 1*(1), 71.

Claassen, D. O., & Torres-Russotto, D. (2018, January 23). CSF microRNA in patients with Huntington disease. *Neurology, 90*(4), 151–152.

Clarke, & O'Malley, C. D. (1971). The human brain and spinal cord: A historical study illustrated by writings from antiquity to the twentieth century. *Isis, 62*(1) 205–206.

Deisenhammer, F. (2015). The history of cerebrospinal fluid. In F. Deisenhammer, F. Sellebjerg, C. Teunissen, & H. Tumani (Eds.), *Cerebrospinal fluid in clinical neurology* (p. 3). Springer.

Dichev, V., Kazakova, M., & Sarafian, V. (2020, July 28). YKL-40 and neuron-specific enolase in neurodegeneration and neuroinflammation. *Reviews in the Neurosciences, 31*(5), 539–553.

Disatnik, M.-H., Joshi, A., Saw, U., Nay, L., Shamloo, M., Leavitt, B. R., Qi, X., & Rosen. (2016). Daria Mochly- potential biomarkers to follow the progression and treatment response of Huntington's disease. *The Journal of Experimental Medicine, 213*(12), 2655–2669.

Dohil, R., Cabrera, B. L., Gangoiti, J. A., Barshop, B. A., & Rioux, P. (2014, April). Pharmacokinetics of cysteamine bitartrate following intraduodenal delivery. *Fundamental & Clinical Pharmacology, 28*(2), 136–143.

Vecchio, D. (2017). The history of cerebrospinal fluid analysis in multiple sclerosis: A great development over the last centuries. *Journal of Brain Disorders, 1*(1), 35–37.

Ellrichmann, G., Reick, C., Saft, C., & Linker, R. A. (2013). The role of the immune system in Huntington's disease. *Journal of Immunology Research*, 11 pages.

Enna, S. J., Stern, L. Z., Wastek, G. J., & Yamamura, H. I. (1977). Cerebrospinal fluid gamma-aminobutyric acid variations in neurological disorders. *Archives of Neurology, 34*, 683–685.

Ferradelschi, M., Romano, S., Giglio, S., Romano, C., Morena, E., Mechelli, R., Annibali, V., Ubaldi, M., Buscarinu, M. C., Umeton, R., Sani, G., Vecchione, A., Salvetti, M., & Ristori, G. (2021, May). Circulating hsa-miR-323b-3p in Huntington's disease: A pilot study. *Frontiers in Neurology, 12*, 657973.

Fodale, V., Boggio, R., Daldin, M., Cariulo, C., Spiezia, M. C., Byrne, L. M., Leavitt, B. R., Wild, E. J., Macdonald, D., Weiss, A., & Bresciani, A. (2017). Validation of ultrasensitive mutant huntingtin detection in human cerebrospinal fluid by single molecule counting immunoassay. *Journal of Huntington's Disease, 6*(4), 349–361.

Fodale, V., Pintauro, R., Daldin, M., Spiezia, M. C., Macdonald, D., & Bresciani, A. (2022). Quantifying huntingtin protein in human cerebrospinal fluid using a novel polyglutamine length-independent assay. *Journal of Huntington's Disease, 11*(3), 291–305.

Fodale, V., Pintauro, R., Daldin, M., et al. (2020). Analysis of mutant and total huntingtin expression in Huntington's disease murine models. *Scientific Reports, 10*, 22137.

Gaetani, L., Blennow, K., Calabresi, P., et al. (2019). Neurofilament light chain as a biomarker in neurological disorders. *Journal of Neurology, Neurosurgery, and Psychiatry, 90*, 870–881.

García Ruiz, P. J., Mena, M. A., Sanchez Bernardos, V., Díaz Neira, W., Gimenez Roldan, S., Benitez, J., & García de Yebenes, J. (1995, February). Cerebrospinal fluid homovanillic acid is reduced in untreated Huntington's disease. *Clinical Neuropharmacology, 18*(1), 58–63.

González, Q., Vedo, A., Sánchez, M., & González García, S. (2018, April). Cerebrospinal fluid total protein reference intervals derived from 20 years of patient data. *Journal of Laboratory and Precision Medicine, 3*, 4.

Hadju, S. I. (2003). A note from history: Discovery of the cerebrospinal fluid. *Annals of Clinical & Laboratory Science Summer, 33*(3), 334–336.

Harris, C. A., Miranda, A. F., Tanguay, J. J., Boegman, R. J., Beninger, R. J., & Jhamandas, K. (1998). Modulation of striatal quinolinate neurotoxicity by elevation of endogenous brain kynurenic acid. *British Journal of Pharmacology, 124*, 391–399.

Hellem, M. N. N., Cheong, R. Y., Tonetto, S., Vinther-Jensen, T., Hendel, R. K., Larsen, I. U., Nielsen, T. T., Hjermind, L. E., Vogel, A., Budtz-Jørgensen, E., Petersén, Å., & Nielsen, J. E. (2022, June). Decreased CSF oxytocin relates to measures of social cognitive impairment in Huntington's disease patients. *Parkinsonism & Related Disorders, 99*, 23–29.

Herbowski, L. (2018, May). Massa versus Haller: Priority of the cerebrospinal fluid discovery. *Neurologia Medico-Chirurgica (Tokyo), 58*(5), 225–227.

Herbowski, L. (2013). The maze of the cerebrospinal fluid discovery. *Anatomy Research International, 2013*, 596027.

Herman, S., Niemelä, V., Emami Khoonsari, P., Sundblom, J., Burman, J., Landtblom, A. M., Spjuth, O., Nyholm, D., & Kultima, K. (2019, March 11). Alterations in the tyrosine and phenylalanine pathways revealed by biochemical profiling in cerebrospinal fluid of Huntington's disease subjects. *Scientific Reports, 9*(1), 4129.

Heyes, M. P., Saito, K., Crowley, J. S., Davis, L. E., Demitrack, M. A., Der, M., Dilling, L. A., Elia, J., Kruesi, M. J., Lackner, A., Larsen, S. A., Lee, K., Leonard, H. L., Markey, S. P., Martin, A., Milstein, S., Mouradian, M. M., Prazantelli, M. R., Quearry, B. J., Salazar, A., Smith, M., Strauss, S. E., Sunderland, T., Swedo, S. W., & Tourtellotte, W. W. (1992, October). Quinolinic acid and Kynurenine pathway metabolism in inflammatory and non-inflammatory neurological disease. *Brain, 115*(5), 1249–1273.

Jeitner, T. M., Bogdanov, M. B., Matson, W. R., Daikhin, Y., Yudkoff, M., Folk, J. E., Steinman, L., Browne, S. E., & Beal, M. F. (2001). Blass J and Cooper AJL N^{ε} -(γ-l-Glutamyl)-l-lysine (GGEL) is increased in cerebrospinal fluid of patients with Huntington's disease. *Journal of Neurochemistry, 79*, 1109–1112.

Khalil, M., Teunissen, C. E., Otto, M., Piehl, F., Sormani, M. P., Gattringer, T., Barro, C., Kappos, L., Comabella, M., Fazekas, F., Petzold, A., Blennow, K., Zetterberg, H., & Kuhle, J. (2018, October). Neurofilaments as biomarkers in neurological disorders. *Nature Reviews. Neurology, 14*(10), 577–589.

Labbadia, J., & Morimoto, R. I. (2013). Huntington's disease: Underlying molecular mechanisms and emerging concepts. *Trends in Biochemical Sciences, 38*, 378.

Landles, C., & Bates, G. P. (2004, October). Huntingtin and the molecular pathogenesis of Huntington's disease. Fourth in molecular medicine review series. *EMBO Reports, 5*(10), 958–963.

Leavitt, B. R., Kordasiewicz, H. B., & Schobel, S. A. (2020, June). Huntingtin-lowering therapies for Huntington disease: A review of the evidence of potential benefits and risks. *JAMA Neurology, 77*(6), 764–772.

Lehtinen, M. K., Bjornsson, C. S., Dymecki, S. M., Gilbertson, R. J., Holtzman, D. M., & Monuki, E. S. (2013, November). The choroid plexus and cerebrospinal fluid: Emerging roles in development, disease, and therapy. *The Journal of Neuroscience, 33*(45), 17553–17559.

Lesort, M., Chun, W., Johnson, G. V., & Ferrante, R. J. (1999, November). Tissue transglutaminase is increased in Huntington's disease brain. *Journal of Neurochemistry, 73*(5), 2018–2027.

Llorens, F., Thüne, K., Tahir, W., et al. (2017). YKL-40 in the brain and cerebrospinal fluid of neurodegenerative dementias. *Molecular Neurodegeneration, 12*, 83.

Lowe, A. J., Sjödin, S., Rodrigues, F. B., Byrne, L. M., Blennow, K., Tortelli, R., Zetterberg, H., & Wild, E. J. (2020, August 17). Cerebrospinal fluid endo-lysosomal proteins as potential biomarkers for Huntington's disease. *PLoS One, 15*(8), e0233820.

Lugo-Huitrón, R., Ugalde Muñiz, P., Pineda, B., Pedraza-Chaverrí, J., Ríos, C., & Pérez-de la Cruz, V. (2013). Quinolinic acid: An endogenous neurotoxin with multiple targets. *Oxidative Medicine and Cellular Longevity, 2013*, 104024.

Manyam, B. V., Ferraro, T. N., & Hare, T. A. (1987, April). Isoniazid-induced alteration of CSF neurotransmitter amino acids in Huntington's disease. *Brain Research, 408*(1–2), 125–130.

Manyam, B. V., Giacobini, E., & Colliver, J. A. (1990, August). Cerebrospinal fluid acetylcholinesterase and choline measurements in Huntington's disease. *Journal of Neurology, 237*(5), 281–284.

Manyam, B. V., Katz, L., Hare, T. A., Kaniefski, K., & Tremblay, R. D. (1981, July 10). Isoniazid-induced elevation of CSF GABA levels and effects on chorea in Huntington's disease. *Annals of Neurology, 1*, 35.

Manyam, N. V. B., Hare, T. A., Katz, L., & Glaeser, B. S. (1978). Huntington's disease: Cerebrospinal fluid GABA levels in at-risk individuals. *Archives of Neurology, 35*(11), 728–730.

McGarry, A., Gaughan, J., Hackmyer, C., Lovett, J., Khadeer, M., Shaikh, H., Pradhan, B., Ferraro, T. N., Wainer, I. W., & Moaddel, R. (2020). Cross-sectional analysis of plasma and CSF metabolomic markers in Huntington's disease for participants of varying functional disability: A pilot study. *Scientific Reports* 2020 November 24, *10*(1), 20490 Erratum in: *Scientific Reports* 2021 May, *11*(1), 9947.

Muszyński, P., Groblewska, M., Kulczyńska-Przybik, A., Kułakowska, A., & Mroczko, B. (2017). YKL-40 as a potential biomarker and a possible target in therapeutic strategies of Alzheimer's disease. *Current Neuropharmacology, 15*(6), 906–917.

Niemelä, V., Landtblom, A. M., Blennow, K., & Sundblom, J. (2017, February). Tau or neurofilament light-which is the more suitable biomarker for Huntington's disease? *PLoS One, 12*(2), e0172762.

Niemela, V., Landtblom, A. M., Nyholm, D., Kneider, M., Constantinescu, R., Paucar, M., Svenningsson, P., Abujrais, S., Burman, J., Shevchenko, G., Bergquist, J., & Sundblom, J. (2021, February). Proenkephalin decreases in cerebrospinal fluid with symptom progression of Huntington's disease. *Movement Disorders, 36*(2), 481–491.

Oliveira, J. P. S., Mendes, N. T., Martins, Á. R., Sanvito, W. L., & Luiz, W. (2020). Cerebrospinal fluid: History, collection techniques, indications, contraindications and complications (Líquido cefalorraquidiano: história, técnicas de coleta, indicações, contraindicações e complicações). *J Bras Patol Med Lab, 56*, 1–11.

Orešković, D., Radoš, M., & Klarica, M. (2017, June). Role of choroid plexus in cerebrospinal fluid hydrodynamics. *Neuroscience, 354*, 69–87.

Ou, Z. A., Byrne, L. M., Rodrigues, F. B., Tortelli, R., Johnson, E. B., Foiani, M. S., Arridge, M., De Vita, E., Scahill, R. I., Heslegrave, A., Zetterberg, H., & Wild, E. J. (2021, February). Brain-derived neurotrophic factor in cerebrospinal fluid and plasma is not a biomarker for Huntington's disease. *Scientific Reports, 11*(1), 3481.

Peltier, L. F. (1988, February). The first description of the spinal fluid by Domenico Cotugno, 1775. *Clinical Orthopaedics and Related Research, 227*, 6–9.

Pfalzer, A. C., Yan, Y., Kang, H., Totten, M., Silverman, J., Bowman, A. B., Erikson, K., & Claassen, D. O. (2022, June). Alterations in metal homeostasis occur prior to canonical markers in Huntington disease. *Scientific Reports, 12*(1), 10373.

Quincke, H. (1891). *Ueber Hydrocephalus* (1st ed., pp. 321–339, 5 Abb.).

Quinn, J. P., Kandigian, S. E., Trombetta, B. A., Arnold, S. E., & Carlyle, B. C. (2021, October 27). VGF as a biomarker and therapeutic target in neurodegenerative and psychiatric diseases. *Brain Communications, 3*(4), fcab261.

Reed, E. R., Latourelle, J. C., Bockholt, J. H., Bregu, J., Smock, J., Paulsen, J. S., & Myers, R. H. (2018, January). PREDICT-HD CSF ancillary study investigators. MicroRNAs in CSF as prodromal biomarkers for Huntington disease in the PREDICT-HD study. *Neurology, 90*(4), e264.

Reiber, H. (2001, August). Dynamics of brain-derived proteins in cerebrospinal fluid. *Clinica Chimica Acta, 310*(2), 173–186.

Reiner, A., Dragatsis, I., & Dietrich, P. (2011). Genetics and neuropathology of Huntington's disease. *International Review of Neurobiology, 98*, 325–372.

Rodrigues, F. B., Byrne, L., McColgan, P., Robertson, N., Tabrizi, S. J., Leavitt, B. R., Zetterberg, H., & Wild, E. J. (2016a, October). Cerebrospinal fluid total tau concentration predicts clinical phenotype in Huntington's disease. *Journal of Neurochemistry, 139*(1), 22–25.

Rodrigues, F. B., Byrne, L. M., Lowe, A. J., Tortelli, R., Heins, M., Flik, G., Johnson, E. B., De Vita, E., Scahill, R. I., Giorgini, F., & Wild, E. J. (2021, July). Kynurenine pathway metabolites in cerebrospinal fluid and blood as potential biomarkers in Huntington's disease. *Journal of Neurochemistry, 158*(2), 539–553.

Rodrigues, F. B., Byrne, L. M., McColgan, P., Robertson, N., Tabrizi, S. J., Zetterberg, H., & Wild, E. J. (2016b, September 22). Cerebrospinal fluid inflammatory biomarkers reflect clinical severity in Huntington's disease. *PLoS One, 11*(9), e0163479.

Rodrigues, F. B., Byrne, L. M., Tortelli, R., Johnson, E. B., Wijeratne, P. A., Arridge, M., De Vita, E., Ghazaleh, N., Houghton, R., Furby, H., Alexander, D. C., Tabrizi, S. J., Schobel, S., Scahill, R. I., Heslegrave, A., Zetterberg, H., & Wild, E. J. (2020, December 16). Mutant huntingtin and neurofilament light have distinct longitudinal dynamics in Huntington's disease. *Science Translational Medicine, 12*(574), 1–13.

Rodrigues, F. B., et al. (2022). Safety and feasibility of research lumbar puncture in Huntington's disease: The HDClarity cohort and bioresource. *Journal of Huntington's Disease, 11*(1), 59–69.

Roos. (2010). RA Huntington's disease: A clinical review. *Orphanet Journal of Rare Diseases, 5*, 40.

Ross, C. A., Aylward, E. H., Wild, E. J., Langbehn, D. R., Long, J. D., Warner, J. H., Scahill, R. I., Leavitt, B. R., Stout, J. C., Paulsen, J. S., Reilmann, R., Unschuld, P. G., Wexler, A., Margolis, R. L., & Tabrizi, S. J. (2014, April). Huntington disease: Natural history, biomarkers and prospects for therapeutics. *Nature Reviews. Neurology, 10*(4), 204–216.

Scahill, R. I., Zeun, P., Osborne-Crowley, K., Johnson, E. B., Gregory, S., Parker, C., Lowe, J., Nair, A., O'Callaghan, C., Langley, C., Papoutsi, M., McColgan, P., Estevez-Fraga, C., Fayer, K., Wellington, H., Rodrigues, F. B., Byrne, L. M., Heselgrave, A., Hyare, H., Sampaio, C., Zetterberg, H., Zhang, H., Wild, E. J., Rees, G., Robbins, T. W., Sahakian, B. J., Langbehn, D.,

& Tabrizi, S. J. (2020, June). Biological and clinical characteristics of gene carriers far from predicted onset in the Huntington's disease Young Adult Study (HD-YAS): A cross-sectional analysis. *Lancet Neurology, 19*(6), 502–512.

Schraen-Maschke, S., Sergeant, N., Dhaenens, C. M., Bombois, S., Deramecourt, V., Caillet-Boudin, M. L., Pasquier, F., Maurage, C. A., Sablonnière, B., Vanmechelen, E., & Buée, L. (2008, August). Tau as a biomarker of neurodegenerative diseases. *Biomarkers in Medicine, 2*(4), 363–384.

Schwarcz, R., Tamminga, C. A., Kurlan, R., & Shoulson, I. (1988). Cerebrospinal fluid levels of quinolinic acid in Huntington's disease and schizophrenia. *Annals of Neurology, 24*, 580–582.

Silajdžić, E., & Björkqvist, M. (2018). A critical evaluation of wet biomarkers for Huntington's disease: Current status and ways forward. *Journal of Huntington's Disease, 7*(2), 109–135.

Southwell, A. L., Smith, S. E., Davis, T. R., Caron, N. S., Villanueva, E. B., Xie, Y., Collins, J. A., Ye, M. L., Sturrock, A., Leavitt, B. R., Schrum, A. G., & Hayden, M. R. (2015, July). Ultrasensitive measurement of huntingtin protein in cerebrospinal fluid demonstrates increase with Huntington disease stage and following brain huntingtin suppression. *Scientific Reports, 15*(5), 12166.

Soylu-Kucharz, R., Sandelius, Å., Sjögren, M., Blennow, K., Wild, E. J., Zetterberg, H., & Björkqvist, M. (2017, October). Neurofilament light protein in CSF and blood is associated with neurodegeneration and disease severity in Huntington's disease R6/2 mice. *Scientific Reports, 7*(1), 14114.

Szejko, N., Picón, C., García-Caldentey, J., de Yebenes, J. G., Alvarez-Cermeño, J. C., Villar, L. M., & López-Sendón Moreno, J. L. (2018, August 31). Quantification of the light subunit of neurofilament protein in cerebrospinal fluid of Huntington's disease patients. *PLoS Currents, 10*, 1–10.

Tabrizi, S. J., Flower, M. D., Ross, C. A., & Wild, E. J. (2020, October). Huntington disease: New insights into molecular pathogenesis and therapeutic opportunities. *Nature Reviews. Neurology, 16*(10), 529–546.

Tabrizi, S. J., Leavitt, B. R., Landwehrmeyer, G. B., Wild, E. J., Saft, C., Barker, R. A., Blair, N. F., Craufurd, D., Priller, J., Rickards, H., Rosser, A., Kordasiewicz, H. B., Czech, C., Swayze, E. E., Norris, D. A., Baumann, T., Gerlach, I., Schobel, S. A., Paz, E., Smith, A. V., & Bennett, C. F. (2019, June). Lane RM; Phase 1–2a IONIS-HTTRx study site teams. Targeting Huntingtin expression in patients with Huntington's disease. *The New England Journal of Medicine, 380*(24), 2307–2316.

Tubbs, R. S., Loukas, M., Shoja, M. M., Shokouhi, G., & Oakes, W. J. (2008). François Magendie (1783–1855) and his contributions to the foundations of neuroscience and neurosurgery. *Journal of Neurosurgery, 108*(5), 1038–1042.

Van Raamsdonk, J. M., Pearson, J., Bailey, C. D., Rogers, D. A., Johnson, G. V., Hayden, M. R., & Leavitt, B. R. (2005, October). Cystamine treatment is neuroprotective in the YAC128 mouse model of Huntington disease. *Journal of Neurochemistry, 95*(1), 210–220.

Wang, L., & Zhang, L. (2020, April). Circulating exosomal miRNA as diagnostic biomarkers of neurodegenerative diseases. *Frontiers in Molecular Neuroscience, 15*(13), 53.

Wang, S. Y., Chen, W., Xu, W., Li, J. Q., Hou, X. H., Ou, Y. N., Yu, J. T., & Tan, L. (2019). Neurofilament light chain in cerebrospinal fluid and blood as a biomarker for neurodegenerative diseases: A systematic review and meta-analysis. *Journal of Alzheimer's Disease, 72*(4), 1353–1361.

Weir, D. W., Sturrock, A., & Leavitt, B. R. (2011, June). Development of biomarkers for Huntington's disease. *Lancet Neurology, 10*(6), 573–590.

Wild, E. J., Boggio, R., Langbehn, D., Robertson, N., Haider, S., Miller, J. R., Zetterberg, H., Leavitt, B. R., Kuhn, R., Tabrizi, S. J., Macdonald, D., & Weiss, A. (2015, May). Quantification of mutant huntingtin protein in cerebrospinal fluid from Huntington's disease patients. *The Journal of Clinical Investigation, 125*(5), 1979–1986.

Xiang, Y., Xin, J., Le, W., & Yang, Y. (2020, October). Neurogranin: A potential biomarker of neurological and mental diseases. *Frontiers in Aging Neuroscience, 12*, 584743.

Extracellular Vesicles as Possible Sources of Huntington's Disease Biomarkers

Hanadi Ananbeh and Helena Kupcova Skalnikova

Abbreviations

AAV5	Adeno-associated virus serotype 5
Aβ	Amyloid β
AD	Alzheimer's disease
APP	Amyloid precursor protein
ALS	Amyotrophic lateral sclerosis
ASOs	Antisense oligonucleotides
BBB	Blood brain barrier
CNS	Central nervous system
CSF	Cerebrospinal fluid
EVs	Extracellular vesicles
ex-miRNAs	Exosomal micro RNAs
HD	Huntington's disease
hsiRNAs	Hydrophobically modified siRNAs
$hsiRNA^{HTT}$	hsiRNA targeting huntingtin mRNA
HTT	Huntingtin protein
HTT	Human gene coding huntingtin protein
ILVs	Intraluminal vesicles

H. Ananbeh
Laboratory of Applied Proteome Analyses and Research Center PIGMOD, Institute of Animal Physiology and Genetics of the Czech Academy of Sciences, Libechov, Czech Republic

H. Kupcova Skalnikova (✉)
Laboratory of Applied Proteome Analyses and Research Center PIGMOD, Institute of Animal Physiology and Genetics of the Czech Academy of Sciences, Libechov, Czech Republic

Institute of Biochemistry and Experimental Oncology, First Faculty of Medicine, Charles University, Prague, Czech Republic
e-mail: helena.skalnikova@lf1.cuni.cz

E. A. Thomas, G. M. Parkin (eds.), *Biomarkers for Huntington's Disease*, Contemporary Clinical Neuroscience, https://doi.org/10.1007/978-3-031-32815-2_3

lncRNA	Long non-coding RNA
ISEV	International Society for Extracellular Vesicles
KI	Knock-in
KI-HD	Knock-in pig model of Huntington's disease
mHTT	Mutant huntingtin
miHTT	miRNA targeting human *HTT*
mRNA	Messenger RNA
MVB	Multivesicular body
NDs	Neurodegenerative diseases
ONTs	Oligonucleotide therapeutics
PD	Parkinson's disease
PolyQ	Polyglutamine repeats
REST	RE1-Silencing Transcription Factor
RISC	RNA Induced Silencing Complex
RVG	Rabies virus glycoprotein
siRNA	Small interfering RNA
SOD1	Superoxide dismutase 1
TgHD	Transgenic pig model for Huntington's disease

1 Introduction

Extracellular vesicles (EVs; Fig. 1) are nanosized particles enveloped by a lipid bilayer membrane. They are produced basically by all cell types and are released from cells into the extracellular space and body fluids. EVs are present, for example, in plasma, serum, urine, saliva, breast milk, amniotic fluid, seminal fluid, nasal and bronchial lavage fluid, cerebrospinal fluid (CSF) and bile, as well as in the supernatants of *in vitro* cultured cells (Brennan et al., 2020; Kalra et al., 2016; Kupcova Skalnikova et al., 2019; Yáñez-Mó et al., 2015).

History and Significance of Extracellular Vesicles

EVs were initially observed by Chargaff and West in 1946 during their studies on blood clotting. They discovered a particular fraction with high clotting capability that sediments at 31,000 g, and they suggested that this fraction might contain the thromboplastic agent, as well as a variety of small fragments of blood cells (Chargaff & West, 1946). Twenty years later, electron microscopy images of minute particles obtained from plasma and serum by high speed centrifugation were published by Wolf. Such particles were referred as 'platelet dust' as they contained several platelet proteins. They were also rich in phospholipids (Wolf, 1967). Later on, in 1971, additional electron microscopy images were published for these particles and they were described as microparticles obtained from platelet free plasma. Moreover,

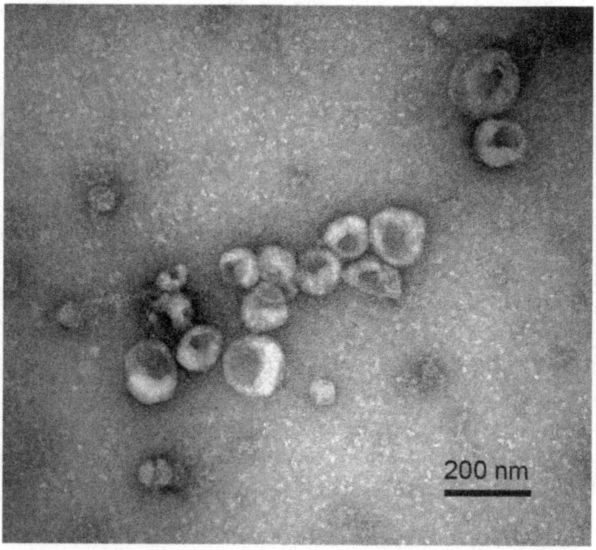

Fig. 1 Small extracellular vesicles isolated from blood plasma of HD patient and visualized by transmission electron microscopy. Uranylacetate contrasting; scalebar 200 nm

basic information about their composition, such as the presence of cholesterol, phospholipids, the platelet contractile protein trombosthenin, and ATPase activity were provided for the first time (Crawford, 1971). Early observations by electron microscopy also detected vesicle-like structures that were present in the alveolar space and released from alveolar cells (Sun, 1966), and matrix vesicles that have been identified in the matrix of hypertrophic cartilage of the bone (Anderson, 1969; Bonucci, 1967). Moreover, the early observation of EVs included reports of "virus like particles" and membrane enveloped particles in biofluids of cancer patients and healthy controls (Levine et al., 1967; Prince & Adams, 1966; Seman et al., 1971), rectal adenoma tissue (De Broe et al., 1975), thyroid gland (Nunez et al., 1974), bovine serum (Benz & Moses, 1974; Dalton, 1975), and seminal plasma (Stegmayr & Ronquist, 1982). The presence of multivesicular bodies (MVBs), i.e. late endosomes whose content can be released from cells as EVs, were initially reported by Nunez et al. in 1974 (Nunez et al., 1974). The fusion of the MVB membrane with the cell membrane leading to release of the EVs was reported in detail in 1983 by Pan and Johnstone, in a study of differentiation of immature sheep reticulocytes (Pan & Johnstone, 1983). Later in 1987, Johnstone et al. reported vesicle formation during the reticulocyte maturation in more detail. Such vesicles were responsible for the transferrin receptor removal during this stage of erythrocyte development, and the vesicles originating from luminal vesicles of MVBs were named 'exosomes' (Johnstone et al., 1987). For the following decade, EVs had been mostly considered as a way to remove unwanted molecules and cellular waste, and largely stood as a side scientific interest.

Since late 1990s, however, important biological and physiological roles of EVs started to emerge, including roles in antigen presentation (Raposo et al., 1996) and cell to cell communication. A major milestone in EV research was met in 2007, by the discovery that EVs contain mRNA that can be transferred to recipient cells and translated to protein, modifying the recipient cell phenotype (Valadi et al., 2007). Later on, it was revealed that EVs are involved in a wide range of physiological and pathological processes, such as blood coagulation, cell proliferation, inflammation and immune responses, reproduction, and also in nervous system intercellular communication, neural cell survival and neurite growth (Yáñez-Mó et al., 2015; Colombo et al., 2014; Busatto et al., 2021; Bahmani & Ullah, 2022; Mathieu et al., 2019; Basso & Bonetto, 2016).

Recent advances in high-throughput techniques have led to identification of various bioactive molecules carried by EVs, such as lipids, proteins, small coding and non-coding nucleic acids, and metabolites (Colombo et al., 2014; Vandendriessche et al., 2020; Wang & Zhang, 2020). Such molecular EV cargo reflects the cytoplasmic signatures of the cell of origin. Due to this fact, and due to their availability from body fluids, attention has risen in EVs, mainly exosomes, as potential sources for novel non-invasive biomarker discovery for wide range of human disorders including neurodegenerative diseases (NDs), several types of cancers, cardiovascular diseases, and many other disorders (Beetler et al., 2022; Takeuchi & Nagai, 2022).

EV Types and Recommendations by the International Society for Extracellular Vesicles

Based on the mechanism of their cellular biogenesis, EVs are classified into three main subtypes: exosomes, microvesicles (also called microparticles or ectosomes), and apoptotic bodies (Kalra et al., 2016; Yáñez-Mó et al., 2015). Exosomes are the smallest (30–150 nm) and the most studied type of EVs. Exosomes originate in the endocytic pathway, i.e. in the intraluminal vesicles of MVBs, which are released from cells by the fusion of the MVB membrane with the plasma membrane (Fig. 2). Microvesicles (diameter 100–1000 nm) are derived directly from the plasma membrane by membrane budding and are released in response to specific stimuli, like ATP level variation (Fig. 2) (Basso & Bonetto, 2016; Clancy et al., 2021; Colombo et al., 2014). The apoptotic bodies are the largest type of EVs (0.5–2 μm in diameter) and are formed by plasma membrane blebbing during the programmed cell death (Fig. 2) (Basso & Bonetto, 2016; Battistelli & Falcieri, 2020). The International Society for Extracellular Vesicles (ISEV) recommends to use the superordinate term “extracellular vesicles” for all studies, where the particular vesicle subtype (i.e. exosome, microvesicle, apoptotic body) cannot be identified (Witwer & Théry, 2019).

Various isolation techniques are being used to obtain EVs from samples leading to various EV yields and purity. The most frequently used method is ultracentrifugation, which can be combined with gradient or density cushion ultracentrifugation to increase the purity of isolated EVs (Roux et al., 2020). Other frequently used techniques are based on EV precipitation, size-exclusion chromatography, immunoaffinity capture by surface markers and many other principles. The isolation

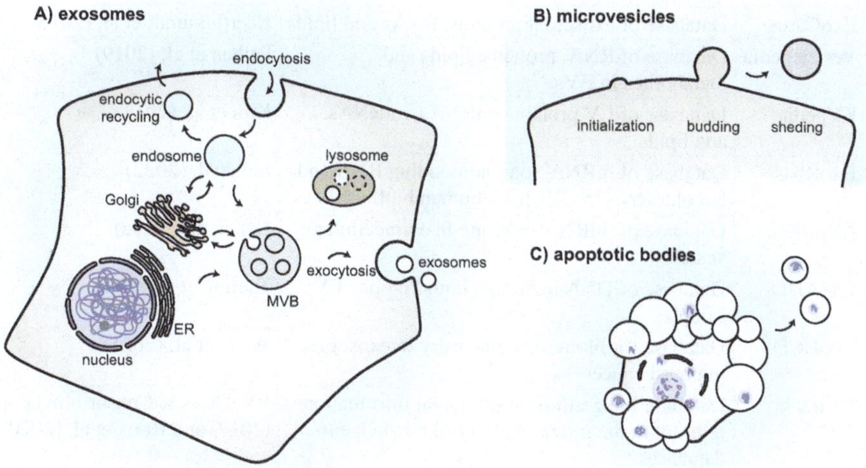

Fig. 2 Biogenesis of extracellular vesicles. (**a**) Exosomes originate from endocytic pathway. Early endosomes are formed by endocytosis. By inward budding of the endosome membrane, intraluminal vesicles (ILVs) appear inside of the endosome, giving rise to multivesicular body (MVB). The ILVs can be directed to the lysosome for degradation or released from cells in the form of exosomes. (**b**) Microvesicles arise directly from the plasma membrane after remodeling of the cytoskeleton and lipid and protein rearrangement, which leads to changes in membrane rigidity and enable outward budding of the membrane and shedding of microvesicles. (**c**) Apoptotic bodies are formed as blebs of cells undergoing apoptosis. (Adapted from Han et al., 2022, Vassileff et al., 2020; Yáñez-Mó et al., 2015)

technique or combination of techniques should be carefully selected prior following studies, such as molecular profiling of the EV cargo, to obtain EVs in required purity and minimize contamination from other molecules present in the original sample. The quality and purity of isolated EVs should be precisely analyzed to enable correct interpretation of the results of subsequent studies. Guidelines for minimal experimental requirements have been published by the ISEV (Lötvall et al., 2014; Théry et al., 2018; Witwer et al., 2021). Several databases have collected the results of studies on the molecular composition of EVs (Table 1).

2 Clinical Significance of EVs in Huntington's Disease

Among neurodegenerative diseases, EVs have been intensively studied in frequently occurring disorders, such as Alzheimer's disease (AD) and Parkinson's disease (PD) (Li et al., 2019; Vandendriessche et al., 2020; Quiroz-Baez et al., 2020; Marchetti et al., 2020; Xia et al., 2022; Natale et al., 2022). However, in Huntington's

Table 1 Databases collecting results from extracellular vesicle research

Database	Description	References
ExoCarta	Database of exosomal proteins, RNAs and lipids	Keerthikumar et al. (2016)
Vesiclepedia	Database of RNA, proteins, lipids and metabolites in EVs	Pathan et al. (2019)
EVpedia	Database of EV proteins, mRNAs, miRNAs, and lipids	Kim et al. (2013)
ExoRbase	Database of mRNA, long non-coding RNA and circular RNA in EVs from human biofluids	Lai et al. (2022)
EVmiRNA	Database of miRNA profiling in extracellular vesicles	Liu et al. (2019a)
EV-ADD	Database of DNA in human liquid biopsy EV samples	Tsering et al. (2022)
ExoBCD	Database for biomarker discovery in exosomes in breast cancer	Wang et al. (2021)
EVtrack	Database integrating experimental information relevant to the interpretation of knowledge in databases	EV-TRACK Consortium et al. (2017) and Roux et al. (2020)

disease (HD), the research of EVs, particularly in the context of biomarker discovery, is at its beginning (Table 2). Therefore, in this chapter we will discuss in broader context the roles of EVs in HD pathogenesis and prognosis, as well as their potential use as sources of diagnostic biomarkers and therapeutic vehicles for HD.

EVs in HD Pathogenesis

Intercellular Spreading of Misfolded Proteins in HD

The accumulation of misfolded proteins is the distinctive hallmark of NDs, including AD, PD, HD, amyotrophic lateral sclerosis (ALS), and prion diseases. The accumulated proteins form various types of inclusions such as amyloid β (Aβ) and tau deposits in AD, α-synuclein aggregates called Lewy bodies in PD, and mutant superoxide dismutase 1 (SOD1) aggregates in ALS (Ananbeh et al., 2021). In HD, due to a mutation in the gene coding the huntingtin protein (HTT), a mutant huntingtin protein (mHTT) with an elongated polyglutamine (polyQ) sequence is produced. The mHTT protein accumulates and gives rise to nuclear and cytoplasmic inclusions (Arrasate & Finkbeiner, 2012; Jarosińska & Rüdiger, 2021). The protein aggregates exhibit cytotoxic effects. Neurons are particularly vulnerable to such toxicity. Misfolded proteins can also be transferred to neighboring cells in a prion like pattern (Aguzzi & Rajendran, 2009; Busatto et al., 2021).

In the case of HD, propagation of mHTT between cells has been well documented both *in vitro* and *in vivo*. Initial experiments with the addition of synthetic polyQ peptides (i.e. peptides with elongated polyglutamine sequence to mimic the N-terminal part of mHTT), showed polyQ peptide internalization into mammalian

Table 2 Publications with relation to EVs/exosomes as sources of HD biomarkers

Purpose of study	Sample type	Organism/ model	Method of EVs isolation	EV markers	References
Incorporation of mHTT exon 1 RNA and protein into EVs and transfer to recipient cells	Conditioned medium	Human embryonic kidney 293T cells	Ultracentrifugation	ALIX	Zhang et al. (2016)
Platelet-derived EVs as possible sources of HD biomarkers	Blood plasma	Human (HD patients and healthy controls)	Gradient centrifugation	CD41, TSG101, ALIX	Denis et al. (2018)
Proteomic study of urine-derived EVs in PD with relation to HD	Urine	Human (PD patients and healthy controls)	Ultracentrifugation	CD63, TSG101	Wang et al. (2019)
Proteomic study of EV roles in mouse and human cell models of HD pathogenesis	Conditioned medium	HD knock-in mouse-derived striatal cells; neural cells derived from human HD iPS cells	Size exclusion chromatography	Tetraspanins	Tartaglia et al. (2022) (EHDN conference abstract)
Monitoring HTT and mHTT content in plasma-derived EVs	Blood plasma	Human (HD patients and healthy controls); minipig (TgHD, KI-HD, and wild-type)	Ultracentrifugation and gradient ultracentrifugation	CD9, TSG101, ALIX	Ananbeh et al. (2022)

Abbreviations: *TgHD* transgenic pig model of HD, *KI-HD* knock-in pig model of HD

cells cultured *in vitro,* and the formation of nuclear and cytoplasmic aggregates within cells (Ren et al., 2009; Yang et al., 2002). Later studies confirmed direct cell-to-cell transmission of mHTT between *in vitro* cultured glioma cells, neuronal cells and primary neurons (Costanzo et al., 2013; Herrera et al., 2011), *ex vivo* cortico-striatal slice model, as well as *in vivo* in an HD mouse model (Pecho-Vrieseling et al., 2014). The spreading and aggregation of mHTT into genetically normal neural tissue was also reported in three HD patients undergoing fetal striatal tissue transplantation where the mHTT aggregates were translocated into the extracellular matrix of the transplanted tissue (Cicchetti et al., 2014). Several mechanisms for intercellular transport of mHTT or its aggregates have been suggested, including spreading by tunneling nanotubes (Costanzo et al., 2013), by phagocytic glia (Pearce et al., 2015), through endocytic uptake (Babcock & Ganetzky, 2015; Pecho-Vrieseling et al., 2014) or by direct penetration of plasma membranes (Masnata &

Cicchetti, 2017). Targeting mHTT spreading is a candidate for future therapeutic interventions (Masnata & Cicchetti, 2017).

Extracellular Vesicles in Spreading of mHTT

Recently, EVs, particularly exosomes, became recognized as particles with the capacity of spreading misfolded proteins in the central nervous system (CNS) in NDs (Cheng et al., 2020; Guo & Lee, 2014; Vandendriessche et al., 2021; Vella et al., 2016). Obviously, the neurotoxic proteins can be selectively integrated into the intraluminal vesicles of MVBs and thereafter released into the extracellular environment within exosomes (Vella et al., 2008). Several studies reported the transfer of the neurotoxic proteins such as prions (Coleman et al., 2012; Vingtdeux et al., 2012), amyloid precursor protein (APP), phosphorylated tau and α-synuclein (Vella et al., 2016) across the CNS via exosomes thereby contributing to the spread of the disease (Bellingham et al., 2012; Properzi et al., 2015). Beside exosomes, extrusion of very large (~4 μm) membrane-enveloped vesicles was observed in neurons from *Caenorhabditis elegans*. Such vesicles were suggested to function in clearing protein aggregates and dysfunctional organelles and also being able to transfer protein aggregates among cells (Melentijevic et al., 2017).

In the last decade, information also appears about spreading of mHTT between cells mediated by EVs, both at the RNA and protein level. Incorporation of RNA coding mHTT into EVs, and its transmission to cells, was studied by Zhang et al. They co-cultured human embryonic kidney 293T cells with *HTT*-exon 1 polyQ-GFP construct, which led to the formation of EVs containing both expanded glutamine repeat-coding mRNA and polyQ-GFP protein. Such vesicles could be taken up by mouse striatal neural cells leading to increases in the polyQ-GFP RNA and GFP protein levels. Interestingly, in experiments with differentiated mouse striatal cell lines, the cell line expressing mRNA encoding mHTT with expanded polyQ (Q111/Q111) incorporated higher amounts of this RNA into EVs, compared to cells expressing normal repeat lengths (Q7/Q7) (Zhang et al., 2016). Additional work was performed by Jeon et al., who studied dissemination of mHTT pathology between cells by exosomes *in vitro* and *in vivo*. The *in vitro* incubation of normal murine neurons with exosomes produced by fibroblasts derived from a severe juvenile HD patient carrying 143 CAG repeats (HD143F cells) led to the formation of intracellular mHTT aggregates in mouse neurons. Similarly, implantation of HD143F-derived exosomes into the ventricle of neonatal mice triggered the manifestation of HD phenotypes, with neurological symptoms and detectable mHTT in mouse striatal neurons (Jeon et al., 2016). Taken together, these studies suggest the involvement of EVs in HD pathogenesis by mHTT transfer. On the other hand, EVs may participate in mHTT removal from cells (Deng et al., 2017). The potential for mHTT to integrate into EVs is consistent with previous findings by Velier et al. showing mHTT association with membranes and vesicles in fibroblasts from HD patients (Velier et al., 1998). More studies are necessary to confirm such results and analyze the principles and regulations of the molecular sorting of mHTT into EVs.

Disruption of Vesicular Transport and EV Production in HD

Although the physiological cellular functions of HTT are not completely understood, its role as an integrator of vesicular trafficking and transport has been known for a long time (Caviston & Holzbaur, 2009; Velier et al., 1998). Pathological conditions, such as downregulation of endogenous HTT or expression of mHTT, affect vesicular transport (Caviston & Holzbaur, 2009; Schweitzer et al., 2009), which can have consequences for several other cellular processes, particularly autophagy and exosome release. Autophagy is a process used for cellular component degradation, where cellular components, including organelles targeted for degradation, are enveloped by a lipid bilayer membrane, transported and fused with the lysosome. Autophagy and exosome biogenesis are overlapping pathways (Li et al., 2023). HTT has been shown to act as a scaffold protein that promotes initiation of autophagy and selective recruitment of cargo into autophagosomes (Ochaba et al., 2014; Rui et al., 2015). Currently it is not clear if the HTT standing at the initiation of (autophagic) vesicle formation can be subsequently incorporated into exosomes. Functional autophagy, particularly functional MVB formation, is necessary for HTT degradation (Rusten & Simonsen, 2008). The presence of mHTT leads to impaired autophagy (Croce & Yamamoto, 2019; Šonský et al., 2021; Tabrizi et al., 2020) which may further facilitate the accumulation of mHTT and aggregate formation.

There are several clues, that mHTT can also affect EV biosynthetic pathways. In the HD140Q knock-in mouse model, it has been shown that exosome secretion in the brain decreases with age and correlates with the accumulation of mHTT aggregates (Hong et al., 2017). In addition, decreased production of EVs in HD was documented in a study of primary astrocytes from mouse HD models, in which lower numbers of exosomes were secreted in HD mice compared to astrocytes from wild-type mice (Hong et al., 2017). Decreased numbers of Alix-positive exosomes were also detected in brains of HD patients (Perez-Gonzalez et al., 2023). Disruption in EV homeostasis was further detected in a transcriptomic and proteomic study of peripheral tissues of HD patients (Neueder et al., 2022a) and the disruption of endocytic pathway and EV production in HD is also evident from several other proteomic studies of EVs (see section "Protein composition of EVs in HD" for details).

EVs in the Search for HD Biomarkers

Practical, reliable, and non-invasive biomarkers are currently needed for the monitoring of HD in the pre-symptomatic stage, to understand disease progression, and to estimate the efficiency of applied therapeutic interventions (Ananbeh et al., 2022; Busatto et al., 2021; Denis et al., 2018). EVs have become extremely attractive as a potential source of biomarkers of human diseases (Kalluri & LeBleu, 2020; Lee et al., 2019; Wu & Shen, 2020) for several reasons. First, EVs carry biological cargo, such as proteins, nucleic acids, metabolites, lipids, and even some organelles that are derived from the parent cells and might reflect disease status (Nieland et al., 2021). Second, the EV biological cargo is protected from degradation by the double layer membrane (Ananbeh et al., 2021; Nieland et al., 2021). Third, EV availability in various body fluids, such as blood, urine, and saliva, makes them easily accessible with non-invasive or minimally-invasive interventions (Battistelli & Falcieri, 2020; Brennan et al., 2020; Yáñez-Mó et al., 2015). Fourth, the ability of neural cell-derived EVs to cross the blood brain barrier (BBB) (Basso & Bonetto, 2016; Busatto et al., 2021) allows the use of peripheral fluids to provide information about neurons and neural cells. Fifth, with importance to miRNA studies, exosomes are highly enriched with miRNAs compared to their source cells and cell-free blood (Cheng et al., 2014).

Blood is the body fluid of primary choice in the majority of biomarker studies, as it is easily obtainable, and due to its circulation throughout the whole body, it contains products of various cell types (Kupcova Skalnikova et al., 2019). However, the EV content in blood is highly influenced by sampling technique and sample processing. For this reason, the ISEV members published procedural details to improve reliability and reproducibility of EV molecular analyses (Coumans et al., 2017; Roux et al., 2020). In blood plasma and serum, the major population of EVs are platelet-derived EVs (Espinosa-Parrilla et al., 2019; Foster et al., 2016; Palviainen et al., 2020; Zhang et al., 2022), while tissue derived EVs represent only a minor part (Xu et al., 2016). Neural cell-derived EVs can be enriched from plasma/serum by immuno-capturing using antibodies against surface proteins expressed on neural cells, such as L1-CAM (Mustapic et al., 2017), which is frequently used in combination with EV precipitation by the ExoQuick reagent. Such techniques have been used for quantification of pathological proteins (e.g. Aβ, tau forms, α-synuclein) in neural-derived plasma EVs in AD and PD, where levels of pathological proteins could predict AD development and could support the PD diagnosis as biomarkers (Fiandaca et al., 2015; Goetzl et al., 2015; Gomes et al., 2022; Liu et al., 2022; Nila et al., 2022).

Platelet-Derived EVs in the Search for HD Biomarkers

Besides the major function of platelets in bleeding prevention, platelets also play roles in pathological conditions including NDs (Espinosa-Parrilla et al., 2019; Pluta & Ułamek-Kozioł, 2019). Platelets and platelet-derived EVs have been investigated as a possible source of biomarkers for NDs, such as AD (Hosseinzadeh et al., 2018; Odaka et al., 2021; Villar-Vesga et al., 2020) and PD (Tomlinson et al., 2015). In AD, platelet-derived EVs were more abundant in serum of AD patients compared to controls (Odaka et al., 2021; Villar-Vesga et al., 2020).

In HD, Denis et al. studied number of platelet-derived EVs in platelet-free plasma in 59 HD patients and 54 age- and sex-matched controls. However, no differences have been found in the number of EVs, either in resting state or after platelet activation. In addition, there were no correlations between the number of platelet derived EVs and the age of the patients, the number of CAG repeats, or the disease stage (Denis et al., 2018).

Importantly, platelets show high expression of proteins related to NDs, such as APP and tau protein (Espinosa-Parrilla et al., 2019; Pluta & Ułamek-Kozioł, 2019). In HD, mHTT is ubiquitously expressed in all body cells, including blood cells (Denis et al., 2019; Weiss et al., 2012). Among blood elements, mHTT is particularly present in platelets, as the ratio of mHTT/HTT protein content is higher in platelets compared to erythrocytes and mononuclear cells (Denis et al., 2019). In a study of mHTT content in platelet-derived EVs, the EVs were isolated from plasma of HD patients by high speed centrifugation and mHTT was detected by highly sensitive 2B7-MW1 Singulex assay, which is able to detect mHTT concentrations as low as 4 pg/ml (Denis et al., 2018). Such assay uses a combination of HTT N-terminal (2B7) and polyQ (MW1) –specific antibodies and was previously used to detect mHTT in CSF, blood and saliva (Fodale et al., 2017; Parkin et al., 2023; Wild et al., 2015). Interestingly, despite high mHTT protein content in platelets, mHTT was undetectable in platelet-derived EVs both in pre-manifest and manifest HD patient samples (Denis et al., 2018). The study of Denis et al. suggests that, despite high mHTT content in platelets, neither the number of platelet-derived EVs nor the mHTT content in such EVs can be used as biomarkers in HD (Denis et al., 2018). Nevertheless, it cannot be excluded that other mHTT species, undetectable by 2B7-MW1 assay, are present in platelet-derived EVs.

Protein Composition of EVs in HD

The stability of EV proteins in body fluids and the variation in their proteomic composition reflecting the status of source cells, make EV proteins an ideal target for diagnostic and prognostic biomarker research in various diseases (Vinaiphat & Sze, 2022; Mitchell et al., 2022; Valencia et al., 2022). Results from proteomic studies of EVs in NDs frequently indicate endolysosomal and autophagic impairment affecting the biogenesis and molecular composition of EVs (Gallart-Palau et al., 2020; Quiroz-Baez et al., 2020; Vinaiphat & Sze, 2022). In PD, the proteome of urinary EVs was studied in 28 PD patients and 22 controls. Urine was chosen as it is a non-invasively collectable body fluid and to evaluate its usefulness in detection of neurological disease-linked proteins. EVs were isolated by differential centrifugation and ultracentrifugation. The proteomic analysis revealed enrichment of proteins involved in vesicle transport, localization, and vesicle biogenesis, documented by the identification of 25 different Rab proteins. Bioinformatic analysis by PANTHER and DAVID pathway classifications identified an enrichment of proteins linked to neurological diseases, including PD, AD, and HD. The results of this study suggest that urine derived EVs are a non-invasively accessible source for biomarker discovery with specific promise to NDs (Wang et al., 2019).

In HD, proteomic study of EVs secreted by striatal cells derived from HD knock-in mice and from neural cells derived from HD patient iPS cells were performed by Tartaglia et al. Small EVs were isolated from conditioned medium by size exclusion chromatography and submitted to quantitative proteomic analysis. Their preliminary results suggest that the presence of mHTT in neural cells may reprogram the biogenesis and secretion of small EVs of endosomal origin during neuronal differentiation (Tartaglia et al., 2022) and further supports endosomal pathway disruption in HD.

Regarding HTT and mHTT proteins themselves, their presence and potential forms (fragments) in plasma-derived EVs were studied in HD pig models and HD patients (Ananbeh et al., 2022). The EVs were isolated by ultracentrifugation from blood plasma of transgenic (TgHD) and knock-in HD (KI-HD) pigs and wild-type control siblings (in total n = 28 for TgHD and n = 16 for KI-HD model), as well as from plasma of 14 HD patients and 4 controls. The HTT/mHTT forms were analyzed by western blot using HTT N-terminal and polyQ-specific antibodies. In EV samples, full length endogenous HTT/mHTT bands were detected at the molecular weight of ~360 kDa. In addition, in the TgHD pig model the expression of the transgene (N-terminal fragment of human HTT protein with elongated polyQ; ~110 kDa) and its fragment (~70 kDa) were detected in plasma-derived EVs of TgHD pigs and were not present in wild-type sibling samples. Co-isolation of HTT/mHTT with EVs was further confirmed by density gradient ultracentrifugation, where HTT/mHTT co-isolated within the same density fraction as EVs. Comparison of total sum of all HTT/mHTT band intensities in EVs from both pig models and HD patients and their respective controls indicated an increase of the total amount of HTT forms in EVs from blood plasma of all three HD groups compared to controls; however, the difference was statistically significant only in the TgHD model. Additional studies on larger cohorts are necessary to investigate whether HTT/mHTT in plasma-derived EVs has potential as a candidate biomarker for HD (Ananbeh et al., 2022).

RNA Composition of EVs in HD

EVs contain various forms of RNA species, including intact messenger RNA (mRNA) and mRNA fragments, long non-coding RNA (lncRNA), micro RNA (miRNA), circular RNAs (circRNAs), small nucleolar RNA (snoRNAs), small nuclear RNAs (snRNAs), piwi-interacting RNA, ribosomal RNA (rRNA) and fragments of tRNA-, vault-RNA and Y-RNA (Kim et al., 2017; Yáñez-Mó et al., 2015). Some transcripts are selectively incorporated into EVs, while some are specifically excluded from EVs. However, the regulation of the sorting processes remains unknown (Kim et al., 2017). Around 10% of the circulating miRNAs are secreted in exosomes (Ho et al., 2022).

There is no data available in HD regarding the possible roles of EV RNAs as biomarkers. Nonetheless, several research groups are currently working on this research topic and preliminary information appears in conference abstracts. Such initial studies include RNA analysis of blood plasma-derived EVs (Neueder et al., 2022b), including analysis of small non-coding RNAs and miRNAs in free form in plasma and associated with plasma EVs (Herrero-Lorenzo et al., 2022), and RNAs from CSF-derived EVs (von Einem et al., 2022).

Among the various RNAs, the miRNAs are highly important targets in biomarker discovery studies because of their crucial regulatory roles in gene expression and cellular communication and presence in EVs (Wang & Zhang, 2020). In the CNS, miRNAs are reported to manage a wide range of cellular processes from normal development to disease and regeneration (Tsujimura et al., 2022) and their expression levels are widely distinct between HD patients and healthy individuals (Johnson et al., 2008). The free miRNAs circulating in blood have been frequently investigated as biomarkers for early diagnosis and monitoring of ND progression (Wang & Zhang, 2020). In patients with NDs, expression of specific miRNAs varied in body fluids such as blood and interstitial fluids (Ban et al., 2017; Wang & Zhang, 2020). There are several peripheral circulating miRNAs with potential to become diagnostic biomarkers for NDs, such as elevated miR-206 in AD (Kenny et al., 2019), decreased miR-29a and miR-29c in PD (Bai et al., 2017), increased miR-100-5p and decreased miR-330-3p and miR-641 in HD (Díez-Planelles et al., 2016), and increased miR-338-3p in ALS (De Felice et al., 2014).

The exosomal miRNAs (ex-miRNAs) are attractive targets in diagnostic biomarker research in NDs and other diseases from several reasons. First, the ex-miRNAs are more stable than free miRNAs, as the double layer EV membrane protects them from degradation by nucleases, which are abundant in body fluids (Cheng et al., 2014; Wang & Zhang, 2020). Second, they are more accessible than cellular miRNA and can be isolated noninvasively from various body fluids (e.g. blood) (Kumar & Reddy, 2016). Third, they have distinct expression patterns that might be interrupted in the pathogenesis and neurodegeneration of the CNS during exosome biogenesis, therefore, ex-miRNAs can be more disease-specific than cellular and free circulating miRNAs (Chen et al., 2015, 2017; Xia et al., 2019; Wang et al., 2022). Finally, exosomes have the ability to cross the BBB and CNS derived ex-miRNAs might be thus available from periphery (Hornung et al., 2020). There are several ex-miRNAs candidates for NDs biomarkers, such as decreased miR-342-3p, miR-125a-5p, miR-125b-5p, miR-451a, miR-23a-3p, and miR-126-3p in AD (Barbagallo et al., 2020; Lugli et al., 2015; Palaniswamy et al., 2020), increased miR-331-5p, miR-22, miR-23a, and miR-24 in PD (Barbagallo et al., 2020; Ghafouri-Fard et al., 2022; Yao et al., 2018).

In HD, there is no available information about miRNAs in blood- or CSF-derived EVs. Alternatively, circulating miRNAs in blood and CSF have been investigated in HD patients. One study identified six CSF miRNAs (miR-135b-3p, miR-520f-3p, miR-4317, miR-3928-5p, miR-140-5p, and miR-8082) that were significantly increased in CSF in prodromal HD gene mutation carriers compared to controls (Reed et al., 2018). In another study, 168 circulating miRNAs were found altered in the plasma of symptomatic patients with HD, and 13 of them (miR-877-5p, miR-223-3p, miR-223-5p, miR-30d-5p, miR-128, miR-22-5p, miR-222-3p, miR-338-3p, miR-130b-3p, miR-425-5p, miR-628-3p, miR-361-5p, and miR-942) were significantly increased in HD patients plasma compared to controls (Díez-Planelles et al., 2016) (Table 3). Interestingly, the levels of miR-100-5p, miR-641 and miR-330-5p were significantly altered in a later stage of the disease, and the miR-100-5p fold change was significantly higher in the later stage of the disease

Table 3 Selected miRNAs with diagnostic and therapeutic potential in HD

miRNA	Sample type	Organism/ HD model	Expression level in HD	Role in HD pathology	References
miR-22	Brain	Mouse	Up-regulated	Inhibits apoptosis and HD-related mRNA expression Has a potential protective effect on neurons and has been confirmed to delay the progression of HD	Ehrnhoefer et al. (2009) and Jovicic et al. (2013)
miR-27a	Brain	Mouse	Up-regulated	Reduces mHTT aggregation by modulating multidrug resistance protein-1 function	Ban et al. (2017)
miR-34b, miR-1285	Plasma	Human	Up-regulated	Affect mHTT cytoplasmic distribution and toxicity *in vitro*	Gaughwin et al. (2011)
miR-196a, miR-196b	Brain, plasma	Human	Up-regulated	Improve mitochondrial morphology and function in HD cells Suppress the expression of mHTT at the mRNA and protein levels	Chang et al. (2017), Cheng et al. (2013), Hoss et al. (2015a), Packer et al. (2008), Petry et al. (2022), and Tan et al. (2015)
miR-146a	Cell line	Mouse striatum-derived STHdh cell line	Down-regulated	Targets TATA box binding protein and is involved in HD pathogenesis in several pathways	Ghose et al. (2011) and Sinha et al. (2010, 2011)
miR-214	Cell line	Mouse striatum-derived STHdh cell line	Up-regulated	Downregulates the endogenous expression of *Htt* at mRNA and protein level	Bucha et al. (2015), Ghatak and Raha 2018, and Sinha et al. (2011)
miR-207, miR-448, miR-669c, miR-18a*	Brain	Mouse	Up-regulated	ND	Lee et al. (2011)

(continued)

Table 3 (continued)

miRNA	Sample type	Organism/ HD model	Expression level in HD	Role in HD pathology	References
miR-877-5p, miR-223-3p, miR-223-5p, miR-30d-5p, miR-22-5p, miR-222-3p, miR-338-3p, miR-130b-3p, miR-425-5p, miR-628-3p, miR-361-5p, miR-942, miR-100-5p	Plasma	Human	Up-regulated	Upregulated in HD patient blood plasma Have a role in cholesterol metabolism	Díez-Planelles et al. (2016)
miR-100, miR-16, miR-151-3p, miR-451, miR-27b, miR-92a	Brain	Human	Up-regulated	ND	Martí et al. (2010)
miR-135b-3p, miR-140-5p, miR-520f-3p, miR-3928-5p, miR-4317, miR-8082	CSF	Human	Up-regulated	ND	Reed et al. (2018)
miR-10b-5p, miR-486-5p	Brain, plasma	Human	Up-regulated	Target expression of brain-derived neurotrophic factor (BDNF)	Hoss et al. (2015a, b) and Petry et al. (2022)
miR-9, miR-9*	Brain, peripheral leukocytes	Human	Down-regulated	Have a role in transcriptional dysregulation and target the neuron restrictive silencer factor (NRSF)	Chang et al. (2017), Johnson et al. (2008), Johnson and Buckley (2009), and Packer et al. (2008)
miR-124, miR-124a	Brain	Human, mouse	Down-regulated	Slows down disease progression by affecting neuronal differentiation and development	Chang et al. (2017), Johnson et al. (2008), and Johnson and Buckley 2009, and Liu et al. (2015)

(continued)

Table 3 (continued)

miRNA	Sample type	Organism/ HD model	Expression level in HD	Role in HD pathology	References
miR-132	Brain	Human, mouse	Down-regulated	Potentially disrupts mRNA regulation and neuronal functions	Chang et al. (2017), Chen et al. (2018), Ehrnhoefer et al. (2009), Johnson et al. (2008), Langfelder et al. (2018), Lee et al. (2011), and Petry et al. (2022)
miR-122-5p, miR-330-3p, miR-641	Plasma	Human	Down-regulated	Have a role in cholesterol metabolism	Díez-Planelles et al. (2016)
miR-212, miR-128, and miR-218	Brain	Mouse	Down-regulated	Associated with neuronal development, differentiation and survival	Díez-Planelles et al. (2016), Ehrnhoefer et al. (2009), Langfelder et al. (2018), Lee et al. (2011), Packer et al. (2008), and Petry et al. (2022)
miR-128, miR-222, miR-139-3p, miR-382, miR-483-3p, miR-433	Brain	Human	Down-regulated	ND	Martí et al. (2010)
miR-28*, miR-29c, miR-128, miR-138, miR-222, miR-344, miR-466h and miR-674*	Brain	Mouse, Rat	Down-regulated	Have a role in transcriptional dysregulation and mitochondrial dysfunction	Ehrnhoefer et al. (2009), Jovicic et al. (2013), Lee et al. (2011), and Petry et al. (2022)

Abbreviations: *ND* not determined, *mHtt* mutant huntingtin, *HD* Huntington's disease

compared to the early stage of the disease as well as controls (Díez-Planelles et al., 2016). This study suggests that the miR-100-5p expression levels might be affected by disease stage, making them a candidate for disease monitoring biomarker.

The miRNA miR-124, plays a crucial role in neurogenesis and it is the most abundant miRNA in the adult brain, where it is predominantly expressed by neurons and microglia (Cao et al., 2007; Liu et al., 2019b). In HD, decreased levels of miR-124 have been reported in HD patient brains (Johnson et al., 2008; Packer et al., 2008), mouse striatal cells STHdh^{Q111}/Hdh^{Q111} and the striatum of the R6/2

HD mouse (Das et al., 2013). Reduction of miR-124 in plasma was also observed in PD patients (Angelopoulou et al., 2019). On the other hand, miR-124 levels in serum-derived exosomes were elevated after stroke with ischemic brain damage (Ji et al., 2016). Delivery of miR-124 to the CNS by exosomes has therapeutic potential in both HD and stroke (Lee et al., 2017; Liu et al., 2019b).

Therapeutic Potential of EVs in HD

EVs can cross the BBB, therefore they are considered as promising therapeutic drug carriers into the CNS to specific regions or cells (Xia et al., 2019, 2022). In NDs, a few preclinical trials using EVs are taking place in AD and PD (Kumar et al., 2020; Muhammad, 2021; Yu et al., 2021). In HD, drug development strategies are currently focused on targeting HTT with the aim to lower mHTT and relieve its pathogenic consequences by reducing *HTT* RNA and/or HTT protein expression (Tabrizi et al., 2020). Various strategies are used to downregulate *HTT* expression, such as antisense oligonucleotides (ASOs), small interfering RNA (siRNA), and miRNAs that are targeting mHTT production pathway (Evers et al., 2018; Miniarikova et al., 2016; Tabrizi et al., 2019). As a type of EVs, exosomes showed efficiency in the delivery of therapeutic oligonucleotides (miRNA and siRNA) targeting *HTT* expression (Biscans et al., 2018; Didiot et al., 2016; Lee et al., 2017; Wu et al., 2018). Exosome-based miRNA delivery also express potential to modify BBB integrity (Zhao & Zlokovic, 2017).

EV Potential in Delivery of siRNA-Based Therapies

Oligonucleotide therapeutics (ONTs) are a novel class of drugs targeting RNA or DNA to prevent the expression of proteins responsible for disease phenotypes. The siRNAs are a class of ONTs that induce gene silencing by targeting complementary mRNA via RNA Induced Silencing Complex (RISC) (Biscans et al., 2018; Didiot et al., 2016). Exosome-mediated delivery of siRNA into transgenic HD mouse models was shown to significantly reduce HTT protein expression (Wu et al., 2018). Modified exosomes expressing the neuron specific rabies viral glycoprotein (RVG) peptide loaded with siRNA targeting human *HTT* exon 1 transcript were intravenously injected to normal mice and BACHD and N171-82Q transgenic mice every two days for 2 weeks. Results indicated that the siRNA was efficiently delivered into the mouse brain and that the siRNA-RVG exosomes significantly lowered human HTT mRNA and protein levels in transgenic mice up to 46% and 54%, respectively. In addition, an improvement in rotarod performance was observed in N171-82Q mice receiving siRNA-RVG exosomes (Wu et al., 2018). This study demonstrates the therapeutic potential of *HTT*-siRNA RVG exosomes in HD.

Hydrophobically modified siRNAs (hsiRNAs) are asymmetric, chemically modified ONTs that enhance stability and improve cellular internalization (Didiot et al.,

2018). Loading RNA into small EVs still represents a bottleneck for their clinical applications as delivery vesicles for therapeutic RNA (Biscans et al., 2018). A cholesterol moiety that is covalently conjugated to siRNAs can make loading of siRNAs to exosomes more effective, resulting in thousands of RNAs copies per vesicle (Biscans et al., 2018; Didiot et al., 2016, 2018). Exosome mediated delivery of TEG-Cholesterol conjugated hsiRNAs targeting *HTT* mRNA (hsiRNA^{HTT}) were efficiently internalized into mice primary cortical neurons resulting in dose dependent silencing of HTT mRNA and protein (Didiot et al., 2016). Unilateral infusion of hsiRNA^{HTT} loaded exosomes into mouse striatum resulted in bilateral oligonucleotide distribution and statistically significant silencing of *HTT* mRNA. This study suggests that exosomes increase the stability of hsiRNAs, promote *in vivo* bilateral spreading of hsiRNAs, and enter a powerful neuronal uptake pathway (Didiot et al., 2016). In another study, conjugation of hsiRNAs with various lipids, fatty acids, sterols and vitamins was tested to improve packing of hsiRNAs into exosomes and the ability to silence *HTT* mRNA in primary murine cortical neurons. The results showed that hydrophobicity drives loading of lipid-conjugated hsiRNAs into EVs (Biscans et al., 2018). Vitamin E-conjugated siRNA facilitated siRNA loading into exosomes and productive RNA delivery to neurons (Biscans et al., 2018).

EV Potential in Delivery of miRNA-Based Therapies

Therapeutic miRNAs are oligonucleotides with the length around 22 nt that regulate target gene expression directly by increasing or decreasing particular mRNA levels, or indirectly by mRNA activation or enhancement of endogenous repair mechanism in the brain (Dong & Cong, 2021). Several miRNAs, such as miR-214, miR-150, miR-146a and miR-125b, target the human and mouse *HTT* gene and are able to reduce *HTT* gene expression in mouse cells (Sinha et al., 2011). mHTT expression can be also reduced by engineered miRNAs. For example, in HD minipig models, the injection of adeno-associated virus serotype vector 5 carrying engineered miRNA targeting human *HTT* (AAV5-miHTT) significantly reduced the *HTT* mRNA and protein in all brain regions transduced by AAV5-miHTT (Evers et al., 2018; Keskin et al., 2019; Vallès et al., 2021). Similar results were reported in Q175 HD mice (Spronck et al., 2019), HD rats (Miniarikova et al., 2017; Spronck et al., 2021) and non-human primates (*Macaca fascicularis*) (Spronck et al., 2021).

Interestingly, EVs demonstrated the capability to transport miHTT into new cells and thus together with axonal transport further participate in the widespread biodistribution of miHTT therapeutic construct. In a study by Morais et al., HD patient iPS cell-derived neurons were transduced by AAV5-miHTT construct to produce EVs enriched in the miHTT construct. When such cells were co-cultured in a contactless transwell system with naïve HD patient iPSC-derived neurons as recipient cells, the miHTT packed EVs were taken up by the naïve recipient cells. The EV transfer of miHTT construct was functional, as the miHTT was active and was able to downregulate *HTT* expression in recipient cells (Morais et al., 2022).

The miR-124, was selected in the majority of exosome-based miRNA delivery studies in NDs because it is highly expressed in all brain regions, has a critical role in regulation of neuronal differentiation (Liu et al., 2015; Sun et al., 2015) and is one of the most suppressed miRNAs in HD models (Das et al., 2013; Ghafouri-Fard et al., 2022; Liu et al., 2015). The exosome-based therapeutic delivery of miR-124 in a HD mouse model was developed by Lee et al. miR-124 was overexpressed in the HEK293 cell line to produce exosomes packed with miR-124, and such exosomes with high level of miR-124 were injected into the striatum of R6/2 transgenic HD mice. This resulted in the downregulation of RE1-Silencing Transcription Factor (REST), which is the key target protein of miR-124. However, no significant behavioral improvement was observed after 1 week of miR-124 delivery, and longer time intervals were not studied (Lee et al., 2017). Because of the moderate therapeutic efficiency of miR-124, Lee et al. suggested the increment of miRNAs dose packed into exosomes, and recommended other miRNAs such as miR-9, miR-22, miR-125b, miR-146a, miR-150, and miR-214 as therapeutic candidates to be delivered by exosomes, which might have greater therapeutic efficacy than miR-124 (Lee et al., 2017). In another study, the induction of miR-27a in HD cell resulted in reduction of mHTT aggregates, which indicates that the miR-27a might be additional promising therapeutic molecule for HD (Ban et al., 2017).

Therapeutic Potential of EVs Derived from Glia and Stem Cells

EVs produced by glia and stem cells may carry neuroprotective factors that prevent neuron degeneration in presence of mHTT. The injection of astrocytic exosomes into the striatum of 140Q KI mice decreased the density of mHTT aggregates (Hong et al., 2017). Similarly, exosomes derived from adipose stem cells that are known to secrete several neuroprotective factors, were shown to reduce the accumulation of mHTT aggregates, mitochondrial dysfunction and cell apoptosis in *in vitro* cultured neuronal cells (Lee et al., 2016).

Altogether, despite currently being studied only *in vitro* or in experimental animals, EVs, mainly exosomes, are believed to provide future therapeutic possibilities for the currently incurable HD (Ananbeh et al., 2021; Morais et al., 2022; Zhang et al., 2021).

3 Conclusions

HD is an inherited, fatal and incurable ND caused by the abnormal expansion of polyQ repeats in HTT protein. Currently, there are several therapeutic interventions based on gene therapies that are being developed to prevent or delay the disease progress. Nonetheless, the efficacy of developing therapies is currently unpredictable and biomarkers are required to provide information about the effects of applied therapies, and in possible postponing or slowing down disease progression. EVs have emerged as promising novel, noninvasive biomarker sources for HD and other NDs, mainly because of their ability to cross the BBB and their availability from body fluids. However, only very limited information is currently available about the molecular composition of EVs in HD related to biomarker discovery.

Various molecules, such as proteins, RNA, lipids, and metabolites are packed into EVs and EVs can transfer such cargo, including mHTT, between cells. In this way, EVs can participate in the spreading of mHTT. On the other hand, EVs may also participate in toxic protein removal from cells. EVs have also therapeutic potential because of their ability to transfer trophic and neuroprotective factors to the CNS. EVs can also be engineered to carry therapeutic oligonucleotides, such as miRNA and siRNA to target HTT/mHTT expression. Overall, EV research in HD is in the beginning stages and majority of work remains to be done to elucidate whether EV may find their diagnostic or therapeutic applications in the future.

Acknowledgements This study was supported by the Czech Science Foundation (project 19-01747S) and by the project National institute for cancer research (Programme EXCELES, project LX22NPO5102) funded by the European Union – Next Generation EU.

Conflict of Interest The authors declare that they have no conflict of interest.

References

Aguzzi, A., & Rajendran, L. (2009, December 24). The transcellular spread of cytosolic amyloids, prions, and prionoids. *Neuron, 64*(6), 783–790.

Ananbeh, H., Vodicka, P., & Kupcova, S. H. (2021, April 15). Emerging roles of exosomes in Huntington's disease. *International Journal of Molecular Sciences, 22*(8), 4085.

Ananbeh, H., Novak, J., Juhas, S., Juhasova, J., Klempir, J., Doleckova, K., et al. (2022, May 17). Huntingtin co-isolates with small extracellular vesicles from blood plasma of TgHD and KI-HD pig models of Huntington's disease and human blood plasma. *International Journal of Molecular Sciences, 23*(10), 5598.

Anderson, H. C. (1969, April). Vesicles associated with calcification in the matrix of epiphyseal cartilage. *The Journal of Cell Biology, 41*(1):59–72.

Angelopoulou, E., Paudel, Y. N., & Piperi, C. (2019, December). miR-124 and Parkinson's disease: A biomarker with therapeutic potential. *Pharmacological Research, 150*, 104515.

Arrasate, M., & Finkbeiner, S. (2012, November). Protein aggregates in Huntington's disease. *Experimental Neurology, 238*(1), 1–11.

Babcock, D. T., & Ganetzky, B. (2015, September 29). Transcellular spreading of huntingtin aggregates in the Drosophila brain. *Proceedings of the National Academy of Sciences of the United States of America, 112*(39), E5427–E5433.

Bahmani, L., & Ullah, M. (2022, June 21). Different sourced extracellular vesicles and their potential applications in clinical treatments. *Cell, 11*(13), 1989.

Bai, X., Tang, Y., Yu, M., Wu, L., Liu, F., Ni, J., et al. (2017, July 14). Downregulation of blood serum microRNA 29 family in patients with Parkinson's disease. *Scientific Reports, 7*(1), 5411.

Ban, J.-J., Chung, J.-Y., Lee, M., Im, W., & Kim, M. (2017, June 24). MicroRNA-27a reduces mutant hutingtin aggregation in an in vitro model of Huntington's disease. *Biochemical and Biophysical Research Communications, 488*(2), 316–321.

Barbagallo, C., Mostile, G., Baglieri, G., Giunta, F., Luca, A., Raciti, L., et al. (2020, May). Specific signatures of serum miRNAs as potential biomarkers to discriminate clinically similar neurodegenerative and vascular-related diseases. *Cellular and Molecular Neurobiology, 40*(4), 531–546.

Basso, M., & Bonetto, V. (2016). Extracellular vesicles and a novel form of communication in the brain. *Frontiers in Neuroscience, 10*, 127.

Battistelli, M., & Falcieri, E. (2020, January 20). Apoptotic bodies: Particular extracellular vesicles involved in intercellular communication. *Biology, 9*(1), 21.

Beetler, D. J., Di Florio, D. N., Bruno, K. A., Ikezu, T., March, K. L., Cooper, L. T., et al. (2022, November 28). Extracellular vesicles as personalized medicine. *Molecular Aspects of Medicine, 91*, 101155.

Bellingham, S. A., Guo, B. B., Coleman, B. M., & Hill, A. F. (2012). Exosomes: Vehicles for the transfer of toxic proteins associated with neurodegenerative diseases? *Frontiers in Physiology, 3*, 124.

Benz, E. W., & Moses, H. L. (1974, June). Small, virus-like particles detected in bovine sera by electron microscopy. *Journal of the National Cancer Institute, 52*(6), 1931–1934.

Biscans, A., Haraszti, R. A., Echeverria, D., Miller, R., Didiot, M.-C., Nikan, M., et al. (2018, June). Hydrophobicity of lipid-conjugated siRNAs predicts productive loading to small extracellular vesicles. *Molecular Therapy, 26*(6), 1520–1528.

Bonucci, E. (1967, September). Fine structure of early cartilage calcification. *Journal of Ultrastructure Research, 20*(1), 33–50.

Brennan, K., Martin, K., FitzGerald, S. P., O'Sullivan, J., Wu, Y., Blanco, A., et al. (2020, January 23). A comparison of methods for the isolation and separation of extracellular vesicles from protein and lipid particles in human serum. *Scientific Reports, 10*(1), 1039.

Bucha, S., Mukhopadhyay, D., & Bhattacharyya, N. P. (2015, October 2). Regulation of mitochondrial morphology and cell cycle by microRNA-214 targeting Mitofusin2. *Biochemical and Biophysical Research Communications, 465*(4), 797–802.

Busatto, S., Morad, G., Guo, P., & Moses, M. A. (2021, September). The role of extracellular vesicles in the physiological and pathological regulation of the blood-brain barrier. *FASEB BioAdvances, 3*(9), 665–675.

Cao, X., Pfaff, S. L., & Gage, F. H. (2007, March 1). A functional study of miR-124 in the developing neural tube. *Genes & Development, 21*(5), 531–536.

Caviston, J. P., & Holzbaur, E. L. F. (2009, April). Huntingtin as an essential integrator of intracellular vesicular trafficking. *Trends in Cell Biology, 19*(4), 147–155.

Chang, K.-H., Wu, Y.-R., & Chen, C.-M. (2017, December 19). Down-regulation of miR-9* in the peripheral leukocytes of Huntington's disease patients. *Orphanet Journal of Rare Diseases* [Internet] [cited 2018 Jan 18], *12*. Available from: https://www.ncbi.nlm.nih.gov/pmc/articles/PMC5737985/

Chargaff, E., & West, R. (1946, November). The biological significance of the thromboplastic protein of blood. *The Journal of Biological Chemistry, 166*(1), 189–197.

Chen, F., Du, Y., Esposito, E., Liu, Y., Guo, S., Wang, X., et al. (2015, December). Effects of focal cerebral ischemia on exosomal versus serum miR126. *Translational Stroke Research, 6*(6), 478–484.

Chen, Y., Song, Y., Huang, J., Qu, M., Zhang, Y., Geng, J., et al. (2017). Increased circulating exosomal miRNA-223 is associated with acute ischemic stroke. *Frontiers in Neurology, 8*, 57.

Chen, D., Hu, S., Wu, Z., Liu, J., & Li, S. (2018). The role of MiR-132 in regulating neural stem cell proliferation, differentiation and neuronal maturation. *Cellular Physiology & Biochemistry. International Journal of Experimental Cellular Physiology, Biochemistry and Pharmacology, 47*(6), 2319–2330.

Cheng, P.-H., Li, C.-L., Chang, Y.-F., Tsai, S.-J., Lai, Y.-Y., Chan, A. W. S., et al. (2013, August 8). miR-196a ameliorates phenotypes of Huntington disease in cell, transgenic mouse, and induced pluripotent stem cell models. *American Journal of Human Genetics, 93*(2), 306–312.

Cheng, L., Sharples, R. A., Scicluna, B. J., & Hill, A. F. (2014). Exosomes provide a protective and enriched source of miRNA for biomarker profiling compared to intracellular and cell-free blood. *Journal of Extracellular Vesicles, 3*, 23743.

Cheng, L., Vella, L. J., Barnham, K. J., McLean, C., Masters, C. L., & Hill, A. F. (2020, June 4). Small RNA fingerprinting of Alzheimer's disease frontal cortex extracellular vesicles and their comparison with peripheral extracellular vesicles. *Journal of Extracellular Vesicles, 9*(1), 1766822.

Cicchetti, F., Lacroix, S., Cisbani, G., Vallières, N., Saint-Pierre, M., St-Amour, I., et al. (2014, July). Mutant huntingtin is present in neuronal grafts in huntington disease patients: Transfer of mutant Huntingtin to normal tissue. *Annals of Neurology, 76*(1), 31–42.

Clancy, J. W., Schmidtmann, M., & D'Souza-Schorey, C. (2021, June). The ins and outs of microvesicles. *FASEB BioAdvances, 3*(6), 399–406.

Coleman, B. M., Hanssen, E., Lawson, V. A., & Hill, A. F. (2012, October). Prion-infected cells regulate the release of exosomes with distinct ultrastructural features. *FASEB Journal: Official Publication of the Federation of American Societies for Experimental Biology, 26*(10), 4160–4173.

Colombo, M., Raposo, G., & Théry, C. (2014). Biogenesis, secretion, and intercellular interactions of exosomes and other extracellular vesicles. *Annual Review of Cell and Developmental Biology, 30*, 255–289.

Costanzo, M., Abounit, S., Marzo, L., Danckaert, A., Chamoun, Z., Roux, P., et al. (2013, August 15). Transfer of polyglutamine aggregates in neuronal cells occurs in tunneling nanotubes. *Journal of Cell Science, 126*(16), 3678–3685.

Coumans, F. A. W., Brisson, A. R., Buzas, E. I., Dignat-George, F., Drees, E. E. E., El-Andaloussi, S., et al. (2017, May 12). Methodological guidelines to study extracellular vesicles. *Circulation Research, 120*(10), 1632–1648.

Crawford, N. (1971, July). The presence of contractile proteins in platelet microparticles isolated from human and animal platelet-free plasma. *British Journal of Haematology, 21*(1), 53–69.

Croce, K. R., & Yamamoto, A. (2019, February 1). A role for autophagy in Huntington's disease. *Neurobiology of Disease, 122*, 16–22.

Dalton, A. J. (1975, May). Microvesicles and vesicles of multivesicular bodies versus 'virus-like' particles. *Journal of the National Cancer Institute, 54*(5), 1137–1148.

Das, E., Jana, N. R., & Bhattacharyya, N. P. (2013, July 26). MicroRNA-124 targets CCNA2 and regulates cell cycle in STHdhQ111/HdhQ111 cells. *Biochemical and Biophysical Research Communications, 437*(2), 217–224.

De Broe, M., Wieme, R., & Roels, F. (1975, December 13). Letter: Membrane fragments with koinozymic properties released from villous adenoma of the rectum. *Lancet (London England), 2*(7946), 1214–1215.

De Felice, B., Annunziata, A., Fiorentino, G., Borra, M., Biffali, E., Coppola, C., et al. (2014, October). miR-338-3p is over-expressed in blood, CFS, serum and spinal cord from sporadic amyotrophic lateral sclerosis patients. *Neurogenetics, 15*(4), 243–253.

Deng, J., Koutras, C., Donnelier, J., Alshehri, M., Fotouhi, M., Girard, M., et al. (2017, April 19). Neurons export extracellular vesicles enriched in cysteine string protein and misfolded protein cargo. *Scientific Reports* [Internet]. [cited 2021 Feb 11], *7*. Available from: https://www.ncbi.nlm.nih.gov/pmc/articles/PMC5430488/

Denis, H. L., Lamontagne-Proulx, J., St-Amour, I., Mason, S. L., Weiss, A., Chouinard, S., et al. (2018, November). Platelet-derived extracellular vesicles in Huntington's disease. *Journal of Neurology, 265*(11), 2704–2712.

Denis, H. L., Lamontagne-Proulx, J., St-Amour, I., Mason, S. L., Rowley, J. W., Cloutier, N., et al. (2019, March). Platelet abnormalities in Huntington's disease. *Journal of Neurology, Neurosurgery, and Psychiatry, 90*(3), 272–283.

Didiot, M.-C., Hall, L. M., Coles, A. H., Haraszti, R. A., Godinho, B. M., Chase, K., et al. (2016, October). Exosome-mediated delivery of hydrophobically modified siRNA for Huntingtin mRNA silencing. *Molecular Therapy, 24*(10), 1836–1847.

Didiot, M.-C., Haraszti, R. A., Aronin, N., & Khvorova, A. (2018). Loading of extracellular vesicles with hydrophobically modified siRNAs. *Methods in molecular biology (Clifton, N.J.), 1740*, 199–214.

Díez-Planelles, C., Sánchez-Lozano, P., Crespo, M. C., Gil-Zamorano, J., Ribacoba, R., González, N., et al. (2016, June 1). Circulating microRNAs in Huntington's disease: Emerging mediators in metabolic impairment. *Pharmacological Research, 108*, 102–110.

Dong, X., & Cong, S. (2021). MicroRNAs in Huntington's disease: Diagnostic biomarkers or therapeutic agents? *Frontiers in Cellular Neuroscience, 15*, 705348.

Ehrnhoefer, D. E., Butland, S. L., Pouladi, M. A., & Hayden, M. R. (2009). Mouse models of Huntington disease: Variations on a theme. *Disease Models & Mechanisms, 2*(3–4), 123–129.

Espinosa-Parrilla, Y., Gonzalez-Billault, C., Fuentes, E., Palomo, I., & Alarcón, M. (2019, July 2). Decoding the role of platelets and related MicroRNAs in aging and neurodegenerative disorders. *Frontiers in Aging Neuroscience* [Internet]. [cited 2021 Feb 11], *11*. Available from: https://www.ncbi.nlm.nih.gov/pmc/articles/PMC6614495/

Evers, M. M., Miniarikova, J., Juhas, S., Vallès, A., Bohuslavova, B., Juhasova, J., et al. (2018, September). AAV5-miHTT gene therapy demonstrates broad distribution and strong human mutant huntingtin lowering in a Huntington's disease minipig model. *Molecular Therapy, 26*(9), 2163–2177.

EV-TRACK Consortium, Van Deun, J., Mestdagh, P., Agostinis, P., Akay, Ö., Anand, S., et al. (2017, February 28). EV-TRACK: Transparent reporting and centralizing knowledge in extracellular vesicle research. *Nature Methods, 14*(3), 228–232.

Fiandaca, M. S., Kapogiannis, D., Mapstone, M., Boxer, A., Eitan, E., Schwartz, J. B., et al. (2015, June). Identification of pre-clinical Alzheimer's disease by a profile of pathogenic proteins in neurally-derived blood exosomes: A case-control study. *Alzheimer's & Dementia. Journal of the Alzheimer's Association, 11*(6), 600–607.e1.

Fodale, V., Boggio, R., Daldin, M., Cariulo, C., Spiezia, M. C., Byrne, L. M., et al. (2017). Validation of ultrasensitive mutant huntingtin detection in human cerebrospinal fluid by single molecule counting immunoassay. *Journal of Huntington's Disease, 6*(4), 349–361.

Foster, B. P., Balassa, T., Benen, T. D., Dominovic, M., Elmadjian, G. K., Florova, V., et al. (2016). Extracellular vesicles in blood, milk and body fluids of the female and male urogenital tract and with special regard to reproduction. *Critical Reviews in Clinical Laboratory Sciences, 53*(6), 379–395.

Gallart-Palau, X., Guo, X., Serra, A., & Sze, S. K. (2020, May 8). Alzheimer's disease progression characterized by alterations in the molecular profiles and biogenesis of brain extracellular vesicles. *Alzheimer's Research & Therapy, 12*(1), 54.

Gaughwin, P. M., Ciesla, M., Lahiri, N., Tabrizi, S. J., Brundin, P., & Björkqvist, M. (2011, June 1). Hsa-miR-34b is a plasma-stable microRNA that is elevated in pre-manifest Huntington's disease. *Human Molecular Genetics, 20*(11), 2225–2237.

Ghafouri-Fard, S., Khoshbakht, T., Hussen, B. M., Taheri, M., Ebrahimzadeh, K., & Noroozi, R. (2022). The emerging role of long non-coding RNAs, microRNAs, and an accelerated epigenetic age in Huntington's disease. *Frontiers in Aging Neuroscience, 14*, 987174.

Ghatak, S., & Raha, S. (2018, September 3). Beta catenin is regulated by its subcellular distribution and mutant huntingtin status in Huntington's disease cell STHdhQ111/HdhQ111. *Biochemical and Biophysical Research Communications, 503*(1), 359–364.

Ghose, J., Sinha, M., Das, E., Jana, N. R., & Bhattacharyya, N. P. (2011). Regulation of miR-146a by RelA/NFkB and p53 in STHdh(Q111)/Hdh(Q111) cells, a cell model of Huntington's disease. *PLoS One, 6*(8), e23837.

Goetzl, E. J., Boxer, A., Schwartz, J. B., Abner, E. L., Petersen, R. C., Miller, B. L., et al. (2015, July 7). Altered lysosomal proteins in neural-derived plasma exosomes in preclinical Alzheimer disease. *Neurology, 85*(1), 40–47.

Gomes, P., Tzouanou, F., Skolariki, K., Vamvaka-Iakovou, A., Noguera-Ortiz, C., Tsirtsaki, K., et al. (2022, December). Extracellular vesicles and Alzheimer's disease in the novel era of Precision Medicine: Implications for disease progression, diagnosis and treatment. *Experimental Neurology, 358*, 114183.

Guo, J. L., & Lee, V. M. Y. (2014, February). Cell-to-cell transmission of pathogenic proteins in neurodegenerative diseases. *Nature Medicine, 20*(2), 130–138.

Han, Q.-F., Li, W.-J., Hu, K.-S., Gao, J., Zhai, W.-L., Yang, J.-H., et al. (2022, November 1). Exosome biogenesis: Machinery, regulation, and therapeutic implications in cancer. *Molecular Cancer, 21*(1), 207.

Herrera, F., Tenreiro, S., Miller-Fleming, L., & Outeiro, T. F. (2011, February 11). Visualization of cell-to-cell transmission of mutant huntingtin oligomers. *PLOS Currents, 3*, RRN1210.

Herrero-Lorenzo, M., Pérez, J., Escaramís, G., Pérez-González, R., Kulisevsky, J., Gámez-Valero, A., et al. (2022, September 1). D10 plasma extracellular small RNAS are early deregulated in Huntington's disease and correlate with cognitive symptoms. *Journal of Neurology, Neurosurgery, and Psychiatry, 93*(Suppl 1), A23–A24. BMJ Publishing Group Ltd.

Ho, P. T. B., Clark, I. M., & Le, L. T. T. (2022, June 28). MicroRNA-based diagnosis and therapy. *International Journal of Molecular Sciences, 23*(13), 7167.

Hong, Y., Zhao, T., Li, X.-J., & Li, S. (2017, September 27). Mutant huntingtin inhibits αB-crystallin expression and impairs exosome secretion from astrocytes. *The Journal of Neuroscience, 37*(39), 9550–9563.

Hornung, S., Dutta, S., & Bitan, G. (2020). CNS-derived blood exosomes as a promising source of biomarkers: Opportunities and challenges. *Frontiers in Molecular Neuroscience, 13*, 38.

Hoss, A. G., Labadorf, A., Latourelle, J. C., Kartha, V. K., Hadzi, T. C., Gusella, J. F., et al. (2015a, March 1). miR-10b-5p expression in Huntington's disease brain relates to age of onset and the extent of striatal involvement. *BMC Medical Genomics* [Internet]. [cited 2018 Jan 21], *8*. Available from: https://www.ncbi.nlm.nih.gov/pmc/articles/PMC4349621/

Hoss, A. G., Lagomarsino, V. N., Frank, S., Hadzi, T. C., Myers, R. H., & Latourelle, J. C. (2015b, December). Study of plasma-derived miRNAs mimic differences in Huntington's disease brain. *Movement Disorders, the Official Journal of Movement Disorder Society, 30*(14), 1961–1964.

Hosseinzadeh, S., Noroozian, M., Mortaz, E., & Mousavizadeh, K. (2018, February). Plasma microparticles in Alzheimer's disease: The role of vascular dysfunction. *Metabolic Brain Disease, 33*(1), 293–299.

Jarosińska, O. D., & Rüdiger, S. G. D. (2021). Molecular strategies to target protein aggregation in Huntington's disease. *Frontiers in Molecular Biosciences, 8*, 769184.

Jeon, I., Cicchetti, F., Cisbani, G., Lee, S., Li, E., Bae, J., et al. (2016). Human-to-mouse prion-like propagation of mutant huntingtin protein. *Acta Neuropathologica (Berl.), 132*(4), 577–592.

Ji, Q., Ji, Y., Peng, J., Zhou, X., Chen, X., Zhao, H., et al. (2016). Increased brain-specific MiR-9 and MiR-124 in the serum exosomes of acute ischemic stroke patients. *PLoS One, 11*(9), e0163645.

Johnson, R., & Buckley, N. J. (2009). Gene dysregulation in Huntington's disease: REST, microRNAs and beyond. *Neuromolecular Medicine, 11*(3), 183–199.

Johnson, R., Zuccato, C., Belyaev, N. D., Guest, D. J., Cattaneo, E., & Buckley, N. J. (2008, March). A microRNA-based gene dysregulation pathway in Huntington's disease. *Neurobiology of Disease, 29*(3), 438–445.

Johnstone, R. M., Adam, M., Hammond, J. R., Orr, L., & Turbide, C. (1987, July 5). Vesicle formation during reticulocyte maturation. Association of plasma membrane activities with released vesicles (exosomes). *The Journal of Biological Chemistry, 262*(19), 9412–9420.

Jovicic, A., Zaldivar Jolissaint, J. F., Moser, R., Silva Santos, M. de F., & Luthi-Carter, R. (2013, January 17). MicroRNA-22 (miR-22) overexpression is neuroprotective via general anti-apoptotic effects and may also target specific Huntington's disease-related mechanisms. *PLoS ONE* [Internet]. [cited 2018 Jan 18], *8*(1). Available from: https://www.ncbi.nlm.nih.gov/pmc/articles/PMC3547907/

Kalluri, R., & LeBleu, V. S. (2020, February 7). The biology, function, and biomedical applications of exosomes. *Science, 367*(6478), eaau6977.

Kalra, H., Drummen, G. P. C., Mathivanan, S. (2016, February 6). Focus on extracellular vesicles: Introducing the next small big thing. *International Journal of Molecular Sciences* [Internet]. [cited 2019 Oct 11], *17*(2). Available from: https://www.ncbi.nlm.nih.gov/pmc/articles/PMC4783904/

Keerthikumar, S., Chisanga, D., Ariyaratne, D., Al Saffar, H., Anand, S., Zhao, K., et al. (2016, February 22). ExoCarta: A web-based compendium of exosomal cargo. *Journal of Molecular Biology, 428*(4), 688–692.

Kenny, A., McArdle, H., Calero, M., Rabano, A., Madden, S. F., Adamson, K., et al. (2019, November 13). Elevated plasma microRNA-206 levels predict cognitive decline and progression to dementia from mild cognitive impairment. *Biomolecules, 9*(11), 734.

Keskin, S., Brouwers, C. C., Sogorb-Gonzalez, M., Martier, R., Depla, J. A., Vallès, A., et al. (2019, December 13). AAV5-miHTT lowers huntingtin mRNA and protein without off-target effects in patient-derived neuronal cultures and astrocytes. *Molecular Therapy – Methods & Clinical Development, 15*, 275–284.

Kim, D.-K., Kang, B., Kim, O. Y., Choi, D.-S., Lee, J., Kim, S. R., et al. (2013). EVpedia: An integrated database of high-throughput data for systemic analyses of extracellular vesicles. *Journal of Extracellular Vesicles, 2*.

Kim, K. M., Abdelmohsen, K., Mustapic, M., Kapogiannis, D., & Gorospe, M. (2017, July). RNA in extracellular vesicles. *Wiley Interdisciplinary Reviews RNA, 8*(4), e1413.

Kumar, S., & Reddy, P. H. (2016, September). Are circulating microRNAs peripheral biomarkers for Alzheimer's disease? *Biochimica et Biophysica Acta, 1862*(9), 1617–1627.

Kumar, A., Zhou, L., Zhi, K., Raji, B., Pernell, S., Tadrous, E., et al. (2020, December 25). Challenges in biomaterial-based drug delivery approach for the treatment of neurodegenerative diseases: Opportunities for extracellular vesicles. *International Journal of Molecular Sciences* [Internet]. [cited 2021 Feb 11], *22*(1). Available from: https://www.ncbi.nlm.nih.gov/pmc/articles/PMC7795247/

Kupcova Skalnikova, H., Bohuslavova, B., Turnovcova, K., Juhasova, J., Juhas, S., Rodinova, M., et al. (2019, April 25). Isolation and characterization of small extracellular vesicles from porcine blood plasma, cerebrospinal fluid, and seminal plasma. *Proteomes* [Internet]. [cited 2020 Jul 3], *7*(2). Available from: https://www.ncbi.nlm.nih.gov/pmc/articles/PMC6630935/

Lai, H., Li, Y., Zhang, H., Hu, J., Liao, J., Su, Y., et al. (2022, January 7). exoRBase 2.0: An atlas of mRNA, lncRNA and circRNA in extracellular vesicles from human biofluids. *Nucleic Acids Research, 50*(D1), D118–D128.

Langfelder, P., Gao, F., Wang, N., Howland, D., Kwak, S., Vogt, T. F., et al. (2018, January 11). MicroRNA signatures of endogenous Huntingtin CAG repeat expansion in mice. *PLoS One, 13*(1), e0190550.

Lee, S.-T., Chu, K., Im, W.-S., Yoon, H.-J., Im, J.-Y., Park, J.-E., et al. (2011, January 1). Altered microRNA regulation in Huntington's disease models. *Experimental Neurology, 227*(1), 172–179.

Lee, M., Liu, T., Im, W., & Kim, M. (2016, August 1). Exosomes from adipose-derived stem cells ameliorate phenotype of Huntington's disease in vitro model. *The European Journal of Neuroscience, 44*(4), 2114–2119.

Lee, S.-T., Im, W., Ban, J.-J., Lee, M., Jung, K.-H., Lee, S. K., et al. (2017, January). Exosome-based delivery of miR-124 in a Huntington's disease model. *Journal of Movement Disorders, 10*(1), 45–52.

Lee, S., Mankhong, S., & Kang, J.-H. (2019, April 8). Extracellular vesicle as a source of Alzheimer's biomarkers: Opportunities and challenges. *International Journal of Molecular Sciences* [Internet]. [cited 2019 Nov 5], *20*(7). Available from: https://www.ncbi.nlm.nih.gov/pmc/articles/PMC6479979/

Levine, P. H., Horoszewicz, J. S., Grace, J. T., Chai, L. S., Ellison, R. R., & Holland, J. F. (1967, October). Relationship between clinical status of leukemic patients and virus-like particles in their plasma. *Cancer, 20*(10), 1563–1577.

Li, T.-R., Wang, X.-N., Sheng, C., Li, Y.-X., Li, F. Z.-T., Sun, Y., et al. (2019, December). Extracellular vesicles as an emerging tool for the early detection of Alzheimer's disease. *Mechanisms of Ageing and Development, 184*, 111175.

Li, R., Hu, Q., Shi, X., Luo, Z., & Shao, D. (2023). Crosstalk between exosomes and autophagy in spinal cord injury: Fresh positive target for therapeutic application. *Cell and Tissue Research, 391*(1), 1–17.

Liu, T., Im, W., Mook-Jung, I., & Kim, M. (2015, May). MicroRNA-124 slows down the progression of Huntington's disease by promoting neurogenesis in the striatum. *Neural Regeneration Research, 10*(5), 786–791.

Liu, T., Zhang, Q., Zhang, J., Li, C., Miao, Y.-R., Lei, Q., et al. (2019a, January 8). EVmiRNA: A database of miRNA profiling in extracellular vesicles. *Nucleic Acids Research, 47*(D1), D89–D93.

Liu, X., Feng, Z., Du, L., Huang, Y., Ge, J., Deng, Y., et al. (2019b, December 23). The potential role of microRNA-124 in cerebral ischemia injury. *International Journal of Molecular Sciences, 21*(1), 120.

Liu, W.-L., Lin, H.-W., Lin, M.-R., Yu, Y., Liu, H.-H., Dai, Y.-L., et al. (2022, November). Emerging blood exosome-based biomarkers for preclinical and clinical Alzheimer's disease: A meta-analysis and systematic review. *Neural Regeneration Research, 17*(11), 2381–2390.

Lötvall, J., Hill, A. F., Hochberg, F., Buzás, E. I., Di Vizio, D., Gardiner, C., et al. (2014, December 22). Minimal experimental requirements for definition of extracellular vesicles and their functions: A position statement from the International Society for Extracellular Vesicles. *Journal of Extracellular Vesicles* [Internet]. [cited 2019 Nov 5], *3*. Available from: https://www.ncbi.nlm.nih.gov/pmc/articles/PMC4275645/

Lugli, G., Cohen, A. M., Bennett, D. A., Shah, R. C., Fields, C. J., Hernandez, A. G., et al. (2015). Plasma exosomal miRNAs in persons with and without Alzheimer disease: Altered expression and prospects for biomarkers. *PLoS One, 10*(10), e0139233.

Marchetti, B., Leggio, L., L'Episcopo, F., Vivarelli, S., Tirolo, C., Paternò, G., et al. (2020, June 21). Glia-derived extracellular vesicles in Parkinson's disease. *Journal of Clinical Medicine, 9*(6), 1941.

Martí, E., Pantano, L., Bañez-Coronel, M., Llorens, F., Miñones-Moyano, E., Porta, S., et al. (2010, November). A myriad of miRNA variants in control and Huntington's disease brain regions detected by massively parallel sequencing. *Nucleic Acids Research, 38*(20), 7219–7235.

Masnata, M., & Cicchetti, F. (2017). The evidence for the spread and seeding capacities of the mutant huntingtin protein in in vitro systems and their therapeutic implications. *Frontiers in Neuroscience, 11*, 647.

Mathieu, M., Martin-Jaular, L., Lavieu, G., & Théry, C. (2019, January). Specificities of secretion and uptake of exosomes and other extracellular vesicles for cell-to-cell communication. *Nature Cell Biology, 21*(1), 9–17.

Melentijevic, I., Toth, M. L., Arnold, M. L., Guasp, R. J., Harinath, G., Nguyen, K. C., et al. (2017, February 16). C. elegans neurons jettison protein aggregates and mitochondria under neurotoxic stress. *Nature, 542*(7641), 367–371.

Miniarikova, J., Zanella, I., Huseinovic, A., van der Zon, T., Hanemaaijer, E., Martier, R., et al. (2016, March). Design, characterization, and lead selection of therapeutic miRNAs targeting Huntingtin for development of gene therapy for Huntington's disease. *Molecular Therapy—Nucleic Acids, 5*(3), e297.

Miniarikova, J., Zimmer, V., Martier, R., Brouwers, C. C., Pythoud, C., Richetin, K., et al. (2017, October). AAV5-miHTT gene therapy demonstrates suppression of mutant huntingtin aggregation and neuronal dysfunction in a rat model of Huntington's disease. *Gene Therapy, 24*(10), 630–639.

Mitchell, M. I., Ma, J., Carter, C. L., & Loudig, O. (2022, July 10). Circulating exosome cargoes contain functionally diverse cancer biomarkers: From biogenesis and function to purification and potential translational utility. *Cancers, 14*(14), 3350.

Morais, R. D. V. S., Sogorb-González, M., Bar, C., Timmer, N. C., Van der Bent, M. L., Wartel, M., et al. (2022, September 3). Functional intercellular transmission of miHTT via extracellular vesicles: An in vitro proof-of-mechanism study. *Cells, 11*(17), 2748.

Muhammad, S. A. (2021, March). Are extracellular vesicles new hope in clinical drug delivery for neurological disorders? *Neurochemistry International, 144*, 104955.

Mustapic, M., Eitan, E., Werner, J. K., Berkowitz, S. T., Lazaropoulos, M. P., Tran, J., et al. (2017, May 22). Plasma extracellular vesicles enriched for neuronal origin: A potential window into brain pathologic processes. *Frontiers in Neuroscience* [Internet]. [cited 2019 Nov 4], *11*. Available from: https://www.ncbi.nlm.nih.gov/pmc/articles/PMC5439289/

Natale, F., Fusco, S., & Grassi, C. (2022 November 27). Dual role of brain-derived extracellular vesicles in dementia-related neurodegenerative disorders: Cargo of disease spreading signals and diagnostic-therapeutic molecules. *Translational Neurodegeneration, 11*(1), 50.

Neueder, A., Kojer, K., Hering, T., Lavery, D. J., Chen, J., Birth, N., et al. (2022a, September 7). Abnormal molecular signatures of inflammation, energy metabolism, and vesicle biology in human Huntington disease peripheral tissues. *Genome Biology, 23*(1), 189.

Neueder, A., Nitzschner, P., Wagner, R., Hummel, J., Hoschek, F., Wagner, M., et al. (2022b, September 1). D17 extracellular vesicles biology in huntington disease. *Journal of Neurology, Neurosurgery, and Psychiatry, 93*(Suppl 1), A26–A26. BMJ Publishing Group Ltd.

Nieland, L., Morsett, L. M., Broekman, M. L. D., Breakefield, X. O., & Abels, E. R. (2021, March). Extracellular vesicle-mediated bilateral communication between glioblastoma and astrocytes. *Trends in Neurosciences, 44*(3), 215–226.

Nila, I. S., Sumsuzzman, D. M., Khan, Z. A., Jung, J. H., Kazema, A. S., Kim, S. J., et al. (2022, December). Identification of exosomal biomarkers and its optimal isolation and detection method for the diagnosis of Parkinson's disease: A systematic review and meta-analysis. *Ageing Research Reviews, 1*(82), 101764.

Nunez, E. A., Wallis, J., & Gershon, M. D. (1974, October). Secretory processes in follicular cells of the bat thyroid. 3. The occurrence of extracellular vesicles and colloid droplets during arousal from hibernation. *The American Journal of Anatomy, 141*(2), 179–201.

Ochaba, J., Lukacsovich, T., Csikos, G., Zheng, S., Margulis, J., Salazar, L., et al. (2014, November 25). Potential function for the Huntingtin protein as a scaffold for selective autophagy. *Proceedings of the National Academy of Sciences, 111*(47), 16889–16894.

Odaka, H., Hiemori, K., Shimoda, A., Akiyoshi, K., & Tateno, H. (2021, March). Platelet-derived extracellular vesicles are increased in sera of Alzheimer's disease patients, as revealed by Tim4-based assays. *FEBS Open Bio, 11*(3), 741–752.

Packer, A. N., Xing, Y., Harper, S. Q., Jones, L., & Davidson, B. L. (2008, December 31). The bifunctional microRNA miR-9/miR-9* regulates REST and CoREST and is downregulated in Huntington's disease. *Journal of Neuroscience: The Official Journal of the Society for Neuroscience, 28*(53), 14341–14346.

Palaniswamy, R., Sevugan, K., & Sampathkumar, S. A. (2020, June 6). Molecular signatures in exosomes as diagnostic markers for neurodegenerative disorders. *Annals of Alzheimer's and Dementia Care, 4*(1), 012–017.

Palviainen, M., Saraswat, M., Varga, Z., Kitka, D., Neuvonen, M., Puhka, M., et al. (2020). Extracellular vesicles from human plasma and serum are carriers of extravesicular cargo-implications for biomarker discovery. *PLoS One, 15*(8), e0236439.

Pan, B. T., & Johnstone, R. M. (1983, July). Fate of the transferrin receptor during maturation of sheep reticulocytes in vitro: Selective externalization of the receptor. *Cell, 33*(3), 967–978.

Parkin, G. M., Corey-Bloom, J., Snell, C., Smith, H., Laurenza, A., Daldin, M., et al. (2023, January). Salivary Huntingtin protein is uniquely associated with clinical features of Huntington's disease. *Scientific Reports, 19*(13), 1034.

Pathan, M., Fonseka, P., Chitti, S. V., Kang, T., Sanwlani, R., Van Deun, J., et al. (2019, January 8). Vesiclepedia 2019: A compendium of RNA, proteins, lipids and metabolites in extracellular vesicles. *Nucleic Acids Research, 47*(D1), D516–D519.

Pearce, M. M. P., Spartz, E. J., Hong, W., Luo, L., & Kopito, R. R. (2015, April). Prion-like transmission of neuronal huntingtin aggregates to phagocytic glia in the Drosophila brain. *Nature Communications, 13*(6), 6768.

Pecho-Vrieseling, E., Rieker, C., Fuchs, S., Bleckmann, D., Esposito, M. S., Botta, P., et al. (2014, August). Transneuronal propagation of mutant huntingtin contributes to non-cell autonomous pathology in neurons. *Nature Neuroscience, 17*(8), 1064–1072.

Perez-Gonzalez, R., Martinez-Horta, S., Riva-Asensio, E., Perez-Perez, J., Garcia-Forn, M., Perez-Navarro, E., & Kulisevsky, J. (2023). Altered levels of brain extracellular vesicle subpopulations in Huntington's disease. *Journal of Neurology, Neurosurgery & Psychiatry*. https://jnnp.bmj.com/content/93/Suppl_1/A27.1

Petry, S., Keraudren, R., Nateghi, B., Loiselle, A., Pircs, K., Jakobsson, J., et al. (2022, July 22). Widespread alterations in microRNA biogenesis in human Huntington's disease putamen. *Acta Neuropathologica Communications, 10*(1), 106.

Pluta, R., & Ułamek-Kozioł, M. (2019). Lymphocytes, platelets, erythrocytes, and exosomes as possible biomarkers for Alzheimer's disease clinical diagnosis. *Advances in Experimental Medicine and Biology, 1118*, 71–82.

Prince, A. M., & Adams, W. R. (1966, August). Virus-like particles in human plasma and serum: Role of platelet lysosomes. *Journal of the National Cancer Institute, 37*(2), 153–166.

Properzi, F., Ferroni, E., Poleggi, A., & Vinci, R. (2015, November 12). The regulation of exosome function in the CNS: Implications for neurodegeneration. Swiss Medical Weekly [Internet]. [cited 2021 Jan 28]. Available from: http://doi.emh.ch/smw.2015.14204

Quiroz-Baez, R., Hernández-Ortega, K., & Martínez-Martínez, E. (2020). Insights into the proteomic profiling of extracellular vesicles for the identification of early biomarkers of neurodegeneration. *Frontiers in Neurology, 11*, 580030.

Raposo, G., Nijman, H. W., Stoorvogel, W., Liejendekker, R., Harding, C. V., Melief, C. J., et al. (1996, March 1). B lymphocytes secrete antigen-presenting vesicles. *The Journal of Experimental Medicine, 183*(3), 1161–1172.

Reed, E. R., Latourelle, J. C., Bockholt, J. H., Bregu, J., Smock, J., Paulsen, J. S., et al. (2018, January 23). MicroRNAs in CSF as prodromal biomarkers for Huntington disease in the PREDICT-HD study. *Neurology, 90*(4), e264–e272.

Ren, P.-H., Lauckner, J. E., Kachirskaia, I., Heuser, J. E., Melki, R., & Kopito, R. R. (2009, February). Cytoplasmic penetration and persistent infection of mammalian cells by polyglutamine aggregates. *Nature Cell Biology, 11*(2), 219–225.

Roux, Q., Van Deun, J., Dedeyne, S., & Hendrix, A. (2020). The EV-TRACK summary add-on: Integration of experimental information in databases to ensure comprehensive interpretation of biological knowledge on extracellular vesicles. *Journal of Extracellular Vesicles, 9*(1), 1699367.

Rui, Y.-N., Xu, Z., Patel, B., Chen, Z., Chen, D., Tito, A., et al. (2015, March). Huntingtin functions as a scaffold for selective macroautophagy. *Nature Cell Biology, 17*(3), 262–275.

Rusten, T. E., & Simonsen, A. (2008, May 1). ESCRT functions in autophagy and associated disease. *Cell Cycle (Georgetown, Texas), 7*(9), 1166–1172.

Schweitzer, J. K., Krivda, J. P., & D'Souza-Schorey, C. (2009). Neurodegeneration in Niemann-Pick type C disease and Huntingtons disease: Impact of defects in membrane trafficking. *Current Drug Targets, 10*(7), 653–665.

Seman, G., Gallager, H. S., Lukeman, J. M., & Dmochowski, L. (1971, December). Studies on the presence of particles resembling RNA virus particles in human breast tumors, pleural effusions, their tissue cultures, and milk. *Cancer, 28*(6), 1431–1442.

Sinha, M., Ghose, J., Das, E., & Bhattarcharyya, N. P. (2010, June 4). Altered microRNAs in STHdhQ111/HdhQ111 cells: miR-146a targets TBP. *Biochemical and Biophysical Research Communications, 396*(3), 742–747.

Sinha, M., Ghose, J., & Bhattarcharyya, N. P. (2011). Micro RNA -214,-150,-146a and-125b target Huntingtin gene. *RNA Biology, 8*(6), 1005–1021.

Šonský, I., Vodička, P., Vodičková Kepková, K., & Hansíková, H. (2021, October). Mitophagy in Huntington's disease. *Neurochemistry International, 1*(149), 105147.

Spronck, E. A., Brouwers, C. C., Vallès, A., de Haan, M., Petry, H., van Deventer, S. J., et al. (2019, March). AAV5-miHTT gene therapy demonstrates sustained Huntingtin lowering and functional improvement in Huntington disease mouse models. *Molecular Therapy – Methods & Clinical Development, 16*(13), 334–343.

Spronck, E. A., Vallès, A., Lampen, M. H., Montenegro-Miranda, P. S., Keskin, S., Heijink, L., et al. (2021, January 20). Intrastriatal administration of AAV5-miHTT in non-human primates and rats is well tolerated and results in miHTT transgene expression in key areas of Huntington disease pathology. *Brain Sciences, 11*(2), 129.

Stegmayr, B., & Ronquist, G. (1982). Promotive effect on human sperm progressive motility by prostasomes. *Urological Research, 10*(5), 253–257.

Sun, C. N. (1966, June). Lattice structures and osmiophilic bodies in the developing respiratory tissue of rats. *Journal of Ultrastructure Research, 15*(3), 380–388.

Sun, Y., Luo, Z.-M., Guo, X.-M., Su, D.-F., & Liu, X. (2015). An updated role of microRNA-124 in central nervous system disorders: A review. *Frontiers in Cellular Neuroscience, 9*, 193.

Tabrizi, S. J., Leavitt, B. R., Landwehrmeyer, G. B., Wild, E. J., Saft, C., Barker, R. A., et al. (2019, June 13). Targeting Huntingtin expression in patients with Huntington's disease. *The New England Journal of Medicine, 380*(24), 2307–2316.

Tabrizi, S. J., Flower, M. D., Ross, C. A., & Wild, E. J. (2020, October). Huntington disease: New insights into molecular pathogenesis and therapeutic opportunities. *Nature Reviews. Neurology, 16*(10), 529–546.

Takeuchi, T., & Nagai, Y. (2022, July). Emerging roles of extracellular vesicles in polyglutamine diseases: Mutant protein transmission, therapeutic potential, and diagnostics. *Neurochemistry International, 157*, 105357.

Tan, L., Yu, J.-T., & Tan, L. (2015). Causes and consequences of microRNA dysregulation in neurodegenerative diseases. *Molecular Neurobiology, 51*(3), 1249–1262.

Tartaglia, N. R., Farina, F., Martin-Jaular, L., Loew, D., Dingli, F., Brouillet, E., et al. (2022, September 1). D18 small extracellular vesicles in response to mutant huntingtin. *Journal of Neurology, Neurosurgery, and Psychiatry, 93*(Suppl 1), A26–A26. BMJ Publishing Group Ltd.

Théry, C., Witwer, K. W., Aikawa, E., Alcaraz, M. J., Anderson, J. D., Andriantsitohaina, R., et al. (2018). Minimal information for studies of extracellular vesicles 2018 (MISEV2018): A position statement of the International Society for Extracellular Vesicles and update of the MISEV2014 guidelines. *Journal of Extracellular Vesicles, 7*(1), 1535750.

Tomlinson, P. R., Zheng, Y., Fischer, R., Heidasch, R., Gardiner, C., Evetts, S., et al. (2015, April). Identification of distinct circulating exosomes in Parkinson's disease. *Annals of Clinical Translational Neurology, 2*(4), 353–361.

Tsering, T., Li, M., Chen, Y., Nadeau, A., Laskaris, A., Abdouh, M., et al. (2022, October). EV-ADD, a database for EV-associated DNA in human liquid biopsy samples. *Journal of Extracellular Vesicles, 11*(10), e12270.

Tsujimura, K., Shiohama, T., & Takahashi, E. (2022, October 9). microRNA biology on brain development and neuroimaging approach. *Brain Sciences, 12*(10), 1366.

Valadi, H., Ekström, K., Bossios, A., Sjöstrand, M., Lee, J. J., & Lötvall, J. O. (2007, June). Exosome-mediated transfer of mRNAs and microRNAs is a novel mechanism of genetic exchange between cells. *Nature Cell Biology, 9*(6), 654–659.

Valencia, J., Ferreira, M., Merino-Torres, J. F., Marcilla, A., & Soriano, J. M. (2022, September 29). The potential roles of extracellular vesicles as biomarkers for Parkinson's disease: A systematic review. *International Journal of Molecular Sciences, 23*(19), 11508.

Vallès, A., Evers, M. M., Stam, A., Sogorb-Gonzalez, M., Brouwers, C., Vendrell-Tornero, C., et al. (2021, April 7). Widespread and sustained target engagement in Huntington's disease minipigs upon intrastriatal microRNA-based gene therapy. *Science Translational Medicine, 13*(588), eabb8920.

Vandendriessche, C., Bruggeman, A., Van Cauwenberghe, C., & Vandenbroucke, R. E. (2020, November 15). Extracellular vesicles in Alzheimer's and Parkinson's disease: Small entities with large consequences. *Cells, 9*(11), E2485.

Vandendriessche, C., Balusu, S., Van Cauwenberghe, C., Brkic, M., Pauwels, M., Plehiers, N., et al. (2021, August 23). Importance of extracellular vesicle secretion at the blood-cerebrospinal fluid interface in the pathogenesis of Alzheimer's disease. *Acta Neuropathologica Communications, 9*(1), 143.

Vassileff, N., Cheng, L., & Hill, A. F. (2020, December 11). Extracellular vesicles – Propagators of neuropathology and sources of potential biomarkers and therapeutics for neurodegenerative diseases. *Journal of Cell Science, 133*(23), jcs243139.

Velier, J., Kim, M., Schwarz, C., Kim, T. W., Sapp, E., Chase, K., et al. (1998, July). Wild-type and mutant huntingtins function in vesicle trafficking in the secretory and endocytic pathways. *Experimental Neurology, 152*(1), 34–40.

Vella, L. J., Greenwood, D. L. V., Cappai, R., Scheerlinck, J.-P. Y., & Hill, A. F. (2008, August 15). Enrichment of prion protein in exosomes derived from ovine cerebral spinal fluid. *Veterinary Immunology and Immunopathology, 124*(3–4), 385–393.

Vella, L. J., Hill, A. F., & Cheng, L. (2016, February 6). Focus on extracellular vesicles: Exosomes and their role in protein trafficking and biomarker potential in Alzheimer's and Parkinson's disease. *International Journal of Molecular Sciences* [Internet]. [cited 2019 Nov 5], *17*(2). Available from: https://www.ncbi.nlm.nih.gov/pmc/articles/PMC4783907/

Villar-Vesga, J., Henao-Restrepo, J., Voshart, D. C., Aguillon, D., Villegas, A., Castaño, D., et al. (2020). Differential profile of systemic extracellular vesicles from sporadic and familial Alzheimer's disease leads to neuroglial and endothelial cell degeneration. *Frontiers in Aging Neuroscience, 12*, 587989.

Vinaiphat, A., & Sze, S. K. (2022, September). Proteomics for comprehensive characterization of extracellular vesicles in neurodegenerative disease. *Experimental Neurology, 355*, 114149.

Vingtdeux, V., Sergeant, N., & Buée, L. (2012). Potential contribution of exosomes to the prion-like propagation of lesions in Alzheimer's disease. *Frontiers in Physiology* [Internet]. [cited 2021 Jan 28], *3*. Available from: http://journal.frontiersin.org/article/10.3389/fphys.2012.00229/abstract

von Einem, B., Abdelmoez, A., Lewerenz, J., Landwehrmeyer, G. B., & Neueder, A. (2022, September 1). D19 purification of extracellular vesicles from human control and HTT mutant carrier CSF and quantitative analysis of their RNA content. *Journal of Neurology, Neurosurgery, and Psychiatry, 93*(Suppl 1), A26–A27. BMJ Publishing Group Ltd.

Wang, L., & Zhang, L. (2020, April). Circulating exosomal miRNA as diagnostic biomarkers of neurodegenerative diseases. *Frontiers in Molecular Neuroscience, 15*(13), 53.

Wang, S., Kojima, K., Mobley, J. A., & West, A. B. (2019, July). Proteomic analysis of urinary extracellular vesicles reveal biomarkers for neurologic disease. *eBioMedicine, 45*, 351–361.

Wang, X., Chai, Z., Pan, G., Hao, Y., Li, B., Ye, T., et al. (2021, May 20). ExoBCD: A comprehensive database for exosomal biomarker discovery in breast cancer. *Briefings in Bioinformatics, 22*(3), bbaa088.

Wang, Z.-Y., Wen, Z.-J., Xu, H.-M., Zhang, Y., & Zhang, Y.-F. (2022). Exosomal noncoding RNAs in central nervous system diseases: Biological functions and potential clinical applications. *Frontiers in Molecular Neuroscience, 15*, 1004221.

Weiss, A., Träger, U., Wild, E. J., Grueninger, S., Farmer, R., Landles, C., et al. (2012, October 1). Mutant huntingtin fragmentation in immune cells tracks Huntington's disease progression. *The Journal of Clinical Investigation, 122*(10), 3731–3736.

Wild, E. J., Boggio, R., Langbehn, D., Robertson, N., Haider, S., Miller, J. R. C., et al. (2015, May). Quantification of mutant huntingtin protein in cerebrospinal fluid from Huntington's disease patients. *Journal of Clinical Investigation American Society for Clinical Investigation, 125*(5), 1979–1986.

Witwer, K. W., & Théry, C. (2019). Extracellular vesicles or exosomes? On primacy, precision, and popularity influencing a choice of nomenclature. *Journal of Extracellular Vesicles, 8*(1), 1648167.

Witwer, K. W., Goberdhan, D. C., O'Driscoll, L., Théry, C., Welsh, J. A., Blenkiron, C., et al. (2021, December). Updating MISEV: Evolving the minimal requirements for studies of extracellular vesicles. *Journal of Extracellular Vesicles, 10*(14), e12182.

Wolf, P. (1967, May). The nature and significance of platelet products in human plasma. *British Journal of Haematology, 13*(3), 269–288.

Wu, J., & Shen, Z. (2020, October). Exosomal miRNAs as biomarkers for diagnostic and prognostic in lung cancer. *Cancer Medicine, 9*(19), 6909–6922.

Wu, T., Yu, M., Zhang, L., Chen, X., & Pei, Z. (2018, September 1). I02 systemic injection of exosomal sirna significantly reduced huntingtin expression in transgenic mice of Huntington's disease. *Journal of Neurology, Neurosurgery, and Psychiatry, 89*(Suppl 1), A88–A89. BMJ Publishing Group Ltd.

Xia, X., Wang, Y., Huang, Y., Zhang, H., Lu, H., & Zheng, J. C. (2019, December). Exosomal miRNAs in central nervous system diseases: Biomarkers, pathological mediators, protective factors and therapeutic agents. *Progress in Neurobiology, 183*, 101694.

Xia, X., Wang, Y., & Zheng, J. C. (2022, December 12). Extracellular vesicles, from the pathogenesis to the therapy of neurodegenerative diseases. *Translational Neurodegeneration, 11*(1), 53.

Xu, R., Greening, D. W., Zhu, H.-J., Takahashi, N., & Simpson, R. J. (2016, April 1). Extracellular vesicle isolation and characterization: Toward clinical application. *The Journal of Clinical Investigation, 126*(4), 1152–1162.

Yáñez-Mó, M., Siljander, P. R.-M., Andreu, Z., Zavec, A. B., Borràs, F. E., Buzas, E. I., et al. (2015). Biological properties of extracellular vesicles and their physiological functions. *Journal of Extracellular Vesicles, 4*, 27066.

Yang, W., Dunlap, J. R., Andrews, R. B., & Wetzel, R. (2002, November 1). Aggregated polyglutamine peptides delivered to nuclei are toxic to mammalian cells. *Human Molecular Genetics, 11*(23), 2905–2917.

Yao, Y.-F., Qu, M.-W., Li, G.-C., Zhang, F.-B., & Rui, H.-C. (2018, August). Circulating exosomal miRNAs as diagnostic biomarkers in Parkinson's disease. *European Review for Medical and Pharmacological Sciences, 22*(16), 5278–5283.

Yu, Y., Hou, K., Ji, T., Wang, X., Liu, Y., Zheng, Y., et al. (2021, January 29). The role of exosomal microRNAs in central nervous system diseases. *Molecular and Cellular Biochemistry, 476*, 2111.

Zhang, X., Abels, E. R., Redzic, J. S., Margulis, J., Finkbeiner, S., & Breakefield, X. O. (2016, April 1). Potential transfer of polyglutamine and CAG-repeat RNA in extracellular vesicles in Huntington's disease: Background and evaluation in cell culture. *Cellular and Molecular Neurobiology, 36*(3), 459–470.

Zhang, N., He, F., Li, T., Chen, J., Jiang, L., Ouyang, X.-P., et al. (2021). Role of exosomes in brain diseases. *Frontiers in Cellular Neuroscience, 15*, 743353.

Zhang, X., Takeuchi, T., Takeda, A., Mochizuki, H., & Nagai, Y. (2022). Comparison of serum and plasma as a source of blood extracellular vesicles: Increased levels of platelet-derived particles in serum extracellular vesicle fractions alter content profiles from plasma extracellular vesicle fractions. *PLoS One, 17*(6), e0270634.

Zhao, Z., & Zlokovic, B. V. (2017, July). Remote control of BBB: A tale of exosomes and microRNA. *Cell Research, 27*(7), 849–850.

Saliva as a Relevant Biofluid for Huntington's Disease Biomarker Research

Steven W. Granger and Elizabeth A. Thomas

Abbreviations

BDNF	Brain-derived neurotrophic factor
CAR	Cortisol awakening response
CNS	Central nervous system
CRP	C-reactive protein
CSF	Cerebral spinal fluid
DST	Dexamethasone suppression test (DST)
GCF	Gingival crevicular fluid
HADS	Hospital Anxiety and Depression Scale
HD	Huntington's disease
HPA	Hypothalamic pituitary adrenal
HTT	Huntingtin
HTT	Huntington gene
IL-1B	Interleukin-1beta
IL-6	Interleukin-6
MMSE	Mini-Mental State Examination
MOCA	Montreal Cognitive Assessment
SDMT	Symbol Digit Modalities Test
SMC	Single Molecular Counting
TFC	Total Functional Capacity

S. W. Granger
Salimetrics, LLC, Carlsbad, CA, USA

E. A. Thomas (✉)
Department of Neurobiology and Behavior, Institute for Interdisciplinary Salivary Bioscience Research, University of California, Irvine, CA, USA

Department of Neuroscience, The Scripps Research Institute, La Jolla, CA, USA
e-mail: eathoma1@uci.edu

E. A. Thomas, G. M. Parkin (eds.), *Biomarkers for Huntington's Disease*, Contemporary Clinical Neuroscience, https://doi.org/10.1007/978-3-031-32815-2_4

TMS	Total Motor Symptom
UA	Uric acid
UHDRS	Unified Huntington's Disease Rating Scale

1 Introduction

Studies using traditional biofluids, such as cerebral spinal fluid (CSF) and blood (plasma or serum), have led to important advances in biomarker research for neurodegenerative diseases. However, there is a need for additional biomarkers that can be measured in easily-accessible fluids. Saliva represents an alternative, non-invasive biofluid that has received growing attention over the past decade (Chojnowska et al., 2018; Granger et al., 2012; Kaczor-Urbanowicz et al., 2017; Song et al., 2023; Vining & McGinley, 1987; Yoshizawa et al., 2013). Landmark studies in the early 80 s first revealed that stress-related and reproductive hormones (i.e. cortisol, testosterone, estradiol and progesterone) could be measured in saliva and were associated with corresponding levels found in the blood (Granger et al., 1999; Perogamvros et al., 2010; Vining & McGinley, 1987; Vining et al., 1983; Wang et al., 1981). These studies paved the way for numerous additional studies on saliva, notably the discovery of the salivary proteome in the early 2000s, revealing ~1000 proteins to be present in saliva (Denny et al., 2008; Hu et al., 2005; Yan et al., 2009), although current estimates of the salivary proteome are now over 3000 (Lau et al., 2021). In addition, saliva contains DNA, RNA, miRNA, extracellular vesicles and metabolites, which also have utility as biomarkers. The integration of omics methods has allowed the accurate detection and quantification of many types of molecules in saliva, which have led to the emerging field of saliva-omics (Zhang et al., 2014). While saliva has been utilized for more than two decades as a diagnostic matrix for several pathological conditions, such as celiac disease (Lenander-Lumikari et al., 2000), diabetes mellitus (Belazi et al., 1998; Lopez et al., 2003), rheumatoid arthritis (Helenius et al., 2005), HIV (Holmstrom et al., 1990; Matsuda et al., 1993), breast cancer (Kaczor-Urbanowicz et al., 2019; Streckfus & Bigler, 2005) and Sjögren's syndrome (Jung et al., 2021; Ryu et al., 2006), this biofluid has yet to be fully accepted as a tool for biomarker research in neurodegenerative disorders. Nonetheless, studies have begun to emerge that have appreciated the unique value that saliva can provide as a biofluid for neurodegenerative diseases (Ashton et al., 2019; Goldoni et al., 2022).

As discussed in other chapters throughout this book, there is a critical need to identify biomarkers for Huntington's Disease (HD). HD is an inherited, progressive neurodegenerative disorder characterized by motor dysfunction, uncontrolled chorea, cognitive decline and psychiatric disturbances (Huntington Disease Collaborative Research Group, 1993). In the U.S. about 30,000 people have been diagnosed with HD and another 150,000 are at risk for developing the disease, given its autosomal dominant nature of inheritance. Despite great strides in our understanding of the molecular and pathophysiological features of the disease, there are currently no satisfactory treatment options for patients although the number of compounds in clinical trials is growing every year. In this context, accessible biomarkers are greatly needed in order to select patients who will most likely benefit from a particular trial, predict trial endpoints and to track outcomes of novel therapeutic approaches. Aside from clinical trials, HD biomarkers would serve additional utility to predict disease onset and track disease progression and severity; both of these features would also have enormous value for patients and their caregivers to allow planning for lifestyle changes and changing outcomes.

This chapter will review published studies from the literature that provide precedence for the use of saliva for biomarker research in HD. Although this field is still in its early stages, analysis of biomarkers in saliva has the potential to transform the medical landscape and to, overall, improve patient care.

2 Saliva as a Non-invasive Biofluid

CSF collection and blood sampling are invasive techniques, which limit their clinical utility as biofluids. With its non-invasive nature, saliva sampling represents an alternative biofluid that has received increased attention over the past decade as a novel matrix for biomarker discovery. It is now appreciated that saliva can provide an assessment of the biological state of the body and can reflect changes in health status, making it an attractive biofluid for many purposes.

Saliva Composition and Secretion

Whole saliva is a complex fluid resulting from secretion of the three major salivary glands, the parotid, submandibular and sublingual glands, as well as hundreds of minor salivary glands, which are located throughout the oral cavity and aerodigestive tract (Baum, 1993; Pedersen et al., 2018). Each gland is comprised of secretory acinar cells connecting a system of ducts by which saliva flows into the oral cavity. Although saliva is mostly composed of water, it also contains electrolytes, proteins, glycoproteins, nucleic acids, hormones, lipids and metabolites (Yan et al., 2009; Zhang et al., 2016), all of which make up a variety of biomolecules that can serve as biomarkers for disease states. Specifically, saliva represents a rich source of proteins, with over 3000 identified in the salivary proteome by mass spectrometry. Interestingly, there is an estimated overlap of 1245 proteins that are present in both blood and saliva fluids (Lau et al., 2021), and it is also known that different saliva glands can express distinct protein signatures, with whole saliva representing the most extensive array of proteins (Fig. 1).

Saliva plays essential roles in many oral functions, including the digestion of carbohydrates and lipids, lubricating the oral cavity to facilitate swallowing, and facilitating the processes of taste and speech. Saliva also helps to protect the teeth, tooth enamel and gums from bacteria and to promote oral health by washing away

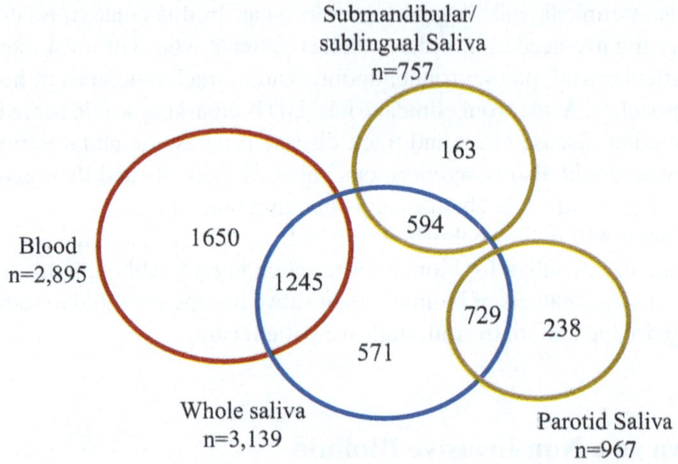

Fig. 1 Overlap of proteins present in blood, whole saliva and specific salivary glands. Proteins listed have been manually reviewed and annotated by UniProt and have been identified through tandem mass spectrometry (MS/MS) experiments using samples collected from healthy subjects. Information taken from www.salivaryproteome.org. The three main salivary glands are the submandibular, sublingual and parotid glands, as indicated

food particles and neutralizing acids. The pH of saliva in a healthy individual ranges from 6.6 to 7.6, which aids in this buffering capacity.

Unstimulated saliva is the basal level of saliva production made by an individual. Secretion is controlled by both the parasympathetic and sympathetic nervous systems, with saliva produced by sympathetic innervation being thicker and containing more mucins, while saliva stimulated by parasympathetic innervation is more clear and watery-like (de Almeida Pdel et al., 2008). Saliva production, or flow-rate, can vary according to many factors; however, the average unstimulated flow rate is about 0.3–0.4 mL/min which results in the production of approximately 1–1.5 liters of saliva every day. Aside from mastication, (i.e. the chewing of food), other factors, such as gum, lemon/citrus or wax can all stimulate saliva production, which can result in five to ten-fold increase in the saliva secretion rate (i.e. up to 1.5–6.0 ml/min) (Lee & Linden, 1992). Because this dramatic increase in flow rate can alter saliva composition, the basal, unstimulated form of saliva is the preferred matrix for biomarker research studies (Song et al., 2023).

Advantages of Saliva Collection

The use of saliva for biomarker research has several advantages over more traditional fluids for several reasons (Table 1). First, saliva collection is non-invasive, painless, and relatively stress-free compared to CSF or blood sampling, which can

Table 1 Advantages and disadvantages of saliva as a biofluid

Advantages	Disadvantages
• Simple and safe • Non-invasive • Improved compliance from subjects • Reduced risk of infection compared to invasive methods using needles • Easier for subjects who have difficulties with blood collection (i.e., children, elderly, etc.) • Conducive to repeat sampling • Does not require immediate sample processing • Cost effective • Can be collected in any setting	• Complex matrix; may interfere with measurements • Low concentrations of markers; may require highly sensitive methods for detection • Interferences from food, smoking, and periodontal diseases • Differences in saliva collection methods • Uncertainty over origin of some analytes

be uncomfortable and anxiety-causing, especially in elderly patients. The ease of collection also helps for older subjects who physically cannot give blood. Secondly, whole saliva collection requires no specially trained personnel, hence it can be collected in any setting. The lack of the need for a phlebotomist also provides a cost-effective benefit, as hiring phlebotomists for clinical research can be expensive. Thirdly, saliva does not need to be processed immediately, unlike blood, which requires centrifugation within a few hours of collection. This feature also allows for sample collection in any setting, as the sample can be frozen immediately and processed at a later time. It is worth mentioning here that the ability to collect a biological specimen in any setting comes with its own set of advantages, which include sample collection in remote locations, from individuals who live far from a centralized laboratory or those who have substantial disease progression making it difficult to come into a clinic. Fourth, saliva is stable and can be stored for long periods, making it an excellent resource for retrospective studies. Fifth, saliva tests are safer than blood tests with regards to the risk for hepatitis and HIV, given that needles are not used (Campo et al., 2006; Wormwood et al., 2015). This feature can also help with people who have "trypanophobia", which is a fear of needles, preventing them from getting injections or blood draws. Finally, for some molecules, such as cortisol, the form of the substance present in saliva represents the unbound, fraction, hence can provide a measure of the biologically active form of the analyte (Schwartz et al., 1998; Vining et al., 1983). Overall, these many advantages improve patient compliance and facilitate the collection of repeat samples, including over long periods of time, all of which would benefit use in clinical trials.

Like any method or procedure, there are drawbacks to the use of saliva for biomarker studies and research. These are covered further below, but include factors such as low concentrations of analytes, matrix effects for reliable quantification, differences in collection methods and potential interferences from food, smoking and periodontal disease (Table 1).

Origins of Salivary Analytes

There are several mechanisms by which proteins, hormones or other factors, can enter into saliva fluids (Fig. 2). First, salivary glands themselves are known to express a rich array of proteins, including several neurodegenerative disease proteins, such as the Huntingtin (HTT) protein, the amyloid beta precursor protein and tau (Conrad et al., 2002; Marques Sousa & Humbert, 2013; Oh & Turner, 2006). Next, due to the dense beds of capillaries surround the salivary glands, many blood constituents can enter into saliva. Components in blood can enter into saliva via three main mechanisms: (1). passive intracellular diffusion; (2). active transport through secretory cells of the glands; and (3). filtration, which can occur through the tight junctions between cells. Serum-like components can also enter the oral cavity via outflow from gingival crevicular fluid (GCF), which is found in the gingival sulcus (the space between a tooth and gum tissue) (Fig. 2). Hence, salivary levels of a biomarker can sometimes reflect the overall circulating levels in the body. Entry of molecules from blood may be particularly relevant to neurodegenerative diseases, given that leakage or breakdown of the blood brain barrier is known to occur in neurodegenerative diseases and this could account for how CNS proteins end up in the blood, and then subsequently find their way into saliva.

Cellular constituents of saliva can also contribute to molecules present in saliva. Notably, buccal cells, which are epithelial cells found inside of the cheek, slough off and are continually present in the oral cavity. In addition, several types of white blood cells, (i.e., granulocytes, monocytes and lymphocytes), are found in whole saliva, and molecules expressed in these cells will also be found in saliva. One issue of note is that proportions of these cell types can vary across individuals, which might confound data interpretation.

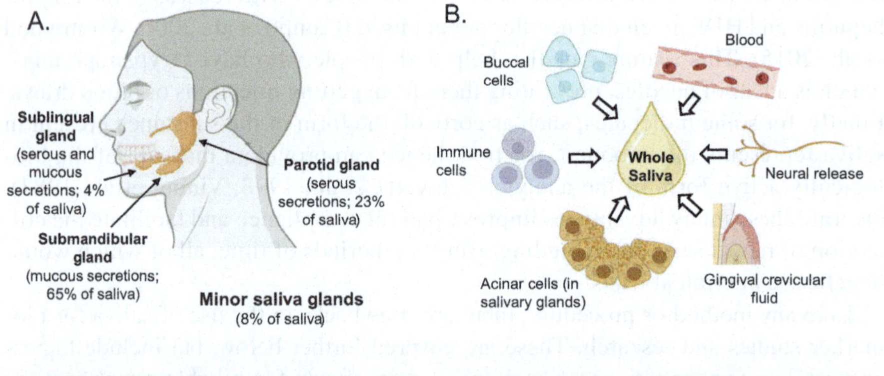

Fig. 2 Summary of the saliva glands and their contribution to saliva secretion (**a**) and summary of sources of saliva analytes. A. Depiction of the three major salivary glands, the parotid, submandibular and sublingual glands, as well minor salivary glands, which are located throughout the oral cavity and aerodigestive tract. (**b**). Factors can enter into saliva via many routes. These routes include analytes expressed in buccal, acinar and immune cells and those factors transported from blood or neural release. This figure was generated in part using BioRender, www.biorender.com

Finally, levels of biomarkers in saliva might possibly reflect neuronal release from nerves innervating the salivary glands, or via the glymphatic clearance system (Jessen et al., 2015). Parasympathetic innervation to the submandibular glands is achieved by the superior salivatory nucleus via the VIIth cranial nerve, a branch of the facial nerve, while sympathetic innervation of the salivary glands originates from the superior cervical ganglion, part of the paravertebral chain (Silvers & Som, 1998).

It will be important for future investigations to investigate the different origins of salivary biomarkers, not only for understanding the fundamental mechanisms involved in the transportation of these proteins in disease states, but also for controlling the effects of potential covariables if salivary biomarkers are to be used in the clinic.

3 Salivary Biomarkers in HD

The mutation responsible for the development of HD is an abnormal CAG repeat expansion in the *HTT* gene (Gusella et al., 1983; Huntington Disease Collaborative Research Group, 1993). This gene encodes the huntingtin protein (HTT), which, in its normal form, is involved in a wide range of cellular functions including gene regulation, vesicle transport and energy metabolism (Tabrizi et al., 2020). Pathogenesis in HD arises largely from the expression of the mutant form of the HTT protein, mutant HTT (mHTT), leading to the formation of insoluble protein aggregates which are the primary pathological hallmark of HD. Accordingly, an obvious candidate biomarker in HD is the disease protein itself, HTT, and studies have investigated salivary levels of this protein in HD saliva (Table 2). The presence of mHTT causes the disruption of many physiological pathways, including gene expression, oxidative stress, mitochondrial abnormalities, autophagy disturbances, metabolic abnormalities and immune system dysfunction (van der Burg et al., 2009). Markers of these pathways and systems have also been investigated in saliva samples (Table 2) and are described below.

Huntingtin Protein in Saliva

Studies using ELISA immunoassays have measured HTT in saliva samples collected from manifest HD patients, premanifest (PM) patients (i.e. those without overt motor symptoms) and normal controls to determine if levels differed according to diagnosis (Corey-Bloom et al., 2018). The immunoassay utilized in this study could not distinguish between the normal and mutant forms of the HTT protein, hence was only able to quantify total HTT (tHTT). Results showed that levels of tHTT were significantly higher in saliva from HD patients compared to matched, normal controls, as well as a non-significant elevation of tHTT in PM patients

Table 2 Summary of saliva biomarker studies in Huntington's disease

Biomarker:	Samples:	Brief findings:	Reference:
Huntingtin	n = 116 subjects: n = 37 HD patients; n = 35 PM subjects; n = 44 controls	Elevated total HTT observed in HD patients compared to controls. Significant correlations observed between salivary total HTT and MMSE, UHDRS and TFC	Corey-Bloom et al. (2018)
Huntingtin	n = 95 subjects: n = 19 HD patients; n = 34 PM subjects; n = 42 controls	Elevated levels of salivary mutant HTT found in HD patients compared to normal controls. Salivary mutant HTT was significantly correlated with age, CAG repeat length and CAP score and salivary total HTT was correlated with clinical measures	Parkin et al. (2023a, b)
Cortisol	n = 57 subjects: n = 20 PM subjects; n = 17 early-HD participants; n = 20 controls	Higher evening cortisol was associated with poorer performance on motor and memory tasks in PM subjects	Shirbin, CA et al. (2013a)
Cortisol	n = 57 subjects: n = 20 PM subjects; n = 17 early-HD participants; n = 20 controls	Increased morning cortisol levels observed in PM patients who had depressive symptoms compared to those without depression. Increased CAR in association with depression in early-HD patients	Shirbin, CA et al. (2013b)
Cortisol	n = 40 subjects: n = 26 PM subjects; n = 14 controls	No significant difference between salivary and hair cortisol between PM patients and controls groups. Hair cortisol was associated with disease onset and disease burden in patients	Cruickshank et al. (2021)
Cortisol	n = 112 subjects: n = 26 PM subjects; n = 58 HD patients, n = 28 controls	Higher cortisol awakening response in PM subjects compared to both HD patients and normal controls	van Duijn et al. (2010)

(continued)

Table 2 (continued)

Biomarker:	Samples:	Brief findings:	Reference:
Uric Acid	n = 178 subjects: n = 49 PM subjects; n = 45 HD patients; n = 84 controls	Salivary levels of uric acid were lower in PM and manifest HD patients compared to controls. Salivary UA levels were significantly correlated with TFC and TMS in male patients only	Corey-Bloom et al. (2020a)
BDNF	n = 81 subjects; n = 23 PM subjects; n = 38 HD patients; n = 20 controls	Significantly lower levels of salivary BDNF in PM and manifest HD patients compared to control subjects; lower BDNF levels were observed in PM patients who were within a predicted 10 years to disease onset	Gutierrez et al. (2019)
Interleukin-6/ Interleukin 1B	n = 125 subjects: n = 37 HD patients; n = 36 PM subjects; n = 52 healthy controls	Salivary IL-6 and IL-1B levels were significantly correlated with TMS, chorea scores and TFC in HD patients	Corey-Bloom et al. (2020b)
C-Reactive Protein	n = 125 subjects: n = 37 HD patients; n = 36 PM subjects; n = 52 healthy controls	Elevated salivary CRP levels measured in saliva from HD patients compared to normal controls	Corey-Bloom et al. (2020b)

PM Premanifest, *HD* Huntington's disease, *HTT* Huntingtin, *UHDRS* Unified Huntington's Disease Rating Scale, *TMS* Total Motor Symptom, *TFC* Total Functional Capacity, *MMSE* Mini-Mental State Examination, *HADS* Hospital Anxiety and Depression Scale, *MoCA* Montreal Cognitive Assessment, *SDMT* Symbol Digit Modalities Test, *CAR* cortisol awakening response, *BDNF* brain-derived neurotrophic factor; *CAP* CAG repeat product, *IL* interleukin; *CRP* C-reactive protein

compared to normal controls (Corey-Bloom et al., 2018). Salivary tHTT showed significant positive correlations to age in both HD patients and normal controls, but no effects of sex, CAG mutation length, nor age-of-onset were observed (Corey-Bloom et al., 2018). Comparing salivary tHTT to clinical measures revealed significant correlations to the Unified Huntington's Disease Rating Scale (UHDRS) and the Total Functional Capacity (TFC) score, with no significant associations with the other measures (Corey-Bloom et al., 2018).

A follow-up study used more sensitive methodology to specifically distinguish mHTT from tHTT in saliva and plasma samples from manifest and PM patients as well as non-mutation carriers (Parkin et al., 2023a, b). Total HTT was measured in each biofluid using Single Molecular Counting (SMC), an ultrasensitive immunoassay, with sensitivity down to the femtogram/mL level. SMC was utilized with an antibody combinations that specifically recognize only polyglutamine-expanded HTT, or separately, HTT in a polyglutamine-independent manner (i.e., tHTT) (Fodale et al., 2017; Fodale et al., 2022). Levels of mHTT were significantly higher in saliva samples from HD and PM patients compared to normal controls, while tHTT concentrations were lower in HD patients compared to normal controls (Parkin et al., 2023a, b). Further, saliva tHTT levels were found to be positively associated with chorea and total motor score (TMS) in PM individuals, but were negatively associated in patients with manifest HD. One explanation is that the association in manifest HD patients is influenced by other factors related to more severe disease progression, as has been suggested previously (Storey & Beal, 1993). Importantly, this latter paper demonstrated no significant correlation between salivary and plasma levels of tHTT or mHTT, suggesting different processing mechanisms of HTT in these fluids, or an inability of certain forms of HTT in the blood to enter into saliva.

Salivary Cortisol and the Stress Response in Early HD

Dysfunction in the hypothalamic pituitary adrenal (HPA) axis, the stress system of the body, have been linked to learning and memory deficits in a variety of neurodegenerative conditions, including HD (Aziz et al., 2009; Du & Pang, 2015; Hubers et al., 2015). In particular, evidence supports a link between alterations in HPA axis activity with depressive symptoms in HD (Hubers et al., 2015). Depression and cognitive deficits are among the most common non-motor symptoms and typically occur prior to the onset of overt movement dysfunction (Gargiulo et al., 2009). HPA functioning can be assessed by measuring salivary cortisol in several ways, including assessing cortisol levels throughout the day, determining the cortisol awakening response (CAR), measuring the morning rise in cortisol levels, and the dexamethasone suppression test (DST) (van Duijn et al., 2010).

Studies that have investigated salivary cortisol responses have mostly focused on the early stages of the disease (Table 2). In one study, salivary cortisol levels were measured diurnally across a single day in combination with a verbal memory performance task in early-stage HD (early-HD) patients and PM HD patients (Shirbin et al., 2013a). Higher levels of evening cortisol were detected in pre-HD patients with greater severity of motor abnormalities and this was found to be correlated with worse performance on a memory retrieval task (Shirbin et al., 2013a). Further, increased levels of salivary cortisol were observed in PM patients compared to controls, suggesting that HPA dysfunction might represents an early feature of disease progression (Shirbin et al., 2013a). Another study showed that the CAR was statistically different among PM, HD and control participants, although no significant differences were found in evening cortisol levels and post-DST cortisol concentrations across diagnoses. Similar to the previous study, this report indicated disturbed cortisol secretion in HD mutation carriers prior to the onset of motor symptoms (van Duijn et al., 2010).

Other studies have specifically investigated salivary cortisol levels in relation to depressive symptoms in HD. One study found that morning cortisol levels were elevated in PM HD patients who had depressive symptoms compared to those without depression (Shirbin et al., 2013b). Further, this study showed increased cortisol awakening responses in association with depression symptoms in patients with early-stage HD, but not in patients with advanced stage HD (Shirbin et al., 2013b).

Finally, another study showed no difference in salivary cortisol in PM HD patients compared to controls, nor any significant association between salivary cortisol and cognitive, mood or lifestyle factors (Cruickshank et al., 2021). Cortisol measured in hair, however, was associated with measures of disease burden and disease onset in HD patients (Cruickshank et al., 2021). Hair cortisol is thought to reflect chronic stress, compared to salivary levels, which typically reflect acute stress. Overall, these findings implicate salivary cortisol as playing a role, early in disease and particularly associated with depressive symptoms.

Salivary Uric Acid as a Sex-Specific Biomarker

Uric acid (UA), an end-product of purine metabolism, is also an anti-oxidant molecule that is found throughout the human body. Although imbalanced UA levels have a long-standing association with medical conditions, such as gout, hyperuricemia and oxidative stress-related disorders, UA has also been specifically linked to neurodegenerative diseases (Bowman et al., 2010; Fang et al., 2013), whereby oxidative damage, mitochondrial dysfunction and impairment in the electron transport chain have been suggested to have important roles in the degenerative process. Studies carried out on blood samples have demonstrated low levels of UA in association with Alzheimer's disease, Parkinson's disease, multiple sclerosis, and mild cognitive impairment (Chen et al., 2012; Constantinescu & Zetterberg, 2011; Crotty et al., 2017). Accordingly, studies in HD, have reported an association between higher plasma levels of UA and slower HD progression, (Auinger et al., 2010). Importantly, past studies have demonstrated significant correlations between UA levels present in blood compared to those found in saliva (Corey-Bloom et al., 2020a; Wang et al., 2020), and these findings have established the importance of salivary UA to potentially replace urine and blood tests for this analyte (Jaiswal et al., 2021).

To date, only one study has specifically measured UA in PM and manifest HD patients, as well as control subjects, using a simple colorimetric enzymatic reaction kit (Table 1) (Corey-Bloom et al., 2020a). Consistent with other studies on UA (Yang et al., 2019), male patients showed higher UA levels compared to female patients and sex differences persisted when comparing across diagnoses. Salivary UA levels were significantly lower in female PM and manifest HD patients compared to control subjects; however, levels were significantly lower only in male manifest HD patients compared to PM patients and controls (Corey-Bloom et al., 2020a). Further, in male HD patients, salivary UA levels were significantly correlated with TFC and TMS. In contrast, female HD patients only showed a significant correlation between salivary UA levels and disease burden. Studies examining postmortem HD brain also found decreased levels of UA in HD brains compared to controls (Corey-Bloom et al., 2020a), suggesting that decreased levels of UA in the brain, can be reflected in peripheral fluids.

Possible confounds of these studies are the potential effects of diet and exercise on UA levels, which could differ between patients and controls. It is known that UA levels can vary according to diet (Towiwat & Li, 2015; Zgaga et al., 2012) and strenuous exercise have been shown to elevate UA (Gonzalez et al., 2008). Hence, it is intriguing to consider that salivary UA might be a modifiable biomarker to monitor possible clinical interventions related to diet and exercise.

Overall, UA may represent a unique sex-specific marker, which is becoming increasingly important, in light of the appreciation of biological sex differences in HD. For example, biological sex has emerged as a factor that can affect disease outcomes in HD, with women showing worse symptoms and a faster rate of progression (Dorner et al., 2007; Zielonka et al., 2013). Further information on biological sex differences in HD can be found in Chap. 19 of this book.

Brain-Derived Neurotrophic Factor (BDNF), an Early Marker of Disease?

Deficits in brain-derived neurotrophic factor (BDNF), a growth factor associated with neuronal survival, development, and synaptic plasticity has long been associated with the pathophysiology of HD (Greenberg et al., 2009; Lu, 2003). Accordingly, BDNF gained early attention as a potential biomarker of the disease, not only due to its role in the pathophysiology of the disease, but also due to the possibility of tracking therapeutic approaches designed to elevate BDNF levels (Corey-Bloom et al., 2017; Giampa et al., 2013; Kells et al., 2004; Nagahara & Tuszynski, 2011). While blood measures of BDNF could serve this purpose, conflicting literature exists over whether levels of BDNF are altered in the blood of HD patients; one study has reported decreased levels of BDNF transcripts in whole blood of HD patients (Krzyszton-Russjan et al., 2013); another study showed increased BDNF protein in platelets (Betti et al., 2018); and a third study showing no significant differences in levels of BDNF protein in plasma or serum in HD patients (Zuccato et al., 2011) compared to control subjects. Given that levels of this protein are not correlated between blood and saliva (Gutierrez et al., 2019), measures of BDNF in saliva might represent an alternative avenue for BDNF exploration.

Investigations into whether levels of BDNF in saliva are altered in relation to HD diagnosis have been carried out. Levels of this neurotrophic factor were quantified using a modified immunoassay in saliva samples, consisting of manifest HD patients, PM patients and normal controls (Gutierrez et al., 2019). Significantly lower levels of BDNF were observed in saliva from PM and manifest HD patients compared to control subjects (Gutierrez et al., 2019). Salivary BDNF levels in patients, however, did not correlate with clinical scores including MMSE, TFC, TMS or HADS score. Interestingly exploring BDNF levels in relation to the onset of disease symptoms, lower levels of BDNF were observed specifically in PM patients who were within a predicted 10 years to disease onset, as estimated using the Langbehn formula (Long et al., 2017). This approach to identify fluid biomarkers associated with predicted years to onset has become an important tool to identify patients who might benefit the greatest from a particular therapeutic in a clinical trial.

Interleukin-6/Interleukin 1B as a Marker of Disease Severity

Although inflammation is not an initiating factor in HD, growing consensus suggests that inflammatory responses involving astrocytes, microglia, and the peripheral immune system contribute to disease progression, not only in HD, but other neurodegenerative diseases as well (see Chap. 11 of this book). Several past studies have measured proinflammatory cytokines, such as interleukin-6 (IL-6) and transforming growth factor beta 1 (TGF-β1) in plasma samples from HD patients and found elevated levels of both, when compared with controls (Bjorkqvist et al., 2008; Chang et al., 2015; Dalrymple et al., 2007). However, one main drawback of measuring cytokine proteins in blood is that endogenous levels of many cytokines are relatively low. Perhaps surprisingly, several past studies have shown that cytokines are present at higher concentrations in saliva compared to blood (Browne et al., 2013; Byrne et al., 2013; La Fratta et al., 2018; Nam et al., 2019; Parkin et al., 2023a, b). These findings put forth the idea that saliva might represent a better peripheral fluid to gauge inflammatory processes and responses (Parkin et al., 2023a, b).

With respect to HD, past studies have quantified levels of IL-6 and interleukin 1 beta (IL-1B), as well as C-reactive protein (CRP), in saliva samples from HD patients and healthy controls (Corey-Bloom et al., 2020b). Highly sensitive immunoassays, using electrochemiluminescence, revealed elevated salivary levels of IL-6, IL-1B and CRP across different disease groups. Increased levels of IL-6 were found in saliva from HD compared to PM patients, as well as increased salivary IL-1B levels in HD patients vs. controls. Further, salivary IL-6 and IL-1B levels were significantly correlated with TMS and chorea scores and negatively correlated with TFC in all gene mutation carriers (Corey-Bloom et al., 2020b). In control subjects, levels of IL-6 were found to be significantly correlated with Montreal Cognitive Assessment (MoCA) and the Symbol Digit Modalities test (SDMT). Elevated levels of salivary CRP were reportred in HD patients compared to normal controls. Plasma levels of IL-6, IL-1B and CRP were also quantified in this study in the same patients, and only plasma IL-6 was significantly elevated in HD patients, similar to previous studies (Corey-Bloom et al., 2020b). However, unlike that observed in saliva, plasma levels of IL-6 did not show any correlations to any clinical measures in either HD or control subjects, suggesting that there is a unique feature of IL-6 present in saliva, compared to other peripheral fluids.

4 Challenges of Saliva Biomarker Research

Despite the enormous potential for saliva to serve as a novel matrix for biomarker studies, there are several challenges that have hindered its ability to take the forefront in clinical studies (Table 1). The features described below have contributed to the overall variability and lack of reproducibility observed in saliva across analytical evaluations, impeding its clinical potential.

Differences in Saliva Collection Methods

A major drawback in implementation of salivary biomarker data is the lack of standardization of saliva collection methods (Table 1). There are several different techniques utilized to collect saliva, with even newer methods on the horizon. While collection of whole saliva via the passive drool method is the gold standard (Granger et al., 2012), there are many collection devices that employ swabbing material, in which swabs are placed under the tongue or in different parts of the mouth. A main drawback of swabbing methods is that the swabs are made of different types of materials, that can differentially affect analyte concentrations in saliva. Other collection devices, such as those developed by Oragene, contain stabilizers that allow for samples to remain at room temperature, reducing transportation and storage costs. These types of devices are most commonly used for DNA isolation from saliva samples.

Newer collection devices, such as PureSal™ and Saletto™, allow for separation of mucins and cellular components in freshly-collected saliva. These features should improve saliva sample variability and purity, thereby reducing complexity of downstream laboratory procedures. Further, these types of devices will play important roles in the development of point-of-care tests involving saliva, whereby saliva measures are quantified "on the spot", with the need to outsource samples for measurements. A recent example is the use of the PurSal™ device in a POC colorimetric molecular test for SARS-CoV-2 in saliva (Davidson et al., 2021). Overall, standardization of saliva collection techniques is critical if salivary biomarkers are going to reliably be utilized in clinical settings.

Low Levels of Biomarkers in Saliva

Saliva is rich in proteins, hormones and metabolites, however, many of these are present in lower abundance than corresponding levels found in blood. Aside from levels of proteins or molecules secreted by the salivary glands or expressed in cells found in the oral cavity (i.e., buccal epithelial cells and leukocytes), components that arise from blood can be highly diluted by the time they reach saliva. Similarly, molecules that may be present in saliva resulting from neuronal release may also be present only at very low levels. One solution to this problem is to utilize supersensitive methods of detections, which are becoming more available with ultrasensitive methods detecting analyte levels as low as femtograms per mL.

The potentially low levels of biomarkers in saliva can lead to significant consequences if blood is leaked into saliva. It is possible for blood and blood components to leak into saliva in numerous ways. Aside from obvious reasons, such as abrasions or cuts to the cheeks, gums, or tongue, individuals with poor oral health (e.g., gingivitis, open sores, or periodontal disease), those suffering from infectious diseases, or the use of tobacco products, can cause leakage of blood components into saliva. One remedy to account for blood contamination in saliva is to measure transferrin, a plasma glycoprotein that plays a central role in iron metabolism (Kivlighan et al., 2004; Schwartz & Granger, 2004).

State of Oral Health of the Participant

Poor oral health, including dental caries, mouth sores and poor hygiene, as well as periodontitis, have a great potential to confound measures of analytes in saliva. These situations cause increased inflammation in the mouth, including the infiltration of neutrophils, making accurate measures of other proinflammatory cytokines difficult. Additionally, advanced age, access to health care, systemic infection or inflammation, substance vaping and tobacco use, dental work, oral injury, shedding or loosing teeth all can contribute to alterations in the oral health of participants and can affect saliva composition. Salivary total Immunoglobulin G is thought to represent a potential surrogate marker of oral inflammation and immune activity, given its strong associations between Immunoglobulin G and several biologic indices of oral health (Riis et al., 2021).

Medication Considerations

Patients with HD can be taking several different types of medications and these could affect measurements of biomarkers in saliva. In past studies, it has been estimated that ~50% of the HD patients can be taking antidepressants, such as Paxil,

Fluoxetine, Wellbutrin and Zoloft, and these may affect the levels of markers measured in saliva (Corey-Bloom et al., 2018). Medications can have notable effects on the composition of saliva, the transport of serum constituents into oral fluid, and the rate of saliva secretion. For example, antidepressant drugs, antihistamines, antipsychotics, sedatives, methyldopa, and diuretics are all known to potentially cause low saliva volume, or hyposalivation, in many, but not all, patients. The presence of these drugs may not directly reduce saliva production, but may lead to a loss of saliva volume secondary to dehydration. Dehydration can lead to a change in saliva composition; this is the case, for example, of diuretics. In addition, cholinergic agents, such as those used in Parkinson's disease patients, can cause dry mouth or xerostomia. Overall, medication effects can be substantial, and must be considered in statistical analyses, either as potential confounders or stratifying by medication type, when possible.

5 Future Perspectives

The exploration of biomarkers in saliva for HD is a relatively new area of research that holds significant promise for the future. Due to its highly complex mixture of substances originating from multiple local and systemic sources, saliva has the capability to reveal systemic health conditions. This feature supports salivary measures as intrinsically important in their own right, yet also capable of reflecting the presence of disease throughout the body. Expanding our understanding of the markers that can be found in saliva could accelerate the development of non-invasive tools for the management of HD that are more accessible and affordable than current methods.

References

Ashton, N. J., Ide, M., Zetterberg, H., & Blennow, K. (2019). Salivary biomarkers for Alzheimer's disease and related disorders. *Neurology and Therapy, 8*(Suppl 2), 83–94.

Auinger, P., Kieburtz, K., & McDermott, M. P. (2010). The relationship between uric acid levels and Huntington's disease progression. *Movement Disorders, 25*(2), 224–228.

Aziz, N. A., Pijl, H., Frolich, M., van der Graaf, A. W., Roelfsema, F., & Roos, R. A. (2009). Increased hypothalamic-pituitary-adrenal axis activity in Huntington's disease. *The Journal of Clinical Endocrinology and Metabolism, 94*(4), 1223–1228.

Baum, B. J. (1993). Principles of saliva secretion. *Annals of the New York Academy of Sciences, 694*, 17–23.

Belazi, M. A., Galli-Tsinopoulou, A., Drakoulakos, D., Fleva, A., & Papanayiotou, P. H. (1998). Salivary alterations in insulin-dependent diabetes mellitus. *International Journal of Paediatric Dentistry, 8*(1), 29–33.

Betti, L., Palego, L., Unti, E., Mazzucchi, S., Kiferle, L., Palermo, G., et al. (2018). Brain-derived neurotrophic factor (BDNF) and serotonin transporter (SERT) in platelets of patients with mild

Huntington's disease: Relationships with social cognition symptoms. *Archives Italiennes de Biologie, 156*(1–2), 27–39.

Bjorkqvist, M., Wild, E. J., Thiele, J., Silvestroni, A., Andre, R., Lahiri, N., et al. (2008). A novel pathogenic pathway of immune activation detectable before clinical onset in Huntington's disease. *The Journal of Experimental Medicine, 205*(8), 1869–1877.

Bowman, G. L., Shannon, J., Frei, B., Kaye, J. A., & Quinn, J. F. (2010). Uric acid as a CNS antioxidant. *Journal of Alzheimer's Disease, 19*(4), 1331–1336.

Browne, R. W., Kantarci, A., LaMonte, M. J., Andrews, C. A., Hovey, K. M., Falkner, K. L., et al. (2013). Performance of multiplex cytokine assays in serum and saliva among community-dwelling postmenopausal women. *PLoS One, 8*(4), e59498.

Byrne, M. L., O'Brien-Simpson, N. M., Reynolds, E. C., Walsh, K. A., Laughton, K., Waloszek, J. M., et al. (2013). Acute phase protein and cytokine levels in serum and saliva: A comparison of detectable levels and correlations in a depressed and healthy adolescent sample. *Brain, Behavior, and Immunity, 34*, 164–175.

Campo, J., Perea, M. A., del Romero, J., Cano, J., Hernando, V., & Bascones, A. (2006). Oral transmission of HIV, reality or fiction? An update. *Oral Diseases, 12*(3), 219–228.

Chang, K. H., Wu, Y. R., Chen, Y. C., & Chen, C. M. (2015). Plasma inflammatory biomarkers for Huntington's disease patients and mouse model. *Brain, Behavior, and Immunity, 44*, 121–127.

Chen, X., Wu, G., & Schwarzschild, M. A. (2012). Urate in Parkinson's disease: More than a biomarker? *Current Neurology and Neuroscience Reports, 12*(4), 367–375.

Chojnowska, S., Baran, T., Wilinska, I., Sienicka, P., Cabaj-Wiater, I., & Knas, M. (2018). Human saliva as a diagnostic material. *Advances in Medical Sciences, 63*(1), 185–191.

Conrad, C., Vianna, C., Freeman, M., & Davies, P. (2002). A polymorphic gene nested within an intron of the tau gene: Implications for Alzheimer's disease. *Proceedings of the National Academy of Sciences of the United States of America, 99*(11), 7751–7756.

Constantinescu, R., & Zetterberg, H. (2011. Drugs Today (Barc)). Urate as a marker of development and progression in Parkinson's disease, *47*(5), 369–380.

Corey-Bloom, J., Aikin, A. M., Gutierrez, A. M., Nadhem, J. S., Howell, T. L., & Thomas, E. A. (2017). Beneficial effects of glatiramer acetate in Huntington's disease mouse models: Evidence for BDNF-elevating and immunomodulatory mechanisms. *Brain Research, 1673*, 102–110.

Corey-Bloom, J., Fischer, R. S., Kim, A., Snell, C., Parkin, G. M., Granger, D. A., et al. (2020a). Levels of Interleukin-6 in saliva, but not plasma, correlate with clinical metrics in Huntington's disease patients and healthy control subjects. *International Journal of Molecular Sciences, 21*(17), 6363.

Corey-Bloom, J., Haque, A., Aboufadel, S., Snell, C., Fischer, R. S., Granger, S. W., et al. (2020b). Uric acid as a potential peripheral biomarker for disease features in Huntington's patients. *Frontiers in Neuroscience, 14*, 73.

Corey-Bloom, J., Haque, A. S., Park, S., Nathan, A. S., Baker, R. W., & Thomas, E. A. (2018). Salivary levels of total huntingtin are elevated in Huntington's disease patients. *Scientific Reports, 8*(1), 7371.

Crotty, G. F., Ascherio, A., & Schwarzschild, M. A. (2017). Targeting urate to reduce oxidative stress in Parkinson disease. *Experimental Neurology, 298*(Pt B), 210–224.

Cruickshank, T., Porter, T., Laws, S. M., Ziman, M., & Bartlett, D. M. (2021). Hair and salivary cortisol and their relationship with lifestyle, mood and cognitive outcomes in premanifest Huntington's disease. *Scientific Reports, 11*(1), 5464.

Dalrymple, A., Wild, E. J., Joubert, R., Sathasivam, K., Bjorkqvist, M., Petersen, A., et al. (2007). Proteomic profiling of plasma in Huntington's disease reveals neuroinflammatory activation and biomarker candidates. *Journal of Proteome Research, 6*(7), 2833–2840.

Davidson, J. L., Wang, J., Maruthamuthu, M. K., Dextre, A., Pascual-Garrigos, A., Mohan, S., et al. (2021). A paper-based colorimetric molecular test for SARS-CoV-2 in saliva. *Biosensors and Bioelectronics: X, 9*, 100076.

de Almeida Pdel, V., Gregio, A. M., Machado, M. A., de Lima, A. A., & Azevedo, L. R. (2008). Saliva composition and functions: a comprehensive review. *The Journal of Contemporary Dental Practice, 9*(3), 72–80.

Denny, P., Hagen, F. K., Hardt, M., Liao, L., Yan, W., Arellanno, M., et al. (2008). The proteomes of human parotid and submandibular/sublingual gland salivas collected as the ductal secretions. *Journal of Proteome Research, 7*(5), 1994–2006.

Dorner, J. L., Miller, B. R., Barton, S. J., Brock, T. J., & Rebec, G. V. (2007). Sex differences in behavior and striatal ascorbate release in the 140 CAG knock-in mouse model of Huntington's disease. *Behavioural Brain Research, 178*(1), 90–97.

Du, X., & Pang, T. Y. (2015). Is dysregulation of the HPA-axis a core pathophysiology mediating co-morbid depression in neurodegenerative diseases? *Frontiers in Psychiatry, 6*, 32.

Fang, P., Li, X., Luo, J. J., Wang, H., & Yang, X. F. (2013). A double-edged sword: uric acid and neurological disorders. *Brain Disorders & Therapy, 2*(2), 109.

Fodale, V., Boggio, R., Daldin, M., Cariulo, C., Spiezia, M. C., Byrne, L. M., et al. (2017). Validation of ultrasensitive mutant huntingtin detection in human cerebrospinal fluid by single molecule counting immunoassay. *The Journal of Huntington's Disease, 6*(4), 349–361.

Fodale, V., Pintauro, R., Daldin, M., Spiezia, M. C., Macdonald, D., & Bresciani, A. (2022). Quantifying huntingtin protein in human cerebrospinal fluid using a novel polyglutamine length-independent assay. *The Journal of Huntington's Disease, 11*, 1–15.

Gargiulo, M., Lejeune, S., Tanguy, M. L., Lahlou-Laforet, K., Faudet, A., Cohen, D., et al. (2009). Long-term outcome of presymptomatic testing in Huntington disease. *European Journal of Human Genetics, 17*(2), 165–171.

Giampa, C., Montagna, E., Dato, C., Melone, M. A., Bernardi, G., & Fusco, F. R. (2013). Systemic delivery of recombinant brain derived neurotrophic factor (BDNF) in the R6/2 mouse model of Huntington's disease. *PLoS One, 8*(5), e64037.

Goldoni, R., Dolci, C., Boccalari, E., Inchingolo, F., Paghi, A., Strambini, L., et al. (2022). Salivary biomarkers of neurodegenerative and demyelinating diseases and biosensors for their detection. *Ageing Research Reviews, 76*, 101587.

Gonzalez, D., Marquina, R., Rondon, N., Rodriguez-Malaver, A. J., & Reyes, R. (2008). Effects of aerobic exercise on uric acid, total antioxidant activity, oxidative stress, and nitric oxide in human saliva. *Research in Sports Medicine, 16*(2), 128–137.

Granger, D. A., Fortunato, C. K., Beltzer, E. K., Virag, M., Bright, M. A., & Out, D. (2012). Focus on methodology: Salivary bioscience and research on adolescence: an integrated perspective. *Journal of Adolescence, 35*(4), 1081–1095.

Granger, D. A., Schwartz, E. B., Booth, A., & Arentz, M. (1999). Salivary testosterone determination in studies of child health and development. *Hormones and Behavior, 35*(1), 18–27.

Greenberg, M. E., Xu, B., Lu, B., & Hempstead, B. L. (2009). New insights in the biology of BDNF synthesis and release: Implications in CNS function. *The Journal of Neuroscience, 29*(41), 12764–12767.

Gusella, J. F., Wexler, N. S., Conneally, P. M., Naylor, S. L., Anderson, M. A., Tanzi, R. E., et al. (1983). A polymorphic DNA marker genetically linked to Huntington's disease. *Nature, 306*(5940), 234–238.

Gutierrez, A., Corey-Bloom, J., Thomas, E. A., & Desplats, P. (2019). Evaluation of biochemical and epigenetic measures of peripheral brain-derived neurotrophic factor (BDNF) as a biomarker in Huntington's disease patients. *Frontiers in Molecular Neuroscience, 12*, 335.

Helenius, L. M., Meurman, J. H., Helenius, I., Kari, K., Hietanen, J., Suuronen, R., et al. (2005). Oral and salivary parameters in patients with rheumatic diseases. *Acta Odontologica Scandinavica, 63*(5), 284–293.

Holmstrom, P., Syrjanen, S., Laine, P., Valle, S. L., & Suni, J. (1990). HIV antibodies in whole saliva detected by ELISA and western blot assays. *Journal of Medical Virology, 30*(4), 245–248.

Hu, S., Xie, Y., Ramachandran, P., Ogorzalek Loo, R. R., Li, Y., Loo, J. A., & Wong, D. T. (2005). Large-scale identification of proteins in human salivary proteome by liquid chromatography/

mass spectrometry and two-dimensional gel electrophoresis-mass spectrometry. *Proteomics, 5*(6), 1714–1728.

Hubers, A. A., van der Mast, R. C., Pereira, A. M., Roos, R. A., Veen, L. J., Cobbaert, C. M., et al. (2015). Hypothalamic-pituitary-adrenal axis functioning in Huntington's disease and its association with depressive symptoms and suicidality. *Journal of Neuroendocrinology, 27*(3), 234–244.

Huntington Disease Collaborative Research Group, D. (1993). A novel gene containing a trinucleotide repeat that is expanded and unstable on Huntington's disease chromosomes. The Huntington's disease collaborative research group. *Cell, 72*(6), 971–983.

Jaiswal, A., Madaan, S., Acharya, N., Kumar, S., Talwar, D., & Dewani, D. (2021). Salivary uric acid: A noninvasive wonder for clinicians? *Cureus, 13*(11), e19649.

Jessen, N. A., Munk, A. S. F., Lundgaard, I., & Nedergaard, M. (2015). The glymphatic system: A beginner's guide. *Neurochemical Research, 40*(12), 2583–2599.

Jung, J. Y., Kim, J. W., Kim, H. A., & Suh, C. H. (2021). Salivary biomarkers in patients with Sjögren's syndrome-a systematic review. *International Journal of Molecular Sciences, 22*(23).

Kaczor-Urbanowicz, K. E., Martin Carreras-Presas, C., Aro, K., Tu, M., Garcia-Godoy, F., & Wong, D. T. (2017). Saliva diagnostics – Current views and directions. *Experimental Biology and Medicine (Maywood, N.J.), 242*(5), 459–472.

Kaczor-Urbanowicz, K. E., Wei, F., Rao, S. L., Kim, J., Shin, H., Cheng, J., et al. (2019). Clinical validity of saliva and novel technology for cancer detection. *Biochimica Et Biophysica Acta. Reviews on Cancer, 1872*(1), 49–59.

Kells, A. P., Fong, D. M., Dragunow, M., During, M. J., Young, D., & Connor, B. (2004). AAV-mediated gene delivery of BDNF or GDNF is neuroprotective in a model of Huntington disease. *Molecular Therapy, 9*(5), 682–688.

Kivlighan, K. T., Granger, D. A., Schwartz, E. B., Nelson, V., Curran, M., & Shirtcliff, E. A. (2004). Quantifying blood leakage into the oral mucosa and its effects on the measurement of cortisol, dehydroepiandrosterone, and testosterone in saliva. *Hormones and Behavior, 46*(1), 39–46.

Krzyszton-Russjan, J., Zielonka, D., Jackiewicz, J., Kusmirek, S., Bubko, I., Klimberg, A., et al. (2013). A study of molecular changes relating to energy metabolism and cellular stress in people with Huntington's disease: Looking for biomarkers. *Journal of Bioenergetics and Biomembranes, 45*(1–2), 71–85.

La Fratta, I., Tatangelo, R., Campagna, G., Rizzuto, A., Franceschelli, S., Ferrone, A., et al. (2018). The plasmatic and salivary levels of IL-1β, IL-18 and IL-6 are associated to emotional difference during stress in young male. *Scientific Reports, 8*(1), 3031.

Lau, W. W., Hardt, M., Zhang, Y. H., Freire, M., & Ruhl, S. (2021). The human salivary proteome wiki: A community-driven research platform. *Journal of Dental Research, 100*(13), 1510–1519.

Lee, V. M., & Linden, R. W. (1992). An olfactory-submandibular salivary reflex in humans. *Experimental Physiology, 77*(1), 221–224.

Lenander-Lumikari, M., Ihalin, R., & Lahteenoja, H. (2000). Changes in whole saliva in patients with coeliac disease. *Archives of Oral Biology, 45*(5), 347–354.

Long, J. D., Langbehn, D. R., Tabrizi, S. J., Landwehrmeyer, B. G., Paulsen, J. S., Warner, J., & Sampaio, C. (2017). Validation of a prognostic index for Huntington's disease. *Movement Disorders, 32*(2), 256–263.

Lopez, M. E., Colloca, M. E., Paez, R. G., Schallmach, J. N., Koss, M. A., & Chervonagura, A. (2003). Salivary characteristics of diabetic children. *Brazilian Dental Journal, 14*(1), 26–31.

Lu, B. (2003). BDNF and activity-dependent synaptic modulation. *Learning & Memory, 10*(2), 86–98.

Marques Sousa, C., & Humbert, S. (2013). Huntingtin: Here, there, everywhere! *Journal of Huntington's Disease, 2*(4), 395–403.

Matsuda, S., Oka, S., Honda, M., Takebe, Y., & Takemori, T. (1993). Characteristics of IgA antibodies against HIV-1 in sera and saliva from HIV-seropositive individuals in different clinical stages. *Scandinavian Journal of Immunology, 38*(5), 428–434.

Nagahara, A. H., & Tuszynski, M. H. (2011). Potential therapeutic uses of BDNF in neurological and psychiatric disorders. *Nature Reviews. Drug Discovery, 10*(3), 209–219.

Nam, Y., Kim, Y. Y., Chang, J. Y., & Kho, H. S. (2019). Salivary biomarkers of inflammation and oxidative stress in healthy adults. *Archives of Oral Biology, 97*, 215–222.

Oh, Y. S., & Turner, R. J. (2006). Effect of gamma-secretase inhibitors on muscarinic receptor-mediated calcium signaling in human salivary epithelial cells. *American Journal of Physiology. Cell Physiology, 291*(1), C76–C82.

Parkin, G. M., Corey-Bloom, J., Snell, C., Smith, H., Laurenza, A., Daldin, M., et al. (2023a). Salivary Huntington protein is uniquely associated with clinical features of Huntington's disease. *Scientific Reports, 13*(1), 1034.

Parkin, G. M., Kim, S., Mikhail, A., Malhas, R., McMillan, L., Hollearn, M., et al. (2023b). Associations between saliva and plasma cytokines in cognitively normal, older adults. *Aging Clinical and Experimental Research, 35*(1), 117–126.

Pedersen, A. M. L., Sørensen, C. E., Proctor, G. B., Carpenter, G. H., & Ekström, J. (2018). Salivary secretion in health and disease. *Journal of Oral Rehabilitation, 45*(9), 730–746.

Perogamvros, I., Keevil, B. G., Ray, D. W., & Trainer, P. J. (2010). Salivary cortisone is a potential biomarker for serum free cortisol. *The Journal of Clinical Endocrinology and Metabolism, 95*(11), 4951–4958.

Riis, J. L., Ahmadi, H., Hamilton, K. R., Hand, T., & Granger, D. A. (2021). Best practice recommendations for the measurement and interpretation of salivary proinflammatory cytokines in biobehavioral research. *Brain, Behavior, and Immunity, 91*, 105–116.

Ryu, O. H., Atkinson, J. C., Hoehn, G. T., Illei, G. G., & Hart, T. C. (2006). Identification of parotid salivary biomarkers in Sjogren's syndrome by surface-enhanced laser desorption/ionization time-of-flight mass spectrometry and two-dimensional difference gel electrophoresis. *Rheumatology (Oxford), 45*(9), 1077–1086.

Schwartz, E. B., & Granger, D. A. (2004). Transferrin enzyme immunoassay for quantitative monitoring of blood contamination in saliva. *Clinical Chemistry, 50*(3), 654–656.

Schwartz, E. B., Granger, D. A., Susman, E. J., Gunnar, M. R., & Laird, B. (1998). Assessing salivary cortisol in studies of child development. *Child Development, 69*(6), 1503–1513.

Shirbin, C. A., Chua, P., Churchyard, A., Hannan, A. J., Lowndes, G., & Stout, J. C. (2013a). The relationship between cortisol and verbal memory in the early stages of Huntington's disease. *Journal of Neurology, 260*(3), 891–902.

Shirbin, C. A., Chua, P., Churchyard, A., Lowndes, G., Hannan, A. J., Pang, T. Y., et al. (2013b). Cortisol and depression in pre-diagnosed and early stage Huntington's disease. *Psychoneuroendocrinology, 38*(11), 2439–2447.

Silvers, A. R., & Som, P. M. (1998). Salivary glands. *Radiologic Clinics of North America, 36*(5), 941–966. vi.

Song, M., Bai, H., Zhang, P., Zhou, X., & Ying, B. (2023). Promising applications of human-derived saliva biomarker testing in clinical diagnostics. *International Journal of Oral Science, 15*(1), 2.

Storey, E., & Beal, M. (1993). Neurochemical substrates of rigidity and chorea in Huntington's disease. *Brain, 116*(5), 1201–1222.

Streckfus, C., & Bigler, L. (2005). The use of soluble, salivary c-erbB-2 for the detection and post-operative follow-up of breast cancer in women: The results of a five-year translational research study. *Advances in Dental Research, 18*(1), 17–24.

Tabrizi, S. J., Flower, M. D., Ross, C. A., & Wild, E. J. (2020). Huntington disease: New insights into molecular pathogenesis and therapeutic opportunities. *Nature Reviews. Neurology, 16*(10), 529–546.

Towiwat, P., & Li, Z. G. (2015). The association of vitamin C, alcohol, coffee, tea, milk and yogurt with uric acid and gout. *International Journal of Rheumatic Diseases, 18*(5), 495–501.

van der Burg, J. M., Bjorkqvist, M., & Brundin, P. (2009). Beyond the brain: Widespread pathology in Huntington's disease. *Lancet Neurology, 8*(8), 765–774.

van Duijn, E., Selis, M. A., Giltay, E. J., Zitman, F. G., Roos, R. A., van Pelt, H., & van der Mast, R. C. (2010). Hypothalamic-pituitary-adrenal axis functioning in Huntington's disease mutation carriers compared with mutation-negative first-degree controls. *Brain Research Bulletin, 83*(5), 232–237.

Vining, R. F., & McGinley, R. A. (1987). The measurement of hormones in saliva: Possibilities and pitfalls. *Journal of Steroid Biochemistry, 27*(1–3), 81–94.

Vining, R. F., McGinley, R. A., Maksvytis, J. J., & Ho, K. Y. (1983). Salivary cortisol: A better measure of adrenal cortical function than serum cortisol. *Annals of Clinical Biochemistry, 20*(Pt 6), 329–335.

Wang, C., Plymate, S., Nieschlag, E., & Paulsen, C. A. (1981). Salivary testosterone in men: Further evidence of a direct correlation with free serum testosterone. *The Journal of Clinical Endocrinology and Metabolism, 53*(5), 1021–1024.

Wang, Q., Wen, X., & Kong, J. (2020). Recent progress on uric acid detection: A review. *Critical Reviews in Analytical Chemistry, 50*(4), 359–375.

Wormwood, K. L., Aslebagh, R., Channaveerappa, D., Dupree, E. J., Borland, M. M., Ryan, J. P., et al. (2015). Salivary proteomics and biomarkers in neurology and psychiatry. *Proteomics. Clinical Applications, 9*(9–10), 899–906.

Yan, W., Apweiler, R., Balgley, B. M., Boontheung, P., Bundy, J. L., Cargile, B. J., et al. (2009). Systematic comparison of the human saliva and plasma proteomes. *Proteomics. Clinical Applications, 3*(1), 116–134.

Yang, Y., Zhou, W., Wang, Y., & Zhou, R. (2019). Gender-specific association between uric acid level and chronic kidney disease in the elderly health checkup population in China. *Renal Failure, 41*(1), 197–203.

Yoshizawa, J. M., Schafer, C. A., Schafer, J. J., Farrell, J. J., Paster, B. J., & Wong, D. T. (2013). Salivary biomarkers: Toward future clinical and diagnostic utilities. *Clinical Microbiology Reviews, 26*(4), 781–791.

Zgaga, L., Theodoratou, E., Kyle, J., Farrington, S. M., Agakov, F., Tenesa, A., et al. (2012). The association of dietary intake of purine-rich vegetables, sugar-sweetened beverages and dairy with plasma urate, in a cross-sectional study. *PLoS One, 7*(6), e38123.

Zhang, C. Z., Cheng, X. Q., Li, J. Y., Zhang, P., Yi, P., Xu, X., & Zhou, X. D. (2016). Saliva in the diagnosis of diseases. *International Journal of Oral Science, 8*(3), 133–137.

Zhang, Y., Sun, J., Lin, C. C., Abemayor, E., Wang, M. B., & Wong, D. T. (2014). The emerging landscape of salivary diagnostics. *Oral Health and Dental Management, 13*(2), 200–210.

Zielonka, D., Marinus, J., Roos, R. A., De Michele, G., Di Donato, S., Putter, H., et al. (2013). The influence of gender on phenotype and disease progression in patients with Huntington's disease. *Parkinsonism & Related Disorders, 19*(2), 192–197.

Zuccato, C., Marullo, M., Vitali, B., Tarditi, A., Mariotti, C., Valenza, M., et al. (2011). Brain-derived neurotrophic factor in patients with Huntington's disease. *PLoS One, 6*(8), e22966.

Part III
Neuroimaging Biomarkers for Huntington's Disease

Retinal Imaging and Functional Biomarkers of Huntington's Disease

Abera Saeed and Peter van Wijngaarden

Abbreviations

AD	Alzheimer's Disease
AI	Artificial Intelligence
CNS	Central Nervous System
DCL	Diagnostic Confidence Level
ERG	Electroretinography
FAF	Fundus autofluorescence
GCL	Ganglion cell layer
HD	Huntington's Disease
IRL	Inner retinal layer
mfERG	Multifocal electroretinography
mHTT	Mutant Huntingtin protein
mRNFL	Macular retinal nerve fibre layer
MS	Multiple sclerosis
OCT	Optical Coherence Tomography
PD	Parkinson's Disease
pRNFL	Peripapillary retinal nerve fibre layer
RGC	Retinal Ganglion Cells
RPE	Retinal Pigment Epithelium
RPE	Retinal pigment epithelium
TMS	Total Motor Score
UHDRS	Unified Huntington's Disease Rating Scale
VEP	Visual evoked potentials

A. Saeed · P. van Wijngaarden (✉)
Centre for Eye Research Australia, Royal Victorian Eye and Ear Hospital, Melbourne, VIC, Australia

Ophthalmology, Department of Surgery, University of Melbourne, Melbourne, VIC, Australia
e-mail: peterv@unimelb.edu.au

E. A. Thomas, G. M. Parkin (eds.), *Biomarkers for Huntington's Disease*, Contemporary Clinical Neuroscience, https://doi.org/10.1007/978-3-031-32815-2_5

1 Introduction

The eye has been considered a "window to the soul" long before scientists investigated the scientific and clinical relevance of this idiom. Embryologically the retina and the brain have a shared origin. The retina is comprised of specialised neurons, glial cells and a rich vascular supply. Visual signals are relayed to the brain via retinal ganglion cells (RGC), whose axons converge to form the optic nerve. The eye has distinctive cell populations, immune responses, and surface molecules comparable to those of the brain and spinal cord (London et al., 2013). Several well-defined neurodegenerative diseases of the brain and spinal cord, including Huntington's disease (HD), have ocular manifestations, and ocular symptoms frequently present before other clinical signs of these disorders (Colligris et al., 2018; Mazur-Michałek et al., 2022).

HD is an autosomal dominant neurodegenerative disorder caused by a mutation resulting in a trinucleotide (CAG) repeat expansion in the Huntingtin (HTT) gene. It has a monogenic mode of inheritance, complete penetrance, and there is a correlation between the age of onset and the CAG repeat length (Roos, 2010). HD has two phases: premanifest and manifest. According to currently accepted criteria, HD is formally diagnosed in a person who (1) carries a known CAG-expanded allele of the HD gene or has a family history of HD and (2) develops motor symptoms that are "unequivocal signs of HD" as defined in the Diagnostic Confidence Level (DCL) of the Unified Huntington's Disease Rating Scale (UHDRS) (Reilmann et al., 2014). Within the UHDRS, motor symptoms are assessed using the Total Motor Score (TMS). The DCL offers levels of confidence ranging from 0 (no motor abnormalities suggestive of HD) to 4 ($\geq$99% likelihood of being due to HD), with a score of 4 describing motor onset or 'manifest' HD based on a clinical examination (McColgan & Tabrizi, 2018; Reilmann et al., 2014). However, other non-motor clinical abnormalities can emerge gradually over many years during the 'premanifest' phase of the disease, which can begin at least 12–15 years before the formal diagnosis (Paulsen et al., 2006; The Huntington Study Group PHAROS Investigators*, 2006). Redefining the diagnostic criteria of HD remains an ongoing area of research (McColgan & Tabrizi, 2018). To this end, a biological classification of Huntington's disease, known as the Huntington's disease integrated staging system, has recently been proposed. Staging (stages 0–3) is based on biological, clinical and functional assessments (Tabrizi et al., 2022).

Visual dysfunction and perceptual disturbances are well documented in both the premanifest and manifest stages of HD (Blekher et al., 2006; Golding et al., 2006; Lasker & Zee, 1997). In fact, early studies of HD mutation carriers noted that increased blinking, slowness of horizontal saccades, a decrease in saccade speed and an increase in saccade latency were common clinical findings (Lasker & Zee, 1997). Paulus et al. were amongst the first to suggest retinal involvement in HD based on a functional test, the foveal blue test light, using a Maxwellian view system (Paulus et al., 1993). They observed abnormally elevated thresholds in HD patients, suggesting retinal dysfunction. Other studies have also indicated that retinal changes may be apparent well before motor symptoms of the disease (Helmlinger et al., 2002; Mazur-Michałek et al., 2022). These early findings have piqued interest in ocular biomarkers of HD.

Retinal imaging technology is rapidly advancing and can provide detailed information about the structure, function, and even the molecular composition of retinal tissue in humans *in vivo* (Christinaki et al., 2022). Most importantly, this information can be obtained rapidly and non-invasively. These developments have enabled the identification of novel biomarkers of a number of neurodegenerative diseases including multiple sclerosis (MS) (Costello & Burton, 2018; Lambe et al., 2020) and Alzheimer's disease (AD) (Hadoux et al., 2019; van Wijngaarden et al., 2020). Advances in next-generation computational techniques for image analysis, such as artificial intelligence (AI) algorithms, have further enhanced the potential of retinal imaging as a promising tool and a source of biomarker discovery. An array of retinal imaging modalities depicted in Fig. 1 have been comprehensively reviewed elsewhere (Christinaki et al., 2022; Kashani et al., 2021) and will be explored in the context of HD in this chapter. At present there are no effective treatments to prevent HD, halt its progression, or delay its onset, but as the future holds promise for disease-modifying strategies, there is a need for reliably accessible disease progression biomarkers, as well as markers to evaluate therapeutic interventions. This chapter will explore the retinal manifestations of HD, the status of retinal imaging and functional biomarkers of HD, as well as the potential of these methods for the detection and monitoring of the disease in the future.

The retina is the innermost, light- sensitive layer of the eye, responsible for receiving and processing visual stimuli. It transmits electrical signals to the brain for the perception of vision. During embryonic development, the retina forms as an out-pouching of the neuroectoderm of the diencephalon, called the optic vesicle, which undergoes invagination to form the optic cup. The inner wall of the optic cup gives rise to the retina (Fig. 2a). The mature retina is a layered structure comprised

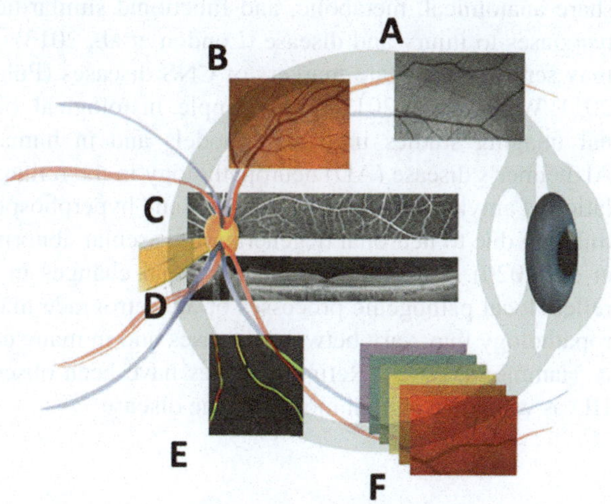

Fig. 1 Depiction of a variety of retinal imaging methods. From top: (**a**) infrared and (**b**) colour retinal photography, (**c**) optical coherence tomography angiography (OCT-A), (**d**) optical coherence tomography (OCT), (**e**) retinal oximetry, (**f**) hyperspectral imaging. (Adapted from: (Christinaki et al., 2022))

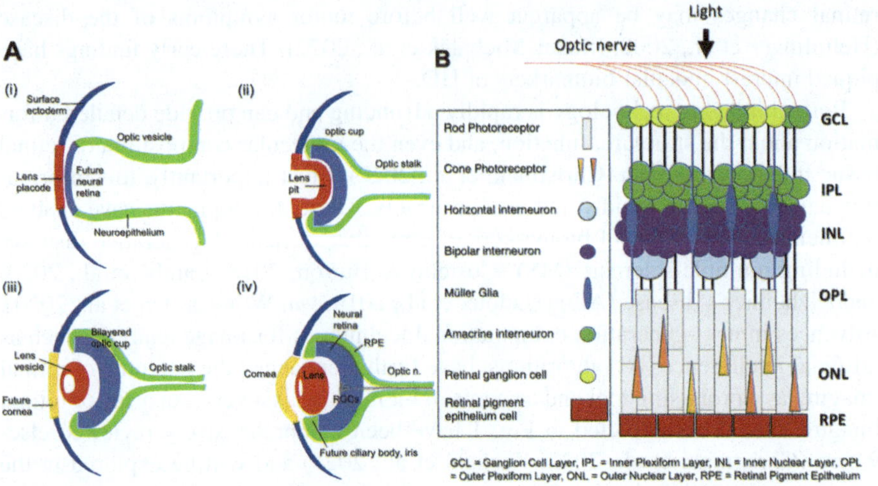

Fig. 2 (**a**) Embryological development of the eye. (i) The optic vesicle forms as an extension of the neural tube, initiating development of the eye. (ii) Invagination of the distal part of the optic vesicle forms the bilayered optic cup. This is accompanied by an invagination of the overlying surface ectoderm to form the lens pit. Constriction of the proximal portion of the optic vesicle gives rise to the optic stalk, a precursor of the optic nerve. (iii) Formation of the lens vesicle is accompanied by the development of the cornea from surface ectoderm and migratory waves of neural crest cells. (iv) The outer layer of the optic cup gives rise to the retinal pigment epithelium (RPE), whereas inner layer gives rise to the neural retina. (**b**) Schematic representation of the retinal microstructure. Retinal ganglion cell (RGC) axons travel in the optic nerve to synapse on neurons in the brain

of several types of neurons, glial cells and rich networks of blood vessels. RGCs constitute the relay neurons of the retina and their axons coalesce to form the optic nerve (Fig. 2b). Due to their common embryological origins, the retina and brain share anatomical, metabolic, and functional similarities, and they exhibit similar responses to injury and disease (London et al., 2013). As a result, retinal changes may serve as surrogate markers of CNS diseases (Pula et al., 2011; Saidha et al., 2011; Walter et al., 2012). For example, histological, biochemical, and *in vivo* retinal imaging studies in animal models and in humans have revealed signs of Alzheimer's disease (AD) neuropathology in the retina. These include the accumulation of amyloid-beta (Aβ) aggregates and hyperphosphorylated tau protein, retinal thinning due to neuronal degeneration, vascular abnormalities and gliosis (Mirzaei et al., 2020). The extent to which retinal changes in neurodegenerative diseases reflect local pathogenic processes or are retrograde manifestations of cortical neuropathology may vary between diseases and in many cases is unclear (Jones-Odeh & Hammond, 2015). Retinal changes have been observed in humans affected by HD as well as in animal models of the disease.

Insights from Retinal Histopathology

The Huntingtin gene encodes the Huntingtin protein which is involved in several critical cellular processes such as metabolism, protein turnover, regulation of gene expression, endocytosis, intracellular trafficking and cytoskeleton dynamics (Schulte & Littleton, 2011). In HD the CAG DNA repeat expansion in the Huntingtin gene is translated into a polyglutamine (polyQ) stretch in the mutant Huntingtin protein. Mutant Huntingtin protein interferes with a range of intracellular functions and accumulates as intranuclear inclusions, leading to neuronal malfunction and cell death (Huang et al., 2016). This process manifests as progressive degeneration of the basal ganglia, with the loss of striatal neurons resulting in atrophy of the caudate and putamen and enlargement of the lateral ventricles (Rosser & Dunnett, 2007). This progressive degeneration results in the hallmark motor, cognitive and psychiatric features of HD.

In the study of HD over the years, several animal models have been developed to recapitulate the genetic and phenotypical aspects of the disease (Gómez-Jaramillo et al., 2022). Experimental genetic models of the disease have been developed in fruit flies (Drosophila), mice and more recently in rats (Helmlinger et al., 2002; Jackson et al., 1998; Johnson et al., 2014). Evidence of retinal manifestations of HD in these animal models and in human post-mortem tissues are mixed (Table 1).

Drosophila Model

A transgenic Drosophila model of HD shows evidence of photoreceptor degeneration resembling the neuronal degeneration that is seen in the human brain in HD (Jackson et al., 1998). In this model, amino-terminal fragments of human Huntingtin gene containing tracts of 2, 75 and 120 polyglutamine residues were expressed in photoreceptor neurons under the control of an eye-specific expression construct. The age of onset of neurodegeneration and its severity correlated with the repeat length. Interestingly, and in contrast to findings in the brains of people with HD, no nuclear inclusions or protein aggregates were observed in the retinas of these transgenic flies (Jackson et al., 1998).

Mouse Model Studies

In keeping with the findings in Drosophila, retinal neurodegeneration has been described in a number of HD mouse models (Gómez-Jaramillo et al., 2022; Menalled & Chesselet, 2002; Rubinsztein, 2002). The R6 line (R6/1 & R6/2) was the first transgenic mouse model of HD to be developed and it continues to be widely used today (Helmlinger et al., 2002). These mice have an elongated exon 1 of the human HD gene (~115 CAG repeats in R6/1 mice and ~ 145–150 repeats in R6/2 mice) under the control of the human Huntingtin gene promoter. Both strains express a

Table 1 Summary of histological retinal changes observed in animal models and in humans with Huntington's disease

Retina Location	Histological findings	Drosophila	R6/1 mice	R6/2 mice	HdhQ150 mice	Rat	Human
Whole retina	Prescence of white spots	x	Yes – (Helmlinger et al., 2002)	Yes – (Helmlinger et al., 2002)	x	x	Ø changes – (Petrasch-Parwez et al., 2005)
Neuronal layers (RGC, INL, ONL)	Nuclear mHTT aggregates	Ø aggregates – (Jackson et al., 1998)	Aggregates – (Helmlinger et al., 2002)	Aggregates – (Helmlinger et al., 2002; Ragauskas et al., 2014; Li et al., 2013)	Aggregates – (Young et al., 2013)	x	
Amacrine cells	Cell loss	x	x	x	x	Yes – (Johnson et al., 2014)	
	Nuclear mHTT aggregates	x	Aggregates – (Helmlinger et al., 2002)	Aggregates – (Petrasch-Parwez et al., 2004; Helmlinger et al., 2002)	x	x	
Müller cells	Prescence of gliosis	x	Gliosis (Batcha et al., 2012)	Ø gliosis – (Young et al., 2013)	Ø gliosis – (Young et al., 2013)	x	
Horizontal cells	Nuclear mHTT aggregates	x	x	Aggregates – (Petrasch-Parwez et al., 2004)	x	Aggregates – (Johnson et al., 2014)	
Outer Nuclear Layer (ONL)	Thinning/cell loss	x	Ø thinning – (Batcha et al., 2012)	Thinning – (Petrasch-Parwez et al., 2004; Mazur-Michałek et al., 2022)	x	x	
	Disorganisation, abnormal morphology	x	Yes – (Helmlinger et al., 2002)	Yes – (Helmlinger et al., 2002; Mazur-Michałek et al., 2022)	x	x	
Photoreceptors	Thinning/cell loss	Thinning – (Jackson et al., 1998)	Ø thinning – (Batcha et al., 2012)	Thinning – (Helmlinger et al., 2002; Mazur-Michałek et al., 2022)	Ø thinning (Young et al., 2013)	x	
	Disorganisation, abnormal cell morphology	Yes – (Jackson et al., 1998)	Yes – (Batcha et al., 2012)	Yes – (Ragauskas et al., 2014; Mazur-Michałek et al., 2022; Petrasch-Parwez et al., 2004)	x	x	

Yes: Presence of histological finding, Ø: Absence of histological findings, X: No data/ not studied
mHTT mutant huntingtin protein

Huntingtin protein fragment. R6/2 mice exhibit a highly aggressive, swiftly progressive form of HD, much like the juvenile form of HD in humans (Cepeda et al., 2010). Overt phenotypes are evident as early as 9 weeks of age and mortality occurs within 10–13 weeks (Cepeda et al., 2010; Gómez-Jaramillo et al., 2022). By contrast, the R6/1 line displays similar deficits, but over a more protracted timescale. Accordingly, symptom onset is delayed (13–21 weeks), and lifespans are substantially longer (32–40 weeks).

Histological examinations of R6/2 mice at various phases of disease progression reveal structural abnormalities in the retina (Table 1). Mutant Huntingtin protein expression levels in R6/2 retinas have been shown to be comparable to those in the brain (Helmlinger et al., 2002). Furthermore, studies have shown that the overexpression of mutant Huntingtin protein in these mice led to the accumulation of protein aggregates in all three neuronal layers of the retina, particularly in the retinal ganglion cell (RGC) layer as early as 8–10 weeks of age (Helmlinger et al., 2002; Ragauskas et al., 2014). These aggregates were found to be predominantly nuclear however some cytoplasmic aggregates were also observed in the plexiform layers (Helmlinger et al., 2002; Petrasch-Parwez et al., 2004). Helmlinger et al. and Li et al. observed patchy cell loss and disorganisation of the outer nuclear layer (ONL) that progressed with disease duration in the R6/2 transgenic line (Helmlinger et al., 2002; Li et al., 2013). A fundus examination of these living transgenic mice revealed diffuse retinal white spots, an appearance that is suggestive of retinal dysplasia (Helmlinger et al., 2002). In another R6/2 mouse study, Huntingtin protein has been identified as an important component of a large variety of cilia, including the photoreceptor cilium (Karam et al., 2015). Accordingly, toxic expansion of polyglutamine in Huntingtin, as occurs in HD, has been correlated with structural and functional deficits in R6/2 photoreceptor cilia and these changes are postulated to contribute to photoreceptor degeneration (Karam et al., 2015).

Histological studies of R6/1 mice show similar retinal pathology to that observed in R6/2 mice, but with a later onset and slower progression (Batcha et al., 2012; Helmlinger et al., 2002). Mutant Huntingtin aggregates were observed across all three neuronal layers of the retina. Despite large and numerous aggregates in the RGCs and inner nuclear layer (INL), the inner retina was preserved (Helmlinger et al., 2002). Importantly, R6/1 mice showed no evidence of ONL thinning (Batcha et al., 2012). Furthermore, Batcha et al. identified Müller cell gliosis and the beginning of photoreceptor degeneration by 13 weeks (Batcha et al., 2012). Müller cell bodies typically reside in the INL, and gliosis occurs when these cells are placed under stress. Of note, gliosis and neurodegeneration in the striatum are prominent features of HD in the brain, indicating that these shared features between eye and brain may serve as the basis of retinal biomarkers of the disease (Vonsattel, 2008).

Other mouse lines, including HdhQ150 mice, the so-called 'knock in' or 'full length' model (as they express full length Huntingtin, as opposed to fragments of the protein), have also exhibited retinal structural and functional changes. Knock-in genetic models are considered to more accurately represent human HD because the mutation causing HD (polyQ expansion) is introduced into the orthologous full-length gene in its native genomic context (Cepeda et al., 2010). As a case in point, the HdhQ150 model was created by inserting 150 expanded CAG repeats into the murine Huntingtin gene. As a result, these rodents have two copies of the Huntingtin gene, with either one (heterozygous) or two (homozygous) modified copies. These transgenes are under the control of the endogenous murine Huntingtin promoter and consequently these mice have a longer premanifest phase than is the case for R6/2 mice. No retinal neurodegeneration was evident at 12 weeks, as is seen in R6/2 mice, however nuclear inclusions were found in all retinal neuronal cell types (Young et al., 2013).

Rat Studies

An immunohistochemistry study of mutant HD rat models revealed Huntingtin accumulation in horizontal cells, accompanied by the loss of amacrine cells (Johnson et al., 2014). This rat model is considered by some to resemble the human disease more closely as the rats have fewer CAG repeat expansions (60–75) than is the case for R6 mice and this is less likely to cause 'off-target' effects (Johnson et al., 2014). The preferential effect on inner retinal neurons in this model hints that human HD may be associated with degeneration of the inner retina. However, this appears to be the only study to examine the retina in a rat model of HD (published as an abstract). Consequently, the generalisability of these findings is uncertain and further studies using this model are required.

Human Studies

Surprisingly, there is only 1 published study that describes post-mortem analysis of human retinal tissue in HD (Petrasch-Parwez et al., 2005). The study was of the retina of a man who died at 69 years of age after a 28-year history of clinically established HD. Genetic analysis confirmed a high number (47) of CAG repeats, consistent with the clinical diagnosis. Interestingly, in this study the retina was devoid of microscopic, macroscopic or ultrastructural changes (Petrasch-Parwez et al., 2005). Fundus photography revealed no signs of degeneration; electron microscopy showed normal photoreceptor nuclei; paraffin sections immunostained for N-terminal Huntingtin and ubiquitin revealed no aggregates or inclusions and cresyl-violent stained sections exhibited regular nuclear layers in the retina (Petrasch-Parwez et al., 2005). In contrast, brain examination demonstrated significant cortical atrophy and immunohistochemical analysis of brain sections revealed abundant nuclear inclusions and aggregates in the cingulate gyrus relative to age-matched HD-free controls. Given the scarcity of human post-mortem data in HD and the disparity between human and animal models, further human post-mortem studies are warranted. Considering the weight of evidence from transgenic animal models and the knowledge that people with other polyglutamine neurological disorders, such as spinocerebellar ataxia 7 (SCA7), develop extensive retinal degeneration, there is a cogent basis for further studies (Michalik et al., 2004).

2 Retinal Imaging Biomarkers of Huntington's Disease

Changes in Peripapillary Retinal Nerve Fibre Layer Thickness

Reductions in retinal thickness have been reported for a wide range of human neurodegenerative diseases including Alzheimer's disease, Parkinson's disease, and multiple sclerosis and this has served as a driver for clinical studies of retinal thickness in HD. Optical coherence tomography (OCT) is a widely available non-invasive imaging method that enables the measurement of retinal layer thickness with precision. The technique is analogous to ultrasound except that it uses near infrared light instead of sound waves (Kashani et al., 2021). Broad bandwidth light is directed at the retina and reflected light is received by a detector (interferometer). The optical path length of light reaching the detector is determined using low-coherence interferometry. In brief, interference patterns generated by light reflected from the retina is analysed to generate a depth-resolved image (Fig. 3). This allows quantitative measurements of retinal thickness, choroidal thickness, and even retinal layer thickness with axial resolution in the range of microns (Asanad et al., 2019). First generation time-domain OCT (TD-OCT) devices have been largely supplanted by faster and higher resolution spectral domain (SD-OCT) and swept-source (SS-OCT) machines.

Retinal OCT parameters have been identified as potential biomarkers of HD (Table 2) and a total of 9 studies of people with HD have reported OCT findings. It is noteworthy that all but one of these studies had fewer than 30 participants. The most consistent finding was thinning of the retinal nerve fibre layer (RNFL), the inner-most layer of the retina that is comprised of retinal ganglion cell (RGC) axons. Thinning was most commonly detected in the peripapillary RNFL (pRNFL), adjacent to the optic nerve, and this appeared to be most pronounced in the temporal peripapillary quadrant (Gatto et al., 2018; Gulmez Sevim et al., 2019; Kersten et al., 2015; Mazur-Michałek et al., 2022; Svetozarskiy et al., 2020).

Whilst mean pRNFL thickness appeared to be reduced in people with HD compared to control participants in most studies (Andrade et al., 2016; Gatto et al., 2018; Gulmez Sevim et al., 2019; Svetozarskiy et al., 2020), the difference was only statistically significant in one study of 13 people with premanifest HD compared to controls. Unfortunately, the exact magnitude of the difference between groups was not provided in this study (Mazur-Michałek et al., 2022). In contrast, a study of 24 people with premanifest (~16 years before disease onset) HD, found that average pRNFL (23.33 ± 2.91 μm) was slightly thicker than in control ($n = 38$) participants (21.26 ± 1.91 μm, $p = 0.002$) (Schmid et al., 2022). Similar observations have been reported in OCT studies of people with early Alzheimer's disease, where it has been postulated that retinal thickening may be due to inflammation in earlier disease stages, before thinning occurs in advanced stages due to neurodegeneration (Ngolab et al., 2019). However, in this study of HD, there was no genetic confirmation of HD status, the sample size was small, and the control group was not a matched control cohort but was retrospectively identified from an existing

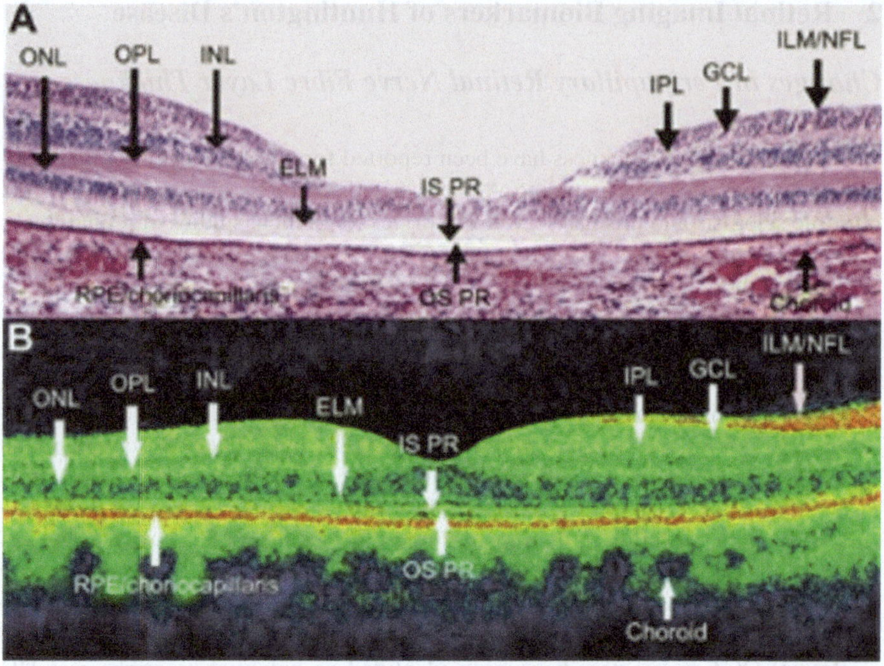

Fig. 3 OCT image and histological comparison of the human retina. (**a**) Histology of a representative control micrograph stained with hematoxylin and eosin. (**b**) OCT image depicts retinal layers relative to the corresponding retinal layers on histology above. Note, the histological micrograph and the OCT image are both centred on the fovea, a depression at the centre of the macula – the area of the retina that is specialised for detailed vision. ONL outer nuclear layer, OPL outer plexiform layer, INL inner nuclear layer, ELM external limiting membrane, IS PR photoreceptor inner segments, OS PR photoreceptor outer segments, IPL inner plexiform layer GCL ganglion cell layer, RPE retinal pigment epithelium, ILM inner limiting membrane, NFL nerve fibre layer. (Source: (Fraser & Lueck, 2021))

database (Schmid et al., 2022). Accordingly, the generalisability of these findings is uncertain. More consistent data is seen when pRNFL quadrants are considered separately.

In a study of 26 people with genetically-confirmed HD, pRNFL thinning in the temporal quadrant was significant compared to age- and sex-matched (n = 29) control participants (62.3 ± 7.3 vs. 69.8 ± 10.8 μm, p = 0.005) (Kersten et al., 2015). Within the HD group, there was no difference in measurements for those with pre-manifest or manifest HD. Accordingly, there does not appear to be a strong association between disease stage and extent of temporal pRNFL thinning. It is possible that change occurs early and stabilises, however these data are derived from cross-sectional studies with small participant numbers, making it difficult to identify temporal associations. Studies by Sevim et al. (n = 15) (Gulmez Sevim et al., 2019), Gatto et al. (n = 14) (Gatto et al., 2018) and Mazur-Michelek et al. (n = 13) (Mazur-Michałek et al., 2022) all identified temporal pRNFL thinning in people with HD. In contrast, a study of 25 people with HD, found no significant RNFL thinning in the

Table 2 Summary of retinal imaging and retinal functional biomarker studies in human subjects with Huntington's disease

Study Author (Year)	Modality	Biomarkers		Participants						Genetically confirmed[#] HD (Y/N)	Premanifest (a) / manifest HD (b) participants	Confounding eye diseases excluded (Y/N) **
				HD			Control					
		HD vs control	No significant difference	n	Mean age (SD)	M: F	n	Mean age (SD)	M: F			
Kersten et al. (2015)	**SD- OCT**	Temporal pRNFL ↓	Total macular thickness Total macular volume	26[d]	52.0 (10.2)	13:13	29[d]	50.7 (10.1)	11:18	Y	20 a 6 b	Y
Andrade et al. (2016)	**SD-OCT**	Average macular choroidal ↓ Central choroid ↓ Inferior choroid ↓	Average pRNFL thickness Superior pRNFL Inferior pRNFL Nasal pRNFL Total macular thickness Peripapillary choroidal thickness	8[d]	49.1 (9.4)	3:5	8[d]	49.8 (10.8)	*	*	8 b (See note (i))	Y
Gatto et al. (2018)	**SD-OCT**	Temporal pRNFL ↓ Superior pRNFL ↓	Inferior pRNFL thickness Nasal pRNFL thickness Total macular thickness Average pRNFL thickness	14[d]	48.1 (16.9)	5:9	13[d]	*	*	Y	27 b (See note (ii))	Y
Gulmez Sevim et al. (2019)	**SD-OCT**	Temporal pRNFL ↓ mRNFL ↓ GCL ↓ RPE ↓ IPL & OPL ↓ INL ↓ ONL ↑	Average pRNFL thickness Superior pRNFL thickness Inferior pRNFL thickness Nasal pRNFL thickness	15[^]	48^*	9:6	15[^]	48^*	*	Y	15 b	Y

(continued)

Table 2 (continued)

Study Author (Year)	Modality	Biomarkers: HD vs control	Biomarkers: No significant difference	Participants: HD n	HD Mean age (SD)	HD M: F	Control n	Control Mean age (SD)	Control M: F	Genetically confirmed[#] HD (Y/N)	Premanifest (a) / manifest HD (b) participants	Confounding eye diseases excluded (Y/N) **
Svetozarskiy et al. (2020)	**SD-OCT**	Temporal pRNFL ↓ Superior pRNFL ↓ Inferior pRNFL ↓ Nasal pRNFL ↓ Sub foveal choroid ↓ GCC layer ↓	–	60[s]	37.6 (10.2)	*	31[s]	37.3 (10.8)	*	Y	29 a 31 b	Y
Di Maio et al. (2021)	**SD-OCT**	Central choroidal↓	GCC thickness Average pRNFL thickness Superior pRNFL thickness Inferior pRNFL thickness	16[d]	57.3 (10.2)	13:3	13[d]	55.1 (11.2)	8:5	*	5 a 11 b	Y
	OCT-A	–	Superficial and deep plexus capillary density									
Amini et al. (2022)	**SD-OCT**	Inferior pRNFL ↓ Macular RNFL ↓	Superior pRNFL Temporal pRNFL Nasal pRNFL	23[d]	49.8 (12.3)	14:11	25[d]	44.3 (12.3)	10:15	Y	25 b	Y
	OCT-A	–	Superficial and deep plexus capillary density									
Schmid et al. (2022)	**SD-OCT**	Average pRNFL ↑	GCL GC + IPL	24[d]	39.7 (9.5)	10:14	38[−]	35.8 (12.2)	14:24	*	24 a	Y

Mazur-Michałek et al. (2022)	**SD- OCT**	Temporal pRNFL ↓ Superior pRNFL ↓ Inferior pRNFL ↓ Average RNFL ↓	Nasal pRNFL GCC thickness	13[d]	43.1 (6.7)	*	14[d]	39.9 (8.7)	*	Y	13 a	Y
Knapp et al. (2018)	**ERG**	Reduced ERG amplitudes	Latency	1	25	1:0	19	28.9 (7.9)	*	Y	1 a	Y
	mfERG	Reduced P1 amplitudes	Latency									
Pearl et al. (2017)	**ERG**	Increased ERG amplitudes	Latency	18	53.1 (*)	9:9	10	54.1 (*)	5:5	Y	See note (iii)	Y

ERG electroretinogram, *mfERG* multifocal ERG, *OCT-A* optical coherence tomography-angiography, *SD-OCT* spectral domain optical coherence tomography, *GCL* ganglion cell layer, *IPL* inner plexiform layer, *ONL* outer nuclear layer, *OPL* outer plexiform layer, *RNFL* retinal nerve fibre layer, *mRNFL* macular RNFL, *pRNFL* peripapillary RNFL, *GCC (ganglion cell complex)* mRNFL+GCL + IPL, *RPE* retinal pigment epithelium

genetic testing is central to the diagnosis of HD however some studies have not specified if genetic testing was performed

↓: reduced thickness; ↑: increased thickness

*Unspecified – information not provided

^*Median age provided only

^*Sevim* et al. *reported 15 eyes of 15 HD patients and 15 healthy controls were analysed however data tables showed n = 30. The number documented in the body of text is referenced here*

**Confounding disease excluded based on thorough ophthalmological exams including visual acuity, intraocular pressure measurement, and dilated fundus examination and history

[s]Represents that single eyes from participants were analysed in the study

[d]Represents that both eyes from participants were analysed in the study

[-]The control cohort is not a matched control cohort and was retrospectively identified from an existing database

Notes:

(i) Andrade et al. (2016) specified that 'all patients had mild to moderate HD' but did not clearly define the basis of definition i.e., CAG length, UHDRS score etc. Given that 7 of 8 patients were on antichorea drugs this has been interpreted as manifest HD patients

(ii) Gatto et al. (2018) classified HD patients according to the TFC (total functional capacity) scale – three in stage I, six in stage II and five in stage III

(iii) Pearl et al. (2017) did not classify HD participants as premanifest or manifest. A UHDRS mean score of 28.3 and a range of 0–70 was provided

temporal peripapillary region, relative to age- and sex-matched control participants (n = 25), but did identify thinning in the inferior region in those with HD (Amini et al., 2022). Another study of 8 people with manifest HD compared to age- and sex-matched controls (n = 8), found no differences in thickness between the groups in any of the RNFL quadrants (Andrade et al., 2016). Overall, whilst findings are mixed, the bulk of the evidence is indicative of an association between temporal quadrant pRNFL thinning and HD.

Preferential loss of retinal ganglion cell axons in the temporal peripapillary region is a pattern shared with mitochondrial diseases (Sitarz et al., 2012). Accordingly, it is possible that mitochondrial dysfunction may underlie this change in HD. Mitochondrial dysfunction has been documented in HD in biochemical, imaging and molecular studies [35, 36] and mutant Huntingtin protein has been shown to impair mitochondrial trafficking (Orr et al., 2008). Furthermore, ultrastructural studies of cortical biopsies obtained from people with juvenile or adult-onset HD have also shown abnormal mitochondrial morphology (Quintanilla & Johnson, 2009). Ganglion cells are particularly susceptible to mitochondrial dysfunction due to their significant bioenergetic demands and as RGC axons in the temporal peripapillary region (papillo-macular bundle) are thinner in calibre than those elsewhere they are considered to be more prone to bioenergetic compromise (Yu-Wai-Man et al., 2016). In keeping with this notion, in mouse models of HD, cone photoreceptors appear to be lost prior to rods and this may be attributed in part to differences in bioenergetic vulnerability (Batcha et al., 2012; Ingram et al., 2020). Similar discrepancies in the extent and location of retinal (pRNFL) thinning have also been reported in Alzheimer's disease (AD) (Svetozarskiy et al., 2020). It has been suggested that variations in case definitions and disease stages between studies may account for these discrepant findings in AD, and the same may also apply to studies of HD. Equally, as RNFL measurements can vary with age, gender, and comorbid eye disease, such as glaucoma, it is important that future OCT studies are appropriately controlled and adequately powered. Longitudinal studies to document changes in RNFL in relation to disease stage will also be important to validate these apparent associations and to contextualise the role of RNFL changes as biomarkers of HD.

Macular Changes in Huntington's Disease

The macula is the specialised region of the retina that subserves high resolution and colour vision. Changes in the macular RNFL (mRNFL) are known to occur early on in a range of neurodegenerative diseases and are increasingly used as a proxy of disease progression in multiple sclerosis (Torbus et al., 2022). Whilst several studies have failed to demonstrate changes in total macular thickness in HD participants (Andrade et al., 2016; Gatto et al., 2018; Kersten et al., 2015), retinal sublayer analysis has revealed significant thinning of all inner retinal layers including the mRNFL in one study (Gulmez Sevim et al., 2019). In this study of 15 people with manifest, genetically confirmed HD and 15 age- and sex-matched controls OCT analysis revealed significant mRNFL thinning (HD 25.95 ± 3.69 μm vs control 30.02 ± 2.34 μm, $p < 0.001$). Several other OCT changes, including thinning of the ganglion cell layer (GCL), inner plexiform layer (IPL), inner nuclear layer (INL) and outer plexiform layer (OPL) as well as reduction in the retinal pigment epithelium (RPE) layer and outer macular volume were also observed. Interestingly, an increase in outer nuclear layer (ONL) thickness was also reported in this study and might be attributed to swelling of cone cells, which has been observed in glaucoma (Fan et al., 2011; Nork et al., 2000). Macular OCT parameters including the mRNFL, GCL and IPL, collectively known as the retinal ganglion cell complex (GCC), have been explored as biomarkers in 4 HD studies. Of these, 3 studies reported weak evidence for lower thickness of the GCC or GC-IPL in people with premanifest HD compared to control participants (Table 2). In contrast, Svetozarskiy et al. demonstrated GCC thinning in people with premanifest (n = 29, mean GCC = 85.36 μm vs 93.8 μm, p = 0.001) and manifest (n = 31, mean GCC = 84.7 μm vs 93.8 μm, p = 0.001) HD compared to controls (n = 31). Given these mixed findings, further studies are needed to clarify the role macular measures may play as biomarkers of HD.

OCT Parameters as Biomarkers of Functional Impairment

Studies on patients with multiple sclerosis indicate that retinal layer thinning on OCT may correlate with disease severity (Gordon-Lipkin et al., 2007; Sepulcre et al., 2007; Toledo et al., 2008). Similar studies have been conducted in HD and, although mixed, the evidence suggests that RNFL thinning may be associated with neurological impairment in HD. For example, total mean RNFL thickness significantly correlated with total functional capacity scores among 14 people with genetically confirmed HD and 13 matched controls (Gatto et al., 2018). In a study of 26 people with HD and 29 controls, temporal pRNFL thickness was found to be negatively correlated with disease duration ($R^2 = -0.51$, $p = 0.04$) (Kersten et al., 2015). In the same study, macular volume was significantly and inversely correlated with both duration of disease ($R^2 = -0.71$, $p = 0.002$) and motor severity score ($R^2 = -0.56$, $p = 0.01$) (Kersten et al., 2015). Notably, these correlations increased in strength following exclusion of premanifest participants from the analysis (Kersten et al., 2015). In a study of 15 people with motor-manifest HD compared to age- and sex-matched controls (n = 15), the average mRNFL thickness was negatively correlated with disease duration ($R^2 = -0.406$, $p = 0.026$), CAG repeats ($R^2 = -0.392$, $p = 0.032$) and burden score (comprising CAG repeat length and age) ($R^2 = -0.436$, $p = 0.016$) (Gulmez Sevim et al., 2019). This correlation was also observed with the thickness of other sublayers including the GCL, IPL, INL and OPL. The GCL thickness was identified as the strongest OCT marker for determining disease progression ($R^2 = -0.648$, $p < 0.001$) which may suggest earlier involvement of the ganglion cell bodies in HD than their axons, which comprise the RNFL

(Gulmez Sevim et al., 2019). This is also in keeping with histological studies of R6 transgenic mice which indicate that mutant Huntingtin aggregates are most numerous in the GCL (Helmlinger et al., 2002). Similar associations have been reported between clinical measures of HD severity and one or more total macular thickness measures in other studies (Andrade et al., 2016).

Retinal Vasculature in Huntington's Disease

Several studies have found that cerebral microvascular alterations and blood-brain barrier (BBB) disruption may play a role in HD pathogenesis (Chan et al., 2021; Drouin-Ouellet et al., 2015). In particular, Drouin-Ouellet et al. found that mutant Huntingtin protein aggregates are present in all major components of the neurovascular unit including in the basal lamina, endothelial cells, smooth muscle, astrocytes, oligodendrocytes and microglia in both transgenic HD mice models (R6/2) and in people with HD (Cisbani & Cicchetti, 2012). These changes were accompanied by increases in blood vessel density, reduced blood vessel diameter and BBB leakage in the striatum of R6/2 mice and humans. Lin and co-workers found that these changes occurred in the first weeks of age, prior to the onset of behavioural symptoms in HD mouse models (R6/2) (Lin et al., 2013). Since the inner blood-retinal-barrier (iBRB) is similar to the BBB, and as the retinal microvasculature is similar to that of the brain, it is possible that vascular changes described in the brains of humans in HD and in mouse models of the disease have correlates in the retina and choroid- the rich vascular network that lies underneath the retina and supplies outer retinal cells.

In keeping with this, studies using enhanced depth imaging SD-OCT have consistently identified an association between HD and reduced central choroidal thickness (Andrade et al., 2016; Di Maio et al., 2021; Svetozarskiy et al., 2020). This was observed in both manifest and premanifest HD patient populations. Choroidal thinning may be an indicator of vascular impairment affecting the brain as seen in Alzheimer's disease (AD) (Salobrar-Garcia et al., 2020; Villa, 2021). It has been speculated that choroidal thinning identified as a biomarker for AD may be related to a reduced perfusion possibly accompanied by atrophic changes in this vascular network caused by local Aβ deposition, a process similar to that described for cerebral vascular system (Villa, 2021). Interestingly, in two separate studies of 8 and 16 people with manifest HD, choroidal thickness did not correlate with disease severity assessed by the UHDRS motor score (Andrade et al., 2016; Di Maio et al., 2021). Conversely, in a study of 60 people with genetically-confirmed HD (29 premanifest, 31 manifest) there was a significant correlation between choroidal thickness and motor score ($R^2 = -0.62$, $p < 0.001$) (Svetozarskiy et al., 2020). Importantly, this study utilised manual choroidal thickness measurements as opposed to standardised protocols, which introduces the potential for bias. Further studies are needed to clarify choroidal changes in HD and to establish the specificity of these changes as this will be key to determining their utility as biomarkers of the disease. Indocyanine green angiographic studies and OCT-angiography may also help to clarify whether choroidal morphological changes are associated with choroidal hypoperfusion, and may establish whether the choriocapillaris or larger choroidal vessels are primarily affected (Amini et al., 2022; Andrade et al., 2016).

Changes in the cerebral vasculature of mouse models and in humans with HD have prompted studies of the retinal vasculature (Amini et al., 2022; Di Maio et al., 2021). Optical coherence tomography-angiography (OCT-A) is a variant of OCT that uses motion contrast to generate depth-resolved maps of the retinal circulation. Di Maio et al. (2021) used OCT-A to examine 16 people with HD at various stages of the disease and found no differences in superficial and deep retinal capillary plexus densities relative to age-matched controls (superficial $p = 0.894$, deep $p = 0.169$). Similarly, no OCT-A differences were found by Amini et al. (2022) in a study of 25 people with manifest, genetically-confirmed HD compared to matched controls (superficial $p = 0.211$, deep $p = 0.896$). Retinal vascular abnormalities - telangiectasia and peripheral intraretinal haemorrhages- were reported in the case of a 25-year-old woman with juvenile HD, who underwent annual eye examinations for 10 years, however the association of these changes with HD are uncertain (Campa, 2020). Further research is needed to characterise retinal and choroidal vascular changes in HD.

Rod and Cone Dysfunction in Huntington's Disease

Changes in retinal function have been extensively studied in a wide array of neurodegenerative eye and brain diseases. Electroretinography (ERG) is a means of measuring the electrical activity of the retina in response to light stimuli. ERG is performed by placing surface electrodes on the eye and eyelids while light stimuli are provided to one or both eyes. Specific stimulation and analysis techniques enable the interrogation of retinal cell class-specific function. Several groups have examined the potential of ERG as a biomarker of HD in mouse models and in humans affected by the disease. ERG recordings from HD mouse models (R6/1, R6/2) have shown progressive retinal dysfunction, involving both the rod and cone pathways (Batcha et al., 2012; Helmlinger et al., 2002; Li et al., 2013; Ragauskas et al., 2014). These electrophysiological changes appear to precede histological changes in HD mouse models, indicating that neural dysfunction may precede structural loss and hinting at the potential of functional tests as means to monitor disease progression. Studies using transgenic HD model mice (R6/1 & R6/2) have shown that cone dysfunction precedes rod dysfunction (Batcha et al., 2012; Helmlinger et al., 2002; Li et al., 2013; Ragauskas et al., 2014). These cone-deficits, indicated by reduced phototopic ERG response amplitudes, were evident at the same age as behavioural and motor dysfunction, but earlier than detectable retinal histological changes (Li et al., 2013).

To the best of our knowledge, only two ERG studies have investigated retinal function in people with HD and the findings of these studies are mixed (Table 2)

(Knapp et al., 2018; Pearl et al., 2017). Electroretinography of a 25-year-old man with premanifest HD showed results consistent with those of the transgenic mouse models (Knapp et al., 2018). Amongst other findings, dark- and light-adapted full field ERG a- and b-waves had reduced amplitudes, suggestive of photoreceptor dysfunction, despite normal retinal structure on OCT. Additionally, this individual had reduced scotopic oscillatory potentials, indicative of dysfunction of the amacrine interneurons, as well as reduced amplitude of the P1 wave on multifocal ERG, indicative of inner retinal dysfunction (bipolar, Muller and amacrine cell). Given that these findings were based on studies of a single participant, their relevance to HD is unclear. A study of 18 people with genetically-confirmed, manifest HD showed significantly increased b-wave amplitudes in dark- and light-adapted conditions compared to controls (n = 10) (Pearl et al., 2017). Furthermore, increased cone activity in response to a low intensity light stimuli was associated with a greater number of CAG repeats (Pearl et al., 2017). Despite differences in the findings of these two small studies, they indicate that HD may be associated with changes in retinal function (Silverstein et al., 2020). These differences may be explained in part by disease stage and medication use, as some medications prescribed to HD patients, such as haloperidol, have been shown to alter ERG amplitudes in non-human primates (Christinaki et al., 2022). Additional studies with larger sample sizes are required to clarify the extent of these retinal functional changes in HD and to characterise changes over time and with disease (Dhalla et al., 2019; Silverstein et al., 2020).

Visual Evoked Potentials in Huntington's Disease

Visual evoked potential (VEP) recording is another means of evaluating visual function that is utilised primarily for disorders of the optic nerve and brain. Electrical potentials in the visual cortex elicited by exposure of the eye to a light stimulus are recorded on the surface of the scalp, in a similar manner to electroencephalography. Evaluation of VEP waveform amplitudes and latencies provides information about the function of the afferent visual pathway (Kashani et al., 2021).

Several studies of VEP responses in HD have yielded relatively consistent findings. Most of the published studies involve people with HD and their at-risk offspring. VEP amplitudes were found to be reduced in people with HD compared to healthy controls (Ellenberger et al., 1978; Hennerici et al., 1985; Lawson et al., 1984; Oepen et al., 1981). VEP amplitudes were typically normal in at-risk offspring, however reduced amplitudes were observed in some instances (Hennerici et al., 1985). VEP amplitude reduction can be indicative of retinal ganglion cell loss, and findings from OCT studies indicate that temporal RNFL thinning may be a feature of HD (Klistorner & Graham, 2021; Wang et al., 2021). In contrast, VEP latencies were normal in people with HD and in their children. Given that most of these studies were conducted before genetic testing for HD was routinely available, the genetic profiles of these study participants are unknown. The practical utility of

VEP in the assessment of HD and other neurodegenerative diseases has been debated, given that abnormalities can be a consequence of changes anywhere along the afferent visual pathway. Nevertheless, VEP recording is increasingly used as a measure of disease progression and response to treatment in clinical trials of remyelination therapies in multiple sclerosis, indicating that the technique may have a role to play in HD in future (Klistorner & Graham, 2021; Wang et al., 2021).

3 Conclusions and Future Directions

There is mixed evidence that structural and functional changes occur in the retina in HD and that these changes may be detected using readily available imaging and functional tests. The clinical utility of these retinal biomarkers of HD is yet to be established. This is due, in large part, to mixed study findings. Key knowledge gaps relate to a paucity of human post-mortem data, the limitations of small clinical sample sizes, varying case definitions and clinical stages, as well as variations in imaging protocols and analysis methods. Improved characterisation of the retinal manifestations of HD rests on further post-mortem studies. These studies have the potential to enhance our understanding of the disease and its manifestations and provide the biological underpinnings of *in vivo* biomarkers. These limitations are not unique to HD; recent expert consensus statements have reported that similar challenges apply to the field of retinal imaging biomarkers of Alzheimer's disease (Snyder et al., 2021).

Current HD biomarker studies are limited by small sample sizes and variations in case definition. Of all published studies, the largest included 60 people with HD at varying stages of disease progression (Svetozarskiy et al., 2020). Furthermore, definitions of 'premanifest' and 'manifest' HD varied amongst studies. Whilst most studies utilised the standard UHDRS classification criteria, others stratified HD cases into stages based on total functional capacity (TFC). Almost one in five studies included in this chapter did not clearly define the clinical stage of participants with HD (Andrade et al., 2016; Gatto et al., 2018; Pearl et al., 2017) (Table 2). Future studies should be adequately powered and include a diversity of participants for age, gender, ethnicity. Furthermore, disease stages should be described in accordance with accepted standards and ideally supported by a genetic diagnosis. In addition, longitudinal studies would be most helpful to inform the causality of the observed associations as well as helping to establish temporal associations between retinal imaging findings and disease stage.

Retinal nerve fibre layer thickness, as measured by OCT, has emerged as one of the most promising potential biomarkers for HD. Unfortunately, current studies on RNFL changes in HD have yielded mixed results. Wide variations in retinal imaging protocols and analysis methods pose significant barriers to inter-study comparisons. Of the 9 OCT studies, 2 did not report on whether retinal imaging was performed in a standard, valid or reliable way with predefined protocols or scan quality thresholds (Svetozarskiy et al., 2020). Some studies did not utilise age- and sex-matched control cohorts for comparison or did not specify the average age or gender ratio in the control cohort (Table 2). This is important as RNFL values are known to vary with age (age-related decline) (Parikh et al., 2007) and sex (higher values are typical in women) (Li et al., 2020). An additional barrier to comparison between studies stems from the use of different imaging devices. Scan patterns, segmentation algorithms, and outcome measures vary considerably between device manufacturers (Kashani et al., 2021). Moreover, even within the same commercial system, segmentation algorithms may change over time, and consequently, the consistency of measurements made with successive generations of the same device may differ subtly but significantly (Chan et al., 2019). Therefore, when comparing studies, it is essential to consider the OCT system and scan parameters utilised.

To overcome the above OCT limitations, a recent consensus paper has called for the harmonisation, standardisation, and replication of imaging techniques to ensure the comparability of study outcomes (Snyder et al., 2021). The interpretation of OCT scans for HD biomarkers requires validated consensus criteria, such as the OSCAR-IB (Tewarie et al., 2012) and APOSTEL (Cruz-Herranz et al., 2016) consensus criteria used for multiple sclerosis. These criteria have facilitated inter-study comparability for multiple sclerosis and would be advantageous for HD.

Electroretinography parameters are emerging as promising biomarkers of HD. A good biomarker is one that not only is specific for a disease state but is scalable, practical, and easily accessible. Unfortunately, current conventional ERG protocols require relatively invasive procedures, including pupillary dilation and the use of corneal or conjunctival lens electrodes (Carpi & Tomei, 2006). Furthermore, interpretation of data requires expertise from ophthalmologists or visual scientists and the size of equipment, lack of transportability and need for trained technicians to perform the test limits the accessibility of this technology. However recent improvements in ERG are making the technique more accessible for general use. For example, new, handheld portable ERG devices and advances in non-invasive electrodes will make this technology more practical and accessible in the future.

References

Amini, E., Moghaddasi, M., Habibi, S. A. H., Azad, Z., Miri, S., Nilforushan, N., et al. (2022). Huntington's disease and neurovascular structure of retina. *Neurological Sciences, 43*(10), 5933–5941.

Andrade, C., Beato, J., Monteiro, A., Costa, A., Penas, S., Guimarães, J., et al. (2016). Spectral-domain optical coherence tomography as a potential biomarker in Huntington's disease. *Movement Disorders, 31*(3), 377–383.

Asanad, S., Ross-Cisneros, F. N., Nassisi, M., Barron, E., Karanjia, R., & Sadun, A. A. (2019). The retina in Alzheimer's disease: Histomorphometric analysis of an ophthalmologic biomarker. *Investigative Ophthalmology & Visual Science, 60*(5), 1491–1500.

Batcha, A. H., Greferath, U., Jobling, A. I., Vessey, K. A., Ward, M. M., Nithianantharajah, J., et al. (2012). Retinal dysfunction, photoreceptor protein dysregulation and neuronal remodelling in the R6/1 mouse model of Huntington's disease. *Neurobiology of Disease, 45*(3), 887–896.

Blekher, T., Johnson, S. A., Marshall, J., White, K., Hui, S., Weaver, M., et al. (2006). Saccades in presymptomatic and early stages of Huntington disease. *Neurology, 67*(3), 394–399. Wolters Kluwer Health, Inc. on behalf of the American Academy of Neurology.

Campa, C. (2020). Huntingtons disease: A new evidence of neurovascular dysfunction. *Journal of the Neurological Sciences* [Internet]. Elsevier; 2020 Sep 15 [cited 2023 Feb 23];416. Available from: https://www.jns-journal.com/article/S0022-510X(20)30331-2/fulltext

Carpi, F., & Tomei, F. (2006). Non-invasive electroretinography. *Biomedicine & Pharmacotherapy, 60*(8), 375–379.

Cepeda, C., Cummings, D. M., André, V. M., Holley, S. M., & Levine, M. S. (2010). Genetic mouse models of Huntington's disease: Focus on electrophysiological mechanisms. *ASN Neuro, 2*(2), AN20090058. SAGE Publications Inc.

Chan, V. T. T., Sun, Z., Tang, S., Chen, L. J., Wong, A., Tham, C. C., et al. (2019). Spectral-domain OCT measurements in Alzheimer's disease: A systematic review and meta-analysis. *Ophthalmology, 126*(4), 497–510.

Chan, S. T., Mercaldo, N. D., Kwong, K. K., Hersch, S. M., & Rosas, H. D. (2021). Impaired cerebrovascular reactivity in Huntington's disease. *Frontiers in Physiology, 12*, 663898.

Christinaki, E., Kulenovic, H., Hadoux, X., Baldassini, N., Van Eijgen, J., De Groef, L., et al. (2022). Retinal imaging biomarkers of neurodegenerative diseases. *Clinical & Experimental Optometry, 105*(2), 194–204.

Cisbani, G., & Cicchetti, F. (2012). An in vitro perspective on the molecular mechanisms underlying mutant huntingtin protein toxicity. *Cell Death & Disease, 3*(8), e382. Nature Publishing Group.

Colligris, P., Perez de Lara, M. J., Colligris, B., & Pintor, J. (2018). Ocular manifestations of Alzheimer's and other neurodegenerative diseases: The prospect of the eye as a tool for the early diagnosis of Alzheimer's disease. *Journal of Ophthalmology (Hindawi), 2018*, e8538573.

Costello, F., & Burton, J. M. (2018). Retinal imaging with optical coherence tomography: A biomarker in multiple sclerosis? *Eye Brain, 10*, 47–63.

Cruz-Herranz, A., Balk, L. J., Oberwahrenbrock, T., Saidha, S., Martinez-Lapiscina, E. H., Lagreze, W. A., et al. (2016). The APOSTEL recommendations for reporting quantitative optical coherence tomography studies. *Neurology, 86*(24), 2303–2309.

Dhalla, A., Pallikadavath, S., & Hutchinson, C. V. (2019). Visual dysfunction in Huntington's disease: A systematic review. *Journal of Huntington's Disease, 8*(2), 233–242.

Di Maio, L. G., Montorio, D., Peluso, S., Dolce, P., Salvatore, E., De Michele, G., et al. (2021). Optical coherence tomography angiography findings in Huntington's disease. *Neurological Sciences, 42*(3), 995–1001.

Drouin-Ouellet, J., Sawiak, S. J., Cisbani, G., Lagacé, M., Kuan, W.-L., Saint-Pierre, M., et al. (2015). Cerebrovascular and blood–brain barrier impairments in Huntington's disease: Potential implications for its pathophysiology. *Annals of Neurology, 78*(2), 160–177.

Ellenberger, C., Petro, D. J., & Ziegler, S. B. (1978). The visually evoked potential in Huntington disease. *Neurology, 28*(1), 95. Wolters Kluwer Health, Inc. on behalf of the American Academy of Neurology.

Fan, N., Huang, N., Lam, D. S. C., & Leung, C. K. (2011). Measurement of photoreceptor layer in glaucoma: A spectral-domain optical coherence tomography study. *Journal of Ophthalmology (Hindawi), 2011*, e264803.

Fraser, C., & Lueck, C. J. (2021). Optical coherence tomography: A window to the brain? *Practical Neurology, 21*(4), 313–321. BMJ Publishing Group Ltd.

Gatto, E., Parisi, V., Persi, G., Fernandez Rey, E., Cesarini, M., Luis Etcheverry, J., et al. (2018). Optical coherence tomography (OCT) study in Argentinean Huntington's disease patients. *International Journal of Neuroscience, 128*(12), 1157–1162. Taylor & Francis.

Golding, C. V. P., Danchaivijitr, C., Hodgson, T. L., Tabrizi, S. J., & Kennard, C. (2006). Identification of an oculomotor biomarker of preclinical Huntington disease. *Neurology, 67*(3), 485–487.

Gómez-Jaramillo, L., Cano-Cano, F., González-Montelongo, M. d. C., Campos-Caro, A., Aguilar-Diosdado, M., & Arroba, A. I. (2022). A new perspective on Huntington's disease: How a neurological disorder influences the peripheral tissues. *International Journal of Molecular Sciences, 23*(11), 6089.

Gordon-Lipkin, E., Chodkowski, B., Reich, D. S., Smith, S. A., Pulicken, M., Balcer, L. J., et al. (2007). Retinal nerve fiber layer is associated with brain atrophy in multiple sclerosis. *Neurology, 69*(16), 1603–1609. Wolters Kluwer Health, Inc. on behalf of the American Academy of Neurology.

Gulmez Sevim, D., Unlu, M., Gultekin, M., & Karaca, C. (2019). Retinal single-layer analysis with optical coherence tomography shows inner retinal layer thinning in Huntington's disease as a potential biomarker. *International Ophthalmology, 39*(3), 611–621.

Hadoux, X., Hui, F., Lim, J. K. H., Masters, C. L., Pébay, A., Chevalier, S., et al. (2019). Non-invasive in vivo hyperspectral imaging of the retina for potential biomarker use in Alzheimer's disease. *Nature Communications, 10*, 4227.

Helmlinger, D., Yvert, G., Picaud, S., Merienne, K., Sahel, J., Mandel, J.-L., et al. (2002). Progressive retinal degeneration and dysfunction in R6 Huntington's disease mice. *Human Molecular Genetics, 11*(26), 3351–3359.

Hennerici, M., Hömberg, V., & Lange, H. W. (1985). Evoked potentials in patients with Huntington's disease and their offspring. II. Visual evoked potentials. *Electroencephalography and Clinical Neurophysiology, 62*(3), 167–176.

Huang, W.-J., Chen, W.-W., & Zhang, X. (2016). Huntington's disease: Molecular basis of pathology and status of current therapeutic approaches. *Experimental and Therapeutic Medicine, 12*(4), 1951–1956.

Ingram, N. T., Fain, G. L., & Sampath, A. P. (2020). Elevated energy requirement of cone photoreceptors. *Proceedings of the National Academy of Sciences, 117*(32), 19599–19603. Proceedings of the National Academy of Sciences.

Jackson, G. R., Salecker, I., Dong, X., Yao, X., Arnheim, N., Faber, P. W., et al. (1998). Polyglutamine-expanded human Huntingtin transgenes induce degeneration of drosophila photoreceptor neurons. *Neuron, 21*(3), 633–642.

Johnson, M. A., Gelderblom, H., Rüther, K., Priller, J., & Bernstein, S. L. (2014). Evidence that Huntington's disease affects retinal structure and function. *Investigative Ophthalmology & Visual Science, 55*(13), 1644.

Jones-Odeh, E., & Hammond, C. J. (2015). How strong is the relationship between glaucoma, the retinal nerve fibre layer, and neurodegenerative diseases such as Alzheimer's disease and multiple sclerosis? *Eye (London, England), 29*(10), 1270–1284.

Karam, A., Tebbe, L., Weber, C., Messaddeq, N., Morlé, L., Kessler, P., et al. (2015). A novel function of Huntingtin in the cilium and retinal ciliopathy in Huntington's disease mice. *Neurobiology of Disease, 80*, 15–28.

Kashani, A. H., Asanad, S., Chan, J. W., Singer, M. B., Zhang, J., Sharifi, M., et al. (2021). Past, present and future role of retinal imaging in neurodegenerative disease. *Progress in Retinal and Eye Research, 83*, 100938.

Kersten, H. M., Danesh-Meyer, H. V., Kilfoyle, D. H., & Roxburgh, R. H. (2015). Optical coherence tomography findings in Huntington's disease: A potential biomarker of disease progression. *Journal of Neurology, 262*(11), 2457–2465.

Klistorner, A., & Graham, S. L. (2021). Role of multifocal visually evoked potential as a biomarker of demyelination, spontaneous remyelination, and myelin repair in multiple sclerosis. *Frontiers in Neuroscience* [Internet]. 2021 [cited 2023 Mar 3];15. Available from: https://www.frontiersin.org/articles/10.3389/fnins.2021.725187

Knapp, J., VanNasdale, D. A., Ramsey, K., & Racine, J. (2018). Retinal dysfunction in a presymptomatic patient with Huntington's disease. *Documenta Ophthalmologica, 136*(3), 213–221.

Lambe, J., Saidha, S., & Bermel, R. A. (2020). Optical coherence tomography and multiple sclerosis: Update on clinical application and role in clinical trials. *Multiple Sclerosis, 26*(6), 624–639.

Lasker, A. G., & Zee, D. S. (1997). Ocular motor abnormalities in Huntington's disease. *Vision Research, 37*(24), 3639–3645.

Lawson, E. A., Barrett, G., Kriss, A., & Halliday, A. M. (1984). *P300 and VEPs in Huntington's Chorea. Annals of the New York Academy of Sciences* (Vol. 425, pp. 592–597). New York Academy of Sciences.

Li, M., Yasumura, D., Ma, A. A. K., Matthes, M. T., Yang, H., Nielson, G., et al. (2013). Intravitreal administration of HA-1077, a ROCK inhibitor, improves retinal function in a mouse model of Huntington disease. *PLoS One, 8*(2), e56026. Public Library of Science.

Li, D., Rauscher, F. G., Choi, E. Y., Wang, M., Baniasadi, N., Wirkner, K., et al. (2020). Sex-specific differences in circumpapillary retinal nerve fiber layer thickness. *Ophthalmology, 127*(3), 357–368.

Lin, C.-Y., Hsu, Y.-H., Lin, M.-H., Yang, T.-H., Chen, H.-M., Chen, Y.-C., et al. (2013). Neurovascular abnormalities in humans and mice with Huntington's disease. *Experimental Neurology, 250*, 20–30.

London, A., Benhar, I., & Schwartz, M. (2013). The retina as a window to the brain-from eye research to CNS disorders. Nature reviews. *Neurology, 9*(1), 44–53.

Mazur-Michałek, I., Kowalska, K., Zielonka, D., Leśniczak-Staszak, M., Pietras, P., Szaflarski, W., et al. (2022). Structural abnormalities of the optic nerve and retina in Huntington's disease pre-clinical and clinical settings. *International Journal of Molecular Sciences, 23*(10), 5450. Multidisciplinary Digital Publishing Institute.

McColgan, P., & Tabrizi, S. J. (2018). Huntington's disease: A clinical review. *European Journal of Neurology, 25*(1), 24–34.

Menalled, L. B., & Chesselet, M.-F. (2002). Mouse models of Huntington's disease. *Trends in Pharmacological Sciences, 23*(1), 32–39.

Michalik, A., Martin, J.-J., & Van Broeckhoven, C. (2004). Spinocerebellar ataxia type 7 associated with pigmentary retinal dystrophy. *European Journal of Human Genetics, 12*(1), 2–15.

Mirzaei, N., Shi, H., Oviatt, M., Doustar, J., Rentsendorj, A., Fuchs, D.-T., et al. (2020). Alzheimer's retinopathy: Seeing disease in the eyes. *Frontiers in Neuroscience, 14*, 921.

Ngolab, J., Honma, P., & Rissman, R. A. (2019). Reflections on the utility of the retina as a biomarker for Alzheimer's disease: A literature review. *Neurology and Therapy, 8*(Suppl 2), 57–72.

Nork, T. M., Ver Hoeve, J. N., Poulsen, G. L., Nickells, R. W., Davis, M. D., Weber, A. J., et al. (2000). Swelling and loss of photoreceptors in chronic human and experimental glaucomas. *Archives of Ophthalmology, 118*(2), 235–245.

Oepen, G., Doerr, M., & Thoden, U. (1981). Visual (VEP) and somatosensory (SSEP) evoked potentials in Huntington's chorea. *Electroencephalography and Clinical Neurophysiology, 51*(6), 666–670.

Orr, A. L., Li, S., Wang, C.-E., Li, H., Wang, J., Rong, J., et al. (2008). N-terminal mutant Huntingtin Associates with mitochondria and impairs mitochondrial trafficking. *The Journal of Neuroscience, 28*(11), 2783–2792.

Parikh, R. S., Parikh, S. R., Sekhar, G. C., Prabakaran, S., Babu, J. G., & Thomas, R. (2007). Normal age-related decay of retinal nerve fiber layer thickness. *Ophthalmology, 114*(5), 921–926.

Paulsen, J. S., Hayden, M., Stout, J. C., Langbehn, D. R., Aylward, E., Ross, C. A., et al. (2006 Jun). Preparing for preventive clinical trials: The Predict-HD study. *Archives of Neurology, 63*(6), 883–890.

Paulus, W., Schwarz, G., Werner, A., Lange, H., Bayer, A., Hofschuster, M., et al. (1993). Impairment of retinal increment thresholds in Huntington's disease. *Annals of Neurology, 34*(4), 574–578.

Pearl, J. R., Heath, L. M., Bergey, D. E., Kelly, J. P., Smith, C., Laurino, M. Y., et al. (2017). Enhanced retinal responses in Huntington's disease patients. *Journal of Huntington's Disease, 6*(3), 237–247.

Petrasch-Parwez, E., Habbes, H.-W., Weickert, S., Löbbecke-Schumacher, M., Striedinger, K., Wieczorek, S., et al. (2004). Fine-structural analysis and connexin expression in the retina of a transgenic model of Huntington's disease. *Journal of Comparative Neurology, 479*(2), 181–197.

Petrasch-Parwez, E., Saft, C., Schlichting, A., Andrich, J., Napirei, M., Arning, L., et al. (2005). Is the retina affected in Huntington disease? *Acta Neuropathologica, 110*(5), 523–525.

Pula, J. H., Towle, V. L., Staszak, V. M., Cao, D., Bernard, J. T., & Gomez, C. M. (2011). Retinal nerve fibre layer and macular thinning in spinocerebellar ataxia and cerebellar multisystem atrophy. *Neuroophthalmology, 35*(3), 108–114.

Quintanilla, R. A., & Johnson, G. V. W. (2009). Role of mitochondrial dysfunction in the pathogenesis of Huntington's disease. *Brain Research Bulletin, 80*(4–5), 242–247.

Ragauskas, S., Leinonen, H., Puranen, J., Rönkkö, S., Nymark, S., Gurevicius, K., et al. (2014). Early retinal function deficit without prominent morphological changes in the R6/2 mouse model of Huntington's disease. *PLoS One, 9*(12), e113317. Public Library of Science.

Reilmann, R., Leavitt, B. R., & Ross, C. A. (2014). Diagnostic criteria for Huntington's disease based on natural history. *Movement Disorders, 29*(11), 1335–1341.

Roos, R. A. (2010). Huntington's disease: A clinical review. *Orphanet Journal of Rare Diseases (BioMed Central), 5*(1), 1–8.

Rosser, A. E., & Dunnett, S. B. (2007). 24 – Neural transplantation in Huntington's disease. In: C. Halberstadt, & D. Emerich (Eds.), *Cellular transplantation* [Internet]. Academic Press; 2007 [cited 2023 Feb 28]. p. 417–37. Available from: https://www.sciencedirect.com/science/article/pii/B9780123694157500259

Rubinsztein, D. C. (2002). Lessons from animal models of Huntington's disease. *Trends in Genetics, 18*(4), 202–209.

Saidha, S., Syc, S. B., Durbin, M. K., Eckstein, C., Oakley, J. D., Meyer, S. A., et al. (2011). Visual dysfunction in multiple sclerosis correlates better with optical coherence tomography derived estimates of macular ganglion cell layer thickness than peripapillary retinal nerve fiber layer thickness. *Multiple Sclerosis, 17*(12), 1449–1463.

Salobrar-Garcia, E., Méndez-Hernández, C., de Hoz, R., Ramírez, A. I., López-Cuenca, I., Fernández-Albarral, J. A., et al. (2020). Ocular vascular changes in mild Alzheimer's disease patients: Foveal avascular zone, choroidal thickness, and ONH hemoglobin analysis. *Journal of Personalized Medicine, 10*(4), 231.

Schmid, R. D., Remlinger, J., Abegg, M., Hoepner, R., Hoffmann, R., Lukas, C., et al. (2022). No optical coherence tomography changes in premanifest Huntington's disease mutation carriers far from disease onset. *Brain and Behavior, 12*(6), e2592.

Schulte, J., & Littleton, J. T. (2011). The biological function of the Huntingtin protein and its relevance to Huntington's Disease pathology. *Current Trends in Neurology, 5*, 65–78.

Sepulcre, J., Murie-Fernandez, M., Salinas-Alaman, A., García-Layana, A., Bejarano, B., & Villoslada, P. (2007). Diagnostic accuracy of retinal abnormalities in predicting disease activity in MS. *Neurology, 68*(18), 1488–1494. Wolters Kluwer Health, Inc. on behalf of the American Academy of Neurology.

Silverstein, S. M., Demmin, D. L., Schallek, J. B., & Fradkin, S. I. (2020). Measures of retinal structure and function as biomarkers in neurology and psychiatry. *Biomarkers in Neuropsychiatry, 2*, 100018.

Sitarz, K. S., Chinnery, P. F., & Yu-Wai-Man, P. (2012). Disorders of the optic nerve in mitochondrial cytopathies: New ideas on pathogenesis and therapeutic targets. *Current Neurology and Neuroscience Reports, 12*(3), 308–317.

Snyder, P. J., Alber, J., Alt, C., Bain, L. J., Bouma, B. E., Bouwman, F. H., et al. (2021). Retinal imaging in Alzheimer's and neurodegenerative diseases. *Alzheimer's & Dementia, 17*(1), 103–111.

Svetozarskiy, S., Kopishinskaya, S., & Smetankin, I. (2020). Early retinal degeneration in Huntington's disease based on optical coherence tomography: A case-control study. *Medical Hypothesis, Discovery & Innovation in Optometry, 1*(1), 18–24.

Tabrizi, S. J., Schobel, S., Gantman, E. C., Mansbach, A., Borowsky, B., Konstantinova, P., et al. (2022). A biological classification of Huntington's disease: The Integrated Staging System. *The Lancet Neurology, 21*(7), 632–644. Elsevier.

Tewarie, P., Balk, L., Costello, F., Green, A., Martin, R., Schippling, S., et al. (2012). The OSCAR-IB consensus criteria for retinal OCT quality assessment. *PLoS One, 7*(4), e34823.

The Huntington Study Group PHAROS Investigators*. (2006). At risk for Huntington disease: The PHAROS (Prospective Huntington At Risk Observational Study) cohort enrolled. *Archives of Neurology, 63*(7), 991–996.

Toledo, J., Sepulcre, J., Salinas-Alaman, A., García-Layana, A., Murie-Fernandez, M., Bejarano, B., et al. (2008). Retinal nerve fiber layer atrophy is associated with physical and cognitive disability in multiple sclerosis. *Multiple Sclerosis, 14*(7), 906–912. SAGE Publications Ltd., STM.

Torbus, M., Niewiadomska, E., Dobrakowski, P., Papuć, E., Rybus-Kalinowska, B., Szlacheta, P., et al. (2022). The usefulness of optical coherence tomography in disease progression monitoring in younger patients with relapsing-remitting multiple sclerosis: A single-centre study. *Journal of Clinical Medicine, 12*(1), 93.

van Wijngaarden, P., Hui, F., Dang, D., Ha, J., Fowler, C. J., Villemagne, V. L., et al. (2020). A retinal hyperspectral imaging biomarker for Alzheimer's disease: Preliminary study of the influence of eye diseases on imaging scores. *Alzheimer's & Dementia, 16*(S4), e046625.

Villa, C. (2021). Ocular vascular changes: Choroidal thickness as an early biomarker for Alzheimer's disease? *Journal of Personalized Medicine, 11*(12), 1365.

Vonsattel, J. P. G. (2008). Huntington disease models and human neuropathology: Similarities and differences. *Acta Neuropathologica, 115*(1), 55–69.

Walter, S. D., Ishikawa, H., Galetta, K. M., Sakai, R. E., Feller, D. J., Henderson, S. B., et al. (2012). Ganglion cell loss in relation to visual disability in multiple sclerosis. *Ophthalmology, 119*(6), 1250–1257.

Wang, C., Barton, J., Kyle, K., Ly, L., Barnett, Y., Hartung, H.-P., et al. (2021). Multiple sclerosis: Structural and functional integrity of the visual system following alemtuzumab therapy. *Journal of Neurology, Neurosurgery, and Psychiatry, 92*(12), 1319–1324. BMJ Publishing Group Ltd.

Young, D., Mayer, F., Vidotto, N., Schweizer, T., Berth, R., Abramowski, D., et al. (2013). Mutant huntingtin gene-dose impacts on aggregate deposition, DARPP32 expression and neuroinflammation in HdhQ150 mice. *PLoS One, 8*(9), e75108.

Yu-Wai-Man, P., Votruba, M., Burté, F., La Morgia, C., Barboni, P., & Carelli, V. (2016). A neurodegenerative perspective on mitochondrial optic neuropathies. *Acta Neuropathologica, 132*(6), 789–806.

Positron Emission Tomography (PET) Imaging Biomarkers in Huntington's Disease

Liesbeth Everix, Steven Staelens, and Daniele Bertoglio

Abbreviations

AD	Alzheimer's disease
BBB	Blood-brain-barrier
CB	Cannabinoid
CB_1R	Cannabinoid 1 receptor
CB_2R	Cannabinoid 2 receptor
CNS	Central nervous system
D_1R	Dopamine D1 receptor
D_2R	Dopamine D2 receptor
DAT	Dopamine transporter
DDC	Dopa decarboxylase protein
DTBZ	Dihydrotetrabenazine
FDG	Fluorodeoxyglucose
GABA	Gamma amino butyric acid
GLUT	Glucose transporter
GPCR	G-protein coupled receptor
H_3R	Histamine 3 receptor
HAB	High-affinity binder
HD	Huntington's disease
HDRP	HD-related metabolic covariance pattern
iGluR	Ionotropic receptors

L. Everix · S. Staelens
Molecular Imaging Center Antwerp (MICA), University of Antwerp, Antwerp, Belgium

μNeuro Research Centre of Excellence, University of Antwerp, Antwerp, Belgium

D. Bertoglio (✉)
Molecular Imaging Center Antwerp (MICA), University of Antwerp, Antwerp, Belgium

μNeuro Research Centre of Excellence, University of Antwerp, Antwerp, Belgium

Bio-Imaging Lab, University of Antwerp, Antwerp, Belgium
e-mail: daniele.bertoglio@uantwerpen.be

E. A. Thomas, G. M. Parkin (eds.), *Biomarkers for Huntington's Disease*,
Contemporary Clinical Neuroscience, https://doi.org/10.1007/978-3-031-32815-2_6

IL	Interleukin
LAB	Low affinity binders
LEV	Levetiracetam
MAB	Mixed affinity binder
mGluR	Metabotropic receptor
mHTT	Mutated Huntingtin protein
MMP	Matrix metallopeptidase
MSN	Medium spiny neuron
NHP	Non-human primate
OR	Opioid receptor (δ, κ or μ)
ORL1	Opioid receptor like-1
PDE	Phosphodiesterase
PET	Positron emission tomography
PolyQ	Polyglutamine
pwHD	People with HD
SNP	Single nucleotide polymorphism
SPECT	Single-photon emission computed tomography
SV2	Synaptic vesicle glycoprotein 2
TFC	Total functional capacity
TNFα	Tumor necrosis factor α
TSPO	Translocator protein 18 kDa
UHDRS	Unified Huntington's Disease Rating Scale
VMAT2	Vesicular monoamine transporter type 2

1 Introduction

Positron Emission Tomography

Positron emission tomography (PET) is a non-invasive imaging technique in which the distribution of biomarkers can be visualized through the use of radionuclide-labelled molecules, i.e. radioligands. These structures contain a radioisotope which undergoes radioactive decay to achieve a more stable state. The intensity of the accompanying signal, which originates from the location of the target of the radioligand, represents a molecular target of choice or a biochemical/physiological process. Due to the minimal amount of radioligand required (trace amounts), these biological processes go entirely uninterrupted without pharmacological effects (Zhang et al., 2021). Because of the high sensitivity (picomolar range) and quantitative value of PET, biomarkers can be accurately tracked, making this technique not only versatile but also extremely powerful. In neuropathological diseases in particular, PET plays an indispensable role since non-invasive brain mapping is a must. In this chapter, the focus is on PET imaging biomarkers and their accompanying tracers within Huntington's disease.

Huntington's Disease

Huntington's disease (HD) is a rare, autosomal, and dominantly inherited neurodegenerative disease characterized by the presence of mutated Huntingtin (mHTT) proteins, ubiquitously expressed throughout the human body (Bhide et al., 1996). Within the Huntington gene (chromosome 4), people with HD (pwHD) manifest an expanded CAG sequence, encoding a polyglutamine (polyQ) region in the mHTT protein which facilitates aggregation (McColgan & Tabrizi, 2018). HD is considered fully penetrant from 39 repeats or higher, generally linked to an age of onset of 35–44 years with an expected life expectancy of 15–18 years from disease onset (Caron et al., 1993). Juvenile HD has been reported in individuals with ~64 CAG repeats or higher (Cronin et al., 2019). pwHD undergo progressive neuronal cell loss in the striatum and cortex, causing progressive motor impairments such as gait abnormalities and uncontrolled movements (chorea). Besides motor impairment, HD gives rise to psychiatric symptoms and cognitive impairment (Wilson et al., 2017). HD symptoms can partly be explained by dysfunctional communication and neurodegeneration in the basal ganglia. For example, loss of striatal medium spiny neurons (MSNs) has been connected to a variety of detrimental effects, an example of which is given in Fig. 1 (Bergonzoni et al., 2021). MSNs have been connected to a variety of detrimental effects, an example of which is given in Fig. 1 (Bergonzoni et al., 2021).

Based on the absence or presence of symptoms, pwHD are designated 'premanifest' (no clinical symptoms) or 'manifest' (clinical manifestation of symptoms). The most applied assessment tool of symptomatic status in HD clinical research is the total functional capacity (TFC) scale, a component of the Unified Huntington's Disease Rating Scale (UHDRS). The UHDRS considers motor, cognitive, behavioral and functional manifestation (Beglinger et al., 2010). Notably, underlying pathophysiology or the changes in biomarkers that may precede functional decline are rarely taken into account to evaluate disease progression (Tabrizi et al., 2022). To this day, HD is still incurable with no disease-slowing treatments available.

2 PET-Associated Biomarkers in HD

Imaging biomarkers within HD are mostly investigated to monitor disease progression and follow-up therapeutic efficacy in support of clinical translation (Martí-Martínez & Valor, 2022). In this chapter, frequently studied clinical PET biomarkers (e.g. glucose metabolism, neuroinflammation, PDE10a, dopaminergic markers), as well as underexplored, more recently identified biomarkers (e.g. histamine receptor H_3, mHTT) for HD will be considered. The investigation of biomarkers in a manifest stage is beneficial to gain insight into the timing, durability, and regional therapeutic effects of administered drugs. Moreover, it allows for a quantitative evaluation of the efficacy of disease-modifying therapies, e.g. by gauging the effect of mHTT

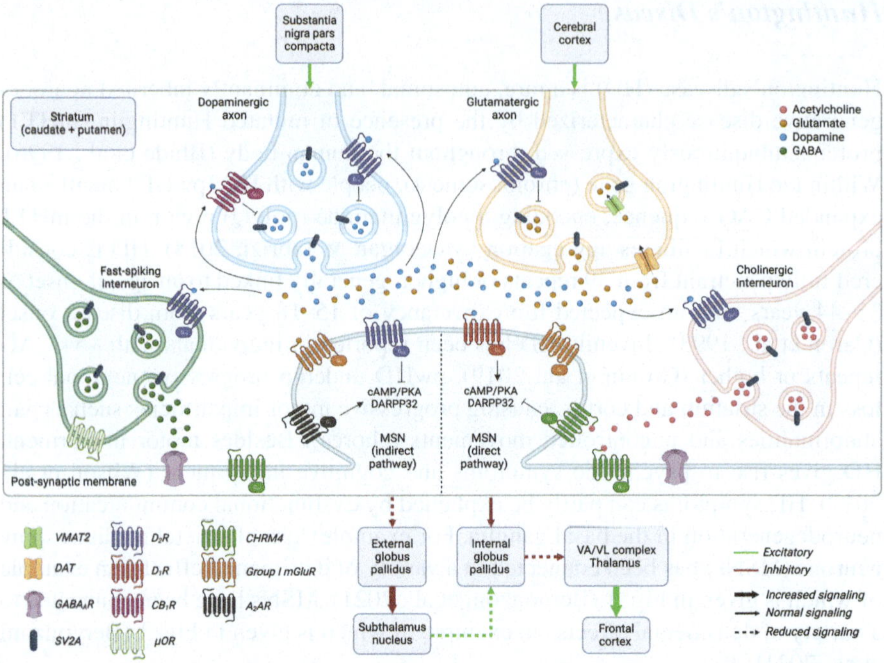

Fig. 1 Signaling pathways in the basal ganglia of people with Huntington's disease. The striatum is an important connection point in the basal ganglia. It is 95% composed of medium spiny neurons (MSN), which are preferentially affected in HD, in part due to the presence of the mutant HTT protein. A wide variety of pathway dysfunctions is seen in HD, involving a complex network of neurotransmitters (e.g. acetylcholine, dopamine, glutamate and GABA) and various proteins and receptors (e.g. opioid receptor, cannabinoid 1 receptor, SV2A, …). Several affected components in HD are depicted on this figure and will be explained further in this chapter. This figure specifically highlights the result of D_2R dysfunction. Under healthy conditions, the striatum receives excitatory dopaminergic and glutamatergic input from the substantia nigra pars compacta and cerebral cortex, respectively. Depending on the receptor type on MSN that is activated (indirect (D_2R) or direct (D_1R) pathway), an inhibitory stimulus is sent to the globus pallidus externus or internus, respectively. In the case of D_2R loss, inhibition of the external globus pallidus is reduced, leading to enhanced inhibition of the subthalamic nucleus and thereby reduced stimulation of the globus pallidus internus. In the case of intact D_1R signaling, inhibition of the globus pallidus internus remains unchanged and thus there is a net reduced output. The reduced communication from the internal globus pallidus to the VA/VL complex is inhibitory, thus results in enhanced signaling to the frontal cortex. In HD, this manifests as the occurrence of chorea. Abbreviations: A_2AR (Adenosine A2 receptor), CB_1R (cannabinoid 1 receptor), CHRM4 (Cholinergic Receptor Muscarinic 4), D_1R (Dopamine D_1 receptor), D_2R (Dopamine D_2 receptor), DAT (Dopamine transporter), GABA (gamma amino butyric acid), $GABA_AR$ (GABA type A receptor), mGluR (metabotropic glutamate receptor), OR (opioid receptor), SV2A (synaptic vesicle protein 2A), VA/VL (ventral anterior/lateral nucleus), VMAT2 (vesicular monoamine transporter 2)

lowering on various symptoms or receptor expression/density. In a premanifest stage, biomarker monitoring may also be valuable to estimate age of onset more accurately. Before onset of symptoms, clinical rating scales such as the UHDRS are inadequate to determine disease severity. Subclinical changes in biomarkers however may already be detectable years before age of onset and are inextricably linked to disease progression. Here, significant emphasis will be put on recent developments in clinical PET imaging of relevant biomarkers (Table 1), and particularly on their translational potential.

Brain Metabolism

Healthy brain function requires the generation of energy in the form of adenosine triphosphate, which is mainly obtained through metabolization of glucose. Glucose is transported past the blood-brain-barrier (BBB) by members of the glucose transporter (GLUT). Deficits in glucose uptake have been connected to neuronal viability and function in a variety of neurodegenerative diseases, including HD. To investigate such deficits, [^{18}F]-fluorodeoxyglucose ([^{18}F]FDG), a glucose analogue that can be transported through the plasma membrane using the GLUT protein family, may be used (Peralta et al., 2019). In people with manifest HD (mHD), [^{18}F]FDG demonstrated a glucose uptake deficit in mHTT-rich areas such as the striatum and cortex. Additionally, bilateral increases in thalamic, occipital, and cerebellar brain regions were reported. Such pattern, termed a HD-related metabolic covariance pattern (HDRP), changes according to the patient's disease progression, as reported by Rees et al. (Ma & Eidelberg, 2007; Rees et al., 2013). Interestingly, the HDRP pattern may indicate HD years before the clinical manifestation of the disease, as regional changes in metabolism may predate anatomical changes. Although glucose metabolism alterations are not limited to HD, [^{18}F]FDG and particularly the HDRP pattern are relevant to aid differential diagnosis between HD and other affected neurodegenerative diseases such as Alzheimer's disease (AD) (Rees et al., 2013).

Neuroinflammation

Neuroinflammation is known to be a component of several neurodegenerative disorders, including HD. It is characterized by reactive gliosis and the production of soluble inflammatory factors in the central nervous system (CNS). Neuroinflammation can be inferred from the detection of the translocator protein 18 kDa (TSPO) (Janssen et al., 2018). TSPO is a highly hydrophobic five-transmembrane-domain protein modestly expressed in resting glial brain cells under normal physiological conditions (Dupont et al., 2017; Zhang et al., 2021). In stress conditions such as neuronal death or oxidative stress, upregulation of TSPO is seen. HD is neuropathologically defined by the formation of intranuclear mHTT inclusions mainly in the striatal, cortical, and hypothalamic regions leading to brain atrophy with an increase in expression of several inflammatory response genes (Crotti et al., 2014; Dupont et al., 2017). In these brain regions, mHTT proteins have been shown to trigger a pathogenic cascade involving oxidative stress and a pro-inflammatory immune

Table 1 Overview of clinical PET studies conducted in people with Huntington's disease

Tracer	Molecular target	Study population	Main findings	Reference
[^{18}F]FDG	Glucose uptake	40 HC + 15 at risk +13 mHD	Glucose hypometabolism in CD and PUT preceded bulk tissue loss.	Kuhl et al. (1982)
[^{18}F]FDG	Glucose uptake	7 HC + 10 mHD	Glucose hypometabolism preceded tissue loss.	Hayden et al. (1986)
[^{18}F]FDG	Glucose uptake	10 HC + 10 early mHD	Abnormal CD and PUT glucose metabolism correlated with functional capacity and motor abnormalities. THA glucose metabolism correlated with dystonia.	Young et al. (1986)
[^{18}F]FDG	Glucose uptake	14 HC + 15 early mHD	Glucose hypometabolism in PUT and CD. CD metabolism correlated with memory and verbal learning.	Berent et al. (1988)
[^{18}F]FDG	Glucose uptake	20 HC + 23 mHD	Reduced rCMRGlc in the STR and CTX, but not in THA or CER. Reduction in CTX linked to severity of clinical symptoms.	Kuwert et al. (1990)
[^{18}F]FDG	Glucose uptake	7 HC + 11 GC	Reduced rCMRGlc and volume in CD. No correlation between volumetric and metabolic changes.	Grafton et al. (1992)
[^{18}F]FDG	Glucose uptake	5 mHD	Fetal striatal allografts may result in reduction of hypometabolism in STR and CTX.	Gaura et al. (2004)
[^{18}F]FDG	Glucose uptake	30 HC + 24 PreHD +47 mHD	Hypometabolism in STR and CTX. Increased CSF volume. Reduction in brain volume. Correlation between white matter volume and predicted age of onset in PreHD.	Ciarmiello et al. (2006)

(continued)

Table 1 (continued)

Tracer	Molecular target	Study population	Main findings	Reference
$[^{18}F]FDG$	Glucose uptake	11 + 12 mHD	Correlation between glucose hypometabolism in major gray matter regions (incl. basal ganglia) and clinical scores.	Squitieri et al. (2009)
$[^{18}F]FDG$	Glucose uptake	4 mHD	Memantine (cognition enhancer) may be protective in HD: CTX and STR metabolism remained unchanged during treatment period.	Hjermind et al. (2011)
$[^{18}F]FDG$	Glucose uptake	21 HC + 17 PreHD +26 mHD	Glucose hypometabolism in CD of mHD. Analysis of CD metabolism may be helpful for age of onset prediction in PreHD.	Ciarmiello et al. (2012)
$[^{18}F]FDG$	Glucose uptake	11 HC + 13 early mHD	Hypometabolism in bilateral frontal, temporal and parietal CTX. CAG repeat size correlated with age of onset, but not clinical progression rate.	Shin et al. (2013)
$[^{18}F]FDG$	Glucose uptake	11 HC + 22 PreHD	Hypometabolism in PUT, CD and PAL. Hypometabolism in PUT correlated with time to conversion from PreHD to mHD.	Herben-Dekker et al. (2014)
$[^{18}F]FDG$	Glucose uptake	1 mHD	Glucose hypometabolism in STR. Hypermetabolism in left THA. Moderate hypometabolism in frontal association areas and supplemental motor areas.	Feil et al. (2018)
$[^{18}F]FDG$	Glucose uptake	1 mHD	Glucose hypometabolism in STR. Progressive cognitive impairment and mild orofacial dyskinesia.	Michels et al. (2019)

(continued)

Table 1 (continued)

Tracer	Molecular target	Study population	Main findings	Reference
[^{18}F]FDG	Glucose uptake	17 HC + 21 PreHD	Glucose hypometabolism in STR correlated with the scaled CAG x age product. No volumetric STR changes.	Hellem et al. (2021)
[^{11}C]PK11195	TSPO	12 HC + 12 PreHD	Increased uptake in CTX, basal ganglia and THA. Somatosensory CTX: increase correlated with higher IL-1β, IL-6, IL-8 and TNF-α.	Politis et al. (2015)
[^{11}C]PBR28	TSPO	6 HC + 1 PreHD +7 mHD	Increased uptake in basal ganglia (PUT, PAL) and occasionally in THA and brainstem. Tracer uptake correlated to brain atrophy.	Lois et al. (2018)
[^{11}C]ER176	TSPO	6 HC + 6 PreHD +6 mHD	Increased uptake in PUT and PAL of mHD correlated with cumulative disease burden and brain volume. PreHD levels were not significantly different from HC or mHD.	Rocha et al. (2021)
[^{11}C]raclopride + [^{11}C]-SCH23390	$D_2R + D_1R$	9 HC + 10 mHD	Severe parallel reduction of D_1R and D_2R in STR, especially in patients with a rigid phenotype.	Turjanski et al. (1995)
[^{11}C]raclopride + [^{11}C]SCH23390	$D_2R + D_1R$	5 HC + 6 at risk +8 PreHD	D_1R and D_2R reduction in 4/8 PreHD subjects, not correlating with CAG repeat size. D_2R reduction in CD and PUT in 1/6 at risk subjects.	Weeks et al. (1996)
[^{11}C]raclopride + [^{11}C]-SCH23390 + [^{11}C] β-CIT	$D_2R + D_1R$ + DAT	5 HC + 5 mHD	D_1R and D_2R reduction in STR correlated with duration of HD. D_1R reduction in CTX. DAT reduction in STR.	Ginovart et al. (1997), Backman et al. (1997)

(continued)

Table 1 (continued)

Tracer	Molecular target	Study population	Main findings	Reference
[^{11}C]raclopride + [^{11}C]SCH23390	$D_2R + D_1R$	5 HC + 4 at risk +8 PreHD +5 mHD	D_1R and D_2R reduction in STR. Correlation with cognitive task performance, especially D_2R reduction.	Lawrence et al. (1998)
[^{11}C]raclopride + [^{11}C]SCH23390	$D_2R + D_1R$	7 HC + 3 at risk +9 PreHD +4 mHD	In PreHD, D_2R reduction in the STR was more prominent than D_1R reduction. In mHD, the opposite was seen with more D1R loss than D_2R.	Andrews et al. (1999)
[^{11}C]raclopride	D_2R	10 HC + 12 mHD	Reduced D_2R availability in the AMY, CTX and STR. No correlation between motor score and STR tracer uptake.	Pavese et al. (2003)
[^{11}C]raclopride	D_2R	11 HC + 18 PreHD	Abnormal D_2R availability in PUT. No significant D_2R availability changes at onset of symptoms. Annual decline was not significantly different compared to controls.	van Oostrom et al. (2009)
[^{11}C]raclopride	D_2R	13 HC + 11 PreHD +16 mHD	Reduced D_2R levels in CTX, mostly in mHD. Subjects with reduced D_2R performed worse on neuropsychological tests.	Pavese et al. (2010)
[^{11}C]DTBZ	VMAT2	64 HC + 19 mHD	Reduced VMAT2 availability in STR, especially in the akinetic-rigid phenotype of HD.	Bohnen et al. (2000)
[^{18}F]-FP-CIT	DAT	1 early/late mHD	Decreased uptake in basal ganglia in early mHD, with an additional decrease in the PUT in late mHD.	Chun (2022)

(continued)

Table 1 (continued)

Tracer	Molecular target	Study population	Main findings	Reference
[^{18}F]JNJ42259152	PDE10a	11 HC + 5 mHD	Strongly reduced tracer uptake in CD and PUT. No correlation between uptake and clinical disease severity measures.	Ahmad et al. (2014)
[^{18}F]MNI-659	PDE10a	9 HC + 3 PreHD +8 mHD	Reduced tracer uptake in basal ganglia, strongly correlating with UHDRS score.	Russell et al. (2014)
[^{11}C]IMA107	PDE10a	12 HC + 12 early PreHD	In PreHD 25 years away from age of onset, PDE10a changes were bidirectional: reduced in STR and PAL, but increased in THA.	Niccolini et al. (2015)
[^{18}F]MNI-659	PDE10a	11 HC + 2 PreHD +6 early mHD	PDE10a reduction in basal ganglia (PUT, CD and to a limited extent in the PAL). Annual PDE10a reduction in HD was at least 7-fold that of controls.	Russell et al. (2016)
[^{11}C]IMA107	PDE10a	12 HC + 12 early PreHD	PDE10a loss in the insular CTX and occipital fusiform gyrus, with no detectable brain atrophy.	Wilson et al. (2016)
[^{18}F]MNI659	PDE10a	10 early HD + 15 late PreHD +19 mHD	PDE10A loss very early in HD, preceding loss of D1R/D2R and atrophy. STR and PAL PDE10a declined in PreHD and appeared to plateau in mHD.	Fazio et al. (2020)
[^{18}F]MK-9470	CB_1R	14 HC + 20 mHD	Reduced CB_1R availability in the gray matter of the CRB, CER, and brainstem, ranging from ~15% (CER) to ~25% (frontal CTX).	Van Laere et al. (2010)
[^{18}F]CPFPX	A_1AR	36 HC + 7 early PreHD +6 late PreHD +8 mHD	A_1AR levels in the CER, CD, THA and AMY were at supranormal levels in PreHD, normal at age of onset, and subnormal in the mHD.	Matusch et al. (2014)

(continued)

Table 1 (continued)

Tracer	Molecular target	Study population	Main findings	Reference
[^{11}C]diprenorphine	OR	9 HC + 5 mHD	Reduced OR availability in the CD and PU. ORs may be less vulnerable to the disease processes than dopamine markers.	Weeks et al. (1997)
[^{11}C]flumazenil + [^{18}F]FDG	GABA+ Glucose uptake	6 HC + 6 mHD	Reduced GABA receptor availability in CD and glucose metabolism in CD, PUT and THA. No GABA receptor loss in PUT and THA.	Holthoff et al. (1993)
[^{18}F]FDG + L-[^{18}F]-fluorodopa + [^{11}C] NMSP	Glucose uptake + DDC + D_2R	34 HC + 1 mHD	No alterations in DDC levels in the STR. Reduced D_2R availability in the STR and hypometabolism in the STR and frontal CTX.	Leenders et al. (1986)
[^{11}C]flumazenil + [^{18}F] FDG + [^{11}C]raclopride	GABA receptor + D_2R + Glucose uptake	11/15/20 HC + 13 PreHD +10 mHD	Inverse correlation between GABA receptor and D_2R availability in the PUT (mHD), likely due to a compensatory GABA receptor upregulation. Not seen in PreHD.	Künig et al. (2000)
[^{18}F]FDG + [^{11}C] raclopride + [^{11}C] SCH22390 +	Glucose uptake + D_2R + D_1R	10 HC + 7 mHD	Fetal striatal allografts in HD follow-up: widespread reductions in glucose uptake and STR D_2R, but not D_1R. No correlation between PET results and clinical function.	Furtado et al. (2005)
[^{18}F]FDG + [^{11}C]Raclopride	Glucose uptake + D_2R	FDG: 12 HC + 12/12 PreHD (RAC)	Progressive STR D_2R decline correlated with metabolic changes. Increased THA metabolism, which decreased to subnormal levels at symptom onset.	Feigin et al. (2007)
[^{18}F]FDG + [^{11}C]Raclopride	Glucose uptake + D_2R	32/12 HC + 12 PreHD +14 early mHD	D_2R reduction in the CD and PUT in PreHD. Metabolic progression. Network changes provided information on HD progression.	Tang et al. (2013)

(continued)

Table 1 (continued)

Tracer	Molecular target	Study population	Main findings	Reference
[^{11}C](R)-PK11195 + [^{11}C]raclopride	TSPO + D_2R	10/10 HCs + 11 mHD	Reduced TSPO availability in the STR, correlating with D_2R reduction in the STR, UHDRS score and CAG index. TSPO increase in CTX.	Pavese et al. (2006)
[^{11}C](R)-PK11195 + [^{11}C]raclopride	TSPO + D_2R	8/8 HCs + 8 PreHD +8 mHD	Reduced D_2R and increased TSPO in STR, BNST, AMY and HYP in PreHD. In mHD, D_2R loss and TSPO elevation became more widespread.	Politis et al. (2011)
[^{11}C]raclopride + [^{18}F]FDG	D_2R + Glucose uptake	43 HC + 10 PreHD +8 early mHD	Hypometabolism in CD and PUT D_2R loss in PUT (PreHD). Both correlated to CAG repeat size, but not age.	Antonini et al. (1996), Antonini et al. (1998)
[^{11}C](R)-PK11195 + [^{11}C]raclopride	TSPO + D_2R	10/9 HC + 10 PreHD +9 mHD	Reduced D_2R and increased TSPO in HYP, correlating with each other but not with tracer uptake in the STR.	Politis et al. (2008)
[^{11}C](R)-PK11195 + [^{11}C]raclopride	TSPO + D_2R	10 HC + 9 mHD	D_2R loss in basal ganglia and diffuse TSPO elevation.	Grecchi et al. 2014)
[^{11}C]UCB-J + [^{18}F]FDG	SV2A + Glucose uptake	15 HC + 7 PreHD +11 mHD	SV2A loss in PUT, CD, PAL, CER, CTX and hypometabolism in CD and PUT in mHD (PreHD: only in CD and PUT). SV2A loss in the PUT correlated with motor impairment.	Delva et al. (2022b)

Abbreviations: *A_1AR* A_1 adenosine receptor, *AMY* Amygdala, *BNST* bed nucleus of the stria terminalis, *CB_1R* cannabinoid receptor 1, *CD* Caudate, *CER* Cerebellum, *CRB* Cerebrum, *CSF* Cerebrospinal fluid, *CTX* Cortex, *D_1R* Dopamine D_1 receptor, *D_2R* Dopamine D_2 receptor, *DAT* Dopamine transporter, *DDC* Dopa decarboxylase, *DTBZ* dihydrotetrabenazine, *GABA* Gamma-aminobutyric acid, *GC* gene carrier, *HC* healthy control, *HD* Huntington's disease, *HDRP* HD-related metabolic covariance pattern, *HYP* Hypothalamus, *mHD* Manifest stage of HD, *NMSP* N-methylspiperone, *OR* Opioid receptor, *PAL* Pallidum, *PDE10a* phosphodiesterase 10a, *PUT* Putamen, *PreHD* Premanifest stage of HD, *RAC* Raclopride, *rCMRGlc* regional metabolic rate of glucose consumption, *STR* Striatum, *SV2A* Synaptic vesicle protein 2A, *THA* Thalamus, *UHDRS* Unified Huntington's Disease Rating Scale, *VMAT2* vesicular monoamine transporter 2

response, leading to the upregulation of TSPO (Rocha et al., 2021). Such detrimental cascade is further reinforced by neuronal dysfunction/death with chronic upregulation of pro-inflammatory mediators, including interleukin 1β (IL-1β), tumor necrosis factor α (TNFA), interleukin 6 (IL-6), interleukin 8 (IL-8), and matrix metallopeptidase 9 (MMP-9) (Politis et al., 2015; Lois et al., 2018).

The prototypical PET tracer for TSPO, namely [^{11}C]PK11195, was developed more than 3 decades ago (1984). The compound PK11195, an isoquinoline carboxamide, was the first selective antagonist for TSPO with the R-enantiomer having the highest affinity for TSPO (Zhang et al., 2021). The first clinical PET study involving [^{11}C]PK11195 in pwHD was published in 2006 by Pavese et al. (Pavese et al., 2006). This study revealed significant increases in TSPO in striatal and cortical regions, correlating with disease severity in symptomatic pwHD (Pavese et al., 2006). Indeed, the correlation between striatal binding and motor symptom severity reflected the extent of neuronal loss and density of reactive microglia (Dupont et al., 2017). Interestingly, in people with premanifest HD (PreHD), microglial activation in the striatum and cognitive regions as measured by [^{11}C]PK11195 correlated with the 5-year probability of HD onset (Politis & Piccini, 2012; Dupont et al., 2017). Furthermore, an inverse correlation between increasing striatal [^{11}C]PK11195 binding and decreasing striatal dopamine D_2 receptor binding (as measured by [^{11}C]Raclopride PET) was reported by Politis et al. (Politis et al., 2008). In addition to striatal and cortical TSPO elevation, increased neuroinflammation was detected in the thalamus and hypothalamus, but not the hippocampus, as reported by Simmons et al. (Simmons et al., 2018). Although still widely used in (pre)clinical research, [^{11}C]PK11195 is characterized by several limitations, including a short half-life, relatively low brain-uptake, a poor metabolic profile, and high levels of non-specific binding resulting in a high signal-to-noise ratio. In an attempt to overcome these shortcomings, a second generation of TSPO tracers was developed, including [^{11}C]DAA1106, [^{18}F]FEDAA1106, [^{11}C]PBR111, [^{18}F]PBR06, [^{11}C]PBR28 and [^{11}C]ER176, amongst others (for review see Zhang et al., 2021). As of today, only two of these tracers, namely [^{11}C]PBR28 and [^{11}C]ER176, have been clinically studied in HD. [^{11}C]PBR28 has first been investigated in pre- and early symptomatic HD gene carriers, showing evidence of increased putamen and pallidum binding, as well as increased basal ganglia in PreHD compared to healthy controls. Importantly, [^{11}C]PBR28 binding was subject to a large interindividual variability that could be predicted by TSPO single nucleotide polymorphism (SNP) *rs6971* expression (Lois et al., 2018) resulting in TSP high-affinity binders (HABs, Ala/Ala), mixed affinity binders (MABs, Ala/Thr), and low affinity binders (LABs, Thr/Thr)(Owen et al., 2012; Zhang et al., 2021). [^{11}C]ER176 is a TSPO tracer that is less sensitive to polymorphisms with no brain-penetrant radiometabolites (Rocha et al., 2021). In mHD, [^{11}C]ER176 showed increased uptake in the globus pallidus and the putamen compared to healthy controls, which correlated to increased microglial activation, cumulative disease burden and reduced volume (Rocha et al., 2021).

Dopaminergic System

The dopaminergic system is involved in activities such as motor control, motivation, reward, cognitive function, maternal behaviors, and reproductive behaviors (Klein et al., 2019). Dopaminergic signaling pathways are regulated through the neurotransmitter dopamine (Klein et al., 2019). Dysfunction of dopaminergic signaling is often seen in neurodegenerative diseases, including HD (Klein et al., 2019). In HD, both pre- and postsynaptic terminals may be affected. The following dopaminergic biomarkers will be considered: vesicular monoamine transporter type 2 (VMAT2), dopamine transporter (DAT) and the dopa decarboxylase protein (DDC) (all presynaptic), as well as the dopamine D_1 receptor and dopamine D_2 receptor (both postsynaptic).

Presynaptic Markers

VMAT2

VMAT2 is a protein involved in the transport of monoamines into synaptic vesicles (Roussakis & Piccini, 2015). Inhibition of VMAT2 causes a depletion of neuroactive monoamines such as serotonin, norepinephrine, and dopamine. Numerous studies have documented successful use of VMAT2 inhibitors in multiple hyperkinetic disorders, including tardive dyskinesia, Tourette's syndrome, and HD (Tarakad & Jimenez-Shahed, 2018). In HD, inhibition of VMAT2 through tetrabenazine and deutetrabenazine, two US FDA-approved VMAT2 inhibitors, has resulted in marked improvement of chorea severity (23.5% and 36.3%, respectively)(Dorsey et al., 2011; Frank et al., 2016; Tarakad & Jimenez-Shahed, 2018). PET imaging with [^{11}C]dihydrotetrabenazine ([^{11}C]DTBZ), which selectively binds to VMAT2, has revealed reduced striatal VMAT2 levels in pwHD compared to controls (Bohnen et al., 2000).

DAT

The DAT is located on presynaptic nigrostriatal neurons. This transporter actively takes up extracellular dopamine in the synaptic cleft to regulate dopamine levels and enable a response to the next presynaptic dopamine release pulse. In the presynaptic neuron, dopamine may be restocked in vesicles or may be catabolized (Booth et al., 2015). Given that there is a decrease in DAT-carrying MSNs decrease over the course of HD, the DAT has been proposed as a potential biomarker. A significant number of single-photon emission computed tomography (SPECT) and PET tracers have been developed for DAT imaging. While most have been researched in the context of in Parkinson's disease, some were investigated for HD, including [^{11}C]β-CIT and [^{18}F]-FP-CIT (Backman et al., 1997; Ginovart et al., 1997; Palermo et al., 2021; Chun, 2022). In mHD patients, accumulation of [^{11}C]β-CIT was markedly reduced in the caudate and putamen compared to controls. Moreover, the mean influx rate was significantly reduced in the caudate (~53%) and putamen (~55%). Regrettably, kinetic studies of [^{11}C]β-CIT have shown signs of irreversible binding, making it not suitable for *in vivo* imaging (Ginovart et al., 1997). Nevertheless, DAT data correlated strongly to cognitive decline as measured by episodic memory and trail making tests, among others (Backman et al., 1997). Recently, another DAT PET tracer '[^{18}F]-FP-CIT' has been tried in a case study HD. Decreased uptake was seen in the basal ganglia in early mHD, with an additional reduction in the putamen in late mHD (Chun, 2022).

DOPA Decarboxylase

DCC is an enzyme involved in the uptake and decarboxylation of amine precursors in peripheral tissues. In neurodegenerative and psychiatric disorders, perturbations in DDC expression have been found (Guenter & Lenartowski, 2016). DDC availability can be evaluated using L-[^{18}F]fluorodopa, which is an L-dopa analogue. In a case study of HD, Leenders et al. showed a normal striatal uptake of L-[^{18}F]fluorodopa despite decreased dopamine receptor D_2 binding (Leenders et al., 1986). Conversely, Martin et al. reported decreased L-[^{18}F]fluorodopa binding in the caudate (Martin & Hayden, 1987). Notably, these studies were conducted when PET imaging was still in its infancy. More research on the DDC with improved PET imaging protocols and data processing are indicated to elucidate its role in HD.

Postsynaptic Markers

Dopamine Receptors

Dopamine receptors play an essential role in daily life functions, particularly in movement, emotions, and the reward system in the brain. They are abundantly expressed in the CNS, including on MSNs. Given the progressive loss of MSNs during HD, dopamine receptors were proposed as a biomarker to evaluate disease progression (Politis et al., 2008). In the dopamine receptor family, 5 distinct members have been identified and classified into two subcategories: D_1-like receptors (D_1, D_5) and D_2-like receptors (D_2, D_3, D_4) (Elsinga et al., 2006). The dopamine D_1 receptor (D_1R) and dopamine D_2 receptor (D_2R) are abundantly expressed in the striatum. To a far lesser extent, their expression is also seen in the cortex (D_1R) or cortex, amygdala, hippocampus and thalamus (D_2R) (Elsinga et al., 2006; van Oostrom et al., 2009; Wei et al., 2018). The D_3, D_4 and D_5 subtypes are less prevalent overall. Functionally, the D_1R is responsible for initiation of movement by disinhibition of the thalamus (direct pathway). Conversely, the D_2R is responsible for reinforcement of thalamic inhibition (indirect pathway), leading to cessation of movement (Koch & Raymond, 2019). In the setting of HD, the D_2R is affected earlier than D_1R, resulting in the characteristic Huntington's chorea (van Oostrom et al., 2009; Koch & Raymond, 2019).

D_1R and D_2R density PET studies have been extensively conducted for a variety of neurodegenerative diseases, primarily with [^{11}C]SCH23390 (D_1R) or [^{11}C]raclopride ($D_{2/3}R$) (Elsinga et al., 2006).

Using these PET tracers, Turjanski et al. measured a severe parallel reduction in striatal D_1R and D_2R in mHD. Although more prominent in patients with muscle rigidity, the decline was notable in all patients irrespective of their predominant phenotype (Turjanski et al., 1995). The importance of D_1R and D_2R was further corroborated by Weeks et al., who showed significant parallel loss of striatal D_1R and D_2R in 50% of asymptomatic pwHD (Weeks et al., 1996). Moreover, a correlation between D_1R and D_2R reduction was also demonstrated here, which was later confirmed in another study by Ginovart et al. (Ginovart et al., 1997). Of note, in latter study, a significant decrease of [^{11}C]β-CIT binding to the presynaptic dopamine transporter was also reported, although this may have been the result of a reduced striatal blood flow (Ginovart et al., 1997). Using serial [^{11}C]SCH23390 and [^{11}C]raclopride PET over the course of 40 months, a respective annual loss of 4.0% and 2.0% has been described in asymptomatic pwHD, with losses up to up to 6.5% ($D_{2/3}R$) and 4.5% (D_1R) in actively progressing patients. In mHD, the balance shifted towards D_1R loss, with an average annual decrease of 5.0% (D_1R) and 3.0% (D_2R) (Andrews et al., 1999). Studies have also been performed with [^{11}C]raclopride alone or [^{11}C]raclopride combined with other PET tracers for HD (e.g., [^{18}F]FDG (glucose), [^{11}C](R)-PK11195 (inflammation), [^{18}F]MNI659 (PDE10A) (Antonini et al., 1996, 1998; Furtado et al., 2005; Pavese et al., 2006; Politis et al., 2008; van Oostrom et al., 2009; Pavese et al., 2010; Grecchi et al., 2014; Fazio et al., 2017). These more recent studies mainly confirm the above findings, but additionally highlight the existence and clinical implications of cortical and hypothalamic $D_{2/3}R$ loss in addition to striatal $D_{2/3}R$ loss (Politis et al., 2008; Pavese et al., 2010). Moreover, they show that striatal D_2R decline can already be detected 19 years before expected onset of symptoms (Politis et al., 2008). A correlation between $D_{2/3}R$ loss and other biomarker fluctuations within HD remains controversial due to of the high inter-patient and regional variability in $D_{2/3}R$ levels. Besides [^{11}C]SCH23390 and [^{11}C]raclopride, no selective dopamine receptor PET tracers have been clinically studied in pwHD.

Phosphodiesterase 10A

Cyclic nucleotide phosphodiesterases (PDEs) are enzymes involved in the regulation of intracellular second messenger protein concentrations (e.g., cAMP or cGMP) by controlling their degradation rates. PDEs are hence regarded as important regulators of cellular signaling, including integration of adenosinergic, dopaminergic and glutaminergic pathways (Niccolini et al., 2015; Erro et al., 2021). Several classes of

PDE enzymes exist, each with specific tissue distributions and certain selectivity for cyclic nucleotides (Erro et al., 2021). On MSNs, which are among the first neurons to be lost in HD, expression of PDE10a is most abundant. Substantial depletion of PDE10a before the onset of motor abnormalities has been reported in HD, highlighting this biomarker's potential as early indicator of disease progression (Ahmad et al., 2014).

Non-invasive monitoring of PDE10a in pwHD has been conducted using the PET radioligands [^{18}F]JNJ42259152, [^{11}C]IMA107 and [^{18}F]MNI-659 (Van Laere et al., 2013; Ahmad et al., 2014; Russell et al., 2014, ; 2016; Niccolini et al., 2015; Wilson et al., 2016; Fazio et al., 2020). Using [^{18}F]JNJ42259152, a significant PDE10a reduction in the putamen (62.6%,) and caudate nucleus (70.7%) was observed in mHD compared to healthy controls. A correlation between PDE10a levels and clinical disease severity measures (TFC scores) was however not found (Ahmad et al., 2014). Regrettably, additional preclinical research provided evidence of a metabolite capable of entering the rat brain, halting further clinical evaluation (Boscutti & Plisson, 2019). Using [^{11}C]IMA107, Niccolini et al. showed a ~33.5% loss of PDE10a in striatonigral/striatopallidal internal and striatopallidal external projecting segments in PreHD 25 years from predicted age of onset. Moreover, a 35% increase was seen in motor thalamic nuclei (Niccolini et al., 2015). Interestingly, the altered balance of PDE10A expression (motor thalamic nuclei over striatopallidal internal projecting segments of the striatum) was strongly associated with the predicted risk of symptomatic conversion (Niccolini et al., 2015). In a similar patient cohort, significantly lower PDE10a expression was also seen in the insular cortex and occipital fusiform gyrus (Wilson et al., 2016). Besides [^{18}F]JNJ42259152 and [^{11}C]IMA107, [^{18}F]MNI-654 and [^{18}F]MNI-659 were put forward as potential PDE10a radioligands. Tracer kinetics were more optimal for [^{18}F]MNI-659, as investigated in healthy volunteers, making this the compound of choice. In pwHD, [^{18}F]MNI-659 uptake was reduced by 47.6%, compared to controls. Additionally, strong correlations were found between disease severity (UHDRS score) and striatal [^{18}F]MNI-659 uptake, burden of disease score, as well as regional atrophy (Russell et al., 2014). Follow-up PET scans were performed in the same HD cohort, showing further declined uptake in the putamen and caudate nucleus in all participants. Mean annualized decline rates in pwHD were reported to be 16.6%, 6.9%, and 5.8% in the caudate, putamen and globus pallidus, respectively, whereas in controls, the reduction remained within 1% in striatal regions (Russell et al., 2016). Latter decline was corroborated by Fazio et al., who reported an estimated PDE10a loss of 8% per decade using the same radioligand ([^{18}F]MNI-659)(Fazio et al., 2017; Fazio et al., 2020). Even so, in mHD with TFC stage 1 and 2, PDE10a levels seemed to reach a plateau (Fazio et al., 2020). Notably, besides the age-related physiological PDE10a reduction, a sex effect was also reported, with generally higher tracer uptake seen in female subjects (Fazio et al., 2017). Together, these studies highlight very early PDE10a changes over the course of HD emphasizing the importance of early intervention, going back to 25 years before onset of symptoms.

Synaptic Vesicle Glycoprotein 2A

The synaptic vesicle glycoprotein 2 (SV2) family are keratan sulfate proteoglycans present in the secretory vesicles of neurons and endocrine cells, where they play a fundamental role in vesicular dynamics (Rossi et al., 2022). The SV2A family has 3 members, namely: SV2A, SV2B and SV2C. These are structurally similar however diverging in function. The dominant isoform SV2A is ubiquitously expressed in all brain regions, with greatest density seen in the basal ganglia and thalamus. SV2B comes second with expression mainly in the cortex and hippocampus (Bajjalieh et al., 1993; Hu et al., 2017). SV2C is least expressed, with expression restricted to older brain areas including the striatum, substantia nigra, pons, medulla oblongata and olfactory bulb, with nearly absent expression levels in the neocortex, hippocampus and thalamus (Hu et al., 2017). Although a role in neurotransmission has been established for SV2 proteins, the entire biological significance of this family is unclear. Their participation in key mechanisms of the vesicular processes is however widely accepted (for review see Rossi et al., 2022).

Synaptic dysfunction has long been suspected to play a major role in HD given the ubiquitous expression of SV2A in virtually all synapses (Delva et al., 2021). In a mouse model of HD, mHTT has proven to associate with synaptic vesicles resulting in impaired synaptic function and behavioral deficits (Smith et al., 2014). Moreover, SV2A loss was also detected in *post-mortem* brain tissue of pwHD (Bertoglio et al., 2021a, b). For *in vivo* assessment of SV2A availability, PET tracers were first developed based on the antiepileptic drug Levetiracetam (LEV), known to specifically bind to SV2A. These LEV-derivatives include [^{11}C]UCB-A, [^{18}F]UCB-H and [^{11}C]UCB-J, of which [^{11}C]UCB-J has showed the most promise for *in vivo* clinical application (Mercier et al., 2014; Nabulsi et al., 2016). *In vivo*, [^{11}C]UCB-J has shown excellent PET properties such as high uptake in gray matter brain areas, fast kinetics, and whole-body distribution in non-human primates (NHPs) (Nabulsi et al., 2016). In healthy humans and NHPs, [^{11}C]UCB-J binding was deemed SV2A-specific, and, in epileptic patients, [^{11}C]UCB-J demonstrated the ability to detect changes in synaptic density (Finnema et al., 2016; Finnema et al., 2018; Rossi et al., 2022). [^{11}C]UCB-J has also been studied in PreHD, where significant SV2A loss could be detected in putamen, caudate, pallidum, cerebellum, parietal, and temporal and frontal cortex. Moreover, SV2A loss in the putamen correlated with motor impairment (Delva et al. 2022a, b). Despite favorable kinetics, good test-retest and reproducibility of binding parameters, [^{11}C]UCB-J is partially limited by its ^{11}C isotope, requiring an on-site cyclotron for production. Hence, efforts shifted to the development of ^{18}F-labeled SV2A radioligands, including [^{18}F]UCB-J, [^{18}F]SynVesT-1 (also referred to as [^{18}F]MNI-1126 or [^{18}F]SDM-8), [^{18}F]SynVesT-2 (also referred to as [^{18}F]SDM-2), and [^{18}F]SDM-16 (Constantinescu et al., 2019; Li et al., 2019a, b; Cai et al., 2020; Rossi et al., 2022; Zheng et al., 2022). Given the potential of these tracers, as seen in healthy volunteers, their application in pwHD can be anticipated (Li et al., 2021; Naganawa et al., 2021).

Cannabinoid System

Cannabinoid (CB) receptors are a class of cell membrane receptors in the G-protein coupled receptor (GPCR) superfamily, located throughout the entire human body (Horne et al., 2013). Within the CB family, there are two recognized subtypes: the CB_1 receptor (CB_1R) and the CB_2 receptor (CB_2R), each with their specific tissue expression pattern (Mackie, 2008). Whereas the CB_2R is mainly located on immune cells, hematopoietic cells and to a limited extent in the brain, the CB_1R is predominantly expressed on GABA (gamma amino butyric acid)-ergic striatal MSNs in the brain with more limited expression in the lungs, liver and kidneys (Mackie, 2008; Pagano et al., 2016). Activation of CBRs is achieved by endocannabinoids or cannabinoids, resulting in suppression of neurotransmitter release to modulate signal transduction pathways (e.g. hormone regulation). Consequently, disruption of CBR signaling has a direct influence on motor behavior, cognition, emotion and food intake, among others (Van Laere et al., 2010; Pagano et al., 2016).

In *post-mortem* brain tissue of pwHD, specifically the basal ganglia, downregulation of CB_1R can be detected. The leading cause of this is hypothesized to be transcriptional repression by mHTT (Van Laere et al., 2010; Blázquez et al., 2011). In advanced mHD, Glass et al. demonstrated loss of CB_1Rs in the substantia nigra down to 97% of control levels. Moreover, the decline in CB_1R availability was noticeable before D_1R and D_2R loss (Glass et al., 2000). *In vivo,* CB_1R PET imaging by Van Laere et al. with the highly selective CB_1R ligand [^{18}F]MK-9470 showed similar results. In a cohort of 20 symptomatic pwHD, reductions in CB_1R availability ranged from ~15% (cerebellum) to ~25% (frontal cortex) compared to controls, with no significant reductions in white matter areas. Such widespread and uniform CB_1R reduction in early symptomatic HD came unexpected given that neuronal loss and mHTT aggregation are mostly restricted to the striatum in early HD. No correlation was found between CB_1R expression and UHDRS motor scores, TFC scores, disease duration, or CAG repeat size (Van Laere et al., 2010). Nevertheless, the available results so far suggest CB_1R dysfunction to start in a presymptomatic stage of HD, stressing the relevance of CB_1R as an early marker.

Adenosinergic System

Adenosine is an important nucleoside for the regulation of the homeostatic bioenergetic network. Tight regulation of adenosine concentration occurs via glial cells through degradation of adenosine 5′-triphosphate (via extracellular ectonucleotidases, e.g. CD73 in the striatum and globus pallidus). Adenosine activates signal transduction cascades through four different receptor subtypes (A_1, A_{2A}, A_{2B} and A_3), all consisting of seven transmembrane domains and belonging to the GPCR family (Blum et al., 2018). Extensive research has been conducted on the A_1 and A_{2A} adenosine receptors (A_1AR and $A_{2A}AR$), being major targets of caffeine. The A_1AR

is widely distributed in brain areas such as the cortex, hippocampus, and cerebellum. The $A_{2A}AR$ is enriched in the basal ganglia and olfactory bulb (Sheth et al., 2014).

In HD, adenosine homeostasis is affected, as evidenced by *post-mortem* data in pwHD (for review see Glass et al., 2000; Blum et al., 2018). Concretely, $A_{2A}R$ downregulation was found in basal ganglia tissue of pwHD in early, intermediate and late disease stages (Glass et al., 2000). In addition, mHTT expression was linked to a significant reduction in transcript levels of endogenous $A_{2A}AR$ (Chiang et al., 2005). Linking A_1R fluctuation to disease progression is however less straightforward. Diverging results were found in mouse models of HD, with the main consensus being that A_1AR is significantly reduced in HD compared to non-HD (Blum et al., 2018). Adenosine analogue PET tracers have been developed to investigate regional A_1AR and $A_{2A}AR$ fluctuations. For $A_{2A}AR$, radioligands have been created (e.g. [^{18}F]MNI-444 and [^{11}C]TMSX), however none have been employed in clinical HD research. For A_1AR on the other hand, the PET ligand [^{18}F]CPFPX has been employed in pwHD. Biphasic alterations in A_1AR levels were found over the course of HD progression (Lee & Chern, 2014; Matusch et al., 2014). In early PreHD, thalamic A_1AR levels were ~ 31% elevated compared to healthy controls, whereas in late PreHD A_1AR levels had returned to normal. In the caudate of mHD, a reduction of ~25% in A_1AR was seen. Additionally, Matusch et al. reported a strong correlation between A_1AR levels and age of onset (Matusch et al., 2014). Further research is indicated to fully unravel the importance of adenosine receptors within HD.

Opioidergic System

Opioids are a class of compounds mostly known for their ability to reduce pain and their addictive properties. There are three classes of endogenous opioids: enkephalins, endorphins, and dynorphins (Reeves et al., 2022). Opioid-binding sites in the CNS as well as in peripheral sensory and automatic nerves have been discovered decades ago and have since been researched extensively. Opioid receptor (OR)-mediated signaling has been studied in the context of pain and analgesia, alcoholism, stress, tolerance, mental illness, behavior, and many others (Henriksen & Willoch, 2008). All ORs belong to the GPCR family and activate inhibitory G-proteins (Al-Hasani & Bruchas, 2011). To date, four subtypes have been identified: the δ-OR (DOR), κ-OR (KOR), μ-OR (MOR) and opioid receptor like-1 (ORL1).

Several *post-mortem* studies have shown that the opioid system is affected in HD. In PreHD, a loss of enkephalin of striatal projection neurons to the external globus pallidus has been reported, whereas in mHD, the proenkephalin/prodynorphin ratio in the caudate, putamen, and internal and external globus pallidus was reduced (Pagano et al., 2016). For PET imaging of the opioid system, the non-selective partial agonist [^{11}C]diprenorphine has often been employed, specifically to study distribution patterns of the DORs, KORs, and MORs. For example, in mHD, Weeks et al. showed decreased [^{11}C]diprenorphine binding in the caudate (~29%) and putamen (~23%) compared to healthy controls (Weeks et al., 1997). Given that this reduction is lower compared to striatal dopaminergic degeneration in a similar population (~60% in mHD), ORs are suggested to be more resistant to the various processes driving HD (Turjanski et al., 1995; Pagano et al., 2016).

GABA/Benzodiazepine System

GABA is an amino acid that serves as an inhibitory neurotransmitter in the brain, where it exerts its function through binding to GABA receptors. Two receptor subtypes have been identified: $GABA_AR$ and $GABA_BR$. Post-synaptic GABA signaling modulates ion channels, hyperpolarizing the cell and inhibiting transmission of action potentials. The complexity of its mechanisms in the CNS is not to be underestimated, evidenced by the multitude of neurologic and psychiatric conditions in which GABA signaling is impaired (Jewett & Sharma, 2022).

In HD, loss of GABA receptor density has been detected in the caudate nucleus, putamen and the frontal cortex in *post-mortem* brain tissue (Künig et al., 2000). GABA release by striatal neurons progressively reduces in HD progression due to the increasing neuronal death (Jewett & Sharma, 2022). To investigate the *in vivo* availability of GABA receptors in regions affected by HD, the radioligand [^{11}C]Flumazenil can be employed. In mHD, [^{11}C]Flumazenil binding was decreased by 17% in the caudate, whereas in the putamen and thalamus no decrease was detectable compared to healthy controls (Holthoff et al., 1993). These results were later corroborated when Künig et al. reported reduced [^{11}C]Flumazenil binding in the caudate with normal putamen values in a similar mHD population (Künig et al., 2000). Interestingly, both studies also reported metabolic disturbances (reduced [^{18}F]FDG uptake) in the putamen and thalamus of their HD subjects, suggesting that alterations in glucose metabolism precede the loss of GABA receptors (Künig et al., 2000; Pagano et al., 2016).

3 Future Perspectives of PET Biomarkers for HD

The following section provides an overview of biomarkers relevant to HD for which PET ligands are available but have not yet been reported in pwHD. These include biomarkers within the histaminergic and glutamatergic systems as well as mHTT aggregates. Given the preclinical evidence, these markers are expected to be brought more into focus in the near future.

Histaminergic System

Histamine is a neuromodulator whose levels vary with a daily rhythm. Peak release is seen during the active, awake cycle and relatively lower levels are seen during sleep phases (Whittaker et al., 2017). Histamine binds to four known GPCRs: H_1, H_2, H_3 and H_4 receptors. The histamine H_3 receptor (H_3R) is predominantly expressed in brain regions involved in cognition or motor function such as the cortex, (hypo) thalamus, hippocampus, amygdala, and the striatum (Moreno-Delgado et al., 2020).

In neuropsychiatric disorders such as AD, Parkinson's disease or HD, the histaminergic system is often disturbed, leading to fragmented sleep patterns, disrupted appetite control or arousal issues, among others (Cheng et al., 2021). In healthy conditions, H_3Rs are ubiquitously expressed in the brain (Jucaite et al., 2013). In *post-mortem* tissue of pwHD, significantly lower H_3R densities were found in the caudate, putamen, and globus pallidus (Jucaite et al., 2013). Given their brain-wide availability and implications in HD, H_3Rs are increasingly proposed as candidate biomarker. Notably, in striatal GABAergic dynorphinergic neurons, D_1R and H_3R may be co-expressed, where they can form molecular complexes termed 'D1R-H3R heteromers'. These complexes are functional in early stages of HD but are increasingly lost with disease progression. Hence, in addition to the H_3R alone, these heteromers were also proposed as novel therapeutic targets in PreHD, specifically to modulate D_1R signaling and ultimately slow progression of HD (Moreno-Delgado et al., 2020). H_3R PET tracers are available and have been applied in humans for brain-wide monitoring of histaminergic changes (e.g. [^{11}C]GSK-189254 (Ashworth et al., 2010, Ashworth et al., 2014) and [^{11}C]AZ-12807110 (Jucaite et al., 2013)), paving the way to application in pwHD.

Glutamatergic System

Glutamate is a major excitatory neurotransmitter in the CNS, playing a key role in integrative brain function and neuronal cell development. Glutamate signaling is exerted through two types of glutamate receptors: ionotropic receptors (iGluR) and metabotropic receptors (mGluR). iGluRs are ligand-gated ion channels for which three subtypes exist: NMDA, AMPA, and kainate receptors. mGluRs are GPCRs for which eight distinct members are known (mGluR 1–8), classified into three groups according to their sequence homology and G-protein coupled specificity (group I-III)(Ribeiro et al., 2010). Group I members (mGluR1, mGluR5) are primarily linked to the activation of phospholipase C, intracellular Ca^{2+} release and neuronal death, whereas group II (mGluR2–3) and group III (mGluR4 and mGluR6–8) members are mostly known to negatively regulate adenylyl cyclase (Francesconi & Duvoisin, 1998; Ribeiro et al., 2011).

Impairment of metabotropic glutamatergic signaling has been reported for a variety of neurological diseases, including HD (Ferraguti et al., 2008). Particularly group I mGluRs are of interest given their expression patterns in the striatum, hippocampus, cortex (mGluR5) and the thalamus and cerebellum (mGluR1) (Bertoglio et al., 2020). Preclinical research in mouse models of HD as well as in *post-mortem* human tissue has revealed mHTT to impair glutamatergic signaling directly (e.g. through interaction with optineurin, a protein known to attenuate mGluR1/5 signaling) and indirectly through disruption of several closely linked signaling pathways (e.g. dopaminergic, cannabinoid)(Ribeiro et al., 2011). Non-invasive PET imaging experiments in HD animals have been performed with [^{11}C]ITDM (mGluR1) and [^{11}C]ABP688 (mGluR5) (Bertoglio et al., 2018, 2020; Verhaeghe et al., 2018). To our knowledge, no mGluR1/5 PET tracers have been evaluated in human subjects so far. Hence, the impairment of mGluR1/5 in pwHD still remains to be demonstrated.

Mutant Huntingtin

The wild-type Huntingtin protein is ubiquitously expressed throughout the human body, with the highest levels seen in the testes and brain, where it is enriched in all neurons and glial cells (Schulte & Littleton, 2011). The accumulation of the mHTT protein is the hallmark of HD, hence mHTT is evidently considered the prime biomarker to assess neuropathological progression. At present, multiple m*HTT*-lowering therapeutic strategies are being explored, most of which employ invasive methods (e.g. antibody-based readouts on biofluid samples) to follow-up the expected decrease in mHTT aggregate load (Tabrizi et al., 2019). While biofluid analyses have proven advantageous in clinical HD research (empowered by their reproducibility, cost and accuracy)(for review see Rodrigues et al., 2018), they lack regional specificity and hence may not fully represent potential therapeutic effects (Rodrigues et al., 2018). Consequently, recent efforts for mHTT detection have shifted to PET imaging, which would provide accurate, region-specific and, moreover, non-invasive quantification of mHTT.

In 2020, the first-in-class mHTT aggregate PET radioligand [^{11}C]CHDI-180R was described by Liu et al. (Liu et al., 2020). [^{11}C]CHDI-180R demonstrated favorable kinetics with excellent brain uptake and elimination in mouse models of HD and healthy rhesus macaques (Liu et al., 2020; Bertoglio et al., 2022). In a NHP model of HD, [^{11}C]CHDI-180R was validated as a viable radioligand for mHTT aggregate visualization, supporting its translation to the clinic (Bertoglio, Weiss et al., unpublished). Of note, *post-mortem* analyses of HD tissue showed non-negligible affinity for Aβ plaques, potentially limiting the applicability of this radioligand. In 2021, Liu et al. reported a new clinical mHTT PET candidate, [^{11}C]CHDI-626, with reduced Aβ plaque/tau-aggregate binding, good affinity for (recombinant) mHTT aggregates and rapid brain exposure and washout (Liu et al., 2021). These characteristics were corroborated *in vivo* by Bertoglio et al., reporting very fast kinetics. However, this hampered quantification in a mouse model of HD (Bertoglio et al. 2021, b). The initial clinical dosimetry study of [^{11}C]CHDI-626 (NCT05224115, *clinicaltrials.gov*) identified accumulation of a brain-penetrant metabolite, making not suitable for clinical application. Conversely, in the same dosimetry study, [^{11}C]CHDI-180R showed promising kinetic properties (Delva et al., 2022a, b). Therefore, clinical evaluation of [^{11}C]CHDI-180R is currently advancing in pwHD (NCT03810898, *clinicaltrials.gov*). Most recently, the novel radioligand [^{18}F]CHDI-650 has been described by Liu et al. (Liu et al., 2023). In healthy NHPs, [^{18}F]CHDI-650 showed a rapid brain uptake and rapid washout, increased metabolic stability, and minimal off-target effects in *post-mortem* human HD tissue (e.g. Aβ plaque/tau-aggregate binding) (Liu et al., 2023). Notably, given these results and the ^{18}F radiolabel on CHDI-650, which is associated with a better half-life, better imaging resolution and therefore expanded utility in a clinical environment compared to ^{11}C, [^{18}F]CHDI-650 studies in pwHD are highly anticipated.

4 Conclusions

In the last decades, tremendous advances have been made in HD research, including in the discovery of affected pathways and the identification of potential therapeutic targets to ameliorate disease progression. The key role of the CAG repeat mutation in HD has been recognized, however the full extent of HD neuropathology reaches beyond the presence of mHTT aggregation, as evidenced by the great deal of signaling pathways and cellular homeostasis components that are affected before the onset of overt symptoms. For example, alterations in brain-wide biomarkers such as glucose uptake and neuroinflammation have been reported, even in the absence of striatal atrophy. Interestingly, PDE10a as well as the dopaminergic marker D_2R showed dysfunction/altered availability dating back decades before the predicted age of onset (e.g. PDE10a (~25 years) or D_2R (~19 years)). Moreover, a variety of key signaling pathways in the CNS have proven to be dysregulated due to decreased receptor availability (e.g. CB_1R, OR, $GABA_AR$) or loss of supporting proteins (e.g. DAT, DDC, SV2A). With the availability of PET radioligands, these alterations can be monitored noninvasively at any stage of disease progression. This is of great importance in clinical research, given that significant attention is now being paid to subclinical indicators of disease progression and the long-term effects of mHTT-lowering (Bertoglio et al., 2022; Marchionini et al., 2022). Besides the biomarkers for which PET imaging has already been validated in the clinic, other potential candidates have been suggested based on preclinical evidence: mGluR group I, H_3R and mHTT. Evidently, a PET tracer targeting mHTT in its soluble form is highly anticipated and would greatly benefit research particularly in the context of mHTT-lowering. Fortunately, this research has gained momentum with the development of several mHTT aggregate-targeting radioligands planned to be tested in pwHD, thus we expect significant progresses in the coming years.

Acknowledgements University of Antwerp, Belgium supported this work through a tenure-track assistant professor position for DB and a full professor position for SS. LE, SS, and DB are members of the μNeuro Research Centre of Excellence at the University of Antwerp. Figure 1 was created with *BioRender.com*.

References

Ahmad, R., Bourgeois, S., Postnov, A., Schmidt, M. E., Bormans, G., Van Laere, K., et al. (2014). PET imaging shows loss of striatal PDE10A in patients with Huntington disease. *Neurology, 82*, 279–281.

Al-Hasani, R., & Bruchas, M. R. (2011). Molecular mechanisms of opioid receptor-dependent signaling and behavior. *Anesthesiology, 115*, 1363–1381.

Andrews, T. C., Weeks, R. A., Turjanski, N., Gunn, R. N., Watkins, L. H., Sahakian, B., et al. (1999). Huntington's disease progression. PET and clinical observations. *Brain, 122*(Pt 12), 2353–2363.

Antonini, A., Leenders, K. L., Spiegel, R., Meier, D., Vontobel, P., Weigell-Weber, M., et al. (1996). Striatal glucose metabolism and dopamine D2 receptor binding in asymptomatic gene carriers and patients with Huntington's disease. *Brain, 119*(Pt 6), 2085–2095.

Antonini, A., Leenders, K. L., & Eidelberg, D. (1998). [11C]raclopride-PET studies of the Huntington's disease rate of progression: Relevance of the trinucleotide repeat length. *Annals of Neurology, 43*, 253–255.

Ashworth, S., Rabiner, E. A., Gunn, R. N., Plisson, C., Wilson, A. A., Comley, R. A., et al. (2010). Evaluation of 11C-GSK189254 as a novel radioligand for the H3 receptor in humans using PET. *Journal of Nuclear Medicine, 51*, 1021–1029.

Ashworth, S., Berges, A., Rabiner, E. A., Wilson, A. A., Comley, R. A., Lai, R. Y., et al. (2014). Unexpectedly high affinity of a novel histamine H(3) receptor antagonist, GSK239512, in vivo in human brain, determined using PET. *British Journal of Pharmacology, 171*, 1241–1249.

Backman, L., Robins-Wahlin, T. B., Lundin, A., Ginovart, N., & Farde, L. (1997). Cognitive deficits in Huntington's disease are predicted by dopaminergic PET markers and brain volumes. *Brain, 120*(Pt 12), 2207–2217.

Bajjalieh, S. M., Peterson, K., Linial, M., & Scheller, R. H. (1993). Brain contains two forms of synaptic vesicle protein 2. *Proceedings of the National Academy of Sciences of the United States of America, 90*, 2150–2154.

Beglinger, L. J., O'Rourke, J. J., Wang, C., Langbehn, D. R., Duff, K., & Paulsen, J. S. (2010). Earliest functional declines in Huntington disease. *Psychiatry Research, 178*, 414–418.

Berent, S., Giordani, B., Lehtinen, S., Markel, D., Penney, J. B., Buchtel, H. A., et al. (1988). Positron emission tomographic scan investigations of Huntington's disease: Cerebral metabolic correlates of cognitive function. *Annals of Neurology, 23*, 541–546.

Bergonzoni, G., Döring, J., & Biagioli, M. (2021). D1R- and D2R-medium-sized spiny neurons diversity: Insights into striatal vulnerability to Huntington's disease mutation. *Frontiers in Cellular Neuroscience, 15*, 628010.

Bertoglio, D., Kosten, L., Verhaeghe, J., Thomae, D., Wyffels, L., Stroobants, S., et al. (2018). Longitudinal characterization of mGluR5 using (11)C-ABP688 PET imaging in the Q175 mouse model of Huntington disease. *Journal of Nuclear Medicine, 59*, 1722–1727.

Bertoglio, D., Verhaeghe, J., Korat, Š., Miranda, A., Wyffels, L., Stroobants, S., et al. (2020). In vitro and in vivo assessment of suitable reference region and kinetic modelling for the mGluR1 Radioligand [(11)C]ITDM in mice. *Molecular Imaging and Biology, 22*, 854–863.

Bertoglio, D., Verhaeghe, J., Miranda, A., Wyffels, L., Stroobants, S., Mrzljak, L., et al. (2021a). Longitudinal preclinical evaluation of the novel radioligand [11C]CHDI-626 for PET imaging of mutant huntingtin aggregates in Huntington's disease. *European Journal of Nuclear Medicine and Molecular Imaging, 49*, 1166–1175.

Bertoglio, D., Verhaeghe, J., Wyffels, L., Miranda, A., Stroobants, S., Mrzljak, L., et al. (2021b). Synaptic vesicle glycoprotein 2A is affected in the CNS of Huntington's disease mice and postmortem human HD brain. *Journal of Nuclear Medicine, 63*, 942–947.

Bertoglio, D., Bard, J., Hessmann, M., Liu, L., Gartner, A., De Lombaerde, S., et al. (2022). Development of a ligand for in vivo imaging of mutant huntingtin in Huntington's disease. *Science Translational Medicine, 14*, eabm3682.

Bhide, P. G., Day, M., Sapp, E., Schwarz, C., Sheth, A., Kim, J., et al. (1996). Expression of normal and mutant huntingtin in the developing brain. *The Journal of Neuroscience, 16*, 5523–5535.

Blázquez, C., Chiarlone, A., Sagredo, O., Aguado, T., Pazos, M. R., Resel, E., et al. (2011). Loss of striatal type 1 cannabinoid receptors is a key pathogenic factor in Huntington's disease. *Brain, 134*, 119–136.

Blum, D., Chern, Y., Domenici, M. R., Buée, L., Lin, C. Y., Rea, W., et al. (2018). The role of adenosine tone and adenosine receptors in Huntington's disease. *Journal of Caffeine and Adenosine Research, 8*, 43–58.

Bohnen, N. I., Koeppe, R. A., Meyer, P., Ficaro, E., Wernette, K., Kilbourn, M. R., et al. (2000). Decreased striatal monoaminergic terminals in Huntington disease. *Neurology, 54*, 1753–1759.

Booth, T. C., Nathan, M., Waldman, A. D., Quigley, A. M., Schapira, A. H., & Buscombe, J. (2015). The role of functional dopamine-transporter SPECT imaging in parkinsonian syndromes, part 1. *AJNR. American Journal of Neuroradiology, 36*, 229–235.

Boscutti, G. E. A. R., & Plisson, C. (2019). PET Radioligands for imaging of the PDE10A in human: Current status. *Neuroscience Letters, 691*, 11–17.

Cai, Z., Li, S., Zhang, W., Pracitto, R., Wu, X., Baum, E., et al. (2020). Synthesis and preclinical evaluation of an (18)F-labeled synaptic vesicle glycoprotein 2A PET imaging probe: [(18) F]SynVesT-2. *ACS Chemical Neuroscience, 11*, 592–603.

Caron, N. S., Wright, G. E. B., & Hayden, M. R. (1993). Huntington disease. GeneReviews((R)). In M.P. Adam, D.B. Everman, G.M. Mirzaa et al. University of Washington, Seattle.

Cheng, L., Liu, J., & Chen, Z. (2021). The histaminergic system in neuropsychiatric disorders. *Biomolecules & Therapeutics, 11*.

Chiang, M. C., Lee, Y. C., Huang, C. L., & Chern, Y. (2005). cAMP-response element-binding protein contributes to suppression of the A2A adenosine receptor promoter by mutant huntingtin with expanded polyglutamine residues. *The Journal of Biological Chemistry, 280*, 14331–14340.

Chun, K. (2022). Dual phase (18)F-FP CIT PET and (99m)Tc- ECD SPECT findings of Huntington's disease. *Radiology Case Reports, 17*, 2460–2463.

Ciarmiello, A., Cannella, M., Lastoria, S., Simonelli, M., Frati, L., Rubinsztein, D. C., et al. (2006). Brain white-matter volume loss and glucose hypometabolism precede the clinical symptoms of Huntington's disease. *Journal of Nuclear Medicine, 47*, 215–222.

Ciarmiello, A., Giovacchini, G., Orobello, S., Bruselli, L., Elifani, F., & Squitieri, F. (2012). 18F-FDG PET uptake in the pre-Huntington disease caudate affects the time-to-onset independently of CAG expansion size. *European Journal of Nuclear Medicine and Molecular Imaging, 39*, 1030–1036.

Constantinescu, C. C., Tresse, C., Zheng, M., Gouasmat, A., Carroll, V. M., Mistico, L., et al. (2019). Development and in vivo preclinical imaging of Fluorine-18-labeled synaptic vesicle protein 2A (SV2A) PET tracers. *Molecular Imaging and Biology, 21*, 509–518.

Cronin, T., Rosser, A., & Massey, T. (2019). Clinical presentation and features of juvenile-onset Huntington's disease: A systematic review. *Journal of Huntington's Disease, 8*, 171–179.

Crotti, A., Benner, C., Kerman, B. E., Gosselin, D., Lagier-Tourenne, C., Zuccato, C., et al. (2014). Mutant huntingtin promotes autonomous microglia activation via myeloid lineage-determining factors. *Nature Neuroscience, 17*, 513–521.

Delva, A., Michiels, L., Koole, M., Van Laere, K., & Vandenberghe, W. (2021). E01 widespread loss of presynaptic terminal marker SV2A in early Huntington disease. *Journal of Neurology, Neurosurgery, and Psychiatry, 92*, A15.

Delva, A., Koole, M., Serdons, K., Bormans, G., Liu, L., Bard, J., et al. (2022a). Biodistribution and dosimetry in human healthy volunteers of the PET radioligands [(11)C]CHDI-00485180-R and [(11)C]CHDI-00485626, designed for quantification of cerebral aggregated mutant huntingtin. *European Journal of Nuclear Medicine and Molecular Imaging, 50*, 48–60.

Delva, A., Michiels, L., Koole, M., Van Laere, K., & Vandenberghe, W. (2022b). Synaptic damage and its clinical correlates in people with early Huntington disease: A PET study. *Neurology, 98*, e83–e94.

Dorsey, R., Biglan, K., Eberly, S., Auinger, P., Brocht, A., Umeh, C. C., et al. (2011). Use of tetrabenazine in Huntington disease patients on antidepressants or with advanced disease: Results from the TETRA-HD study. *PLoS Currents, 3*, Rrn1283.

Dupont, A. C., Largeau, B., Santiago Ribeiro, M. J., Guilloteau, D., Tronel, C., & Arlicot, N. (2017). Translocator Protein-18 kDa (TSPO) positron emission tomography (PET) imaging and its clinical impact in neurodegenerative diseases. *International Journal of Molecular Sciences, 18*.

Elsinga, P. H., Hatano, K., & Ishiwata, K. (2006). PET tracers for imaging of the dopaminergic system. *Current Medicinal Chemistry, 13*, 2139–2153.

Erro, R., Mencacci, N. E., & Bhatia, K. P. (2021). The emerging role of phosphodiesterases in movement disorders. *Movement Disorders, 36*, 2225–2243.

Fazio, P., Schain, M., Mrzljak, L., Amini, N., Nag, S., Al-Tawil, N., et al. (2017). Patterns of age related changes for phosphodiesterase type-10A in comparison with dopamine D(2/3) receptors and sub-cortical volumes in the human basal ganglia: A PET study with (18)F-MNI-659 and (11)C-raclopride with correction for partial volume effect. *NeuroImage, 152*, 330–339.

Fazio, P., Fitzer-Attas, C. J., Mrzljak, L., Bronzova, J., Nag, S., Warner, J. H., et al. (2020). PET molecular imaging of phosphodiesterase 10A: An early biomarker of Huntington's disease progression. *Movement Disorders, 35*, 606–615.

Feigin, A., Tang, C., Ma, Y., Mattis, P., Zgaljardic, D., Guttman, M., et al. (2007). Thalamic metabolism and symptom onset in preclinical Huntington's disease. *Brain, 130*, 2858–2867.

Feil, K., Brendel, M., Hensler, M., Bartenstein, P., Seelos, K., Dieterich, M., et al. (2018). FDG-PET in a case of very late-onset Huntington's disease. *Movement Disorders Clinical Practice, 5*, 227–228.

Ferraguti, F., Crepaldi, L., & Nicoletti, F. (2008). Metabotropic glutamate 1 receptor: Current concepts and perspectives. *Pharmacological Reviews, 60*, 536–581.

Finnema, S. J., Nabulsi, N. B., Eid, T., Detyniecki, K., Lin, S. F., Chen, M. K., et al. (2016). Imaging synaptic density in the living human brain. *Science Translational Medicine, 8*, 348ra396.

Finnema, S. J., Nabulsi, N. B., Mercier, J., Lin, S. F., Chen, M. K., Matuskey, D., et al. (2018). Kinetic evaluation and test-retest reproducibility of [(11)C]UCB-J, a novel radioligand for positron emission tomography imaging of synaptic vesicle glycoprotein 2A in humans. *Journal of Cerebral Blood Flow and Metabolism, 38*, 2041–2052.

Francesconi, A., & Duvoisin, R. M. (1998). Role of the second and third intracellular loops of metabotropic glutamate receptors in mediating dual signal transduction activation. *The Journal of Biological Chemistry, 273*, 5615–5624.

Frank, S., Testa, C. M., Stamler, D., Kayson, E., Davis, C., Edmondson, M. C., et al. (2016). Effect of deutetrabenazine on chorea among patients with Huntington disease: A randomized clinical trial. *JAMA, 316*, 40–50.

Furtado, S., Sossi, V., Hauser, R. A., Samii, A., Schulzer, M., Murphy, C. B., et al. (2005). Positron emission tomography after fetal transplantation in Huntington's disease. *Annals of Neurology, 58*, 331–337.

Gaura, V., Bachoud-Lévi, A. C., Ribeiro, M. J., Nguyen, J. P., Frouin, V., Baudic, S., et al. (2004). Striatal neural grafting improves cortical metabolism in Huntington's disease patients. *Brain, 127*, 65–72.

Ginovart, N., Lundin, A., Farde, L., Halldin, C., Backman, L., Swahn, C. G., et al. (1997). PET study of the pre- and post-synaptic dopaminergic markers for the neurodegenerative process in Huntington's disease. *Brain, 120*(Pt 3), 503–514.

Glass, M., Dragunow, M., & Faull, R. L. (2000). The pattern of neurodegeneration in Huntington's disease: A comparative study of cannabinoid, dopamine, adenosine and GABA(a) receptor alterations in the human basal ganglia in Huntington's disease. *Neuroscience, 97*, 505–519.

Grafton, S. T., Mazziotta, J. C., Pahl, J. J., St George-Hyslop, P., Haines, J. L., Gusella, J., et al. (1992). Serial changes of cerebral glucose metabolism and caudate size in persons at risk for Huntington's disease. *Archives of Neurology, 49*, 1161–1167.

Grecchi, E., Doyle, O. M., Bertoldo, A., Pavese, N., & Turkheimer, F. E. (2014). Brain shaving: Adaptive detection for brain PET data. *Physics in Medicine and Biology, 59*, 2517–2534.

Guenter, J., & Lenartowski, R. (2016). Molecular characteristic and physiological role of DOPA-decarboxylase. *Postępy Higieny i Medycyny Doświadczalnej (Online)*, (70), 1424–1440.

Hayden, M. R., Martin, W. R., Stoessl, A. J., Clark, C., Hollenberg, S., Adam, M. J., et al. (1986). Positron emission tomography in the early diagnosis of Huntington's disease. *Neurology, 36*, 888–894.

Hellem, M. N. N., Vinther-Jensen, T., Anderberg, L., Budtz-Jørgensen, E., Hjermind, L. E., Larsen, V. A., et al. (2021). Hybrid 2-[18F] FDG PET/MRI in premanifest Huntington's disease gene-expansion carriers: The significance of partial volume correction. *PLoS One, 16*, e0252683.

Henriksen, G., & Willoch, F. (2008). Imaging of opioid receptors in the central nervous system. *Brain, 131*, 1171–1196.

Herben-Dekker, M., van Oostrom, J. C., Roos, R. A., Jurgens, C. K., Witjes-Ané, M. N., Kremer, H. P., et al. (2014). Striatal metabolism and psychomotor speed as predictors of motor onset in Huntington's disease. *Journal of Neurology, 261*, 1387–1397.

Hjermind, L. E., Law, I., Jonch, A., Stokholm, J., & Nielsen, J. E. (2011). Huntington's disease: Effect of memantine on FDG-PET brain metabolism? *The Journal of Neuropsychiatry and Clinical Neurosciences, 23*, 206–210.

Holthoff, V. A., Koeppe, R. A., Frey, K. A., Penney, J. B., Markel, D. S., Kuhl, D. E., et al. (1993). Positron emission tomography measures of benzodiazepine receptors in Huntington's disease. *Annals of Neurology, 34*, 76–81.

Horne, E. A., Coy, J., Swinney, K., Fung, S., Cherry, A. E., Marrs, W. R., et al. (2013). Downregulation of cannabinoid receptor 1 from neuropeptide Y interneurons in the basal ganglia of patients with Huntington's disease and mouse models. *The European Journal of Neuroscience, 37*, 429–440.

Hu, Y. W., Xiao, L., Zheng, L., & Wang, Q. (2017). Synaptic vesicle 2C and its synaptic-related function. *Clinica Chimica Acta, 472*, 112–117.

Janssen, B., Vugts, D. J., Windhorst, A. D., & Mach, R. H. (2018). PET imaging of microglial activation—Beyond targeting TSPO. *Molecules, 23*.

Jewett, B. E., & Sharma, S. (2022). *Physiology, GABA*. StatPearls Publishing.

Jucaite, A., Takano, A., Boström, E., Jostell, K. G., Stenkrona, P., Halldin, C., et al. (2013). AZD5213: A novel histamine H3 receptor antagonist permitting high daytime and low nocturnal H3 receptor occupancy, a PET study in human subjects. *The International Journal of Neuropsychopharmacology, 16*, 1231–1239.

Klein, M. O., Battagello, D. S., Cardoso, A. R., Hauser, D. N., Bittencourt, J. C., & Correa, R. G. (2019). Dopamine: Functions, signaling, and association with neurological diseases. *Cellular and Molecular Neurobiology, 39*, 31–59.

Koch, E. T., & Raymond, L. A. (2019). Dysfunctional striatal dopamine signaling in Huntington's disease. *Journal of Neuroscience Research, 97*, 1636–1654.

Kuhl, D. E., Phelps, M. E., Markham, C. H., Metter, E. J., Riege, W. H., & Winter, J. (1982). Cerebral metabolism and atrophy in Huntington's disease determined by 18FDG and computed tomographic scan. *Annals of Neurology, 12*, 425–434.

Künig, G., Leenders, K. L., Sanchez-Pernaute, R., Antonini, A., Vontobel, P., Verhagen, A., et al. (2000). Benzodiazepine receptor binding in Huntington's disease: [11C]flumazenil uptake measured using positron emission tomography. *Annals of Neurology, 47*, 644–648.

Kuwert, T., Lange, H. W., Langen, K. J., Herzog, H., Aulich, A., & Feinendegen, L. E. (1990). Cortical and subcortical glucose consumption measured by PET in patients with Huntington's disease. *Brain, 113*(Pt 5), 1405–1423.

Lawrence, A. D., Weeks, R. A., Brooks, D. J., Andrews, T. C., Watkins, L. H., Harding, A. E., et al. (1998). The relationship between striatal dopamine receptor binding and cognitive performance in Huntington's disease. *Brain, 121*(Pt 7), 1343–1355.

Lee, C. F., & Chern, Y. (2014). Adenosine receptors and Huntington's disease. *International Review of Neurobiology, 119*, 195–232.

Leenders, K. L., Frackowiak, R. S., Quinn, N., & Marsden, C. D. (1986). Brain energy metabolism and dopaminergic function in Huntington's disease measured in vivo using positron emission tomography. *Movement Disorders, 1*, 69–77.

Li, S., Cai, Z., Wu, X., Holden, D., Pracitto, R., Kapinos, M., et al. (2019a). Synthesis and in vivo evaluation of a novel PET radiotracer for imaging of synaptic vesicle glycoprotein 2A (SV2A) in nonhuman primates. *ACS Chemical Neuroscience, 10*, 1544–1554.

Li, S., Cai, Z., Zhang, W., Holden, D., Lin, S. F., Finnema, S. J., et al. (2019b). Synthesis and in vivo evaluation of [(18)F]UCB-J for PET imaging of synaptic vesicle glycoprotein 2A (SV2A). *European Journal of Nuclear Medicine and Molecular Imaging, 46*, 1952–1965.

Li, S., Naganawa, M., Pracitto, R., Najafzadeh, S., Holden, D., Henry, S., et al. (2021). Assessment of test-retest reproducibility of [(18)F]SynVesT-1, a novel radiotracer for PET imaging of synaptic vesicle glycoprotein 2A. *European Journal of Nuclear Medicine and Molecular Imaging, 48*, 1327–1338.

Liu, L., Prime, M. E., Lee, M. R., Khetarpal, V., Brown, C. J., Johnson, P. D., et al. (2020). Imaging mutant huntingtin aggregates: Development of a potential PET ligand. *Journal of Medicinal Chemistry, 63*, 8608–8633.

Liu, L., Johnson, P. D., Prime, M. E., Khetarpal, V., Lee, M. R., Brown, C. J., et al. (2021). [(11)C]CHDI-626, a PET tracer candidate for imaging mutant huntingtin aggregates with reduced binding to AD pathological proteins. *Journal of Medicinal Chemistry, 64*, 12003–12021.

Liu, L., Johnson, P. D., Prime, M. E., Khetarpal, V., Brown, C. J., Anzillotti, L., et al. (2023). Design and evaluation of [(18)F]CHDI-650 as a positron emission tomography ligand to image mutant huntingtin aggregates. *Journal of Medicinal Chemistry, 66*, 641–656.

Lois, C., González, I., Izquierdo-García, D., Zürcher, N. R., Wilkens, P., Loggia, M. L., et al. (2018). Neuroinflammation in Huntington's disease: New insights with (11)C-PBR28 PET/MRI. *ACS Chemical Neuroscience, 9*, 2563–2571.

Ma, Y., & Eidelberg, D. (2007). Functional imaging of cerebral blood flow and glucose metabolism in Parkinson's disease and Huntington's disease. *Molecular Imaging and Biology, 9*, 223–233.

Mackie, K. (2008). Cannabinoid receptors: Where they are and what they do. *Journal of Neuroendocrinology, 20 Suppl 1*, 10–14.

Marchionini, D. M., Liu, J.-P., Ambesi-Impiombato, A., Cox, K., Cirillo, K., Bansal, M., et al. (2022). Benefits of global mutant huntingtin lowering diminish over time in a Huntington's disease mouse model. *bioRxiv, 7*, 2022.2005.2017.492356.

Martí-Martínez, S., & Valor, L. M. (2022). A glimpse of molecular biomarkers in Huntington's disease. *International Journal of Molecular Sciences, 23*.

Martin, W. R., & Hayden, M. R. (1987). Cerebral glucose and dopa metabolism in movement disorders. *The Canadian Journal of Neurological Sciences, 14*, 448–451.

Matusch, A., Saft, C., Elmenhorst, D., Kraus, P. H., Gold, R., Hartung, H. P., et al. (2014). Cross sectional PET study of cerebral adenosine a_1 receptors in premanifest and manifest Huntington's disease. *European Journal of Nuclear Medicine and Molecular Imaging, 41*, 1210–1220.

McColgan, P., & Tabrizi, S. J. (2018). Huntington's disease: A clinical review. *European Journal of Neurology, 25*, 24–34.

Mercier, J., Archen, L., Bollu, V., Carre, S., Evrard, Y., Jnoff, E., et al. (2014). Discovery of heterocyclic nonacetamide synaptic vesicle protein 2A (SV2A) ligands with single-digit nanomolar potency: Opening avenues towards the first SV2A positron emission tomography (PET) ligands. *ChemMedChem, 9*, 693–698.

Michels, S., Buchholz, H. G., Rosar, F., Heinrich, I., Hoffmann, M. A., Schweiger, S., et al. (2019). 18F-FDG PET/CT: An unexpected case of Huntington's disease. *BMC Neurology, 19*, 78.

Moreno-Delgado, D., Puigdellívol, M., Moreno, E., Rodríguez-Ruiz, M., Botta, J., Gasperini, P., et al. (2020). Modulation of dopamine D1 receptors via histamine H3 receptors is a novel therapeutic target for Huntington's disease. *eLife, 9*, e51093.

Nabulsi, N. B., Mercier, J., Holden, D., Carré, S., Najafzadeh, S., Vandergeten, M. C., et al. (2016). Synthesis and preclinical evaluation of 11C-UCB-J as a PET tracer for imaging the synaptic vesicle glycoprotein 2A in the brain. *Journal of Nuclear Medicine, 57*, 777–784.

Naganawa, M., Li, S., Nabulsi, N., Henry, S., Zheng, M. Q., Pracitto, R., et al. (2021). First-in-human evaluation of (18)F-SynVesT-1, a radioligand for PET imaging of synaptic vesicle glycoprotein 2A. *Journal of Nuclear Medicine, 62*, 561–567.

Niccolini, F., Haider, S., Reis Marques, T., Muhlert, N., Tziortzi, A. C., Searle, G. E., et al. (2015). Altered PDE10A expression detectable early before symptomatic onset in Huntington's disease. *Brain, 138*, 3016–3029.

Owen, D. R., Yeo, A. J., Gunn, R. N., Song, K., Wadsworth, G., Lewis, A., et al. (2012). An 18-kDa translocator protein (TSPO) polymorphism explains differences in binding affinity of the PET radioligand PBR28. *Journal of Cerebral Blood Flow and Metabolism, 32*, 1–5.

Pagano, G., Niccolini, F., & Politis, M. (2016). Current status of PET imaging in Huntington's disease. *European Journal of Nuclear Medicine and Molecular Imaging, 43*, 1171–1182.

Palermo, G., Giannoni, S., Bellini, G., Siciliano, G., & Ceravolo, R. (2021). Dopamine transporter imaging, current status of a potential biomarker: A comprehensive review. *International Journal of Molecular Sciences, 22*.

Pavese, N., Andrews, T. C., Brooks, D. J., Ho, A. K., Rosser, A. E., Barker, R. A., et al. (2003). Progressive striatal and cortical dopamine receptor dysfunction in Huntington's disease: A PET study. *Brain, 126*, 1127–1135.

Pavese, N., Gerhard, A., Tai, Y. F., Ho, A. K., Turkheimer, F., Barker, R. A., et al. (2006). Microglial activation correlates with severity in Huntington disease: A clinical and PET study. *Neurology, 66*, 1638–1643.

Pavese, N., Politis, M., Tai, Y. F., Barker, R. A., Tabrizi, S. J., Mason, S. L., et al. (2010). Cortical dopamine dysfunction in symptomatic and premanifest Huntington's disease gene carriers. *Neurobiology of Disease, 37*, 356–361.

Peralta, C., Biafore, F., Depetris, T. S., & Bastianello, M. (2019). Recent advancement and clinical implications of 18FDG-PET in Parkinson's disease, atypical Parkinsonisms, and other movement disorders. *Current Neurology and Neuroscience Reports, 19*, 56.

Politis, M., & Piccini, P. (2012). Positron emission tomography imaging in neurological disorders. *Journal of Neurology, 259*, 1769–1780.

Politis, M., Pavese, N., Tai, Y. F., Tabrizi, S. J., Barker, R. A., & Piccini, P. (2008). Hypothalamic involvement in Huntington's disease: An in vivo PET study. *Brain, 131*, 2860–2869.

Politis, M., Pavese, N., Tai, Y. F., Kiferle, L., Mason, S. L., Brooks, D. J., et al. (2011). Microglial activation in regions related to cognitive function predicts disease onset in Huntington's disease: A multimodal imaging study. *Human Brain Mapping, 32*, 258–270.

Politis, M., Lahiri, N., Niccolini, F., Su, P., Wu, K., Giannetti, P., et al. (2015). Increased central microglial activation associated with peripheral cytokine levels in premanifest Huntington's disease gene carriers. *Neurobiology of Disease, 83*, 115–121.

Rees, E. M., Scahill, R. I., & Hobbs, N. Z. (2013). Longitudinal neuroimaging biomarkers in Huntington's disease. *Journal of Huntington's Disease, 2*, 21–39.

Reeves, K. C., Shah, N., Munoz, B., & Atwood, B. K. (2022). Opioid receptor-mediated regulation of neurotransmission in the brain. *Frontiers in Molecular Neuroscience, 15*, 919773.

Ribeiro, F. M., Paquet, M., Cregan, S. P., & Ferguson, S. S. (2010). Group I metabotropic glutamate receptor signalling and its implication in neurological disease. *CNS & Neurological Disorders Drug Targets, 9*, 574–595.

Ribeiro, F. M., Pires, R. G., & Ferguson, S. S. (2011). Huntington's disease and group I metabotropic glutamate receptors. *Molecular Neurobiology, 43*, 1–11.

Rocha, N. P., Charron, O., Latham, L. B., Colpo, G. D., Zanotti-Fregonara, P., Yu, M., et al. (2021). Microglia activation in basal ganglia is a late event in Huntington disease pathophysiology. *Neurology: Neuroimmunology & Neuroinflammation, 8*.

Rodrigues, F. B., Byrne, L. M., & Wild, E. J. (2018). Biofluid biomarkers in Huntington's disease. In S. V. Precious, A. E. Rosser, & S. B. Dunnett (Eds.), *Huntington's disease* (pp. 329–396).

Rossi, R., Arjmand, S., Bærentzen, S. L., Gjedde, A., & Landau, A. M. (2022). Synaptic vesicle glycoprotein 2A: Features and functions. *Frontiers in Neuroscience, 16*, 864514.

Roussakis, A. A., & Piccini, P. (2015). PET imaging in Huntington's disease. *Journal of Huntington's Disease, 4*, 287–296.

Russell, D. S., Barret, O., Jennings, D. L., Friedman, J. H., Tamagnan, G. D., Thomae, D., et al. (2014). The phosphodiesterase 10 positron emission tomography tracer, [18F]MNI-659, as a novel biomarker for early Huntington disease. *JAMA Neurology, 71*, 1520–1528.

Russell, D. S., Jennings, D. L., Barret, O., Tamagnan, G. D., Carroll, V. M., Caille, F., et al. (2016). Change in PDE10 across early Huntington disease assessed by [18F]MNI-659 and PET imaging. *Neurology, 86*, 748–754.

Schulte, J., & Littleton, J. T. (2011). The biological function of the huntingtin protein and its relevance to Huntington's disease pathology. *Current Trends in Neurology, 5*, 65–78.

Sheth, S., Brito, R., Mukherjea, D., Rybak, L. P., & Ramkumar, V. (2014). Adenosine receptors: Expression, function and regulation. *International Journal of Molecular Sciences, 15*, 2024–2052.

Shin, H., Kim, M. H., Lee, S. J., Lee, K. H., Kim, M. J., Kim, J. S., et al. (2013). Decreased metabolism in the cerebral cortex in early-stage Huntington's disease: A possible biomarker of disease progression? *Journal of Clinical Neurology, 9*, 21–25.

Simmons, D. A., James, M. L., Belichenko, N. P., Semaan, S., Condon, C., Kuan, J., et al. (2018). TSPO-PET imaging using [18F]PBR06 is a potential translatable biomarker for treatment response in Huntington's disease: Preclinical evidence with the p75NTR ligand LM11A-31. *Human Molecular Genetics, 27*, 2893–2912.

Smith, G. A., Rocha, E. M., McLean, J. R., Hayes, M. A., Izen, S. C., Isacson, O., et al. (2014). Progressive axonal transport and synaptic protein changes correlate with behavioral and neuropathological abnormalities in the heterozygous Q175 KI mouse model of Huntington's disease. *Human Molecular Genetics, 23*, 4510–4527.

Squitieri, F., Orobello, S., Cannella, M., Martino, T., Romanelli, P., Giovacchini, G., et al. (2009). Riluzole protects Huntington disease patients from brain glucose hypometabolism and grey matter volume loss and increases production of neurotrophins. *European Journal of Nuclear Medicine and Molecular Imaging, 36*, 1113–1120.

Tabrizi, S. J., Leavitt, B. R., Landwehrmeyer, G. B., Wild, E. J., Saft, C., Barker, R. A., et al. (2019). Targeting huntingtin expression in patients with Huntington's disease. *The New England Journal of Medicine, 380*, 2307–2316.

Tabrizi, S. J., Schobel, S., Gantman, E. C., Mansbach, A., Borowsky, B., Konstantinova, P., et al. (2022). A biological classification of Huntington's disease: The integrated staging system. *Lancet Neurology, 21*, 632–644.

Tang, C. C., Feigin, A., Ma, Y., Habeck, C., Paulsen, J. S., Leenders, K. L., et al. (2013). Metabolic network as a progression biomarker of premanifest Huntington's disease. *The Journal of Clinical Investigation, 123*, 4076–4088.

Tarakad, A., & Jimenez-Shahed, J. (2018). VMAT2 inhibitors in neuropsychiatric disorders. *CNS Drugs, 32*, 1131–1144.

Turjanski, N., Weeks, R., Dolan, R., Harding, A. E., & Brooks, D. J. (1995). Striatal D1 and D2 receptor binding in patients with Huntington's disease and other choreas. A PET study. *Brain, 118*(Pt 3), 689–696.

Van Laere, K., Casteels, C., Dhollander, I., Goffin, K., Grachev, I., Bormans, G., et al. (2010). Widespread decrease of type 1 cannabinoid receptor availability in Huntington disease in vivo. *Journal of Nuclear Medicine, 51*, 1413–1417.

Van Laere, K., Ahmad, R. U., Hudyana, H., Celen, S., Dubois, K., Schmidt, M. E., et al. (2013). Human biodistribution and dosimetry of 18F-JNJ42259152, a radioligand for phosphodiesterase 10A imaging. *European Journal of Nuclear Medicine and Molecular Imaging, 40*, 254–261.

van Oostrom, J. C., Dekker, M., Willemsen, A. T., de Jong, B. M., Roos, R. A., & Leenders, K. L. (2009). Changes in striatal dopamine D2 receptor binding in pre-clinical Huntington's disease. *European Journal of Neurology, 16*, 226–231.

Verhaeghe, J., Bertoglio, D., Kosten, L., Thomae, D., Verhoye, M., Van Der Linden, A., et al. (2018). Noninvasive relative quantification of [(11)C]ABP688 PET imaging in mice versus an input function measured over an arteriovenous shunt. *Frontiers in Neurology, 9*, 516.

Weeks, R. A., Piccini, P., Harding, A. E., & Brooks, D. J. (1996). Striatal D1 and D2 dopamine receptor loss in asymptomatic mutation carriers of Huntington's disease. *Annals of Neurology, 40*, 49–54.

Weeks, R. A., Cunningham, V. J., Piccini, P., Waters, S., Harding, A. E., & Brooks, D. J. (1997). 11C-diprenorphine binding in Huntington's disease: A comparison of region of interest analysis with statistical parametric mapping. *Journal of Cerebral Blood Flow and Metabolism, 17*, 943–949.

Wei, X., Ma, T., Cheng, Y., Huang, C. C. Y., Wang, X., Lu, J., et al. (2018). Dopamine D1 or D2 receptor-expressing neurons in the central nervous system. *Addiction Biology, 23*, 569–584.

Whittaker, D. S., Wang, H. B., Loh, D. H., Cachope, R., & Colwell, C. S. (2017). Possible use of a H3R antagonist for the management of nonmotor symptoms in the Q175 mouse model of Huntington's disease. *Pharmacology Research & Perspectives, 5*.

Wilson, H., Niccolini, F., Haider, S., Marques, T. R., Pagano, G., Coello, C., et al. (2016). Loss of extra-striatal phosphodiesterase 10A expression in early premanifest Huntington's disease gene carriers. *Journal of the Neurological Sciences, 368*, 243–248.

Wilson, H., De Micco, R., Niccolini, F., & Politis, M. (2017). Molecular imaging markers to track Huntington's disease pathology. *Frontiers in Neurology, 8*, 11.

Young, A. B., Penney, J. B., Starosta-Rubinstein, S., Markel, D. S., Berent, S., Giordani, B., et al. (1986). PET scan investigations of Huntington's disease: Cerebral metabolic correlates of neurological features and functional decline. *Annals of Neurology, 20*, 296–303.

Zhang, L., Hu, K., Shao, T., Hou, L., Zhang, S., Ye, W., et al. (2021). Recent developments on PET radiotracers for TSPO and their applications in neuroimaging. *Acta Pharmaceutica Sinica B, 11*, 373–393.

Zheng, C., Holden, D., Zheng, M. Q., Pracitto, R., Wilcox, K. C., Lindemann, M., et al. (2022). A metabolically stable PET tracer for imaging synaptic vesicle protein 2A: Synthesis and preclinical characterization of [(18)F]SDM-16. *European Journal of Nuclear Medicine and Molecular Imaging, 49*, 1482–1496.

Functional and Physiological MRI Measures as Early Biomarkers for Huntington's Disease

Wenzhen Duan

Abbreviations

ASL	Arterial-spin-labeling
BBB	Blood brain barrier
BOLD	Blood-oxygen-level-dependent
CBV	Cerebral Blood Volume
CBVa	Arterial CBV
CBVv	Venous CBV
CMRglc	Glucose metabolism
CMRO2	Cerebral metabolic rate of oxygen
DCE	Dynamic contrast-enhanced
DMN	Default mode network
DSC	Dynamic susceptibility contrast
DTI	Diffusion Tensor Imaging
EPI	Echo planar imaging
fMRI	functional MRI
gluCEST	Chemical Exchange Saturation Transfer of glutamate
GWAS	Genome-wide association studies
HD	Huntington's disease
HTT	Huntingtin
iVASO	Inflow-based vascular-space-occupancy
MRI	Magnetic resonance imaging
MRS	Magnetic Resonance Spectroscopy
MTI	Magnetization Transfer Imaging
NVU	Neurovascular unit

W. Duan (✉)
Division of Neurobiology, Department of Psychiatry and Behavioral Sciences, Department of Neuroscience, Johns Hopkins University School of Medicine, Baltimore, MD, USA
e-mail: wduan2@jhmi.edu

E. A. Thomas, G. M. Parkin (eds.), *Biomarkers for Huntington's Disease*,
Contemporary Clinical Neuroscience, https://doi.org/10.1007/978-3-031-32815-2_7

1 Introduction

The onset of Huntington's disease (HD) typically occurs in the third or fourth decade of life, with a large range in disease onset, for 60–70% of disease onset could be explained by the length of the CAG repeat expansion (Langbehn et al., 2004). Genome-wide association studies (GWAS) have shown that other genetic variants also influence age at onset of HD as well as disease progression, including variants in genes involved in DNA damage repair pathways and sequence variants in the CAG tract (Lee et al., 2015; Moss et al., 2017; Genetic Modifiers of Huntington's Disease Consortium. Electronic address & Genetic Modifiers of Huntington's Disease, 2019). Thus, precision medicine or personalized trial design is crucial for developing effective treatment due to genetic variability and disease heterogeneity. Noninvasive measures of brain changes using functional and physiological MRI in the premanifest phase of HD can provide personalized risk assessments, prognoses, and treatment plans, guide therapeutic interventions, enable personalized monitoring, and assist with family planning decisions, ultimately improving the care and outcomes for individuals with HD (Fig. 1).

Since genetic testing became feasible in 1993, it is possible to identify individuals with risk of HD before the disease onset. The symptom free period of gene carriers prior to inevitable disease onset is referred to as the premanifest phase of the

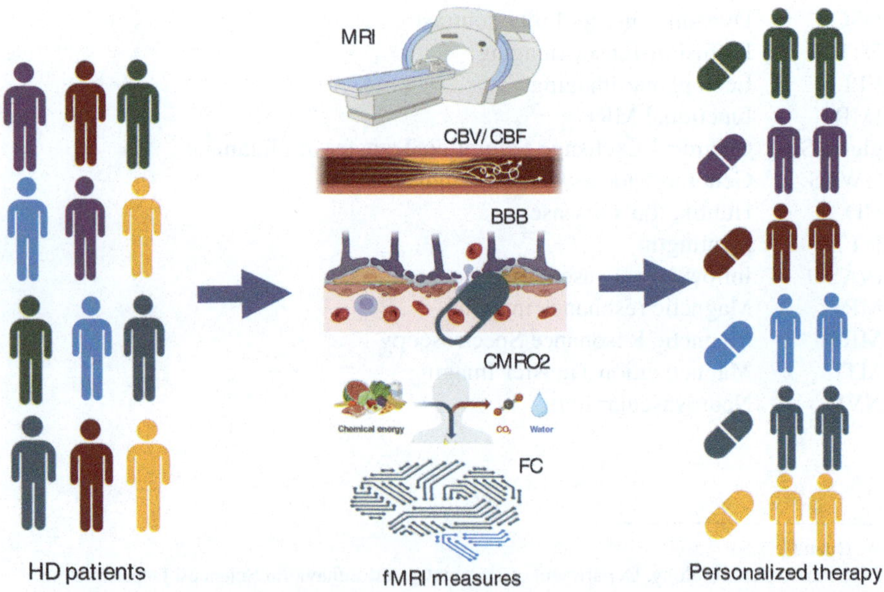

Fig. 1 Proposed functional and physiological MRI measures with prognostic value for precision medicine. Huntington's disease patients can be divided into subgroups by indicated MRI measures. The future clinical trials and treatment would be better to be designed based on the MRI biomarkers. (Figure created using BioRender.com)

disease and historically has been defined by the absence of motor symptoms. However, it has been recognized for some time now that the disease can also begin with cognitive or other behavioral problems (Lemiere et al., 2004; van Duijn et al., 2007).

The pathologic hallmark of HD brain is selective striatal atrophy, with further histological findings of mutant huntingtin (HTT) nuclear inclusions in neurons (Vonsattel et al., 1985). The disadvantage of the histological measures is that it is only available from end-stage HD patients. Non-invasive *in vivo* neuroimaging techniques are critical for characterization of the altered brain activity and function in the premanifest and early manifest HD stages. Initially, MRI studies were conducted in the manifest disease stages. After the widespread introduction of genetic testing, the neuroimaging studies have also expanded to the premanifest HD subjects.

Magnetic resonance imaging (MRI) is a non-invasive tool for examining brain structure, function, and physiology in living human subjects. With advanced MRI technology development, MRI studies in HD has progressed from detecting structural and contrast differences to using more sophisticated techniques such as Diffusion Tensor Imaging (DTI), Magnetization Transfer Imaging (MTI), Magnetic Resonance Spectroscopy (MRS), functional MRI (fMRI), and physiological MRI, to assess a variety of functional and pathophysiological changes in the brain of human HD subjects. Disrupted neuronal function and impaired synaptic transmission inevitably disrupt large scale networks within the brain regions and consequently communication between brain regions starts to break down. In the earlier stages of premanifest disease, alternative brain networks may be accessed to compensate for this initial damage so that brain function can be maintained to a certain extent. Eventually these compensatory networks fail and disease symptoms start to manifest. Functional MRI (fMRI) provides a powerful tool to study these networks within the brain and help to elucidate the impact of toxic mutant huntingtin (mHTT) on brain function.

Functional MRI (fMRI) is an imaging technique used primarily to investigate brain activity during a period of rest (resting state activity) or while performing a specific task (task-based fMRI). fMRI utilizes the neurovascular coupling of neural activity and blood flow (Blood Oxygen Level-Dependent; BOLD) to obtain an indirect measure of neuronal activity in response to performing a task or a function. In studies of both premanifest and early manifest HD, aberrant functional connectivity or task-related regional brain activation patterns were reported in HD subjects.

Small increases in brain activity cause increased metabolic activity, which leads to regional increases in blood flow (hemodynamic response). This causes a local shift in the ratio of oxygenated to deoxygenated blood, enabling the measurement of the BOLD fMRI signal. As this signal is an indirect measure of neuronal activity, a derived hemodynamic response function is convolved with the predicted neuronal response and temporally matched to different aspects of a task or network activity. fMRI uses rapid gradient echo planar imaging (EPI) as this facilitates the acquisition of the large number of images necessary to detect the subtle changes in the BOLD signal. Once acquired, the fMRI images are processed using standard imaging software and a pipeline that includes realignment of all EPI images across a time series; co-registration of the EPI images to a high resolution structural image to improve anatomical specificity; normalization into standard space (for group level comparisons) and smoothing to improve the ratio of signal to noise. The resultant data can then be used for most types of analyses for both task-based and resting state fMRI data (in some cases, not all these steps are required).

By comparing fMRI measures of individuals with HD to healthy controls, we can identify patterns of brain activity that are associated with the HD pathological process. These patterns can be used as biomarkers to study the disease at a premanifest early stage and track disease progression, and hope to assess the treatment response in clinical trials. Besides fMRI measures, physiological MRI, including cerebral blood volume/flow (CBV/CBF), brain oxygen metabolism, and blood brain barrier (BBB) integrity, have also been applied to human brain to distinguish HD from normal subjects.

Individuals who have a family history of HD can undergo genetic testing to determine if they carry the mutant huntingtin gene. fMRI could be used to identify early changes in brain function that occur in these individuals before the onset of clinical symptoms, providing opportunities for earlier intervention and potential disease modifying therapy. fMRI measures could also be used to track changes in brain function over time in individuals with early-stage HD, allowing for the monitoring of disease progression and the evaluation of treatment efficacy. This could potentially allow for more targeted and personalized treatment approaches for individuals with HD.

In this chapter, we will overview the current research progress on functional and physiological MRI measures as potential early biomarkers for HD in both premanifest and early manifest individuals, as well as preclinical HD models. We will also discuss the challenges and limitations of using these MRI measures as biomarkers and future research direction that is needed to develop reliable and sensitive MRI measures for clinical use in HD.

2 Resting State fMRI Studies in HD

Several studies have demonstrated altered functional connectivity patterns in individuals with HD compared to healthy controls. These changes typically involve a reduction in connectivity within and between certain brain regions, particularly those within the basal ganglia and the default mode network (DMN). Studies have also shown that functional connectivity changes in HD are associated with clinical features of the disease, such as motor and cognitive impairments. Additionally, functional connectivity measures have been used to predict disease progression in individuals with HD.

Resting state fMRI studies have investigated changes in functional connectivity between brain regions in HD patients. Several studies have shown that HD patients have altered functional connectivity within and between networks, particularly in the DMN (Wolf et al., 2012; Quarantelli et al., 2013; Werner et al., 2014; McColgan et al., 2017). The DMN is involved in self-referential processing and social cognition and is a network of brain regions that are active during rest and deactivated during task performance. It includes the posterior cingulate cortex, medial prefrontal cortex, and lateral parietal cortex, among other regions. Studies have shown that the DMN is disrupted in individuals with HD, with decreased functional connectivity between the network's core regions. HD individuals also show reduced DMN connectivity, suggesting that this may be an early biomarker of HD (Wolf et al., 2012; Quarantelli et al., 2013; Werner et al., 2014). These changes in functional connectivity have been associated with cognitive and behavioral impairments in HD (Wolf et al., 2012; Quarantelli et al., 2013). Thus, resting-state fMRI measures may have utility as early biomarkers for HD, particularly for monitoring changes in cognitive and behavioral function.

Another study of 18 month longitudinal changes in both functional activation and functional connectivity during working memory in premanifest HD and manifest HD suggest that longitudinal reductions in connectivity over 18 months may represent an early signature of cortico-cortical and cortico-striatal functional disconnectivity in premanifest HD, whereas the concomitant increased cortical and subcortical activation may reflect a compensatory response to the demands for cognitive resources required during task performance (Georgiou-Karistianis et al., 2013). These findings demonstrate that functional imaging modalities have the potential to serve as sensitive methods for the assessment of cortical and subcortical responses to future treatment measures.

A study led by Poudel characterized synchrony in resting state cerebral networks in both premanifest and manifest HD populations and to determine its association with disease burden and neurocognitive functions (Poudel et al., 2014). They found that compared with the control group, the premanifest group showed decreased synchrony in the sensorimotor and dorsal attention networks; decreased level of synchrony in the sensorimotor network was associated with poorer motor performance. Compared with the control group, the manifest HD group showed widespread reduction in synchrony in the dorsal attention network, which was associated with poorer cognitive performance. The posterior putamen and superior parietal cortex were functionally disconnected from the frontal executive network in the manifest compared with control and premanifest groups. Furthermore, the left frontoparietal network showed areas of increased synchrony in the manifest compared with the premanifest HD group. These findings suggest that aberrant synchrony in the sensorimotor and dorsal attention networks may serve as an early signature of neural change in premanifest HD individuals. The altered synchrony in dorsal attention, frontoparietal and cortico-striatal networks may predict the development of specific clinical symptoms in HD.

In addition to its potential as a biomarker for early detection of HD, functional connectivity may also be used to track disease progression and evaluate the effectiveness of treatments. HD patients treated with tetrabenazine, a drug that reduces the amount of dopamine released in the brain, had increased activation in the caudate and putamen during a motor task compared to untreated patients (Paleacu,

2007). In another study, researchers used fMRI to evaluate the effects of the drug latrepirdine on brain function in HD patients. They found that latrepirdine treatment was associated with improvements in neural activity in certain brain regions involved in motor and cognitive functions (Kieburtz et al., 2010). Another study used fMRI to evaluate the effects of cognitive training on brain function in HD patients, and the researchers found that cognitive training was associated with changes in brain activity in regions involved in executive function and working memory (Gargouri et al., 2016). Overall, functional connectivity analysis has shown promise as a biomarker for HD. However, more research is needed to determine its sensitivity and specificity as a diagnostic tool and to identify the most reliable and clinically HD relevant connectivity measures.

On the other hand, longitudinal resting state fMRI analysis in premanifest HD through a 3-year follow-up study revealed no differences in the degree of connectivity change between the groups over a period of 3 years, though a significantly higher rate of striatal atrophy was found in the premanifest HD group compared to controls in the same period (Odish et al., 2015). This study concluded that resting state fMRI lacks sensitivity in detecting changes in functional connectivity in HD gene carriers prior to disease manifestation over a 3-year follow-up period.

3 Task-Based fMRI Studies in HD

In addition to the resting state fMRI, task or stimuli-based connectivity has also been examined in premanifest and early manifest HD subjects. Activation may be elevated or reduced and this varies not only according to disease stage but also the task being presented and the regions of the brain being examined. Consequently, there may appear to be inconsistencies between studies and interpretation is often complex. For example, studies have found that the salience network, which is involved in detecting and responding to salient stimuli, is disrupted in premanifest HD individuals (Harrington et al., 2015). Furthermore, studies of motor networks have shown that premanifest HD individuals exhibit increased activation in the striatum during motor tasks, potentially reflecting compensatory changes in response to early motor dysfunction (Kloppel et al., 2009). Another study used fMRI to measure changes in brain activity in response to a motor task in premanifest HD individuals and found that they had decreased activity in the striatum compared to healthy controls (Braet et al., 2011). Similarly, a study found that premanifest HD individuals had altered brain connectivity in the basal ganglia, another brain region that is affected in HD, compared to healthy controls (Wolf et al., 2014). Moreover, HD patients showed decreased activation in the caudate and putamen, two key components of the striatum, during a visual-spatial working memory task (Rosas et al., 2002).

While much of the research on fMRI as a biomarker for HD has focused on premanifest individuals, there has also been a significant amount of work examining changes in brain function in early manifest individuals. One study investigated changes in brain activity during a motor task in early symptomatic HD patients compared to healthy controls using fMRI (Dumas et al., 2012). The study found that early symptomatic HD patients had reduced activity in the motor cortex and putamen during the motor task compared to healthy controls. Moreover, the changes in brain activity were correlated with motor impairment, indicating that fMRI may be useful for assessing disease severity in early symptomatic HD patients.

Another study by Wolf and colleagues (2014) investigated changes in brain activity during a working memory task in early symptomatic HD patients compared to healthy controls using fMRI. The study found that early symptomatic HD patients had reduced activity in the prefrontal cortex and increased activity in the parietal cortex during the working memory task compared to healthy controls. Moreover, the changes in brain activity were correlated with cognitive impairment, indicating that fMRI may be useful for assessing cognitive dysfunction in early symptomatic HD patients.

A more recent study investigated changes in brain activity during a social cognition task in early symptomatic HD patients compared to healthy controls using fMRI (Aracil-Bolanos et al., 2022). The study found that early symptomatic HD patients had reduced activity in the medial prefrontal cortex, temporal gyrus, and fusiform gyrus during the social cognition task compared to healthy controls. Moreover, the changes in brain activity were correlated with social cognitive impairment, indicating that fMRI may be useful for assessing social cognitive dysfunction in early symptomatic HD patients.

A study used fMRI to investigate changes in brain function in response to a reward processing task in early manifest HD patients (Lahr et al., 2018), and found that HD patients showed decreased activation in the ventral striatum, a region of the brain that is involved in reward processing, compared to healthy controls. Additionally, the study found that HD patients showed increased activation in the prefrontal cortex and the insula, regions of the brain that are involved in emotion regulation, compared to healthy controls.

Overall, these studies suggest that fMRI may be a useful tool for detecting changes in brain function in early symptomatic HD patients, particularly in specific brain regions involved in motor, cognitive, and social processing. However, further research is needed to validate these findings and to explore the potential of fMRI as a biomarker for monitoring disease progression in early symptomatic HD patients. Additionally, future studies may benefit from larger sample sizes, longitudinal designs, and the inclusion of pre-symptomatic HD patients to further elucidate the early changes in brain function associated with HD.

4 Cerebral Blood Volume (CBV)

Cerebral blood volume (CBV) is a physiological measure reflecting the amount of blood in brain tissue (ml blood/100 ml tissue). Under normal physiological conditions, the brain attempts to maintain adequate cerebral blood flow (CBF, in units of ml blood/100 g tissue/minute) by regulating the caliber of arteries and arterioles through an auto-regulatory mechanism (Kuschinsky, 1996). Thus, in early stages of cerebrovascular alterations, CBV can serve as a sensitive indicator for evaluating tissue viability and function. Baseline CBV also demonstrates strong correlations with basal metabolism in the brain (Gonzalez et al., 1995). In addition, CBV is a major modulator of the blood-oxygen-level-dependent (BOLD) effect (Buxton & Frank, 1997; van Zijl et al., 1998) and thus has relevance for the interpretation of the BOLD fMRI signals.

fMRI studies have shown that HD patients exhibit altered total CBV in several brain regions compared to healthy controls. These changes in CBV may reflect early alterations in neural activity that contribute to the development of HD symptoms. The striatum is a key brain region affected in HD, it is composed of several subregions, including the caudate nucleus and the putamen, and involved in motor control, reward processing, and cognitive flexibility. fMRI studies on CBV in HD patients have consistently shown that HD patients exhibit decreased CBV in the striatum compared to healthy controls (Harris et al., 1999; Wolf et al., 2011; Lin et al., 2013). Using dynamic susceptibility contrast (DSC) fMRI, a study reported that HD patients exhibited decreased CBV in the caudate nucleus and putamen compared to healthy controls, and that these changes were associated with motor impairment (Lewandowski et al., 2013). These findings suggest that decreased CBV in the striatum may be a sensitive marker of altered brain metabolism in HD.

Various neuroimaging techniques have been developed to measure CBV. A common requirement of these approaches is the utilization of an endogenous or exogenous tracer to separate signal from the intravascular and extravascular compartments so that CBV (intravascular space) can be quantified. Most CBV imaging approaches measure total CBV, which reflects the sum of signals from the arterial, capillary and venous vessels in the microvasculature. More specific MR techniques have been recently developed to measure CBV in the different compartments of cerebrovascular system, such as arterial CBV (CBVa) or venous CBV (CBVv). In healthy brain parenchyma, CBVa is approximately 20–30% of total CBV at baseline (van Zijl et al., 1998; Piechnik et al., 2008). The relative change in CBVa (in percentage compared to baseline values) during functional stimulation is usually much greater than those in capillaries and venous vessels, which is an important factor to consider when interpreting the BOLD signals.

Arteriolar CBV (CBVa) measures the amount of blood contained within the arterioles, which are the blood vessels supplying oxygen and nutrients to the brain parenchyma. Several studies have investigated the changes in CBVs in HD patients. Drouin-Ouellet et al. (2015) and Hua et al. (2014) both reported significant increase in CBVa in premanifest HD patients compared to healthy controls using two different MRI approaches (Brookes et al., 2007; Hua et al., 2011b). Significantly elevated CBVa values in the frontal cortex strongly correlated with the projected year-to-onset of motor symptoms estimated from the length of the CAG repeat mutation in premanifest HD (Hua et al., 2014) (Fig. 2). The relative magnitude of CBVa changes was greater than that of total CBF and brain volumetric measures in premanifest HD (Hua et al., 2014). Such alterations in arterial vessels have been linked with the

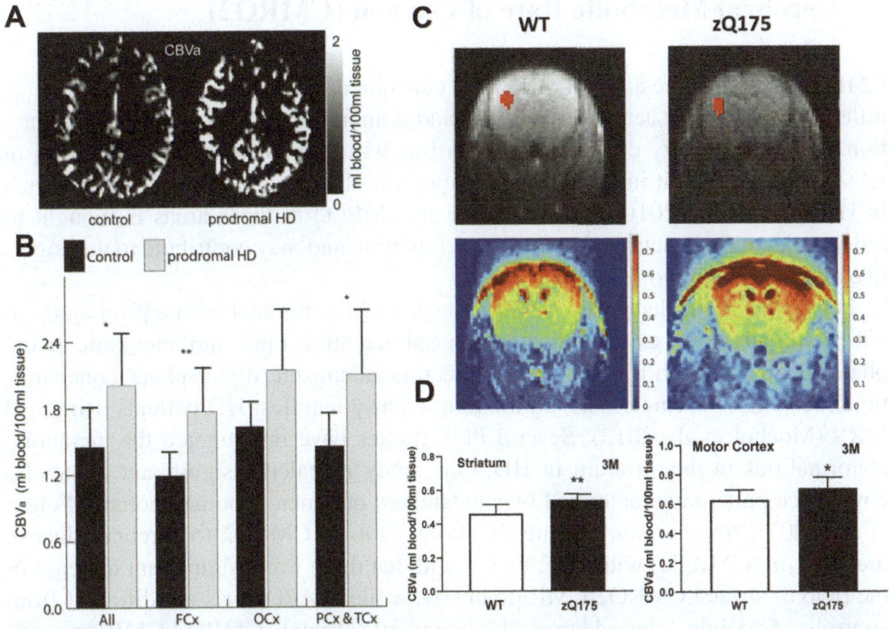

Fig. 2 Elevated arteriolar cerebral blood volume (CBVa) in premanifest Huntington's disease (HD) patients and a mouse model. (**a**, **b**) Shown are typical arteriolar cerebral blood volume (CBVa) maps (**a**) and average gray matter CBVa (**b**) in the whole slice (All), the frontal cortex (FCx), the occipital cortex (OCx), and the parietal and temporal cortices (PCx&TCx). Error bars indicate standard deviations. *p* values were calculated with the Wilcoxon rank-sum test and were adjusted for multiple comparisons. $^*p < 0.05$; $^{**}p < 0.01$ vs Control. (Figure a, b is adopted from Hua et al., 2014). (**c**) Representative CBVa maps calculated from inflow-based vascular-space-occupancy (iVASO) images in zQ175 knock-in HD mouse brains and controls at 3 months of age. The top row shows the raw images, and the red ROIs indicate the quantified brain region-striatum. The bottom row shows the CBVa maps for zQ175 mice and wild type (WT) littermates. (**d**) Quantification of CBVa values in WT or zQ175 HD mice at 3 months of age (3 M) in the striatum and motor cortex. $^*p < 0.05$, $^{**}p < 0.01$ vs WT group by Student's t-tests. n = 8. (Figure c, d is adopted from Liu et al., 2021)

intercellular transportation of mutant huntingtin protein in HD (Drouin-Ouellet et al., 2015). These findings suggest that CBVa changes may be a sensitive marker of early neuropathological signs in HD. Moreover, the study highlights the importance of measuring CBVa changes in multiple brain regions, as CBVa changes were observed not only in the striatum but also in other regions involved in motor and cognitive processes. However, the study has some limitations that need to be considered. First, the sample size was relatively small, which may limit the generalizability of the findings. Second, the study did not include a longitudinal follow-up to investigate the temporal dynamics of CBVa changes in pre-symptomatic HD gene carriers. Future studies with larger sample sizes and longitudinal designs are needed to confirm and extend the findings of this study.

5 Cerebral Metabolic Rate of Oxygen (CMRO2)

CMRO2 measures the amount of oxygen consumed by the brain per unit of time. It reflects the metabolic activity of neurons and is an important indicator of brain function. Deficit in energy consumption is in line with the metabolic abnormalities in HD. There is a deficit in oxygen consumption in the caudate nucleus and putamen in HD (Lou et al., 2016). The reduction in CMRO2 in these areas is thought to reflect a decrease in neuronal activity and function, and may contribute to the cognitive and motor symptoms of HD.

For instance, the decreased CMRO2 response is congruent with a prior study in which impaired inorganic phosphate/phosphocreatine ratio and inorganic phosphate/adenosine triphosphate ratio (reflecting adenosine diphosphate concentrations) responses during visual stimulation in early manifest HD patients using ^{31} P MRS (Mochel et al., 2012). Several PET studies have investigated the metabolic abnormalities in the striatum in HD. One study revealed a significant metabolic covariance pattern characterized by caudate and putamen hypometabolism (Feigin et al., 2001); Another study using the molar ratio of CMRO2 to cerebral glucose metabolism (CMRglc) with PET, they found that there was a significant increase in the ratio of striatal CMRO2/CMRglc in HD, while CMRO2 was not different from controls, CMRglc was decreased. Increased striatal CMRO2/CMRglc with unchanged CMRO2 is inconsistent with a defect in mitochondrial oxidative phosphorylation due to reduced activity of the mitochondrial electron transport system. Because glycolytic metabolism is predominantly astrocytic (Powers et al., 2007), the selective reduction in striatal CMRglc raises the possibility that astrocyte dysfunction may be involved in the pathogenesis of HD (Khakh & Sofroniew, 2014; Khakh & Goldman, 2023). While metabolic alterations are a consistent finding, the nature and origin of these alterations are more controversial. It has long been thought that mitochondrial abnormalities are central, but more recent studies have also suggested the importance of glycolysis, fatty acid metabolism, or other aspects of cellular metabolism (Lou et al., 2016; Dubinsky, 2017; Van Houten, 2019; Consortium, 2020).

In addition, impaired response of cerebral oxygen metabolism to stimulation was reported in premanifest and early manifest HD using a recently developed 3-D-TRiple-acquisition-after-Inversion-Preparation magnetic resonance imaging (MRI) approach (Klinkmueller et al., 2021). These experiments were conducted in 23 HD patients and 16 healthy controls. Decreased occipital cortex CMRO2 responses were observed in premanifest and early manifest HD patients compared to controls, correlating with the CAG-Age Product scores in these patients. The results suggest the potential value of this reduced CMRO2 response during visual stimulation as a biomarker for HD and may illuminate the role of metabolic alterations in the pathophysiology of HD.

Overall, the measurement of CMRO2 can provide important insights into the pathophysiology of HD and may be a useful biomarker for assessing disease severity, predicting the onset of clinical symptoms, and assessing functional impairment in premanifest stage of HD. However, further research and is needed to validate these findings and to explore the potential of CMRO2 as a therapeutic target in HD.

6 Blood Brain Barrier (BBB)

The BBB is centrally positioned within the neurovascular unit (NVU) and is formed by a monolayer of tightly sealed endothelial cells along the vascular tree expressing low paracellular and transcellular permeability (Zlokovic, 2011; Zhao et al., 2015). In the human brain, at the capillary level, BBB integrity is maintained by tight junctions between endothelia cells and pericytes (Armulik et al., 2010; Bell et al., 2010; Daneman et al., 2010). BBB regulates transport of molecules into and out of the central nervous system (CNS), which maintains tightly controlled chemical composition of the neuronal milieu that is required for proper neuronal functioning (Zhao et al., 2015). BBB dysfunction could lead to cellular infiltration and aberrant transport and clearance of CNS molecules (Zlokovic, 2011), which contributes to neurological deficits.

BBB dysfunction is evident in HD (Lin et al., 2013; Drouin-Ouellet et al., 2015; Hsiao et al., 2015). Specifically, HD patients exhibit increased BBB permeability in the caudate nucleus as shown by dynamic contrast-enhanced (DCE)-MRI analysis that associates with disease burden and increased gray matter CBV (Drouin-Ouellet et al., 2015). Histological analysis of BBB breakdown shows perivascular capillary fibrinogen deposition, diminished endothelial tight junction proteins claudin-5 and occludin (Drouin-Ouellet et al., 2015), and reduced pericyte coverage (Hsiao et al., 2015) in HD brains. The ratio of capillaries to arterioles is increased in the HD putamen (Drouin-Ouellet et al., 2015), cortex, and substantia nigra (Lin et al., 2013; Drouin-Ouellet et al., 2015), suggestive of microvascular angiogenesis. Furthermore, mutant huntingtin aggregates accumulate in brain endothelial cells, perivascular macrophages, smooth muscle cells, and vascular basal lamina in HD (Drouin-Ouellet et al., 2015), and in genetically unrelated neural allografts within the brains of patients with advanced HD (Cicchetti et al., 2014), suggesting that cerebral vasculature may contribute spreading of mutant huntingtin. Similarly, HD transgenic mice with about 150 CAG repeats (R6/2 mice) exhibit increased BBB permeability (Drouin-Ouellet et al., 2015), increased microvascular density (Hsiao et al., 2015), and loss of tight junctions that precedes disease onset (Di Pardo et al., 2017).

A non-contrast arterial-spin-labeling (ASL) based MRI technique to assess BBB permeability to water was recently developed in mice (Wei et al., 2023). Using this new MRI technique, a significantly increased BBB permeability to water was evident in zQ175 KI HD mice before exhibiting significant motor deficits (Wei et al., 2023).

These studies suggest that BBB dysfunction may be an important component of the pathophysiology of HD in both human patients and mouse models, it may be a potential target for therapeutic interventions. However, the precise mechanisms underlying BBB dysfunction in HD, as well as its relationship to the clinical symptoms of the disease need to be further characterized.

7 fMRI and Physiological MRI Studies in Preclinical Models of HD

Mouse models have helped elucidate disease pathogenesis, experimental therapeutics, as well as biomarker development in HD (Menalled, 2005; Menalled & Brunner, 2014). While prior focus has been on neurons in HD, it is becoming increasingly apparent that cerebral vascular system may contribute to or at least be reflective of pathogenesis (Drouin-Ouellet et al., 2015; Gray, 2019; Rieux et al., 2021). The validation of potential biomarkers in animal models has aided in the preclinical development of new therapeutic modalities. However, neuroimaging, especially functional and physiological brain imaging methods, has only recently begun to be applied to mouse models due to technical challenges.

Recently, a study utilized inflow-based vascular-space-occupancy (iVASO) MRI which was initially developed for the human brain (Hua et al., 2011a), and reported alterations in CBVa in premanifest HD human brain (Hua et al., 2014). In this study, CBVa was measured in conjunction with behavioral assessments and structural MRI to evaluate the effectiveness of HTT silencing introduced at the premanifest stage longitudinally (Liu et al., 2021). The study demonstrates that lowering HTT in striatal neurons restores CBVa in premanifest zQ175 HD mice (Fig. 2), delays onset, and slows the progression of motor phenotype and striatal atrophy. These data suggest that CBVa measure may be a promising noninvasive biomarker for premanifest HD clinical trials, and that introducing treatment at the premanifest stage has long-lasting benefits in HD. This study has shown for the first time that a sensitive physiological MRI measure responds to an HTT-lowering treatment in the premanifest period and demonstrates the feasibility of detecting arteriolar perfusion changes in the premanifest HD brain with a robust MRI technique that is suitable for longitudinal evaluations of therapeutic efficacy.

The fact that CBVa changes in the striatum emerge over time supports that dysfunction in the basal ganglion circuitry is associated with disease phenotype, though underlying mechanisms that link CBVa and neuronal dysfunction (or compensation for dysfunctional neuronal metabolism) in premanifest HD remain to be explored. These changes indicate that there is a premanifest therapeutic window to test interventions in this HD knock-in mouse model, which is not present in faster progressing mouse models. While no animal model replicates all of the features of HD, the heterozygous zQ175 model offers an alternative system to study functional changes in the premanifest period.

The use of iVASO MRI for CBVa assessment may provide an improved means for the bidirectional translation of findings between animal models and human

HD. Reverse translation of biomarkers to rodents also provides an important tool for evaluating the therapeutic effects of candidate treatment (Parievsky et al., 2012) which is particularly relevant as clinical trials on lowering HTT using different strategies (Tabrizi et al., 2019).

In another study, we revealed perturbations in resting-state brain functional connectivity in the N171-82Q HD mouse model (Li et al., 2017). Notably, we observed a robust association between altered striatal connectivity and striatal atrophy as well as motor performance. In this study we investigated resting-state fMRI signals among functionally connected brain regions and used two complementary analysis approaches, independent components analysis showed that stable network interconnectivity in mouse brain, and seed-based analysis revealed patterns of weakened striatal connections and strengthened striatal to the retrosplenial cortex connections in the HD mice. These disturbances in functional connectivity were strongly associated with selective brain atrophy and motor performance, demonstrating that regional brain connectivity measures may provide a unique window into the structural and functional consequence induced by mutant huntingtin in the brain. These results may bring additional insights into the connections between human and mouse models of HD, e.g., how regionally specific neuronal dysfunction relates to specific phenotype and selective neuropathology. A limitation of this study was the cross-sectional design; further longitudinal studies are needed to reveal dynamic changes of functional connectivity, and determine how the connectivity changes contribute to neuropathology as well as behavioral deficits in HD. The longitudinal studies will also be able to assess modifications of the pattern of the network alterations throughout the disease course, understand its role in the disease progression, and further clarify the mechanisms underlying these network alterations. Such studies will provide more insights into the exact relation between functional connectivity and brain pathology as well as clinical symptoms.

Another recent study conducted using newly developed multimodal MRI techniques by combining anatomical, diffusion weighted, Chemical Exchange Saturation Transfer of glutamate (gluCEST) and magnetization transfer (MT) imaging to monitor changes in the knock-in 140CAG mouse model. The results pointed toward early changes affecting the function and tissue organization of different extra-striatal brain regions, providing new insights on the progression of brain perturbations in this slowly progressive model of HD that could be reminiscent of those occurring in HD human brain (Perot et al., 2022).

8 Challenges and Future Directions

fMRI has shown promise to develop potential biomarkers for HD in both preclinical and clinical applications. However, there are following challenges associated with using fMRI measures as biomarkers.

1). ***Technical challenges***: fMRI data are prone to various technical challenges, including motion artifacts, physiological noise, and scanner instability. These technical challenges can lead to unideal data quality, which can compromise the reliability and validity of fMRI biomarkers.
2). ***Data variability***: There is considerable inter-subject variability in fMRI data, which can be due to several factors such as age, sex, medication use, and comorbidities. This variability can affect the sensitivity and specificity of fMRI biomarkers and limit their generalizability to the larger population.
3). ***Interpretation of fMRI data***: fMRI data are indirect measures of neural function and do not provide information about underlying cellular and molecular mechanisms. Therefore, the interpretation of fMRI data can be challenging and requires careful consideration of the physiological and neurobiological mechanisms that underlie the observed changes in brain activity.
4). ***Standardization of fMRI protocols***: Standardization of fMRI protocols across different research sites is essential for the reliability and reproducibility of fMRI biomarkers. However, there is currently no consensus on the optimal fMRI protocols for HD or other neurological disorders, which can lead to inconsistencies in data acquisition and analysis.
5). ***Longitudinal studies***: Longitudinal studies are essential for assessing the sensitivity and specificity of fMRI biomarkers for detecting changes in disease progression over time. However, conducting longitudinal studies is challenging due to the high cost, patient burden, and technical issues associated with repeated imaging.
6). ***Data analysis***: fMRI data analysis is complex and requires expertise in advanced statistical and computational methods. Data analysis methods are constantly evolving, and it is important to use the most up-to-date methods to ensure the accuracy and reliability of fMRI biomarkers.
7). ***Lack of consensus:*** There are a variety of fMRI measures that can be used to assess brain function, more work needs to be done to examine which measures are most sensitive to HD-related neuropathological changes and which are most appropriate and feasible for use as a biomarker.
8). ***Regulatory challenges***: There are several regulatory challenges associated with using fMRI as a biomarker, including data privacy and confidentiality, informed consent, and ethical considerations. These challenges can affect the recruitment and retention of study participants and limit the availability of data for research.

Addressing these challenges will require collaboration across disciplines and standardization of data acquisition and analysis protocols. With continued research and development, the fMRI and physiological MRI measures will continue to help elucidate the pathophysiology of HD, and may ultimately aid in the development and assessment of new therapies. But there is a need for larger-scale studies to validate the use of each fMRI measure as a biomarker in HD.

To date, several functional and structural imaging approaches have shown promise, but none have been sufficiently confirmed in longitudinal studies. Network analysis is very promising in this regard, though such biomarkers will need to include combination imaging techniques. While early symptomatic fMRI studies have shown promising results, larger studies are needed to confirm these findings and to establish the sensitivity and specificity of fMRI as a biomarker for HD. Currently, the clinical utility of fMRI is still to be established and currently it remains predominantly a research tool. Although sensitive to subtle disease effects, its power to discriminate between groups is relatively weak. Advances in acquisition and further standardization across multiple sites as well as results from ongoing clinical trials will help us to evaluate its future clinical application.

9 Concluding Remarks

In conclusion, fMRI measures show promise as potential biomarkers for detecting early changes in brain functional alterations in HD. Task-related fMRI measures may have utility for monitoring motor and cognitive changes in the disease, while resting-state fMRI measures may have utility for monitoring changes in cognitive and behavioral function. CBVa and CMRO2 measures may have utility for monitoring changes in regional brain metabolism. However, larger scale research in both preclinical models and clinical subjects is needed to fully realize the potential of these fMRI measures as early biomarkers for HD.

Premanifest fMRI studies have provided valuable insights into the neural changes associated with HD in the early stages of the disease before irreversible clinical symptoms develop. fMRI allows for the identification of individuals who are at risk for developing the disease and for the monitoring of disease progression and treatment efficacy. However, challenges mentioned above must be addressed before fMRI can be widely used in clinical practice, including variability in fMRI data, a lack of consensus on which fMRI measures are most sensitive to HD-related changes, and a need for larger-scale studies to validate the use of fMRI as a biomarker in HD.

Acknowledgements This work is supported by National Institute of Health R01NS082338, R01NS124084, R01NS127344, CHDI A16785, and The Bev Hartig Huntington's Disease Foundation (to W.D).

Conflict of Interest Statement The Author declares no conflict of interest.

References

Aracil-Bolanos, I., Martinez-Horta, S., Gonzalez-de-Echavarri, J. M., Sampedro, F., Perez-Perez, J., Horta-Barba, A., Campolongo, A., Izquierdo, C., Gomez-Anson, B., Pagonabarraga, J., & Kulisevsky, J. (2022). Structure and dynamics of large-scale cognitive networks in Huntington's disease. *Movement Disorders, 37*, 343–353.

Armulik, A., Genove, G., Mae, M., Nisancioglu, M. H., Wallgard, E., Niaudet, C., He, L., Norlin, J., Lindblom, P., Strittmatter, K., Johansson, B. R., & Betsholtz, C. (2010). Pericytes regulate the blood-brain barrier. *Nature, 468*, 557–561.

Bell, R. D., Winkler, E. A., Sagare, A. P., Singh, I., LaRue, B., Deane, R., & Zlokovic, B. V. (2010). Pericytes control key neurovascular functions and neuronal phenotype in the adult brain and during brain aging. *Neuron, 68*, 409–427.

Braet, W., Johnson, K. A., Tobin, C. T., Acheson, R., McDonnell, C., Hawi, Z., Barry, E., Mulligan, A., Gill, M., Bellgrove, M. A., Robertson, I. H., & Garavan, H. (2011). fMRI activation during response inhibition and error processing: The role of the DAT1 gene in typically developing adolescents and those diagnosed with ADHD. *Neuropsychologia, 49*, 1641–1650.

Brookes, M. J., Morris, P. G., Gowland, P. A., & Francis, S. T. (2007). Noninvasive measurement of arterial cerebral blood volume using Look-Locker EPI and arterial spin labeling. *Magnetic Resonance in Medicine, 58*, 41–54.

Buxton, R. B., & Frank, L. R. (1997). A model for the coupling between cerebral blood flow and oxygen metabolism during neural stimulation. *Journal of Cerebral Blood Flow and Metabolism, 17*, 64–72.

Cicchetti, F., Lacroix, S., Cisbani, G., Vallieres, N., Saint-Pierre, M., St-Amour, I., Tolouei, R., Skepper, J. N., Hauser, R. A., Mantovani, D., Barker, R. A., & Freeman, T. B. (2014). Mutant huntingtin is present in neuronal grafts in Huntington disease patients. *Annals of Neurology, 76*, 31–42.

Consortium H.D.i. (2020). Bioenergetic deficits in Huntington's disease iPSC-derived neural cells and rescue with glycolytic metabolites. *Human Molecular Genetics, 29*, 1757–1771.

Daneman, R., Zhou, L., Kebede, A. A., & Barres, B. A. (2010). Pericytes are required for blood-brain barrier integrity during embryogenesis. *Nature, 468*, 562–566.

Di Pardo, A., Carrizzo, A., Damato, A., Castaldo, S., Amico, E., Capocci, L., Ambrosio, M., Pompeo, F., De Sanctis, C., Spinelli, C. C., Puca, A. A., Remondelli, P., Maglione, V., & Vecchione, C. (2017). Motor phenotype is not associated with vascular dysfunction in symptomatic Huntington's disease transgenic R6/2 (160 CAG) mice. *Scientific Reports, 7*, 42797.

Drouin-Ouellet, J., Sawiak, S. J., Cisbani, G., Lagace, M., Kuan, W. L., Saint-Pierre, M., Dury, R. J., Alata, W., St-Amour, I., Mason, S. L., Calon, F., Lacroix, S., Gowland, P. A., Francis, S. T., Barker, R. A., & Cicchetti, F. (2015). Cerebrovascular and blood-brain barrier impairments in Huntington's disease: Potential implications for its pathophysiology. *Annals of Neurology, 78*, 160–177.

Dubinsky, J. M. (2017). Towards an understanding of energy impairment in Huntington's disease brain. *Journal of Huntingtons Disease, 6*, 267–302.

Dumas, E. M., van den Bogaard, S. J., Ruber, M. E., Reilman, R. R., Stout, J. C., Craufurd, D., Hicks, S. L., Kennard, C., Tabrizi, S. J., van Buchem, M. A., van der Grond, J., & Roos, R. A. (2012). Early changes in white matter pathways of the sensorimotor cortex in premanifest Huntington's disease. *Human Brain Mapping, 33*, 203–212.

Feigin, A., Leenders, K. L., Moeller, J. R., Missimer, J., Kuenig, G., Spetsieris, P., Antonini, A., & Eidelberg, D. (2001). Metabolic network abnormalities in early Huntington's disease: An [(18) F]FDG PET study. *Journal of Nuclear Medicine, 42*, 1591–1595.

Gargouri, F., Messe, A., Perlbarg, V., Valabregue, R., McColgan, P., Yahia-Cherif, L., Fernandez-Vidal, S., Ben Hamida, A., Benali, H., Tabrizi, S., Durr, A., & Lehericy, S. (2016). Longitudinal changes in functional connectivity of cortico-basal ganglia networks in manifests and premanifest Huntington's disease. *Human Brain Mapping, 37*, 4112–4128.

Genetic Modifiers of Huntington's Disease Consortium. Electronic address g.h.m.h.e. & Genetic Modifiers of Huntington's Disease C. (2019). CAG repeat not polyglutamine length determines timing of Huntington's disease onset. *Cell, 178*(887–900), e14.

Georgiou-Karistianis, N., Poudel, G. R., Dominguez, D. J., Langmaid, R., Gray, M. A., Churchyard, A., Chua, P., Borowsky, B., Egan, G. F., & Stout, J. C. (2013). Functional and connectivity changes during working memory in Huntington's disease: 18 month longitudinal data from the IMAGE-HD study. *Brain and Cognition, 83*, 80–91.

Gonzalez, R. G., Fischman, A. J., Guimaraes, A. R., Carr, C. A., Stern, C. E., Halpern, E. F., Growdon, J. H., & Rosen, B. R. (1995). Functional MR in the evaluation of dementia: Correlation of abnormal dynamic cerebral blood volume measurements with changes in cerebral metabolism on positron emission tomography with fludeoxyglucose F 18. *AJNR. American Journal of Neuroradiology, 16*, 1763–1770.

Gray, M. (2019). Astrocytes in Huntington's disease. *Advances in Experimental Medicine and Biology, 1175*, 355–381.

Harrington, D. L., Rubinov, M., Durgerian, S., Mourany, L., Reece, C., Koenig, K., Bullmore, E., Long, J. D., Paulsen, J. S., Group P.-H.i.o.t.H.S, & Rao, S. M. (2015). Network topology and functional connectivity disturbances precede the onset of Huntington's disease. *Brain, 138*, 2332–2346.

Harris, G. J., Codori, A. M., Lewis, R. F., Schmidt, E., Bedi, A., & Brandt, J. (1999). Reduced basal ganglia blood flow and volume in pre-symptomatic, gene-tested persons at-risk for Huntington's disease. *Brain, 122*(Pt 9), 1667–1678.

Hsiao, H. Y., Chen, Y. C., Huang, C. H., Chen, C. C., Hsu, Y. H., Chen, H. M., Chiu, F. L., Kuo, H. C., Chang, C., & Chern, Y. (2015). Aberrant astrocytes impair vascular reactivity in Huntington disease. *Annals of Neurology, 78*, 178–192.

Hua, J., Qin, Q., Donahue, M. J., Zhou, J., Pekar, J. J., & van Zijl, P. C. (2011a). Inflow-based vascular-space-occupancy (iVASO) MRI. *Magnetic Resonance in Medicine, 66*, 40–56.

Hua, J., Qin, Q., Pekar, J. J., & van Zijl, P. C. (2011b). Measurement of absolute arterial cerebral blood volume in human brain without using a contrast agent. *NMR in Biomedicine, 24*, 1313–1325.

Hua, J., Unschuld, P. G., Margolis, R. L., van Zijl, P. C., & Ross, C. A. (2014). Elevated arteriolar cerebral blood volume in prodromal Huntington's disease. *Movement Disorders, 29*, 396–401.

Khakh, B. S., & Goldman, S. A. (2023). Astrocytic contributions to Huntington's disease pathophysiology. *Annals of the New York Academy of Sciences, 1522*, 42–59.

Khakh, B. S., & Sofroniew, M. V. (2014). Astrocytes and Huntington's disease. *ACS Chemical Neuroscience, 5*, 494–496.

Kieburtz, K., McDermott, M. P., Voss, T. S., Corey-Bloom, J., Deuel, L. M., Dorsey, E. R., Factor, S., Geschwind, M. D., Hodgeman, K., Kayson, E., Noonberg, S., Pourfar, M., Rabinowitz, K., Ravina, B., Sanchez-Ramos, J., Seely, L., Walker, F., Feigin, A., & Huntington Disease Study Group D.I. (2010). A randomized, placebo-controlled trial of latrepirdine in Huntington disease. *Archives of Neurology, 67*, 154–160.

Klinkmueller, P., Kronenbuerger, M., Miao, X., Bang, J., Ultz, K. E., Paez, A., Zhang, X., Duan, W., Margolis, R. L., Zijl, P. C. V., Ross, C. A., & Hua, J. (2021). Impaired response of cerebral oxygen metabolism to visual stimulation in Huntington's disease. *Journal of Cerebral Blood Flow and Metabolism, 41*, 1119–1130.

Kloppel, S., Draganski, B., Siebner, H. R., Tabrizi, S. J., Weiller, C., & Frackowiak, R. S. (2009). Functional compensation of motor function in pre-symptomatic Huntington's disease. *Brain, 132*, 1624–1632.

Kuschinsky, W. (1996). Capillary perfusion in the brain. *Pflügers Archiv, 432*, R42–R46.

Lahr, J., Minkova, L., Tabrizi, S. J., Stout, J. C., Kloppel, S., Scheller, E., & TrackOn, H. D. I. (2018). Working memory-related effective connectivity in Huntington's disease patients. *Frontiers in Neurology, 9*, 370.

Langbehn, D. R., Brinkman, R. R., Falush, D., Paulsen, J. S., Hayden, M. R., & International Huntington's Disease Collaborative G. (2004). A new model for prediction of the age of onset and penetrance for Huntington's disease based on CAG length. *Clinical Genetics, 65*, 267–277.

Lee, J. M., Wheeler, V. C., Chao, M. J., Vonsattel, J. P. G., Pinto, R. M., Lucente, D., Abu-Elneel, K., Ramos, E. M., Mysore, J. S., Gillis, T., MacDonald, M. E., Gusella, J. F., Harold, D., Stone, T. C., Escott-Price, V., Han, J., Vedernikov, A., Holmans, P., Jones, L., Kwak, S., Mahmoudi, M., Orth, M., Landwehrmeyer, G. B., Paulsen, J. S., Dorsey, E. R., Shoulson, I., Myers, R. H., & Dis, G. M. H. (2015). Identification of genetic factors that modify clinical onset of Huntington's disease. *Cell, 162*, 516–526.

Lemiere, J., Decruyenaere, M., Evers-Kiebooms, G., Vandenbussche, E., & Dom, R. (2004). Cognitive changes in patients with Huntington's disease (HD) and asymptomatic carriers of the HD mutation--a longitudinal follow-up study. *Journal of Neurology, 251*, 935–942.

Lewandowski, N. M., Bordelon, Y., Brickman, A. M., Angulo, S., Khan, U., Muraskin, J., Griffith, E. Y., Wasserman, P., Menalled, L., Vonsattel, J. P., Marder, K., Small, S. A., & Moreno,

H. (2013). Regional vulnerability in Huntington's disease: fMRI-guided molecular analysis in patients and a mouse model of disease. *Neurobiology of Disease, 52*, 84–93.

Li, Q., Li, G., Wu, D., Lu, H., Hou, Z., Ross, C. A., Yang, Y., Zhang, J., & Duan, W. (2017). Resting-state functional MRI reveals altered brain connectivity and its correlation with motor dysfunction in a mouse model of Huntington's disease. *Scientific Reports, 7*, 16742.

Lin, C. Y., Hsu, Y. H., Lin, M. H., Yang, T. H., Chen, H. M., Chen, Y. C., Hsiao, H. Y., Chen, C. C., Chern, Y., & Chang, C. (2013). Neurovascular abnormalities in humans and mice with Huntington's disease. *Experimental Neurology, 250*, 20–30.

Liu, H., Zhang, C., Xu, J., Jin, J., Cheng, L., Miao, X., Wu, Q., Wei, Z., Liu, P., Lu, H., van Zijl, P. C. M., Ross, C. A., Hua, J., & Duan, W. (2021). Huntingtin silencing delays onset and slows progression of Huntington's disease: A biomarker study. *Brain, 144*, 3101–3113.

Lou, S., Lepak, V. C., Eberly, L. E., Roth, B., Cui, W., Zhu, X. H., Oz, G., & Dubinsky, J. M. (2016). Oxygen consumption deficit in Huntington disease mouse brain under metabolic stress. *Human Molecular Genetics, 25*, 2813–2826.

McColgan, P., Razi, A., Gregory, S., Seunarine, K. K., Durr, A., Roos R, A. C., Leavitt, B. R., Scahill, R. I., Clark, C. A., Langbehn, D. R., Rees, G., Tabrizi, S. J., & Track On H.D.I. (2017). Structural and functional brain network correlates of depressive symptoms in premanifest Huntington's disease. *Human Brain Mapping, 38*, 2819–2829.

Menalled, L. B. (2005). Knock-in mouse models of Huntington's disease. *NeuroRx, 2*, 465–470.

Menalled, L., & Brunner, D. (2014). Animal models of Huntington's disease for translation to the clinic: Best practices. *Movement Disorders, 29*, 1375–1390.

Mochel, F., N'Guyen, T. M., Deelchand, D., Rinaldi, D., Valabregue, R., Wary, C., Carlier, P. G., Durr, A., & Henry, P. G. (2012). Abnormal response to cortical activation in early stages of Huntington disease. *Movement Disorders, 27*, 907–910.

Moss, D. J. H., Pardinas, A. F., Langbehn, D., Lo, K., Leavitt, B. R., Roos, R., Durr, A., Mead, S., Investigators T.-H., Investigators R, Holmans, P., Jones, L., & Tabrizi, S. J. (2017). Identification of genetic variants associated with Huntington's disease progression: A genome-wide association study. *Lancet Neurology, 16*, 701–711.

Odish, O. F., van den Berg-Huysmans, A. A., van den Bogaard, S. J., Dumas, E. M., Hart, E. P., Rombouts, S. A., van der Grond, J., Roos, R. A., & Group T.-H.I. (2015). Longitudinal resting state fMRI analysis in healthy controls and premanifest Huntington's disease gene carriers: A three-year follow-up study. *Human Brain Mapping, 36*, 110–119.

Paleacu, D. (2007). Tetrabenazine in the treatment of Huntington's disease. *Neuropsychiatric Disease and Treatment, 3*, 545–551.

Parievsky, A., Cepeda, C., & Levine, M. S. (2012). Evidence from the R6/2 mouse model of Huntington's disease for using abnormal brain metabolism as a biomarker for evaluating therapeutic approaches for treatment. *Future Neurology, 7*, 527–530.

Perot, J. B., Celestine, M., Palombo, M., Dhenain, M., Humbert, S., Brouillet, E., & Flament, J. (2022). Longitudinal multimodal MRI characterization of a knock-in mouse model of Huntington's disease reveals early gray and white matter alterations. *Human Molecular Genetics, 31*, 3581–3596.

Piechnik, S. K., Chiarelli, P. A., & Jezzard, P. (2008). Modelling vascular reactivity to investigate the basis of the relationship between cerebral blood volume and flow under CO2 manipulation. *NeuroImage, 39*, 107–118.

Poudel, G. R., Egan, G. F., Churchyard, A., Chua, P., Stout, J. C., & Georgiou-Karistianis, N. (2014). Abnormal synchrony of resting state networks in premanifest and symptomatic Huntington disease: The IMAGE-HD study. *Journal of Psychiatry & Neuroscience, 39*, 87–96.

Powers, W. J., Videen, T. O., Markham, J., McGee-Minnich, L., Antenor-Dorsey, J. V., Hershey, T., & Perlmutter, J. S. (2007). Selective defect of in vivo glycolysis in early Huntington's disease striatum. *Proceedings of the National Academy of Sciences of the United States of America, 104*, 2945–2949.

Quarantelli, M., Salvatore, E., Giorgio, S. M., Filla, A., Cervo, A., Russo, C. V., Cocozza, S., Massarelli, M., Brunetti, A., & De Michele, G. (2013). Default-mode network changes in

Huntington's disease: An integrated MRI study of functional connectivity and morphometry. *PLoS One, 8*, e72159.

Rieux, M., Alpaugh, M., Sciacca, G., Saint-Pierre, M., Masnata, M., Denis, H. L., Levesque, S. A., Herrmann, F., Bazenet, C., Garneau, A. P., Isenring, P., Truant, R., Oueslati, A., Gould, P. V., Ast, A., Wanker, E. E., Lacroix, S., & Cicchetti, F. (2021). Shedding a new light on Huntington's disease: How blood can both propagate and ameliorate disease pathology. *Molecular Psychiatry, 26*, 5441–5463.

Rosas, H. D., Liu, A. K., Hersch, S., Glessner, M., Ferrante, R. J., Salat, D. H., van der Kouwe, A., Jenkins, B. G., Dale, A. M., & Fischl, B. (2002). Regional and progressive thinning of the cortical ribbon in Huntington's disease. *Neurology, 58*, 695–701.

Tabrizi, S. J., Ghosh, R., & Leavitt, B. R. (2019). Huntingtin lowering strategies for disease modification in Huntington's disease. *Neuron, 102*, 899.

van Duijn, E., Kingma, E. M., & van der Mast, R. C. (2007). Psychopathology in verified Huntington's disease gene carriers. *The Journal of Neuropsychiatry and Clinical Neurosciences, 19*, 441–448.

Van Houten, B. (2019). Huntington's disease: Astrocytes shift to fatty acid metabolism. *Trends in Endocrinology and Metabolism, 30*, 575–577.

van Zijl, P. C., Eleff, S. M., Ulatowski, J. A., Oja, J. M., Ulug, A. M., Traystman, R. J., & Kauppinen, R. A. (1998). Quantitative assessment of blood flow, blood volume and blood oxygenation effects in functional magnetic resonance imaging. *Nature Medicine, 4*, 159–167.

Vonsattel, J. P., Myers, R. H., Stevens, T. J., Ferrante, R. J., Bird, E. D., & Richardson, E. P., Jr. (1985). Neuropathological classification of Huntington's disease. *Journal of Neuropathology and Experimental Neurology, 44*, 559–577.

Wei, Z., Liu, H., Lin, Z., Yao, M., Li, R., Liu, C., Li, Y., Xu, J., Duan, W., & Lu, H. (2023). Non-contrast assessment of blood-brain barrier permeability to water in mice: An arterial spin labeling study at cerebral veins. *NeuroImage, 268*, 119870.

Werner, C. J., Dogan, I., Sass, C., Mirzazade, S., Schiefer, J., Shah, N. J., Schulz, J. B., & Reetz, K. (2014). Altered resting-state connectivity in Huntington's disease. *Human Brain Mapping, 35*, 2582–2593.

Wolf, R. C., Gron, G., Sambataro, F., Vasic, N., Wolf, N. D., Thomann, P. A., Saft, C., Landwehrmeyer, G. B., & Orth, M. (2011). Magnetic resonance perfusion imaging of resting-state cerebral blood flow in preclinical Huntington's disease. *Journal of Cerebral Blood Flow and Metabolism, 31*, 1908–1918.

Wolf, R. C., Sambataro, F., Vasic, N., Wolf, N. D., Thomann, P. A., Saft, C., Landwehrmeyer, G. B., & Orth, M. (2012). Default-mode network changes in preclinical Huntington's disease. *Experimental Neurology, 237*, 191–198.

Wolf, R. C., Sambataro, F., Vasic, N., Wolf, N. D., Thomann, P. A., Landwehrmeyer, G. B., & Orth, M. (2014). Longitudinal task-negative network analyses in preclinical Huntington's disease. *European Archives of Psychiatry and Clinical Neuroscience, 264*, 493–505.

Zhao, Z., Nelson, A. R., Betsholtz, C., & Zlokovic, B. V. (2015). Establishment and dysfunction of the blood-brain barrier. *Cell, 163*, 1064–1078.

Zlokovic, B. V. (2011). Neurovascular pathways to neurodegeneration in Alzheimer's disease and other disorders. *Nature Reviews. Neuroscience, 12*, 723–738.

Huntington's disease: An integrated MRI study of functional connectivity and morphometry. PLoS One, 8, e72497.
[illegible], Saint-Pierre, M., [illegible], Denis, H. L., [illegible] S. A., [illegible], A. R., [illegible] A. V., [illegible], Cicchetti, F. (2021). Shedding a new light on Huntington's disease: How blood can both propagate and ameliorate disease pathology. Molecular Psychiatry, 26, 5441–5463.
Rosas, H. D., Liu, A. K., Hersch, S., Glessner, M., Ferrante, R. J., Salat, D. H., van der Kouwe, A., Jenkins, B. G., Dale, A. M., & Fischl, B. (2002). Regional and progressive thinning of the cortical ribbon in Huntington's disease. Neurology, 58, 695–701.
Tabrizi, S. J., Ghosh, R., & Leavitt, B. R. (2019). Huntingtin lowering strategies for disease modification in Huntington's disease. Neuron, 102, 899.
van Duijn, E., Kingma, E. M., & van der Mast, R. C. (2007). Psychopathology in verified Huntington's disease gene carriers. The Journal of Neuropsychiatry and Clinical Neurosciences, 19, 441–448.
[illegible] (2019). Huntington [illegible]. Trends in Endocrinology and Metabolism, 30, 474–479.
van Zijl, P. C., Eleff, S. M., Ulatowski, J. A., Oja, J. M., Uluğ, A. M., Traystman, R. J., & Kauppinen, R. A. (1998). Quantitative assessment of blood flow, blood volume and blood oxygenation effects in functional magnetic resonance imaging. Nature Medicine, 4, 159–167.
Vonsattel, J. P., Myers, R. H., Stevens, T. J., Ferrante, R. J., Bird, E. D., & Richardson, E. P., Jr. (1985). Neuropathological classification of Huntington's disease. Journal of Neuropathology and Experimental Neurology, 44, 559–577.
Wei, Z., Liu, H., Lin, Z., Yao, M., Li, R., Liu, C., Li, Y., Xu, J., Duan, W., & Lu, H. (2023). Non-contrast assessment of blood-brain barrier permeability to water in mice: An arterial spin labeling study at cerebral veins. NeuroImage, 268, 119870.
[illegible] K. (2014). Microstructural changes in Huntington's disease. Human Brain Mapping, 35, [illegible].
Wolf, R. C., Grön, G., Sambataro, F., Vasic, N., Wolf, N. D., Thomann, P. A., Saft, C., Landwehrmeyer, G. B., & Orth, M. (2011). Magnetic resonance perfusion imaging of resting-state cerebral blood flow in preclinical Huntington's disease. Journal of Cerebral Blood Flow & Metabolism, 31, 1908–1918.
Wolf, R. C., Sambataro, F., Vasic, N., Wolf, N. D., Thomann, P. A., Saft, C., Landwehrmeyer, G. B., & Orth, M. (2012). Default-mode network changes in preclinical Huntington's disease. Experimental Neurology, 237, 191–198.
Wolf, R. C., Sambataro, F., Vasic, N., Wolf, N. D., Thomann, P. A., Landwehrmeyer, G. B., & Orth, M. (2014). Longitudinal task-negative network analyses in preclinical Huntington's disease. European Archives of Psychiatry and Clinical Neuroscience, 264, 493–505.
Zhao, Z., Nelson, A. R., Betsholtz, C., & Zlokovic, B. V. (2015). Establishment and dysfunction of the blood-brain barrier. Cell, 163, 1064–1078.
Zlokovic, B. V. (2011). Neurovascular pathways to neurodegeneration in Alzheimer's disease and other disorders. Nature Reviews Neuroscience, 12, 723–738.

Part IV
Multi-omics Approaches to Biomarker Discovery

Metabolomics in Huntington's Disease

Henrik Carlsson, Ida Erngren, and Kim Kultima

Abbreviations

CE	Capillary electrophoresis
CE-MS	Capillary electrophoresis - mass spectrometry
CSF	Cerebrospinal fluid
GC	Gas chromatography
GC-MS	Gas chromatography-mass spectrometry
HD	Huntington's disease
HRMS	High-resolution mass spectrometry
LC	Liquid chromatography
LC-MS	Liquid chromatography-mass spectrometry
MS	Mass spectrometry
NMR	Nuclear magnetic resonance

1 Introduction

During the last two decades, metabolomics has emerged as a primary avenue for biomarker discovery. Metabolomics concerns studies of the metabolome, which is composed of the complete set of small molecules (typically defined as compounds with molecular weights <1.5 kDa) within the biological sample under study (Wishart et al., 2007). In a typical metabolomics experiment, the metabolomic profiles are compared between healthy controls and subjects with disease or under treatment to

Supplementary Information The online version contains supplementary material available at https://doi.org/10.1007/978-3-031-32815-2_8.

H. Carlsson · I. Erngren · K. Kultima (✉)
Department of Medical Sciences, Uppsala University, Uppsala, Sweden
e-mail: kim.kultima@medsci.uu.se

E. A. Thomas, G. M. Parkin (eds.), *Biomarkers for Huntington's Disease*,
Contemporary Clinical Neuroscience, https://doi.org/10.1007/978-3-031-32815-2_8

identify altered metabolites of significance. Using metabolomics, new insights may be gained regarding the biochemical processes involved in disease, which may be used for biomarker applications and, ultimately, new treatment options (Ashrafian et al., 2021).

Metabolomics has been applied in Huntington's disease (HD) studies with promising results for almost two decades. For this chapter, a systematic literature review was performed to identify published data from metabolomics studies in HD in animal and model systems and humans. The PubMed database (Sayers et al., 2022) was utilized to search articles published before March of 2023 comprehensively. The used search criteria combined the term "Huntington's" with the terms "metabolomics", "metabolome", "lipidomics", "lipidome", "metabonomics", and "metabonome". Works not using metabolomics approaches, e.g., analyzing single metabolites and models not including mHtt expression, such as 3-nitropropionic acid-induced neuronal toxicity (Tsang et al., 2009; Chang et al., 2011), were excluded from the literature review. The present chapter contains a comprehensive literature summary describing metabolomics in HD to date. It discusses the potential, limitations, and ways forward for utilizing the HD metabolome in research and biomarker applications.

2 Metabolomics: A Brief Overview

Metabolomics is the study of low-weight molecules (metabolites) comprising the global metabolic network (metabolome) in a biological sample. The number of small metabolites detected so far in the human body (>220,000 metabolites and other small molecules) has dramatically increased in recent years (Wishart et al., 2022b). As the metabolome includes all biochemical pathways' intermediate- and end products, changes caused by various pathophysiological processes will immediately leave traces in the metabolome (Dettmer et al., 2007). The prominence of metabolomics has been fueled by significant technological advances, especially by the increased sensitivity of mass spectrometers commonly used for metabolomic studies. Currently, >10,000 independent spectral features can be measured from a single biological specimen. Under favorable conditions, typically only a third of these features will however be possible to identify (Newgard, 2017). The metabolome is composed of compounds with widely varying chemical properties, e.g., very hydrophilic compounds to highly hydrophobic compounds, and with current analytical techniques, it is impossible to analyze the complete metabolome with a single analytical method.

Several analytical techniques and many different methods have been used for metabolomics applications, with some methods designed to cover as large ranges of metabolites as possible in a single experiment and some more specifically aimed at a subset of metabolites, e.g., lipids. In the search criteria for the literature review for the present chapter, the terms "lipidomics", "lipidome", "metabonomics", and "metabonome" were included. Lipidomics is the study of the lipidome, the complete set of lipids in a biological sample, using methodologies similar to metabolomics (Züllig et al., 2020).

The lipidome may be considered a subset of the metabolome, and general metabolomics methods often detect many lipids. However, specific lipidomics methods may be required to cover the complete lipidome (Carlsson et al., 2022), making such approaches overlapping and sometimes difficult to differentiate. The terms "metabonome" and "metabonomics" is often used synonymously with "metabolome" and "metabolomics", respectively, and to the authors' knowledge, there is no complete agreement regarding the differences. However, "metabonomics" is more associated with nuclear magnetic resonance (NMR) spectroscopy-based methods, whereas "metabolomics" is more associated with mass spectrometry (MS) based methods (Klassen et al., 2017). Furthermore, metabonomics has historically been applied in toxicology, putting more emphasis on exogenous factors than what is typically done in metabolomics, where the focus instead is on endogenous metabolism (Robertson, 2005)

A typical metabolomics workflow is shown in Fig. 1. Briefly, samples are collected based on the aim of the study, followed by appropriate sample preparation for metabolomic profiling. Especially when working with highly metabolically active samples, such as cells and tissues, it is crucial to quench, i.e., rapidly stop, the ongoing metabolism in samples following sample collection (Lu et al., 2017). This is essential in order to produce a sample quantitatively reflective of the endogenous metabolite levels present in the original sample and is often done using quick-freeze methods or liquid nitrogen. Many sample preparation methods used for metabolomics are simple and fast, thus suitable for high-throughput analysis and for analysis of large cohorts (Chetwynd et al., 2017). For biofluids, protein precipitation is often

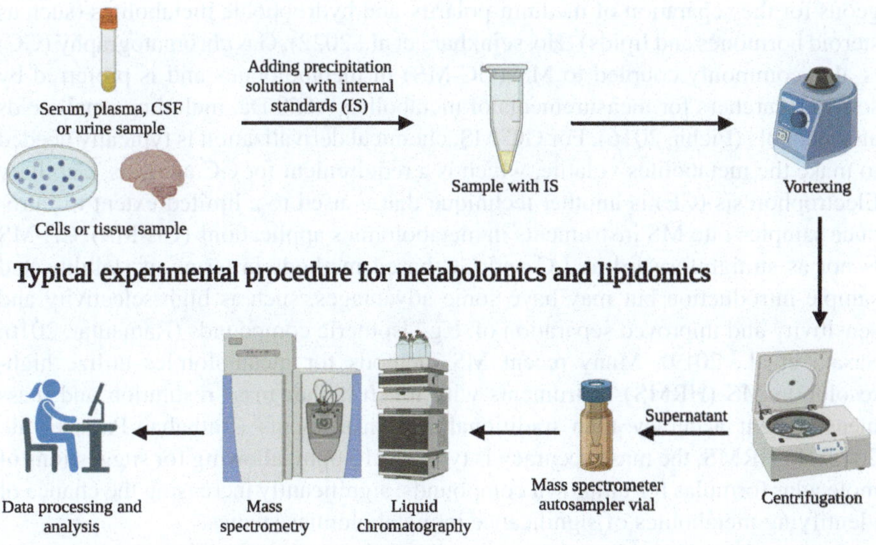

Fig. 1 This illustrates an overview of a typical metabolomics and lipidomics workflow analyzed using LC-MS. For biofluids, protein precipitation is often performed using common solvents such as methanol and acetonitrile that are compatible with MS analysis, whereas tissues are typically homogenized before the extraction of metabolites

performed using common solvents such as methanol and acetonitrile that are compatible with MS analysis, whereas tissues are typically homogenized prior to the extraction of metabolites. To increase coverage and detectability, samples may be cleaned up and concentrated before analysis using approaches such as solid phase extraction and nitrogen gas evaporation (Rico et al., 2014).

The two main analytical techniques used in metabolomics are NMR spectroscopy and MS. MS-based methods are increasingly more common and widespread, accounting for more than 70% of published metabolomics studies (Wishart et al., 2022a). However, NMR spectroscopy has several advantages, including simple sample preparation, an ability to quantify metabolite levels, a high level of experimental reproducibility, and the inherently nondestructive nature of the technique. Modern automated NMR systems can typically process about three times as many samples as typical MS-based metabolomics platforms over 24 hours, primarily because no time-consuming chromatography is needed (Wishart et al., 2022a). A significant disadvantage of NMR spectroscopy is the comparably low sensitivity, with MS-based methods typically achieving detection limits 10 to 100 times better (Emwas et al., 2019).

Regarding MS, the large majority of methods used for metabolomics couple liquid chromatography (LC) to the mass spectrometer (LC-MS) to introduce and separate sample components to the instrument (Perez de Souza et al., 2021). For LC-MS, there are several options regarding the stationary phase of the chromatographic columns used for separation with different advantages. Hydrophilic interaction liquid chromatography (HILIC) is preferable for the separation of polar metabolites (such as amino acids and sugars), whereas reversed-phase chromatography is advantageous for the separation of medium-polarity and hydrophobic metabolites (such as steroid hormones and lipids) (Hosseinkhani et al., 2022). Gas chromatography (GC) is also commonly coupled to MS (GC-MS) in metabolomics and is preferred by some researchers for measurements of metabolites <650 Da, including small acids and alcohols (Fiehn, 2016). For GC-MS, chemical derivatization is typically needed to make the metabolites volatile, which is a requirement for GC analysis. Capillary Electrophoresis (CE) is another technique that is used to a limited extent to introduce samples into MS instruments in metabolomics applications (CE-MS). CE-MS is not as straightforward as LC and GC-based methods in terms of stability and sample introduction but may have some advantages, such as high selectivity and sensitivity and improved separation of, e.g., isomeric compounds (Ramautar, 2016; Sasaki et al., 2019). Many recent MS methods for metabolomics utilize high-resolution MS (HRMS) instruments with much higher mass resolution and mass measurement accuracy than traditional MS instruments (Rathahao-Paris et al., 2015). In HRMS, the mass accuracy is typical <10 ppm, allowing for suggestions of molecular formulas for unknown compounds, significantly increasing the chance of identifying metabolites of significance in metabolomics studies.

Data analysis is an essential aspect of metabolomics. Similar to other 'omics' techniques, the datasets generated in metabolomics are complex and large-scale. Several data analysis tools are typically needed to extract biologically relevant information from metabolomics data, often involving multi-step workflows (Lamichhane et al., 2018; Emami Khoonsari et al., 2019; Novella et al., 2019; Peters et al., 2019). Metabolomics data are typically analyzed using multivariate statistical analysis, such as principal component analysis, to identify differentiating metabolites (Herman et al., 2019a, b). As for other fields where large and complex datasets are commonplace, machine learning and artificial intelligence have been increasingly applied in metabolomics in recent years with promising results (Herman et al., 2018, 2023; Galal et al., 2022). Databases such as the Human Metabolome Database and Metlin (Smith et al., 2005; Wishart et al., 2022b), as well as tools such as MetaboAnalyst (Pang et al., 2021) are essential resources for researchers in metabolomics to interpret metabolomics data.

Metabolomics can be performed using untargeted and targeted approaches. In an untargeted approach, the metabolites are measured in relation to each other, providing relative abundances (Alonso et al., 2015). An untargeted approach can measure the metabolome as thoroughly as the technologies allow. In the discovery phase, this is an advantage as blind exploration will generate unbiased hypotheses and an overview of molecular events (Herman et al., 2019a, b; Peters et al., 2021). Many untargeted metabolomics studies are being performed, comparing various disease states and biological conditions with healthy controls. Once a hypothesis has been generated, one can deploy a targeted approach to verify the proposition. Identification and validation are the primary challenges faced when performing untargeted metabolomics.

In a targeted approach, a defined list of metabolites will be selectively measured, enabling more accurate estimations (Carlsson et al., 2020). By adding chemically labeled internal standards, the measured relative abundances can be converted into concentrations with the help of a calibration curve. The standards are typically labeled by a few heavy hydrogens, deuteriums, which possess one proton and one neutron in their nucleus, whereas the far more common isotope of hydrogen contains only one proton in its nucleus. Hence, the mass of a heavy labeled standard will be increased by the number of hydrogens replaced by deuterium. Carbon-13 labeled and nitrogen-15 labeled internal standards are also commonly used in metabolomics but are typically more expensive.

3 Animal and Cell Model Metabolomics Studies in Huntington's Disease

With a rare disease such as HD, it is challenging to perform metabolomics studies in human subjects considering subject recruitment and sample acquisition while minimizing confounding factors. Acquiring relevant tissue (human post-mortem brains) is difficult, given that tissue needs to be frozen immediately to avoid post-mortem changes. This has led to many animal and model studies in HD metabolomics. Such studies have the advantage of being more easily controlled with fewer sources of variability, which is especially important considering the small sample size typically used. The literature review identified 20 works investigating the HD metabolome in HD's animal and cell model systems. These studies are summarized in Table 1 and discussed below, divided by the animal or cell model used.

Table 1 Summary of published animal and cell model metabolomics studies in Huntington's disease

Organism	Biospecimen	Instrumentation	Key results	Reference
Cell line (mouse)	Striatal cell line model of HD	LC-MS	Network integration analysis of untargeted metabolomics data applied to HD model, identifying dysregulated sphingolipid, steroid, and fatty acid metabolism in HD model	Pirhaji et al. (2016)
Cell line (mouse)	Striatal cell line model of HD	LC-MS	Interactions between HD and neuronal manganese status, e.g. by reduced levels of pantothenic acid and glutathione in HD	Kumar et al. (2015)
Drosophila melanogaster	Fly heads	NMR	HD progression is associated with reduction in NAD and an increase in lactate-to-pyruvate ratio; overall results suggest altered cellular energy homeostasis	Singh et al. (2017)
Drosophila melanogaster	Homogenized Drosophila	NMR	Different metabolomic profiles for the presymptomatic and symptomatic stages of HD, indicate different processes involved	Bertrand et al. (2020)
Mouse	Striatum, frontal cortex, cerebellum, brain stem, muscle, serum, and urine	NMR	Clear differentiation between HD and control metabolomes in blood and urine; increased concentrations of osmolytes, creatine, glutamine, and lactate in HD brain	Tsang et al. (2006)
Mouse	In vivo NMR of the striatum in anesthetized mice	NMR	Evaluation of different clustering analyses, identifying a model able to separate HD from controls and between HD age groups using in vivo NMR data	Nikas and Low (2011)
Mouse	Plasma	Capillary Electrophoresis Mass Spectrometry (CE-MS)	Controls and HD discriminated by 29 compounds, with intracellular signaling the main affected pathway at middle and late disease stage	Pont et al. (2016)
Mouse	Plasma, brown adipose tissue, cerebellum, striatum, liver, white adipose tissue	LC-MS	Metabolome alterations in HD were tissue-specific and most evident in the brain, specifically with an accumulation of lipids in striatum	Carroll et al. (2015)

(continued)

Table 1 (continued)

Organism	Biospecimen	Instrumentation	Key results	Reference
Mouse	Cerebellum, cortex, and striatum	LC-MS	Altered glycerophospholipid profiles in HD, including alterations in phosphatidic acid and lyso-phosphatidic acid	Vodicka et al. (2015)
Mouse	Plasma	LC-MS	Microbiota-gut-brain axis implicated in the pathogenesis of HD through integrated metagenomics and metabolomics	Kong et al. (2021)
Mouse	Plasma and whole brain	CE-MS and LC-MS	Perturbation of histidine metabolism in HD brain, as well as changes in the metabolism of several amino acids and lipid metabolites	Hashimoto et al. (2021)
Mouse	Subcellular fractionation of striatal tissue	LC-MS	HD lipidome altered with changed levels of phosholipids, acylcarnitines, phosphatidic acid and sphingomyelin	Iuliano et al. (2021)
Mouse	Hippocampus	LC-MS	Pre-symptomatic changes in levels of lipid metabolites in the hippocampi	Farzana et al. (2023)
Mouse and human	Mouse: striatum, cortex, and cerebellum. Human: Striatum and cortex	LC-MS	Impaired pyridoxal 5′-phosphate (PLP) dependent metabolism in HD, indicating impaired PLP synthesis in both human and mouse brain	Sorolla et al. (2016)
Rat	CSF and serum	NMR	Mitochondrial energy dysfunction in HD as indicated by several altered metabolites, e.g., succinic acid and N-acetylaspartate	Verwaest et al. (2011)
Sheep	Motor cortex, hippocampus, cerebellum and liver tissue	GC-MS	Tissue-specific alterations in transgenic HD metabolomes, with predominantly amino acids affected in the cerebellum and fatty acids in the liver	Handley et al. (2016)
Sheep	Plasma	LC-MS	Urea cycle and nitric oxide pathways dysregulated at early stages in HD	Skene et al. (2017)
Sheep	Plasma	LC-MS	Metabolites affected both by disease and circadian rhythm in HD, with phosphatidylcholines major contributor to difference between HD and controls	Spick et al. (2023)

(continued)

Table 1 (continued)

Organism	Biospecimen	Instrumentation	Key results	Reference
Yeast	Yeast model of HD	NMR	Alanine, glutamine, glycerol, and valine identified as potential HD biomarkers	Joyner et al. (2010)
Yeast and human	Serum and yeast model of HD	LC-MS	Large overlap in downregulated metabolic pathways between human serum and yeast model, including amino acid and glutathione metabolism	Pradhan et al. (2022)

Cell Lines and Yeast Models

The simplest models used for metabolomics in HD are cell lines and yeast models of HD. Pirhaji et al. performed untargeted metabolomics analysis on the STHdh cell line model of HD using LC-MS (Pirhaji et al., 2016). STHdh cell lines are striatum-derived STHdh cells generated from wild-type and huntingtin knock-in mouse embryos expressing a 111 CAG-expanded huntingtin gene (Menalled, 2005). Comparing the wild-type and the mutant cell lines, 115 significantly differing metabolite features were identified. A network-based approach for integrative analysis of untargeted metabolomics data was used to uncover the molecular context of these metabolite features. This approach identified changes in the metabolism of sphingolipids, fatty acids, and steroids in the STHdh HD model. Kumar et al. also used LC-MS to perform metabolomics on STHdh lines (Kumar et al., 2015). They investigated how neuronal manganese status affected the HD model's metabolome. Manganese is an essential micronutrient in the nervous system, and deficiencies in manganese transport have been implicated in HD pathogenesis. By exposing the STHdh cell lines to varying concentrations of manganese, they observed different metabolomic effects in the HD model compared to controls. The findings demonstrate metabolic interactions between the HD genotype and biologically relevant levels of manganese, strengthening the hypothesis that manganese cellular transport processes are associated with compensatory metabolic responses in HD (Williams et al., 2010).

The metabolomes of yeast models of HD have been described by Joyner et al. and Pradhan et al. (Joyner et al., 2010; Pradhan et al., 2022). Joyner et al. used NMR spectroscopy to compare the metabolomes of yeast models with 25 and 103 CAG repeats, serving as control and HD models, respectively. Eleven metabolites (alanine, acetate, galactose, glutamine, glycerol, histidine, proline, succinate, threonine, trehalose, and valine) were significantly altered in the metabolome of the HD yeast model. After performing correspondence analysis to compare the results with data reported from transgenic mice expressing a mutant huntingtin gene fragment and HD patients, four metabolites were identified as significantly changed in all HD systems: alanine, glutamine, glycerol, and valine. Joyner et al. conclude that these metabolites should be considered potential HD biomarkers. Pradhan et al. used

LC-MS to compare the metabolomes of yeast models with 25 CAG repeats, used as control, and 46, 76, and 103 CAG repeats, corresponding to HD models. The metabolomes of the HD models exhibited significantly altered pathways for amino acids (alanine, aspartate, glutamate, glutamine, arginine, and proline), aminoacyl-tRNA biosynthesis, butanoate, glutathione, and purine metabolism. In the same study, metabolomics of serum from HD patients and healthy controls were performed. The metabolomes of the HD patients and the HD yeast model showed considerable overlap in downregulated metabolic pathways, indicating the applicability of the yeast model.

Drosophila Models

Metabolomics using NMR spectroscopy has been performed on Drosophila models by Singh et al. and Bertrand et al. (Singh et al., 2017; Bertrand et al., 2020). Singh et al. studied HD progression in Drosophila models with 93 and 127 CAG repeats compared with a control with 20 CAG repeats. The metabolomes were studied at 1, 10, and 20 days of age. HD progression was associated with a reduction in nicotinamide adenine dinucleotide and an increase in the lactate-to-pyruvate ratio. Overall, the results suggest an altered cellular energy homeostasis in HD. Bertrand et al. used a Drosophila HD model (93 CAG repeats) to investigate the metabolomes corresponding to presymptomatic and symptomatic stages of HD. Their experiments demonstrated that the metabolomes of 10-day-old flies, corresponding to the presymptomatic stage, and 16-day-old flies, corresponding to the symptomatic stage, highlighted two different sets of altered metabolites. Sixteen discriminatory metabolites were identified and connected by the authors to dysfunctions in energy production in HD, leading to mitochondrial dysfunction and a disruption of glycolysis, tricarboxylic acid, and glutamate–glutamine cycles.

Mouse and Rat Models

Mouse models are the most frequently used HD metabolomics models. Most of these studies concern the mouse plasma and brain metabolome, with a few investigating other tissues and biofluids. The most common mouse model to study the HD metabolome is the R6/2 transgenic mouse model. R6/2 mice express an N-terminal fragment of huntingtin with >150 glutamines (Mangiarini et al., 1996). This corresponds to a CAG repeat size that would cause childhood onset of HD in humans with rapid progression. Therefore, the R6/2 model is the most frequently used for testing therapeutic compounds in preclinical trials (Bates & Hockly, 2003).

The very first study of metabolomics in HD, to the authors' knowledge, was a study of R6/2 transgenic mice by Tsang et al. published in 2006 (Tsang et al., 2006). They used NMR spectroscopy to profile the metabolomes in the brain, muscle, blood, and urine of ten transgenic R6/2 and ten wild-type mice at four, eight, and twelve weeks of age. The brain, blood, and urine metabolomic profiles were differentiated between the R6/2 and wild-type mice. The transgenic mice exhibited increased brain concentrations of osmolytes, creatine, glutamine, and lactate. This was a promising first study in HD metabolomics, demonstrating the potential of omics approaches to detect promising biomarkers.

In a unique study, Nikas and Low used *in vivo* NMR to study the metabolomic profiles of the striatum in anesthetized mice (Nikas & Low, 2011). They compared the striatal metabolomic profiles of R6/2 transgenic mice and that of wild-type mice and were able to separate the groups based on *in vivo* NMR data. Furthermore, they could also separate R6/2 mice aged eight or twelve weeks. Creatine, phosphocreatine, glutamine, and N-acetylaspartate were identified as the most significantly altered metabolites in the HD metabolome and suggested as potential biomarkers.

Capillary Electrophoresis Mass Spectrometry was used by Pont et al. to study the plasma metabolomes of R6/1 transgenic HD mice of varying age (eight, twelve, and 30 weeks, corresponding to the asymptomatic, symptomatic, and terminal disease stages, respectively) compared with wild-type mice as controls (Pont et al., 2016). R6/1 is another mouse HD model with about 115 CAG repeats, compared to about 150 CAG repeats in the R6/2 model, causing a less rapid symptom development (Li et al., 2005). In their untargeted analysis, Pont et al. found 29 tentatively identified metabolites to discriminate between wild-type and R6/1 plasma samples. Intracellular signaling was the most affected metabolic pathway in the transgenic HD mice at the middle (twelve weeks) and late (30 weeks) stage of the disease, agreeing with atrophy and dysfunction of specific neurons, loss of several types of receptors, and altered expression of neurotransmitters.

Kong et al. also investigated the plasma metabolome of R6/1 HD mice using LC-MS (Kong et al., 2021). They combined the metabolomic analysis with investigating the gut microbiome in the same mice. In the integrated analysis, they found several *Bacteroides* species negatively correlated with plasma ATP and pipecolic acid. Kong et al. suggest that the results indicate that the gut microbiome modulates the pathogenesis of HD, potentially via specific altered plasma metabolites that mediate gut-brain signaling.

The single HD metabolomics study to cover the widest variety of tissues was performed by Carroll et al. Using a targeted LC-MS assay, they profiled the metabolomes in plasma, brown adipose tissue, white adipose tissue, cerebellum, striatum, and the liver of Hdh^{Q111} mice as compared to wild-type mice (Carroll et al., 2015). Hdh^{Q111} is another heterozygous HD mouse model with an expanded CAG repeat of about 111 repeats inserted at the proper location in the mouse HTT homolog (Hölter et al., 2013). Tissue-specific metabolomic profiles were found with the most pronounced alterations observed in the brain, with striatum accumulation of a set of lipids, including sphingomyelin, phosphatidylcholine, cholesterol ester, and triglyceride species. With mutant huntingtin expressed widely in all tissues, this study shows how metabolomics may be used to gain insights into how different tissues are metabolically impacted during the disease.

Several research groups have investigated the mouse HD brain metabolome using different approaches (Vodicka et al., 2015; Sorolla et al., 2016; Hashimoto et al., 2021; Iuliano et al., 2021; Farzana et al., 2023). Vodicka et al. performed lipidomic LC-MS profiling of the cerebellum, cortex, and striatum from mice of the HD knock-in model Q140/Q140 as compared to wild-type mice (Vodicka et al., 2015). They observed altered glycerophospholipid profiles in the brain metabolomes of the HD model, including alterations in phosphatidic acid and lysophosphatidic acid.

Hashimoto et al. compared the whole-brain and plasma metabolomes of R6/2 mice and wild-type mice using CE-MS and LC-MS (Hashimoto et al., 2021). A perturbed histidine metabolism was observed in the brain of R6/2 mice, as indicated by decreases in neuroprotective dipeptides and histamine metabolites. They also observed alterations of several amino acids and lipid metabolites in HD, indicating several metabolomic pathways altered by disease, e.g., upregulation of the urea cycle, perturbation of energy homeostasis, and increased oxidative stress.

Iuliano et al. performed untargeted lipidomic profiling of subcellular fractions of striatal tissue in an HD mouse model as compared to controls (knock-in Q175/Q7 HD versus Q7/Q7 mice) (Iuliano et al., 2021). They found the HD lipidome to be significantly altered with changed levels of phospholipids, acylcarnitine, phosphatidic acid, and sphingomyelin species.

Farzana et al. performed a longitudinal spatial mapping of lipid metabolites in the hippocampi of Huntington's disease transgenic mice using the R6/1 model compared to wild-type mice. They investigated the hippocampi lipidomes using LC-MS of mice aged six, eight, ten, twelve, and 16 weeks combined with spatial mapping using MALDI-mass spectrometry imaging. The study showed that the synthesis rates of certain lipid classes increased with age in HD mice, with increased synthesis of lipids involved in adaptation to endoplasmic reticulum stress, thought to be a driver of lipogenesis, in pre-symptomatic HD mice. Pre-symptomatic HD mice also exhibited disturbances in phosphatidylinositol, which may indicate impairment of synaptic plasticity (Farzana et al., 2023).

Studies by Sorolla et al. compared the brain metabolomes of HD mice and human HD patients (Sorolla et al., 2016) using LC-MS. The R6/1 mouse model was used and compared to wild-type mice. Impaired pyridoxal 5′-phosphate (PLP) dependent metabolism was observed in the brains of both transgenic HD mice and HD patients, indicating impaired PLP synthesis. This study is further discussed in section "Brain Studies" of this chapter.

To the authors' knowledge, a single HD metabolomics study using a rat model has been published. This study by Verwaest et al. used NMR to compare the presymptomatic serum and cerebrospinal fluid (CSF) metabolomes in a transgenic rat model of HD (about 51 CAG repeats) and compared that to wild-type rats (Verwaest et al., 2011). N-acetylaspartate, was found to be significantly decreased in the serum of HD rats, accompanied by significantly increased levels of glutamine, succinic acid, glucose, and lactate. The observed metabolic alterations were more profound in serum than in CSF. Verwaest et al. suggests that the results indicate a defect in energy metabolism apparent already in presymptomatic rats, supporting the hypothesis of mitochondrial energy dysfunction in HD.

Sheep Models

A transgenic HD sheep model called OVT73 was developed by Jacobsen et al. (Jacobsen et al., 2010) and has been applied by them (Handley et al., 2016) and others for metabolomics studies (Skene et al., 2017; Spick et al., 2023). OVT73 was developed to allow the study of HD as a late-onset disorder, including the progression of pathogenesis through the pre-manifest period. The OVT73 model does not display overt symptoms of HD but has an altered circadian rhythm and molecular changes characteristic of the early phase disease. OVT73 sheep express a full-length human huntingtin cDNA with 73 CAG repeats and is considered a good model for the study of premanifest HD.

Handley et al. performed extensive metabolomic profiling of the motor cortex, hippocampus, cerebellum, and liver tissue of five-year-old OVT73 sheep and matched controls using GC-MS (Handley et al., 2016). They found tissue-specific alterations in the transgenic HD metabolomes, with predominantly amino acids affected in the cerebellum, including significantly increased levels of phenylalanine and pyroglutamic acid, and fatty acids in the liver, including significantly increased levels of dodecanoic acid, myristic acid, and nicotinic acid. These observed differences indicate a metabolic disruption in the OVT73 model and could provide insights into the presymptomatic stage of the human disease.

Skene et al. studied the plasma metabolome of OVT73 and control-sheep using LC-MS as a model of presymptomatic HD (Skene et al., 2017). Using targeted metabolomic profiling, 89 metabolites were found to differentiate between transgenic sheep and controls. Significant increases in urea, arginine, citrulline, asymmetric and symmetric dimethylarginine, and decreases in sphingolipids were observed in the transgenic HD sheep compared to controls. This was suggested to indicate that both the urea cycle and nitric oxide pathways are dysregulated in the presymptomatic sheep.

The OVT73 model has also been used to study the metabolomic effects of circadian rhythm dysregulation in presymptomatic HD (Spick et al., 2023). Plasma samples were taken hourly across a 27-h period from transgenic and control sheep aged five and seven years and analyzed using a targeted LC-MS approach. Both absolute and rhythmic differences were found between the metabolomes of HD and control sheep. Notable dysregulated metabolites were phosphatidylcholines, amino acids, urea, and threonine. When comparing the results with human metabolomics studies in HD, Spick et al. found similar metabolites altered, confirming the suitability of the OVT73 model.

Enrichment Analysis of Metabolites Altered in Animal and Cell models

To summarize the results of these studies, a list of altered metabolites was compiled based on the metabolites and lipids found in the results sections in each of the studies (Supplementary Table 1). This list was subjected to enrichment analysis using MetaboAnalyst (Pang et al., 2021) (Fig. 2). This method searches several libraries containing approximately 9000 biologically relevant metabolite sets collected primarily from human studies, including >1500 chemical classes. It ranks the most relevant metabolite sets based on the input list of metabolites. The top three metabolite sets identified from the works describing animal and cell model metabolomics studies in HD were arginine biosynthesis, aminoacyl-tRNA biosynthesis, and arginine and proline metabolism, highlighting the potential importance of these and related metabolites in the HD metabolome.

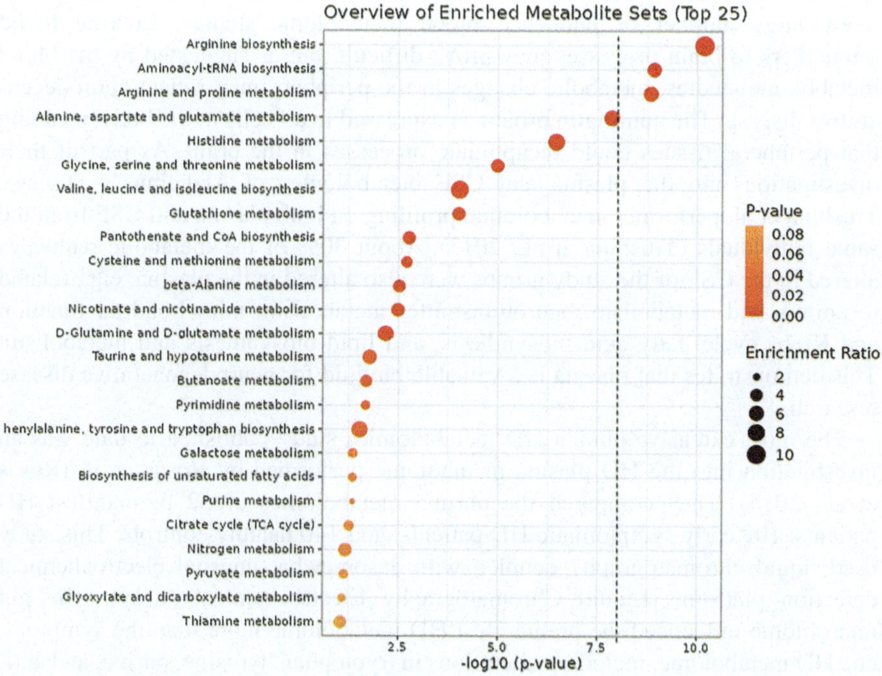

Fig. 2 A metabolite set enrichment analysis of the metabolites altered in published animal and cell model metabolomics studies in Huntington's disease was performed using MetaboAnalyst, using the list of altered metabolites as input. The 25 most significant metabolite sets are shown in the figure, with the dashed line indicating the p-value 0.01 (−log10 = 2), which is here considered a threshold for statistical significance

4 Human Metabolomics Studies in Huntington's Disease

While animal models are essential and valuable for studies of the HD metabolome, human studies are preferred for accuracy and relevance. However, performing metabolomics studies in humans for a rare disease such as HD is challenging, considering the limited number of subjects and tissue availability. As such, many potential confounders may exist, posing challenges to the data analysis (Ghazaleh et al., 2021). The literature review identified 13 works focused on investigations of the HD metabolome in humans. These studies are summarized in Table 2 and discussed below, divided by the biospecimen used.

Plasma and Serum Studies

The collection of blood samples is less invasive than the collection of the other biospecimens relevant for studying the HD metabolome, which is reflected in the relatively large number of published blood metabolome studies. Linking blood biomarkers to brain processes may prove difficult, but as suggested by the blood metabolome studies, metabolic changes in the periphery may reflect neurodegenerative disease. The huntingtin protein is expressed in all cells in the body, meaning that peripheral tissues could recapitulate processes in the brain. As part of their investigation into the plasma and CSF metabolomes of Alzheimer's Disease, Trushina et al. performed metabolomic profiling on both plasma and CSF from the same individuals (Trushina et al., 2013). About 30% of the metabolic pathways altered in the CSF of the study groups were also altered in the plasma, e.g., related to amino acid metabolism, neurotransmitter metabolism, mitochondrial function and Krebs cycle, fatty acid biosynthesis, and lipid biosynthesis and metabolism. This demonstrates that plasma is a valuable biofluid for neurodegenerative disease research.

The most extensive human HD metabolomics study published to date was an investigation into the HD plasma metabolome performed by Rosas et al. (Rosas et al., 2015). They compared the plasma metabolomes of 52 premanifest HD patients, 102 early symptomatic HD patients, and 140 healthy controls. This study used liquid chromatography coupled with a somewhat unusual electrochemical detection platform (Liquid Chromatography Electrochemical Array). The gut microbiome influenced the premanifest HD metabolome more than the symptomatic HD metabolome, including alterations in tryptophan, tyrosine, purine, and antioxidant pathways. Many of these metabolites are derived from the gut microbiome and are related to energetic and oxidative stress. Using multivariate statistical modeling, they showed mutually distinct metabolomic profiles for premanifest and symptomatic HD, suggesting that different processes are involved in the onset and progression of the disease. While gut microbiome-derived metabolites

Table 2 Summary of published human metabolomics studies in Huntington's disease

Biospecimen	Instrumentation	Key results	Reference
Brain (putamen, globus pallidus, substantia nigra, cerebellum, hippocampus, middle temporal gyrus, cingulate gyrus, sensory cortex, motor cortex, middle frontal gyrus, and entorhinal cortex)	GC-MS	Urea significantly elevated in all studied regions of HD brain, with levels sufficient to generate cytotoxicity	Patassini et al. (2015)
Brain (frontal lobe and striatum)	NMR	Accurately discriminating between control and HD, with striatum metabolites (n = 15) more perturbed than frontal lobe metabolites (n = 4) in HD	Graham et al. (2016a, b)
Brain (frontal lobe and striatum)	LC-MS	Seven biochemical pathways altered by HD, with metabolites but-2-enoic acid and uric acid suggested as potential HD biomarkers	Graham et al. (2016a, b)
Brain (frontal lobe and striatum)	LC-MS	Changes in energy metabolism and phospholipid metabolism in HD, as well as significant reductions in neurotransmitters	Graham et al. (2018)
Brain (putamen, globus pallidus, substantia nigra, cerebellum, hippocampus, middle temporal gyrus, cingulate gyrus, sensory cortex, motor cortex, middle frontal gyrus, and entorhinal cortex)	GC-MS	HD metabolome indicates severe regulatory defects in energy-transducing and amino-acid pathways, and urea.	Patassini et al. (2016)
Brain (human and mouse; human: striatum and cortex; mouse: striatum, cortex, and cerebellum)	LC-MS	Impaired pyridoxal 5′-phosphate (PLP) dependent metabolism in HD, indicating impaired PLP synthesis in both human and mouse brain	Sorolla et al. (2016)
CSF	LC-MS	Altered tyrosine metabolism in manifest HD compared to premanifest HD	Herman et al. (2019a, b)
Plasma	Liquid Chromatography Electrochemical Array	Premanifest HD metabolome more influenced by gut microbiome than symptomatic HD metabolome	Rosas et al. (2015)
Plasma	LC-MS	HD and control metabolomes differentiated by several carnitines, amino acids and phosphatidylcholines	Cheng et al. (2016)

(continued)

Table 2 (continued)

Biospecimen	Instrumentation	Key results	Reference
Plasma and CSF	LC-MS	Changes in plasma and CSF metabolomes with worsening diesease; urea cycle, glycine, and serine metabolism suggested as relevant to HD progression	McGarry et al. (2020)
Serum	LC-MS	Eight phosphatidylcholines (PC) associated with disease severity, also associated with four downregulated PC metabolism related genes in same patients	Mastrokolias et al. (2016)
Serum	NMR	Large non-pathogenic CAG repeat sizes in HTT are associated with an unfavorable metabolomic profile despite association with lower BMI	Faquih et al. (2023)
Serum (plus yeast model of HD)	LC-MS	Large overlap in downregulated metabolic pathways between human serum and yeast model, including amino acid and glutatathione metabolism	Pradhan et al. (2022)

differentiated the premanifest metabolome, the symptomatic HD metabolome was influenced by metabolites that reflected mutant huntingtin toxicity and neurodegeneration.

The HD plasma metabolome was also investigated by Cheng et al. (Cheng et al., 2016). They performed both untargeted and targeted analysis using LC-MS. The untargeted study compared the plasma metabolomes between 15 HD patients and 17 controls. The targeted analysis compared 29 HD patients, nine presymptomatic HD patients, and 44 controls. Using both approaches, the HD and control metabolomes were found to differentiate by several carnitines, amino acids, and phosphatidylcholines. To further study the importance of phosphatidylcholines in HD, the expression levels and activities of enzymes responsible for phosphatidylcholine metabolism were examined in a follow-up experiment. PCYT1A was found to be down-regulated with increased activity of phospholipase A2 in HD leukocytes, which was suggested to affect phosphatidylcholine production in HD patients negatively.

In a small but interesting pilot study, McGarry et al. (McGarry et al., 2020) compared the plasma and CSF metabolomes in the same twelve HD patients, ranging from one to twelve years since clinical diagnosis, using a targeted LC-MS approach. Changes in both the plasma and CSF metabolomes were observed with a worsening disease; in plasma, arginine, citrulline, and glycine increased with disease progression, while decreases were observed for serine, cholesterol esters, diacylglycerides, triacylglycerides, phosphatidylcholines, phosphatidylethanolamines, and sphingomyelins. In CSF, increases were observed for NAD+, arginine, saturated long-chain free fatty acids, diacylglycerides, triacylglycerides, and sphingomyelins. The specific diacylglycerides and triacylglyceride species associated with disease progression differed in plasma and CSF. The metabolites highlighted in plasma and CSF indicate different HD-related biological processes in separate compartments.

Regarding HD metabolomics, a relatively large study was performed by Mastrokolias et al. (Mastrokolias et al., 2016). The study included serum samples from 29 presymptomatic and 68 symptomatic HD patients, plus 36 healthy controls. The samples were analyzed using a targeted LC-MS assay. Serine and threonine and one phosphatidylcholine (PC ae C36:0) were significantly altered in HD, and the group also found an additional eight phosphatidylcholines associated with disease severity. The metabolomics data were integrated with previously collected gene expression data from the same patients to investigate phosphatidylcholines' importance further. Four downregulated genes related to phosphatidylcholine metabolism (ALDH1B1, MBOAT1, MTRR, and PLB1) showed significant associations with metabolic alterations. This study illustrates how metabolomics data may be integrated with other datasets to gain new insights.

As mentioned above, in section "Cell Lines and Yeast Models", Pradhan et al. (2022) compared the serum metabolome of HD patients with that of an HD yeast model. Five presymptomatic and eleven symptomatic HD patients, and 15 controls, were studied using a targeted LC-MS method. The serum and yeast HD metabolomes showed overlap in derated metabolic pathways, including alanine, aspartate, and glutamate metabolism. The metabolomes of the HD model exhibited significantly altered pathways for amino acids (alanine, aspartate, glutamate, glutamine, arginine, and proline), aminoacyl-tRNA biosynthesis, butanoate, glutathione, and purine metabolism. In the same study, metabolomics analysis of serum from HD patients and healthy controls was performed. The metabolomes of the HD patients and the HD yeast model showed a large overlap in downregulated metabolic pathways, indicating the applicability of the yeast model.

The most extensive metabolomics study related to HD, although not an HD metabolomics study per se, was performed by (Faquih et al. (2023). This impressive study explored the serum metabolomes of 10228 subjects from three European cohorts with <36 CAG repeats in HTT using NMR. Thirty-six or more CAG repeats cause HD, and this study thus concerns subjects with CAG repeat sizes in the non-pathogenic range. The motivation behind this large study was the findings that repeat sizes below the pathogenic threshold (<36) are associated with BMI (Aziz et al., 2008), suggesting an association between HTT repeat sizes in the non-pathogenic range and metabolic dysregulation. The authors hypothesized that HTT repeat sizes < 36 would be associated with metabolite levels, possibly mediated through reduced BMI. Using a targeted NMR methodology, 145 metabolites were measured in all 10228 samples. Sixty-seven metabolites were associated with non-pathogenic CAG repeat size variations in HTT. Larger HTT CAG repeat sizes were associated with lower BMI and a generally 'unfavorable' metabolomic profile overlapping with those observed in type 2 diabetes, unhealthy adiposity, and coronary artery disease. This study provides important insights into possible metabolic and health effects in subjects with non-pathogenic CAG repeat size variations, a largely unexplored topic.

CSF Studies

Cerebrospinal fluid exports waste products from neural tissues, and supplies the CNS with blood oxygen and nutrition (Segal, 1993). As such, CSF has a vital role in brain metabolism (Pautova et al., 2021). When CNS damage occurs, CSF's molecular composition will be affected. It is commonly accepted that these changes mirror the alterations in the CNS, making CSF a valuable source for monitoring pathological events in the CNS. Various chemical compounds can penetrate the blood–brain barrier into the CSF and subsequently into the brain, affecting the function of the CNS. Since CSF is the closest biological compartment to the CNS (excluding the brain itself), it should reflect the pathological changes occurring in the brain more accurately than in plasma and serum. This makes CSF a valuable biospecimen for metabolomics in HD. The collection of CSF through lumbar punctures is invasive and not without risk for side complications (Doherty & Forbes, 2014), making the number of samples in most CSF metabolomics studies relatively limited. The complexity of CSF sampling combined with the rarity of HD makes the data on the HD CSF metabolome both limited and valuable.

As discussed above, a single animal model study has been performed regarding the HD CSF metabolome. This study by Verwaest et al. compared the presymptomatic serum and CSF metabolomes in a transgenic rat model of HD to wild-type rats (Verwaest et al., 2011). In this study, HD-related metabolic alterations were more evident in the serum metabolome compared with the CSF metabolome. The human HD CSF metabolome has also been studied to a minimal extent. As discussed above, McGarry et al. compared the plasma and CSF metabolomes in the same twelve HD patients (McGarry et al., 2020). Different metabolites were highlighted in plasma and CSF, indicating different biological processes associated with HD in the separate compartments.

Apart from the small study by McGarry et al., the only published and to-date most extensive investigation of the human HD CSF metabolome was performed by Herman et al. (2019b). The CSF metabolomes in 26 mHTT carriers (13 premanifest subjects and 13 manifest HD subjects) and control subjects ($n = 42$) were analyzed using untargeted LC-HRMS. The data analysis revealed that the tyrosine metabolic pathway, including tyrosine, thyroxine, L-DOPA, and dopamine, was significantly altered in manifest HD compared with premanifest HD (Fig. 3). These metabolites demonstrated moderate to strong associations with measures of disease severity and symptoms. Thyroxine, dopamine, and lumichrome concentrations also correlated with the five-year risk of onset in premanifest HD subjects. Lumichrome is part of the riboflavin metabolism, where it is in strict equilibrium with the amount of riboflavin (vitamin B2), an essential compound in the brain that is transported from the

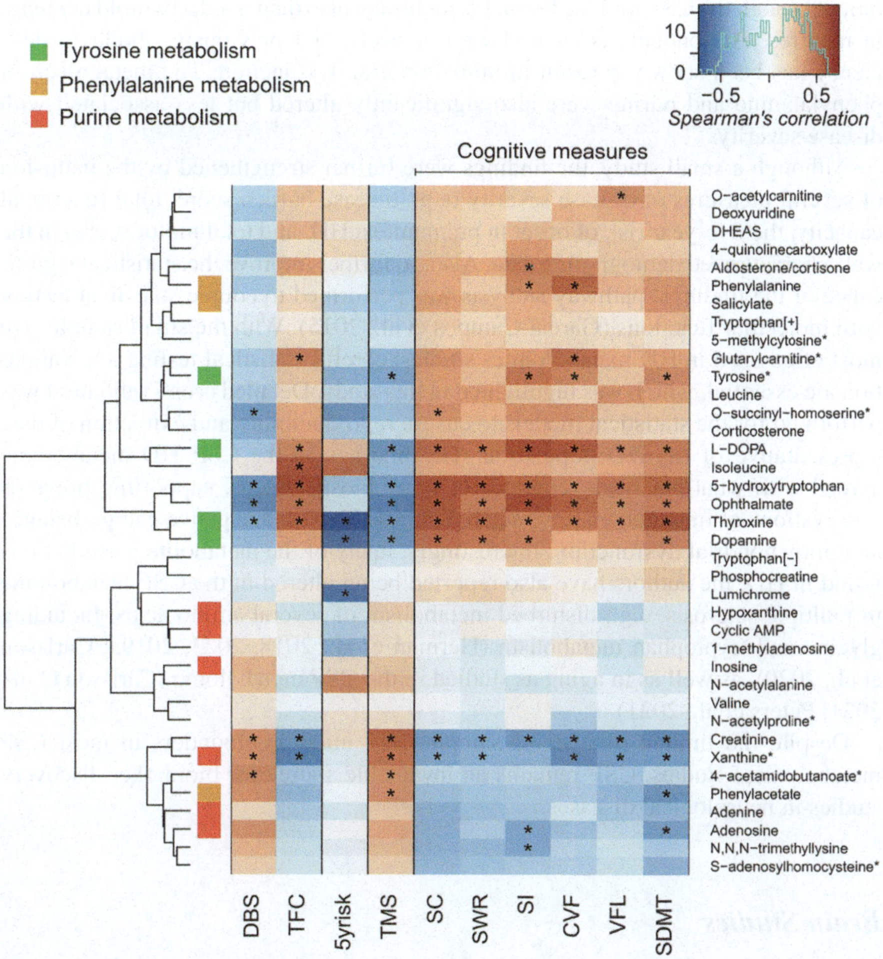

Fig. 3 Associations between CSF metabolites and HD symptoms in 26 mHTT carriers and the five-year risk of onset in the 13 premanifest HD subjects. Tyrosine, thyroxine, L-DOPA, and dopamine, were significantly altered in manifest HD compared with premanifest HD. These metabolites also demonstrated moderate to strong associations with measures of disease severity and symptoms. Thyroxine, dopamine, and lumichrome (part of the riboflavin metabolism, vitamin B2), concentrations also correlated with the five-year risk of onset in premanifest HD subjects. A '*' indicates a corresponding p-value < 0.05. Hierarchical clustering was performed on the correlation values where metabolites from the tyrosine metabolism, phenylalanine metabolism, and purine metabolism have been marked. DBS: disease burden score; TFC: total functional capacity; 5yrisk: the five-year risk of onset; TMS: total motor score; SC: Stroop color; SWR: Stroop word reading; SI: Stroop interference; CVF: category verbal fluency; VFL: verbal fluency letters; SDMT: symbol digit modalities test. Metabolites indicated with a '*' following the metabolite name demonstrated an age dependence which was corrected for in the analysis. (This figure is adapted from Herman et al. (2019b))

blood through the CSF into the brain. Lumichrome also displayed a twofold decrease in manifest HD patients compared with controls, and previously riboflavin deficiency has been shown to result in mitochondrial dysfunction. The metabolism of phenylalanine and purine were also significantly altered but less associated with disease severity.

Although a small study, the findings were further strengthened by the inclusion of several measures of disease severity (e.g., disease burden score, total functional capacity, the five-year risk of onset in premanifest HD, and total motor score) in the well-characterized demographic data. Also, to further improve the statistical significance of the findings, pathway analysis was performed to couple statistical testing with molecular functions (García-Campos et al., 2015). With the small sample size most often used in HD metabolomics studies, careful statistical testing and validation are essential, which was highlighted in this study. Detailed cross-validation was performed for the statistical models to ensure reproducibility and extraction of data representative of the HD population. The findings in the CSF HD metabolome agreed with what has been reported for other biospecimens, supporting previous observations of impaired energy metabolism in HD with a negative energy balance and mitochondrial dysfunction. Interestingly, many of the metabolite perturbations found in HD, the authors have also reported being altered in the CSF metabolome in multiple sclerosis, e.g., disturbed metabolism of several amino acids, including glycine and tryptophan metabolism (Herman et al., 2018, 2023, 2019a; Carlsson et al., 2020), as well as in aging as studied in the CSF metabolome (Carlsson et al., 2021; Peters et al., 2021).

Despite the limited number of samples and many confounders in most CSF metabolomics studies, CSF remains an invaluable source for biomarker discovery studies in neurological diseases.

Brain Studies

For metabolomics in neurodegenerative diseases such as HD, the brain is naturally the most relevant and accurate biospecimen for learning more about the biochemical and molecular processes involved in the disease. This is reflected by six published studies in HD brain metabolomics, the same number as for plasma and serum. These are all exciting works, providing close-to-the-source insights into the disease.

Sorolla et al. compared the brain HD metabolome between humans (eight HD cases and eight controls) and mice (eleven R6/1 transgenic and eleven wild-type mice) (Sorolla et al., 2016), which was mentioned earlier in the chapter (section "Mouse and Rat Models"). In humans, they investigated the striatum and cortex; in mice, the striatum, cortex, and cerebellum were studied. Using a targeted LC-MS approach, they observed impaired pyridoxal 5′-phosphate (PLP) dependent metabolism in the brains of both transgenic HD mice and HD patients. PLP is the active form of vitamin B6 and functions as a coenzyme in many important enzymatic reactions. An impaired PLP synthesis would affect essential pathways, including the synthesis of neurotransmitters like GABA and glutathione metabolism. The authors suggest that PLP treatment should be considered a potential therapy in HD.

In an extensive study, Patassini et al. used GC-MS to measure the brain metabolome in eleven brain regions in brains from nine HD patients and nine matched controls (Patassini et al., 2015). The studied brain regions were: putamen, globus pallidus, substantia nigra, cerebellum, hippocampus, middle temporal gyrus, cingulate gyrus, sensory cortex, motor cortex, middle frontal gyrus, and entorhinal cortex. Although an untargeted metabolomic approach is described, the article reports and focuses on a single metabolite, urea. In all eleven studied brain regions, the urea concentration was significantly elevated. According to the authors, the urea levels observed were sufficient to generate cytotoxicity in the brain. Based on these results, they hypothesize that a defective brain urea metabolism could be essential in the pathogenesis of neurodegeneration in HD, potentially, and osmoregulation or nitrogen metabolism.

In a follow-up study, the same group used GC-MS again to study the same set of samples. Still, this time reporting a more complete metabolomic profile (Patassini et al., 2016) with 63 metabolites altered in the HD brain samples. They again report accumulation of urea in HD, but also metabolites indicating regulatory defects in energy-transducing (e.g., glucose, threitol, glycerol, and ribitol) and amino acid (e.g., lysine, leucine, methionine, isoleucine, and threonine) pathways. Region-specific metabolic signatures were identified by comparing the metabolomic profiles of the eleven brain regions. For example, a distinctive amino-acid signature was observed in HD putamen with significantly increased levels of the aromatic amino acids tryptophan, phenylalanine, and tyrosine, which was not observed in the other brain regions. The region-specific metabolic signatures regions lay a foundation for an improved understanding of how HD cause localized metabolic perturbations in the CNS. According to the authors, the study demonstrates a widespread disruption of brain metabolism, indicating a complex phenotype with evidence of HD being a metabolic brain disease.

Across three publications, Graham et al. used both NMR and LC-MS to profile the striatum and frontal lobe metabolomes of 14 HD patients and 14 control subjects (Graham et al., 2016a, b, 2018). Using NMR, they were able to accurately discriminate between the HD patients and control subjects, with striatum metabolites (n = 15, including, e.g., aspartate, glycine, tyrosine, and urea) and frontal lobe metabolites (n = 4, including leucine, myo-inositol, phenylalanine, and tyrosine) perturbed in HD compared to controls (Graham et al., 2016b). In contrast to Patassini et al., who observed a significant accumulation of urea in eleven brain regions, Graham et al. observed substantially decreased (3.25-fold decrease) urea levels in the HD striatum. However, the striatum was not included among the eleven brain regions studied by Patassini et al. making a direct comparison difficult. It is not uncommon to observe such conflicting results when comparing results from metabolomic studies, highlighting the difficulty in performing reproducible metabolomics studies with limited subjects available.

In the first of two LC-MS studies, Graham et al. used an untargeted approach to profile the same subjects' striatum and frontal lobe metabolomes as in the NMR study (Graham et al., 2016a). A total of seven biochemical pathways were significantly altered in HD; in the striatum: (1) taurine and hypotaurine metabolism, (2) purine metabolism, (3) primary bile acid biosynthesis, and (4) nitrogen metabolism; and in the frontal lobe: (5) arginine and proline metabolism, (6) purine metabolism, and (7) glycine, serine, and threonine metabolism. Two metabolites, but-2-enoic acid, and uric acid, were significantly altered in both the frontal lobe and striatum and were suggested as HD biomarkers by the authors. In their second LC-MS study, they used a targeted metabolomics approach (Graham et al., 2018). Using a commercialized method including a panel of well-characterized metabolites, 185 metabolites were identified and quantified in the samples. Changes in energy metabolism and phospholipid metabolism were observed in HD, and significant reductions in neurotransmitters, e.g., glutamate, accompanied by increased spermidine levels. Based on the results, the authors suggest acylcarnitine and phospholipid metabolism to be critical in HD.

Enrichment Analysis of Metabolites Altered in Humans

As was performed for the animal and cell models metabolomics studies discussed above, a metabolite set enrichment analysis was performed using MetaboAnalyst (Pang et al., 2021) with the compiled list of altered metabolites extracted from the publications as input (Supplementary Table 2) (Fig. 4). The top three metabolite sets identified from the works describing human metabolomics studies in HD were aminoacyl-tRNA biosynthesis, arginine biosynthesis, and alanine, aspartate, and glutamate metabolism.

5 Conclusions, Limitations, and Future Perspectives

Since the first study in HD metabolomics was published in 2006 (Tsang et al., 2006), more than 30 studies using metabolomics approaches to study HD have been published. The large variety of study designs and techniques displayed in these works showcases the heterogeneous nature of metabolomics. Although these differences make direct comparisons between studies complex, some conclusions can be made based on the metabolite set enrichment analysis of metabolites altered in HD performed for this chapter (Figs. 2 and 4). In animal and cell model studies and human studies, arginine biosynthesis and aminoacyl-tRNA biosynthesis are included in the top three enriched metabolite sets, indicating the potential importance of these pathways in HD. Further exploring the metabolite set enrichment analysis finds arginine and proline metabolism, alanine, aspartate and glutamate metabolism, glycine, serine, and threonine metabolism among the top 10 enriched metabolite sets in animal and cell models and humans. This considerable overlap is

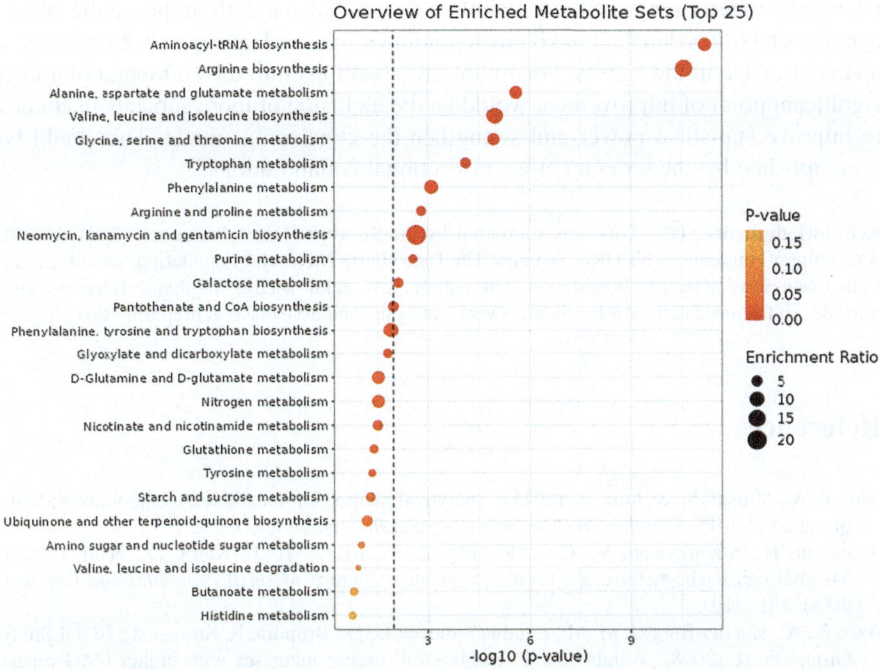

Fig. 4 A metabolite set enrichment analysis of the metabolites altered in published human metabolomics studies in Huntington's disease was performed using MetaboAnalyst, using the list of altered metabolites as input. The 25 most significant metabolite sets are shown in the figure, with the dashed line indicating the p-value 0.01 (−log10 = 2), which is here considered a threshold for statistical significance

a promising indication of the suitability of HD animal and cell models for studying the HD metabolome, with similar results to those of human studies.

A major limitation in all published HD metabolomics studies is the few subjects and samples studied. The often small studies suppress the statistical power and confidence of the findings. This limitation is present in most HD metabolomics studies. It requires clever study designs, bias correction, and addressing potential confounders to achieve the highest statistical confidence possible and to make relevant conclusions. Considering that thousands of potential metabolites may be detected in a single metabolomics study, hundreds to thousands of subjects would ideally be needed to characterize the HD metabolome with high statistical power (Blaise et al., 2016).

In conclusion, a wide range of metabolomics approaches has been applied for studies of the HD metabolome, with promising results for biomarker discovery for HD. The field of metabolomics is constantly improving and evolving, and with new improvements in instruments, analytical methods, and data analysis occurring, this will also benefit studies of the HD metabolome. The use of complementary methods, e.g., several LC-MS methods using different types of chromatography to cover the whole polarity range of the metabolome, would also greatly improve the coverage of metabolites detected in HD metabolomics studies, leading to deeper insights and confidence in the results. For future investigations into the HD metabolome, a significant point of improvement would be the inclusion of more subjects in studies to improve statistical power and strengthen the conclusions made. This could be accomplished by encouraging cross-institutional collaborations.

Acknowledgments This work was supported by the Swedish Research Council (2021-02189), Åke Wiberg Foundation, NEURO Sweden, The Lars Hierta Memorial Foundation, and O. E. and Edla Johanssons Scientific Foundation. The authors also acknowledge Stephanie Herman, Aina Vaivade, and Jenny Jakobsson for their artwork, analysis, and assistance in preparing this chapter.

References

Alonso, A., Marsal, S., & Julià, A. (2015). Analytical methods in untargeted metabolomics: State of the art in 2015. *Frontiers in Bioengineering and Biotechnology, 3*, 23.

Ashrafian, H., Sounderajah, V., Glen, R., Ebbels, T., Blaise, B. J., Kalra, D., et al. (2021). Metabolomics: The stethoscope for the twenty-first century. *Medical Principles and Practice, 30*(4), 301–310.

Aziz, N. A., van der Burg, J. M. M., Landwehrmeyer, G. B., Brundin, P., Stijnen, T., EHDI Study Group, et al. (2008). Weight loss in Huntington disease increases with higher CAG repeat number. *Neurology, 71*(19), 1506–1513.

Bates, G. P., & Hockly, E. (2003). Experimental therapeutics in Huntington's disease: Are models useful for therapeutic trials? *Current Opinion in Neurology, 16*(4), 465–470.

Bertrand, M., Decoville, M., Meudal, H., Birman, S., & Landon, C. (2020). Metabolomic nuclear magnetic resonance studies at presymptomatic and symptomatic stages of Huntington's disease on a drosophila model. *Journal of Proteome Research, 19*(10), 4034–4045.

Blaise, B. J., Correia, G., Tin, A., Young, J. H., Vergnaud, A. C., Lewis, M., et al. (2016). Power analysis and sample size determination in metabolic phenotyping. *Analytical Chemistry, 88*(10), 5179–5188.

Carlsson, H., Abujrais, S., Herman, S., Khoonsari, P. E., Åkerfeldt, T., Svenningsson, A., et al. (2020). Targeted metabolomics of CSF in healthy individuals and patients with secondary progressive multiple sclerosis using high-resolution mass spectrometry. *Metabolomics, 16*(2), 26.

Carlsson, H., Rollborn, N., Herman, S., Freyhult, E., Svenningsson, A., Burman, J., et al. (2021). Metabolomics of cerebrospinal fluid from healthy subjects reveal metabolites associated with ageing. *Metabolites* [Internet], *11*(2). Available from: https://doi.org/10.3390/metabo11020126

Carlsson, H., Vaivade, A., Emami Khoonsari, P., Burman, J., & Kultima, K. (2022). Evaluation of polarity switching for untargeted lipidomics using liquid chromatography coupled to high resolution mass spectrometry. *Journal of Chromatography. B, Analytical Technologies in the Biomedical and Life Sciences, 1195*, 123200.

Carroll, J. B., Deik, A., Fossale, E., Weston, R. M., Guide, J. R., Arjomand, J., et al. (2015). HdhQ111 mice exhibit tissue specific metabolite profiles that include striatal lipid accumulation. *PLoS One, 10*(8), e0134465.

Chang, K. L., New, L. S., Mal, M., Goh, C. W., Aw, C. C., Browne, E. R., et al. (2011). Metabolic profiling of 3-nitropropionic acid early-stage Huntington's disease rat model using gas chromatography time-of-flight mass spectrometry. *Journal of Proteome Research, 10*(4), 2079–2087.

Cheng, M. L., Chang, K. H., Wu, Y. R., & Chen, C. M. (2016). Metabolic disturbances in plasma as biomarkers for Huntington's disease. *The Journal of Nutritional Biochemistry, 31*, 38–44.

Chetwynd, A. J., Dunn, W. B., & Rodriguez-Blanco, G. (2017). Collection and preparation of clinical samples for metabolomics. In A. Sussulini (Ed.), *Metabolomics: From fundamentals to clinical applications* (pp. 19–44). Springer International Publishing.

Dettmer, K., Aronov, P. A., & Hammock, B. D. (2007). Mass spectrometry-based metabolomics. *Mass Spectrometry Reviews, 26*(1), 51–78.

Doherty, C. M., & Forbes, R. B. (2014). Diagnostic lumbar puncture. *The Ulster Medical Journal, 83*(2), 93–102.

Emami Khoonsari, P., Moreno, P., Bergmann, S., Burman, J., Capuccini, M., Carone, M., et al. (2019). Interoperable and scalable data analysis with microservices: Applications in metabolomics. *Bioinformatics, 35*(19), 3752–3760.

Emwas, A. H., Roy, R., McKay, R. T., Tenori, L., Saccenti, E., Gowda, G. A. N., et al. (2019). NMR spectroscopy for metabolomics research. *Metabolites* [Internet], *9*(7). Available from: https://doi.org/10.3390/metabo9070123.

Faquih, T. O., Aziz, N. A., Gardiner, S. L., Li-Gao, R., de Mutsert, R., Milaneschi, Y., et al. (2023). Normal range CAG repat size variations in the HTT gene are associated with an adverse lipoprotein profile partially mediated by body mass index. *Human Molecular Genetics*[Internet]. Available from: https://doi.org/10.1093/hmg/ddad020

Farzana, F., McConville, M. J., Renoir, T., Li, S., Nie, S., Tran, H., et al. (2023). Longitudinal spatial mapping of lipid metabolites reveals pre-symptomatic changes in the hippocampi of Huntington's disease transgenic mice. *Neurobiology of Disease, 176*, 105933.

Fiehn, O. (2016). Metabolomics by gas chromatography-mass spectrometry: Combined targeted and untargeted profiling. *Current Protocols in Molecular Biology, 114*, 30.4.1–30.4.32.

Galal, A., Talal, M., & Moustafa, A. (2022). Applications of machine learning in metabolomics: Disease modeling and classification. *Frontiers in Genetics, 13*, 1017340.

García-Campos, M. A., Espinal-Enríquez, J., & Hernández-Lemus, E. (2015). Pathway analysis: State of the art. *Frontiers in Physiology, 6*, 383.

Ghazaleh, N., Houghton, R., Palermo, G., Schobel, S. A., Wijeratne, P. A., & Long, J. D. (2021). Ranking the predictive power of clinical and biological features associated with disease progression in Huntington's disease. *Frontiers in Neurology, 12*, 678484.

Graham, S. F., Kumar, P., Bahado-Singh, R. O., Robinson, A., Mann, D., & Green, B. D. (2016a). Novel metabolite biomarkers of Huntington's disease as detected by high-resolution mass spectrometry. *Journal of Proteome Research, 15*(5), 1592–1601.

Graham, S. F., Kumar, P. K., Bjorndahl, T., Han, B., Yilmaz, A., Sherman, E., et al. (2016b). Metabolic signatures of Huntington's disease (HD): 1H NMR analysis of the polar metabolome in post-mortem human brain. *Biochim Biophys Acta (BBA), 1862*(9), 1675–1684.

Graham, S. F., Pan, X., Yilmaz, A., Macias, S., Robinson, A., Mann, D., et al. (2018). Targeted biochemical profiling of brain from Huntington's disease patients reveals novel metabolic pathways of interest. *Biochimica et Biophysica Acta - Molecular Basis of Disease, 1864*(7), 2430–2437.

Handley, R. R., Reid, S. J., Patassini, S., Rudiger, S. R., Obolonkin, V., McLaughlan, C. J., et al. (2016). Metabolic disruption identified in the Huntington's disease transgenic sheep model. *Scientific Reports, 6*, 20681.

Hashimoto, M., Watanabe, K., Miyoshi, K., Koyanagi, Y., Tadano, J., & Miyawaki, I. (2021). Multiplatform metabolomic analysis of the R6/2 mouse model of Huntington's disease. *FEBS Open Biology, 11*(10), 2807–2818.

Herman, S., Khoonsari, P. E., Tolf, A., Steinmetz, J., Zetterberg, H., Akerfeldt, T., et al. (2018). Integration of magnetic resonance imaging and protein and metabolite CSF measurements to enable early diagnosis of secondary progressive multiple sclerosis. *Theranostics, 8*(16), 4477–4490.

Herman, S., Åkerfeldt, T., Spjuth, O., Burman, J., Kultima, K. (2019a). Biochemical differences in cerebrospinal fluid between secondary progressive and relapsing–remitting multiple sclerosis. *Cells* [Internet], *8*(2). Available from: https://doi.org/10.3390/cells8020084.

Herman, S., Niemelä, V., Emami Khoonsari, P., Sundblom, J., Burman, J., Landtblom, A. M., et al. (2019b). Alterations in the tyrosine and phenylalanine pathways revealed by biochemical profiling in cerebrospinal fluid of Huntington's disease subjects. *Scientific Reports, 9*(1), 4129.
Herman, S., Arvidsson, M. S., Zjukovskaja, C., Emami, K. P., Svenningsson, A., Burman, J., et al. (2023). Disease phenotype prediction in multiple sclerosis. *iScience, 26*, 106906.
Hölter, S. M., Stromberg, M., Kovalenko, M., Garrett, L., Glasl, L., Lopez, E., et al. (2013). A broad phenotypic screen identifies novel phenotypes driven by a single mutant allele in Huntington's disease CAG knock-in mice. *PLoS One, 8*(11), e80923.
Hosseinkhani, F., Huang, L., Dubbelman, A. C., Guled, F., Harms, A. C., & Hankemeier, T. (2022). Systematic evaluation of HILIC stationary phases for global metabolomics of human plasma. *Metabolites* [Internet], *12*(2). Available from: https://doi.org/10.3390/metabo12020165.
Iuliano, M., Seeley, C., Sapp, E., Jones, E. L., Martin, C., Li, X., et al. (2021). Disposition of proteins and lipids in synaptic membrane compartments is altered in Q175/Q7 Huntington's disease mouse striatum. *Frontiers in Synaptic Neuroscience, 13*, 618391.
Jacobsen, J. C., Bawden, C. S., Rudiger, S. R., McLaughlan, C. J., Reid, S. J., Waldvogel, H. J., et al. (2010). An ovine transgenic Huntington's disease model. *Human Molecular Genetics, 19*(10), 1873–1882.
Joyner, P. M., Matheke, R. M., Smith, L. M., & Cichewicz, R. H. (2010). Probing the metabolic aberrations underlying mutant huntingtin toxicity in yeast and assessing their degree of preservation in humans and mice. *Journal of Proteome Research, 9*(1), 404–412.
Klassen, A., Faccio, A. T., Canuto, G. A. B., da Cruz, P. L. R., Ribeiro, H. C., Tavares, M. F. M., et al. (2017). Metabolomics: Definitions and significance in systems biology. In A. Sussulini (Ed.), *Metabolomics: From fundamentals to clinical applications* (pp. 3–17). Springer International Publishing.
Kong, G., Ellul, S., Narayana, V. K., Kanojia, K., Ha, H. T. T., Li, S., et al. (2021). An integrated metagenomics and metabolomics approach implicates the microbiota-gut-brain axis in the pathogenesis of Huntington's disease. *Neurobiology of Disease, 148*, 105199.
Kumar, K. K., Goodwin, C. R., Uhouse, M. A., Bornhorst, J., Schwerdtle, T., Aschner, M., et al. (2015). Untargeted metabolic profiling identifies interactions between Huntington's disease and neuronal manganese status†. *Metallomics, 7*(2), 363–370.
Lamichhane, S., Sen, P., Dickens, A. M., Hyötyläinen, T., & Orešič, M. (2018). Chapter fourteen - an overview of metabolomics data analysis: Current tools and future perspectives. In J. Jaumot, C. Bedia, & R. Tauler (Eds.), *Comprehensive analytical chemistry* (pp. 387–413). Elsevier.
Li, J. Y., Popovic, N., & Brundin, P. (2005). The use of the R6 transgenic mouse models of Huntington's disease in attempts to develop novel therapeutic strategies. *NeuroRx, 2*(3), 447–464.
Lu, W., Su, X., Klein, M. S., Lewis, I. A., Fiehn, O., & Rabinowitz, J. D. (2017). Metabolite measurement: Pitfalls to avoid and practices to follow. *Annual Review of Biochemistry, 86*, 277–304.
Mangiarini, L., Sathasivam, K., Seller, M., Cozens, B., Harper, A., Hetherington, C., et al. (1996). Exon 1 of the HD gene with an expanded CAG repeat is sufficient to cause a progressive neurological phenotype in transgenic mice. *Cell, 87*(3), 493–506.
Mastrokolias, A., Pool, R., Mina, E., Hettne, K. M., van Duijn, E., van der Mast, R. C., et al. (2016). Integration of targeted metabolomics and transcriptomics identifies deregulation of phosphatidylcholine metabolism in Huntington's disease peripheral blood samples. *Metabolomics, 12*, 137.
McGarry, A., Gaughan, J., Hackmyer, C., Lovett, J., Khadeer, M., Shaikh, H., et al. (2020). Cross-sectional analysis of plasma and CSF metabolomic markers in Huntington's disease for participants of varying functional disability: A pilot study. *Scientific Reports, 10*(1), 20490.
Menalled, L. B. (2005). Knock-in mouse models of Huntington's disease. *NeuroRx, 2*(3), 465–470.
Newgard, C. B. (2017). Metabolomics and metabolic diseases: Where do we stand? *Cell Metabolism, 25*(1), 43–56.

Nikas, J. B., & Low, W. C. (2011). Application of clustering analyses to the diagnosis of Huntington disease in mice and other diseases with well-defined group boundaries. *Computer Methods and Programs in Biomedicine, 104*(3), e133–e147.

Novella, J. A., Emami Khoonsari, P., Herman, S., Whitenack, D., Capuccini, M., Burman, J., et al. (2019). Container-based bioinformatics with pachyderm. *Bioinformatics, 35*(5), 839–846.

Pang, Z., Chong, J., Zhou, G., de Lima Morais, D. A., Chang, L., Barrette, M., et al. (2021). MetaboAnalyst 5.0: Narrowing the gap between raw spectra and functional insights. *Nucleic Acids Research, 49*(W1), W388–W396.

Patassini, S., Begley, P., Reid, S. J., Xu, J., Church, S. J., Curtis, M., et al. (2015). Identification of elevated urea as a severe, ubiquitous metabolic defect in the brain of patients with Huntington's disease. *Biochemical and Biophysical Research Communications, 468*(1–2), 161–166.

Patassini, S., Begley, P., Xu, J., Church, S. J., Reid, S. J., Kim, E. H., et al. (2016). Metabolite mapping reveals severe widespread perturbation of multiple metabolic processes in Huntington's disease human brain. *Biochimica et Biophysica Acta, 1862*(9), 1650–1662.

Pautova, A., Burnakova, N., Revelsky, A. (2021). Metabolic profiling and quantitative analysis of cerebrospinal fluid using gas chromatography-mass spectrometry: Current methods and future perspectives. *Molecules* [Internet], *26*(12). Available from: https://doi.org/10.3390/molecules26123597.

Perez de Souza, L., Alseekh, S., Scossa, F., & Fernie, A. R. (2021). Ultra-high-performance liquid chromatography high-resolution mass spectrometry variants for metabolomics research. *Nature Methods, 18*(7), 733–746.

Peters, K., Bradbury, J., Bergmann, S., Capuccini, M., Cascante, M., de Atauri, P., et al. (2019). PhenoMeNal: Processing and analysis of metabolomics data in the cloud. *Gigascience* [Internet], *8*(2). Available from: https://www.ncbi.nlm.nih.gov/pubmed/30535405

Peters, K., Herman, S., Khoonsari, P. E., Burman, J., Neumann, S., & Kultima, K. (2021). Metabolic drift in the aging nervous system is reflected in human cerebrospinal fluid. *Scientific Reports, 11*(1), 18822.

Pirhaji, L., Milani, P., Leidl, M., Curran, T., Avila-Pacheco, J., Clish, C. B., et al. (2016). Revealing disease-associated pathways by network integration of untargeted metabolomics. *Nature Methods, 13*(9), 770–776.

Pont, L., Benavente, F., Jaumot, J., Tauler, R., Alberch, J., Ginés, S., et al. (2016). Metabolic profiling for the identification of Huntington biomarkers by on-line solid-phase extraction capillary electrophoresis mass spectrometry combined with advanced data analysis tools. *Electrophoresis, 37*(5–6), 795–808.

Pradhan, S. S., Thota, S. M., Rajaratnam, S., Bhagavatham, S. K. S., Pulukool, S. K., Rathnakumar, S., et al. (2022). Integrated multi-omics analysis of Huntington disease identifies pathways that modulate protein aggregation. *Disease Models & Mechanisms* [Internet], *15*(10). Available from: https://doi.org/10.1242/dmm.049492

Ramautar, R. (2016). Chapter one - capillary electrophoresis–mass spectrometry for clinical metabolomics. In G. S. Makowski (Ed.), *Advances in clinical chemistry* (pp. 1–34). Elsevier.

Rathahao-Paris, E., Alves, S., Junot, C., & Tabet, J. C. (2015). High resolution mass spectrometry for structural identification of metabolites in metabolomics. *Metabolomics, 12*(1), 10.

Rico, E., González, O., Blanco, M. E., & Alonso, R. M. (2014). Evaluation of human plasma sample preparation protocols for untargeted metabolic profiles analyzed by UHPLC-ESI-TOF-MS. *Analytical and Bioanalytical Chemistry, 406*(29), 7641–7652.

Robertson, D. G. (2005). Metabonomics in toxicology: A review. *Toxicological Sciences, 85*(2), 809–822.

Rosas, H. D., Doros, G., Bhasin, S., Thomas, B., Gevorkian, S., Malarick, K., et al. (2015). A systems-level "misunderstanding": The plasma metabolome in Huntington's disease. *Annals of Clinical Translational Neurology, 2*(7), 756–768.

Sasaki, K., Sagawa, H., Suzuki, M., Yamamoto, H., Tomita, M., Soga, T., et al. (2019). Metabolomics platform with capillary electrophoresis coupled with high-resolution mass apectrometry for plasma analysis. *Analytical Chemistry, 91*(2), 1295–1301.

Sayers, E. W., Bolton, E. E., Brister, J. R., Canese, K., Chan, J., Comeau, D. C., et al. (2022). Database resources of the national center for biotechnology information. *Nucleic Acids Research, 50*(D1), D20–D26.

Segal, M. B. (1993). Extracellular and cerebrospinal fluids. *Journal of Inherited Metabolic Disease, 16*(4), 617–638.

Singh, V., Sharma, R. K., Athilingam, T., Sinha, P., Sinha, N., & Thakur, A. K. (2017). NMR spectroscopy-based metabolomics of drosophila model of Huntington's disease suggests altered cell energetics. *Journal of Proteome Research, 16*(10), 3863–3872.

Skene, D. J., Middleton, B., Fraser, C. K., Pennings, J. L. A., Kuchel, T. R., Rudiger, S. R., et al. (2017). Metabolic profiling of presymptomatic Huntington's dis202ease sheep reveals novel biomarkers. *Scientific Reports, 7*, 43030.

Smith, C. A., O'Maille, G., Want, E. J., Qin, C., Trauger, S. A., Brandon, T. R., et al. (2005). METLIN: A metabolite mass spectral database. *Therapeutic Drug Monitoring, 27*(6), 747–751.

Sorolla, M. A., Rodríguez-Colman, M. J., Vall-Llaura, N., Vived, C., Fernández-Nogales, M., Lucas, J. J., et al. (2016). Impaired PLP-dependent metabolism in brain samples from Huntington disease patients and transgenic R6/1 mice. *Metabolic Brain Disease, 31*(3), 579–586.

Spick, M., Hancox, T. P. M., Chowdhury, N. R., Middleton, B., Skene, D. J., Morton, A. J. (2023). Metabolomic Analysis of plasma in Huntington's disease transgenic sheep (Ovis aries) reveals progressive circadian rhythm dysregulation. *Journal of Huntington's Disease* [Internet]. Available from: https://doi.org/10.3233/JHD-220552

Trushina, E., Dutta, T., Persson, X. M. T., Mielke, M. M., & Petersen, R. C. (2013). Identification of altered metabolic pathways in plasma and CSF in mild cognitive impairment and Alzheimer's disease using metabolomics. *PLoS One, 8*(5), e63644.

Tsang, T. M., Woodman, B., McLoughlin, G. A., Griffin, J. L., Tabrizi, S. J., Bates, G. P., et al. (2006). Metabolic characterization of the R6/2 transgenic mouse model of Huntington's disease by high-resolution MAS 1H NMR spectroscopy. *Journal of Proteome Research, 5*(3), 483–492.

Tsang, T. M., Haselden, J. N., & Holmes, E. (2009). Metabonomic characterization of the 3-nitropropionic acid rat model of Huntington's disease. *Neurochemical Research, 34*(7), 1261–1271.

Verwaest, K. A., Vu, T. N., Laukens, K., Clemens, L. E., Nguyen, H. P., Van Gasse, B., et al. (2011). 1H NMR based metabolomics of CSF and blood serum: A metabolic profile for a transgenic rat model of Huntington disease. *Biochim Biophys Acta (BBA), 1812*(11), 1371–1379.

Vodicka, P., Mo, S., Tousley, A., Green, K. M., Sapp, E., Iuliano, M., et al. (2015). Mass spectrometry analysis of wild-type and knock-in Q140/Q140 Huntington's disease mouse brains reveals changes in glycerophospholipids including alterations in phosphatidic acid and lysophosphatidic acid. *Journal of Huntington's Disease, 4*(2), 187–201.

Williams, B. B., Li, D., Wegrzynowicz, M., Vadodaria, B. K., Anderson, J. G., Kwakye, G. F., et al. (2010). Disease-toxicant screen reveals a neuroprotective interaction between Huntington's disease and manganese exposure. *Journal of Neurochemistry, 112*(1), 227–237.

Wishart, D. S., Tzur, D., Knox, C., Eisner, R., Guo, A. C., Young, N., et al. (2007). HMDB: The human metabolome database. *Nucleic Acids Research, 35*(Database issue), D521–D526.

Wishart, D. S., Cheng, L. L., Copié, V., Edison, A. S., Eghbalnia, H. R., Hoch, J. C., et al. (2022a). NMR and metabolomics-a roadmap for the future. *Metabolites* [Internet], *12*(8). Available from: https://doi.org/10.3390/metabo12080678

Wishart, D. S., Guo, A., Oler, E., Wang, F., Anjum, A., Peters, H., et al. (2022b). HMDB 5.0: The human metabolome database for 2022. *Nucleic Acids Research, 50*(D1), D622–D631.

Züllig, T., Trötzmüller, M., & Köfeler, H. C. (2020). Lipidomics from sample preparation to data analysis: A primer. *Analytical and Bioanalytical Chemistry, 412*(10), 2191–2209.

Proteomics in Huntington's Disease Biomarker Discovery

Jakub Červenka, Kateřina Budková, Rita Suchá, Petr Vodička, and Eleni Voukali

Abbreviations

2-DE	Two-dimensional gel electrophoresis
AD	Alzheimer's disease
ALDH2	Mitochondrial aldehyde dehydrogenase
APOA4	Apolipoprotein A-IV
BACHD	Bacterial artificial chromosome HD
BASP1	Brain abundant membrane attached signal protein 1
BDNF	Brain-derived neurotrophic factor
CAG	Cytosine-adenine-guanine
CAP	CAG-age product
Cb1	Cannabinoid receptor 1
Ccl5	Chemokine (C-C motif) ligand 5/regulated on activation normal T cell expressed and secreted
CKB	Creatine kinase
CLU	Clusterin
CNS	Central nervous system
Coro1a	Coronin-1A
CPE	Carboxypeptidase E
CRP	C-reactive protein
Cryab	Alpha-crystallin B chain
CSF	Cerebrospinal fluid
DDA	Data-dependent acquisition
DIA	Data-independent acquisition

Supplementary Information The online version contains supplementary material available at https://doi.org/10.1007/978-3-031-32815-2_9.

J. Červenka · K. Budková · R. Suchá · P. Vodička · E. Voukali (✉)
Laboratory of Applied Proteome Analyses, Institute of Animal Physiology and Genetics, Czech Academy of Sciences, Libechov, Czech Republic
e-mail: Voukali@iapg.cas.cz

E. A. Thomas, G. M. Parkin (eds.), *Biomarkers for Huntington's Disease*, Contemporary Clinical Neuroscience, https://doi.org/10.1007/978-3-031-32815-2_9

DIGE	Differential in gel electrophoresis
Dlst	Mitochondrial dihydrolipoamide S-succinyltransferase
DYNC1LI1	Cytoplasmic dynein 1 light intermediate chain 1
ECM	Extracellular matrix
ELISA	Enzyme-linked immunosorbent assay
Eno1	α-Enolase
F2	Prothrombin
Flot2	Flotillin-2
FTL	Ferritin light polypeptide
GABA	Gamma aminobutyric acid
GFAP	Glial fibrillary acidic protein
GluI	Glutamine synthetase
GM2A	Ganglioside GM2 activator
GO	Gene ontology
Got2	Aspartate aminotransferase
GRIA4	Glutamate receptor 4
Gspt2	G1 to S phase transition 2
GSTM2	Glutathione S-transferase mu-2
HD	Huntington's disease
hESCs	Human embryonic stem cells
HP	Haptoglobin
Hspa9	Stress-70 protein
Hypk	Huntingtin interacting protein K
Iapp	Amylin
ICAT	Isotope-coded affinity tag
IL6	Interleukin 6
Ins1/2	Insulin
iPSCs	Induced pluripotent stem cells
IsoHD	Isogenic HD
iTRAQ	Isobaric tags for relative and absolute quantitation
LAP3	Cytosol aminopeptidase
LC	Liquid chromatography
Lep	Leptin
Lmna	Prelamin-A/C
M/z	Mass-to-charge
mHTT	Mutant huntingtin
MOB4	MOB family member 4
MS	Mass spectrometry
MS/MS	Tandem MS
Mtor	Serine/threonine-protein kinase mTOR
NCAM3	Neural cell adhesion molecule 2
NDUFS3	NADH dehydrogenase [ubiquinone] iron-sulfur protein 3, mitochondrial
NFASC	Neurofascin
NfL	Neurofilament light polypeptide
NPCs	Neural progenitor cells

NRXN3	Neurexin 3
OMGP	Oligodendrocyte-myelin glycoprotein
Park7	Parkinson disease protein 7 homolog
PD	Parkinson's disease
Pde10a	Phosphodiesterase 10a
PDYN	Proenkephalin-B
Pebp1	Phosphatidylethanolamine-binding protein 1
PENK	Proenkephalin-A
Pkm2	Pyruvate kinase PKM
PolyQ	Polyglutamine tract
PPFIA3	Liprin-alpha-3
Prdx3	Peroxiredoxin-3
PRM	Parallel reaction monitoring
PTMs	Post-translational modifications
Pyy	Peptide Y
Ran	GTP-binding nuclear protein Ran
Scn4b	Sodium channel subunit 4
SILAC	Stable isotope labelling by amino acids in cell culture
Sod2	Superoxide dismutase [Mn], mitochondrial
SRM	Selected reaction monitoring
STXBP3	Syntaxin-binding protein 3
TMTs	Tandem mass tags
TOF	Time-of-flight
TTR	T_4-binding protein
Vim	Vimentin
YAC	Yeast artificial chromosome

1 Huntington's Disease and the Need for Biomarkers

Huntington's disease (HD) is a progressive lethal neurodegenerative condition with a prevalence of 4.88 per 100,000 (Medina et al., 2022). HD is inherited in an autosomal dominant manner and caused by a single genetic mutation; an expansion of the CAG (cytosine-adenine-guanine) repeats in the huntingtin (*HTT*) gene on chromosome 4. Translation of this gene produces a mutated huntingtin protein (mHTT) with an abnormally long polyglutamine tract (polyQ) at its N-terminal end (The Huntington's Disease Collaborative Research Group, 1993). Expansions longer than 35 CAG repeats lead to HD (Kremer et al., 1994), and the length of the polyQ tract influences the age of disease onset (Rosenblatt et al., 2003, 2006, 2012; Wexler, 2004) (Fig. 1a). HD commonly manifests during the fourth decade of life (Fusilli et al., 2018). The gene-expansion carrier will insidiously and ultimately develop a range of symptoms referred to as Huntington's triad, i.e., choreiform movements, psychiatric problems and cognitive decline (Bates et al., 2015; Ross et al., 2014).

HTT is broadly expressed throughout the central nervous system (CNS) and most tissues (Carroll et al., 2015; Marques Sousa & Humbert, 2013), with widespread subcellular localisation that reflects its pleiotropic functions. HTT protein interacts with a wide range of proteins (Wanker et al., 2019), and participates in many cellular processes, acting as a molecular scaffold and multivalent interaction hub (Guo et al., 2018; Saudou & Humbert, 2016). Two main pathogenic processes have been suggested as a cause of neurodegeneration in HD. The first process involves the loss of HTT function and toxic gain of function of the polyQ repeats, which may disrupt a number of cellular events (Gil & Rego, 2008; Saudou & Humbert, 2016; Zuccato et al., 2001) (Fig. 1b). The second process involves toxic fragments of mHTT released as a result of proteolytic cleavages as post-translational modifications (PTMs) of full-length mHTT (El-Daher et al., 2015). These fragments of mHTT then form large intranuclear inclusions (DiFiglia et al., 1997; Riguet et al., 2021), but the role of these aggregates in neuropathology remains controversial (Kuemmerle et al., 1999; Michalik & Van Broeckhoven, 2003).

Neuropathological features in HD brains are well-characterised. The striatum is the first affected region, but serial atrophic changes occur throughout variable cortical areas, with a loss of brain weight by up to 30% (Aylward et al., 2011; Halliday et al., 1998; Rosas et al., 2008). Striatal volume loss is the hallmark of HD and occurs in a specific temporal manner; cortical thinning is considered an associated pathology (Vonsattel, 2008). Striatal gamma-aminobutyric acid (GABA)-ergic medium spiny neurons are lost, with prominent degenerative changes in dendrites, whereas medium spiny interneurons are not affected (Ferrante et al., 1987; Graveland et al., 1985).

Although the symptomatology of HD is generally well-defined, individual symptoms may vary. Besides CNS pathology, HD patients may suffer from metabolic disturbances, skeletal muscle wasting, weight loss, cardiac failure, testicular atrophy, osteoporosis (van der Burg et al., 2009), and congenital immune dysfunction, which is increasingly recognised as one of the key characteristics of HD and other neurodegenerative diseases (Andre et al., 2016). The HD patient usually succumbs 20 years following symptoms onset due to aspiration pneumonia (Heemskerk & Roos, 2012). Since there is no cure available, HD management is currently best provided by a multidisciplinary team through supportive and symptomatic treatment (Bachoud-Lévi et al., 2019).

A long pre-symptomatic period offers a potential therapeutic window for experimental HD interventions, such as gene-silencing approaches to lower mHTT levels, which are the focus of some clinical trials (Estevez-Fraga et al., 2022; Tabrizi et al., 2022). International consensus has established a set of motor, cognitive and neuroimaging assessments with predictive value for HD onset and progression (Tabrizi et al., 2022). However, during the prodromal stage in the absence of clinical symptoms, reliable biochemical biomarkers would allow more effective monitoring of a given intervention. An ideal biomarker should i) reflect the underlying neuropathology and/or disease progression, ii) show minimal variability between subjects, and iii) be obtainable from a biofluid with minimal invasion (Biomarkers Definitions Working Group, 2001). Molecular changes detected in cerebrospinal fluid (CSF) and blood of HD patients propose many potential biomarkers (Byrne & Wild, 2016; Przybyl et al., 2021; Rodrigues et al., 2018; Silajdžić & Björkqvist, 2018; Tabrizi et al., 2022), but

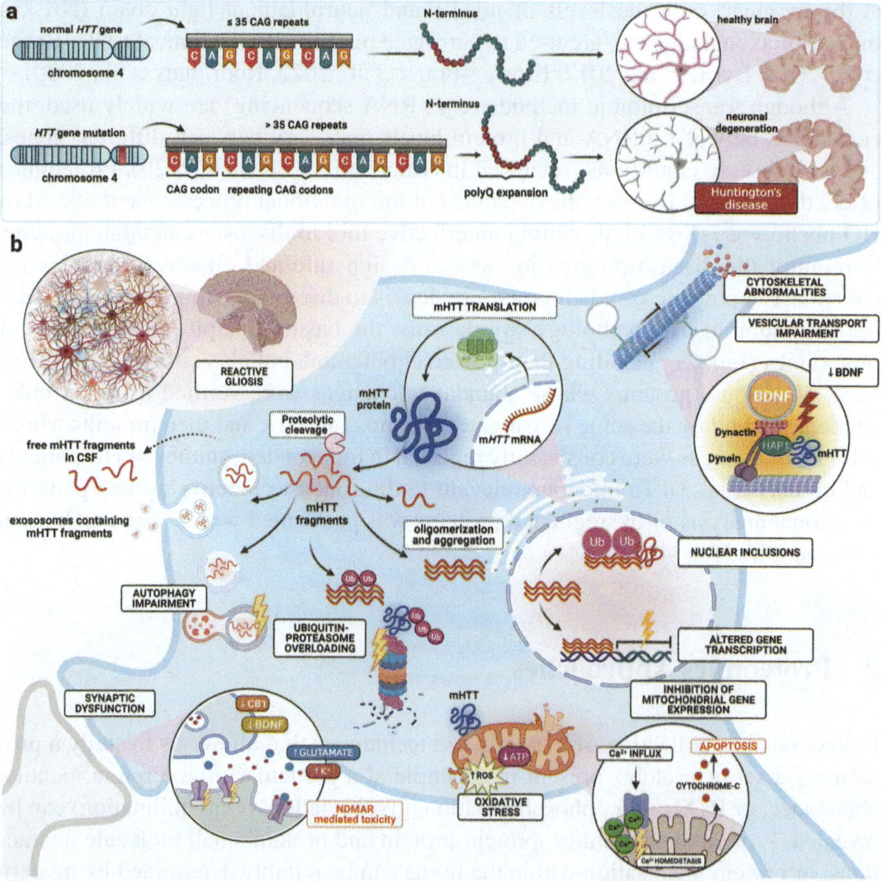

Fig. 1 Pathogenetic mechanisms in Huntington's disease (HD). (**a**) HD is caused by a mutation in the *HTT* gene, which leads to CAG repeats expansion resulting in the production of the toxic mutant huntingtin (mHTT) protein. As a result of the disruption of mHTT functions, degeneration of neurons occurs. (**b**) Dysregulated biological pathways in HD after production of insoluble N-terminal mHTT fragments with a tendency to oligomerize and aggregate. The fragments form nuclear inclusions. Some of the aggregates enter catabolic degradation pathways (ubiquitin-proteasome system and autophagy), which are overwhelmed, abnormal proteins accumulate, and protein homeostasis is disrupted. The effect of mHTT on vesicular transport is closely related to binding to molecular motors and HAP1 protein, transport of vesicles slows down or stops. Synaptic dysfunctions are caused by a combination of many factors, e.g., reduction of brain-derived neurotrophic factor (BDNF) in cortical synapses, downregulation of *BDNF* transcription, switching of synaptic N-methyl-D-aspartate receptor signalling, and the associated excitotoxicity. mHTT affects a range of mitochondrial processes, such as mitochondrial gene expression, resulting in metabolic abnormalities and oxidative stress. mHTT interacts with the outer mitochondrial membrane and reduces the level of Ca^{2+} required for mitochondrial permeability transition pore opening, leading to cytochrome C release and apoptosis. Astrocytes and microglia transit to reactive morphology causing neuroinflammation and gliosis. (The figure was created with BioRender.com)

at the moment, only the levels of mHTT and neurofilament light chain (NfL), a marker of axonal damage, are used as surrogate markers in HD clinical trials (Byrne et al., 2018; Byrne et al., 2017; Estevez-Fraga et al., 2022; Rodrigues et al., 2020).

Although transcriptomic methods (e.g., RNA sequencing) are widely used, the correlation between mRNA and protein levels may vary between different genes, and with disease conditions (reviewed in (Buccitelli & Selbach, 2020). Proteomics offers the potential to study the dynamics of the functional processes at the level of HD phenotypes (Fig. 1), providing an effective tool to discover candidate proteins to monitor the disease progression and to design safe and effective therapies for patients. The purpose of this literature review is to discuss existing proteomic studies recapitulating HD pathology. We describe the basic principles and methods of proteomics (Fig. 2), including their powerful potential, but also current limitations. First, we discuss proteins whose abundance changes were verified using an independent method by the same investigators (Supp. Table 1), and then proteins whose abundance changes were consistently reported in independent studies of HD patients and models (Fig. 3). To integrate relevant findings and to identify protein patterns, functional analysis of dysregulated proteins was performed with a focus on human CNS samples (Fig. 4).

2 Proteomics Approaches

Proteomics is a collection of methods and techniques that allows us to study a proteome – a set of proteins present in a sample at a particular time. Protein identity, abundance, or PTMs (e.g., phosphorylation, glycosylation, or ubiquitination) can be revealed. Protein 3D structures, protein-protein and protein-small molecule interactions, or protein localisation within the tissue can be reliably determined by modern proteomics (Aebersold & Mann, 2016). Proteomics methods can be divided into two categories, the antibody-based, and the methods based on mass spectrometry (MS) (Fig. 2a). However, the division between these two groups is not absolute, as some techniques combine both approaches (e.g., mass cytometry). This subchapter briefly summarises the particular methods enabling global and targeted proteome analyses.

Fig. 2 (continued) reaction monitoring (PRM) and selected reaction monitoring (SRM), require previous knowledge about proteins, i.e., the mass-to-charge (*m/z*) ratios of precursor ions for selected peptides and their retention times. The first quadrupole filters out all other precursor ions than the targeted one, which is sent to the second quadrupole for fragmentation. In PRM, full MS2 spectra of all fragment ions are measured by the third high-resolution mass analyser. For SRM assays, *m/z* ratios of fragment ions are also defined, the third quadrupole filters out all other fragment ions than the targeted one, and several fragment ions are sequentially analysed. Data are collected and integrated for relative or absolute proteins quantification. (The figure was created with BioRender.com)

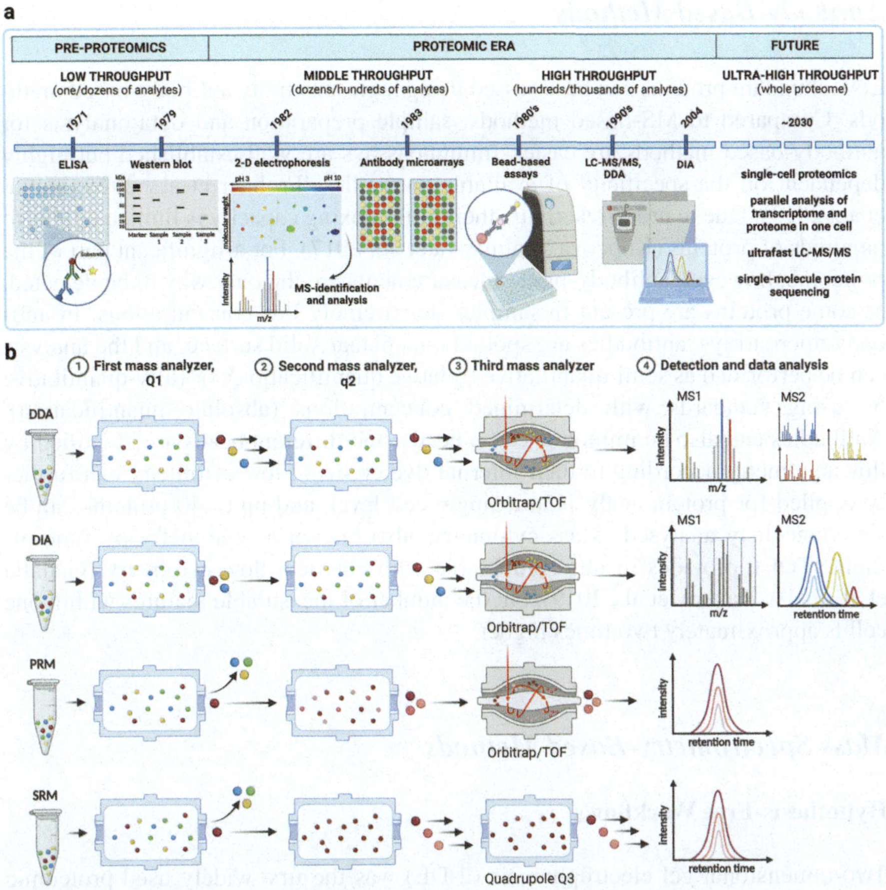

Fig. 2 Evolution of proteomic techniques and the main approaches of bottom-up proteomics. (**a**) The timeline of proteomic technology tools starts with Western blot and enzyme-linked immunosorbent assay (ELISA) that represent protein biology/biochemistry techniques with rather limited multiplexing capabilities. The "true" proteomic methods are 2-D electrophoresis coupled with mass spectrometry (MS), antibody microarrays and bead-based assays. In the early 1990s, liquid chromatography (LC) coupled to tandem MS (LC-MS/MS) instruments with data-dependent acquisition (DDA) approach were introduced. Later, data-independent acquisition (DIA) methods improved quantification reliability and robustness. In the next decade, ultrafast MS-based methods may dominate in proteomics, mainly based on the DIA principle with a focus on single-cell proteomics, eventually combined with transcriptomics. (**b**) Main stages of bottom-up proteomic techniques. Bottom-up MS proteomic experiments begin with the digestion of proteins. All peptides of the proteins (colour balls in tubes) are separated (LC) and ionised before they enter the tandem MS instrument. For both global approaches, DDA and DIA, the complete spectrum of precursor ions (MS1) is measured by high-resolution Orbitrap or time-of-flight (TOF) mass analysers. Then, in the DDA approach, a defined number of precursor ions is sequentially selected by the first quadrupole (Q1) mass analyser, fragmented in the second mass analyser (q2), and product ions are measured in the third mass analyser, generating MS2 spectra. In the DIA approach, the MS1 spectrum is divided into defined windows and all precursor ions within this window are fragmented and analysed together (MS2). When all selected precursor ions in DDA or all windows in DIA are measured, another cycle starts with a new MS1 experiment. Both targeted approaches, parallel

Antibody-Based Methods

Low-abundant proteins can be assessed using high sensitivity antibody-based methods. Compared to MS-based methods, sample preparation and data analysis for antibody-based methods are easier. Immunoassays are well-established but highly dependent on the specificity of available antibodies (Frohner et al., 2020; Hafko et al., 2013). Due to cross-reactivity, their multiplexing capacity is limited to tens to hundreds of proteins (Kupcova Skalnikova et al., 2017). For a significant part of the proteome, however, antibody-based measurements are the only way to be detected, as some proteins are present in samples in extremely low concentrations. In antibody microarrays, antibodies are spotted on a planar solid surface, and the analysis can be performed as semi-quantitative (relative quantification), or fully-quantitative by using standards with determined concentrations (absolute quantification). Antibodies can also be immobilised on beads, and different beads are identified by flow cytometry according to their internal dye or size. Flow cytometry can further be applied for protein analysis at a single-cell level, and up to 40 proteins can be simultaneously analysed. Mass cytometry, also known as cytometry by time-of-flight (TOF), provides results comparable with classical flow cytometry (Gadalla et al., 2019; Ravkov et al., 2019), but the number of measurable features within one cell is approximately two times higher.

Mass Spectrometry-Based Methods

Hypothesis-Free Workflows

Two-dimensional gel electrophoresis (2-DE) was the first widely used proteomic method able to quantify hundreds to thousands of proteins within a sample. Denatured proteins are separated by isoelectric focusing at the first dimension and at the second dimension, proteins are separated by SDS polyacrylamide gel electrophoresis (SDS-PAGE) according to their molecular weight. Proteins are visualised, proteins maps are compared, and differential protein spots are analysed by MS (Thiede et al., 2005). MS instruments are reviewed in (Haag, 2016). 2-DE suffers from lower reproducibility (Zhou et al., 2005), together with high variability between gels, although this issue may be overcome by the employment of differential in gel electrophoresis (DIGE) (Ünlü et al., 1997).

Nowadays, the most widely used MS method is a bottom-up proteomics approach where proteins are cleaved into peptides, separated by liquid chromatography (LC), and measured by the MS instrument. Proteins identification and quantification are then computationally inferred from the peptide data (Zhang et al., 2013). MS-based methods offer very high specificity and unprecedented depth of analysis (more than 10,000 proteins can be quantified simultaneously in one sample) (Meier et al., 2018). On the other hand, analysis of complex samples (e.g., blood plasma) is challenging due to the extremely high concentration range of proteins (Anderson &

Anderson, 2002). For global proteome profiling, data-dependent acquisition (DDA) and data-independent acquisition (DIA) are used. Parallel reaction monitoring (PRM) and selected reaction monitoring (SRM) are used in hypothesis-driven proteomics (Fig. 2b).

DDA, also known as shotgun proteomics, is a global approach for proteome analysis (Aebersold & Mann, 2016). The exact mass-to-charge (*m/z*) ratio of peptides (precursor ions) and their signal intensities are analysed in the MS1 experiment. Precursor ions are then sequentially selected by the first mass analyser in a signal intensity-dependent manner (e.g., 30 precursor ions from the highest to lower signal intensity in MS1) and fragmented in the second mass analyser. The product ions (smaller peptides and amino acid residues from a particular precursor ion, also called fragment ions) are then analysed by the third mass analyser (MS2 experiment). A new cycle starts with the following MS1 experiment, once MS2 experiments of all selected precursor ions are measured. Detected masses of the precursor ions and their product ions are computationally compared with theoretical masses from an *in silico* digested protein database of the analysed organism for protein identification. The major advantage of the DDA method is its high throughput. Without a need for prior knowledge about the sample proteome, hundreds or thousands of proteins can be identified. The main drawback is its partial stochasticity, as precursor ions are selected according to their signal intensity. A particular precursor ion may be detectable, but below the number of precursors selected for fragmentation. This would result in a missing data point that would bias protein quantification.

To reduce the DDA stochasticity, various labelling methods that enable researchers to mix or pool the peptides from several samples into one were developed, e.g., stable isotope labelling by amino acids in cell culture (SILAC) (Ong et al., 2002), stable-isotope dimethyl labelling (Hsu et al., 2003), tandem mass tags (TMTs) (Thompson et al., 2003), isotope-coded affinity tag (ICAT) (Shiio & Aebersold, 2006), or isobaric tags for relative and absolute quantitation (iTRAQ) (Ross et al., 2004). Proteins/peptides are labelled by stable isotopes in cells or small organisms metabolically during their growth, or during sample processing by chemical or enzymatic reactions, and measured by high-resolution MS. The isotope ratios are then used for accurate quantification of proteins.

In the DIA technique, the major difference over the DDA is in the selection of precursor ions for fragmentation. The MS1 spectrum is divided into dozens of windows of pre-specified *m/z* values (e.g., 25 Da – from 760 to 785 *m/z*, and so on, consequently covering the whole range of *m/z* from the MS1 experiment), all precursor ions within the particular window are sent for fragmentation altogether, and resulting product ions are detected in one MS2 experiment. In theory, all peptides are measured and quantified in one analysis, eliminating the DDA stochasticity. DIA method thus provides high throughput for global proteome analysis (over 11,000 proteins can be quantified in one analysis) (Muntel et al., 2019), as well as the possibility for quantification of a selected set of proteins, with high reproducibility and accuracy (Kitata et al., 2022). Proteomic data analysis is possible with machine learning and bioinformatics software tools that are constantly being developed and improved (Neely et al., 2023; Tsiamis et al., 2019).

Hypothesis-Driven Workflows

PRM is, in contrast to DDA and DIA methods, a hypothesis-driven targeted approach suitable for the validation of candidate biomarkers, rather than biomarker discovery. This method requires high resolution and high mass accuracy analysers, i.e., TOF or Orbitrap, at the end of the tandem MS set-up. For PRM analysis, prior knowledge of *m/z* values of peptides from targeted proteins, together with their retention times is essential. Particular precursor ions are selected by the first analyser (quadrupole) and sent to the second quadrupole for fragmentation. The resulting product ions are then monitored by TOF or Orbitrap mass analysers. This leads to very high specificity, sensitivity, and a wide dynamic range (Peterson et al., 2012). The PRM method also enables absolute quantification of targeted analytes, providing that internal standards (peptides labelled with stable isotopes) of known concentrations are used (Gerber et al., 2003). SRM is another targeted method for hypothesis-driven studies. For its implementation, a specific MS instrument consisting of three quadrupoles is required. The first and the third quadrupoles serve as mass filters, while the second quadrupole is used for fragmentation. Prior knowledge of *m/z* values for both precursor and product ions, and their retention time is necessary. The combination of known retention times and pre-defined *m/z* values assures high specificity and sensitivity (Lange et al., 2008; Vidova & Spacil, 2017). The SRM method also enables absolute quantification and provides higher measurement precision than PRM, but a lower dynamic range (Peterson et al., 2012).

3 Dysregulated Candidate Proteins Verified by an Independent Method

The verification by an orthogonal method (e.g., real-time quantitative polymerase chain reaction, Western blotting, etc.) was long considered an inevitable step for the confirmation of 2-DE and quantitative MS results. Since they have undergone imposing development in the last several years – from sample preparation workflows, over instrumentation, to data analyses, MS-based methods have been proven to be highly robust and reproducible (Mehta et al., 2022). For results from the targeted PRM and SRM methods, with several peptides per protein being highly specifically analysed, the requirement for an independent verification with less precise methods is no longer justified (Aebersold et al., 2013). In this part, we discuss findings obtained from global proteome screens, where selected protein changes were followed up in independent experiments. These results are complemented by findings from up-to-date targeted proteomics measurements. Such verified proteins are listed in Supp. Table 1.

Biofluids

Plasma

Presently, human plasma is probably the most preferred biofluid for biomarker discovery. Its proteome has been used to identify changes in the level of proteins that are believed to be associated with blood-brain-barrier leakage, and that reflect many pathological processes. In addition, plasma is accessible and widespread laboratory infrastructure is in place for its analysis; therefore biomarkers that can be measured through a blood test are highly desirable (Rifai et al., 2006). Using two different depletion techniques and 2-DE, Dalrymple et al. identified proteins differentially abundant in the plasma of HD patients from various stages of the disease, including asymptomatic patients. Many of these proteins are regulators of the innate immune system. These findings reflect concurrent inflammation which was confirmed by measuring the levels of interleukin 6 (IL6) in HD patients and the R6/2 transgenic mouse model of HD by the same researchers. The authors proposed clusterin (CLU), an extracellular chaperone glycoprotein that inhibits protein aggregation, as a promising HD biomarker (Dalrymple et al., 2007; Poon et al., 2000; Wyatt et al., 2009). However, further evaluation of candidate biomarkers associated with innate immunity in plasma did not confirm CLU or other protein alterations, but revealed a decrease of C-reactive protein (CRP) in early HD (Silajdžić & Björkqvist, 2018). A later study used SRM assays for analysis of markers of innate immunity, members of the complement system and the CRP in pre-symptomatic and early HD patients. In contrast to the findings from Dalrymple and colleagues, none of the tested proteins showed a difference in any of the disease stages (Rezeli et al., 2014). All these results confirmed significant impact of inter-individual and/or intra-individual variability and support the necessity of validation of these biomarkers. Although the increase in the levels of inflammation markers might be unspecific to HD, substantial evidence indicates that neuroinflammation is one of the underlying pathogenetic mechanisms and clinical trials of immunomodulatory drugs are ongoing (Dickey & La Spada, 2018; Estevez-Fraga et al., 2022).

Cerebrospinal Fluid

Despite being less accessible than plasma, CSF is the primary source for evaluating molecular variations of the CNS. The composition of CSF is dynamic and reflects physiological and pathological changes in the brain (Kroksveen et al., 2011). Fang et al. conducted an integrated analysis from five laboratories employing different MS detection methods and the same CSF samples. Although there was a large inconsistency between labs, the authors identified proteins with abundance changes throughout HD progression. Overall, brain-specific proteins decreased through HD advancement, while neuroinflammatory markers increased (Fang et al., 2009).

Huang et al. studied the CSF proteome from HD patients by 2-DE-based methods. In contrast to Fang et al., they did not use abundant protein depletion steps. They found that prothrombin (F2), apolipoprotein A-IV (APOA4) and haptoglobin (HP) were increased in CSF, but not in the serum of HD patients, and the authors proposed that this increase is CNS-derived (Huang et al., 2011).

A larger proteomic study explored the differences in CSF protein levels of HD gene carriers of different disease stages. After age adjustment, the authors proposed ubiquitin (UBC) as a candidate CSF biomarker. The ubiquitin peaks positively correlated with motor and functional scale scores and with the CAG-age product (CAP) score (Vinther-Jensen et al., 2015). Protein degradation by the ubiquitin-proteasome system and dysfunctions of this system are well-documented in HD and other neurodegenerative diseases (Bennett et al., 2007). So far, the most extensive CSF profiling was recently conducted using label-free MS quantification in a cohort of pre-symptomatic and symptomatic HD patients. This absolute MS quantification method demonstrated that the severity of symptoms negatively correlated with the levels of the opioid precursor proenkephalin-A (PENK), which was decreased also in a separate HD cohort, and the authors proposed PENK as a candidate biomarker for HD progression (Niemela et al., 2021).

Other recent studies have used targeted analysis of candidate protein markers originally identified in earlier HD research (Al Shweiki et al., 2021; Barschke et al., 2022; Caron et al., 2022); in addition, two studies also compared HD patients to patients with other neurodegenerative diseases (Al Shweiki et al., 2021; Barschke et al., 2022). Consistent with findings by Niemela and colleagues (Niemela et al., 2021), Barschke and colleagues reported decreased levels of PENK in the CSF of HD patients, as well as decreased levels of another opioid precursor, proenkephalin-B (PDYN) (Barschke et al., 2022). Caron et al. investigated whether previously reported CSF protein changes (Al Shweiki et al., 2021; Fang et al., 2009; Niemela et al., 2021; Rodrigues et al., 2016; Vinther-Jensen et al., 2016) can be replicated and if candidates from gene expression studies in brains from humans (Corvol et al., 2004; García-Huerta et al., 2020; Hodges et al., 2006) and HD mouse models (Luthi-Carter et al., 2000; Mazarei et al., 2010, 2013; Ryskamp et al., 2017) have the potential to become biomarkers. Using PRM, the levels of pre-specified proteins were evaluated in the CSF from manifest HD patients, and pre-manifest mutation carriers and, consistently with previous studies, PENK, PDYN and NfL levels were altered (Al Shweiki et al., 2021; Niemela et al., 2021; Rodrigues et al., 2016). PENK appeared as the most important protein for the stratification between HD and controls in this study. Also, combinations of biomarkers performed better than individual proteins in discriminating between the various stages of HD (Caron et al., 2022).

Brain

Human Brain Tissues

Neurodegenerative changes in the brains of HD patients occur years before the manifestation of motor symptoms. Only a small number of studies have focused on the proteome of HD human brain samples, possibly due to the difficulty associated with obtaining human samples. These studies used post-mortem brain tissues; therefore, the described changes refer mostly to patients during the terminal stage of HD compared to "healthy" individuals without neurodegenerative disease. Using 2-DE, Sorolla et al. found altered levels of proteins implicated in energy metabolism and antioxidant defence. The striatum was more affected than the cortex. These findings are consistent with the hypothesis of disrupted metabolism and reactive gliosis in HD pathology (Sorolla et al., 2008). A case-control 2-DE study by Schönberger and colleagues described the protein changes in the middle frontal gyrus and visual cortex (Schönberger et al., 2013), and Chen et al. investigated the alterations in substantia nigra using a label-free quantitative analysis. Although there was no overlap with the findings of Sorolla et al., several other proteins implicated in energy metabolism were dysregulated (Chen et al., 2012). More recently, Ratovitski et al. employed quantitative LC-MS/MS in HD superior frontal gyrus and detected, to date, the highest number of protein abundance changes (4,789 proteins) in the examined brains (Ratovitski et al., 2016). Further re-analysis of the dataset utilized by Ratovitski et al. identified changes in extracellular matrix (ECM) caused by HD, such as pronounced alterations in aminoglycan metabolism similarly to Alzheimer's disease (AD) and Parkinson's disease (PD) (Freitas et al., 2021).

Animal Brain Tissues

Transgenic Mice

A variety of animal models that express polyQ-expanded HTT have been developed to mimic the progressive neurological phenotype of HD. Due to genomic similarity and ease of manipulation with relatively low costs, utilization of the most common mammalian disease model, the transgenic mouse, represented an important advance in the study of HD. Despite concerns raised about whether a mouse model can faithfully reproduce human pathology, transgenic HD mice provide an invaluable tool to efficiently investigate HD pathogenesis and therapeutic strategies, and are still preferred in drug screenings and preclinical studies. The well-characterised and most widely used HD mouse model system is R6/2, which expresses exon 1 of the human *HTT* gene with a highly expanded CAG repeat region (Mangiarini et al., 1996). Since the R6/2 model lacks the full-length mHTT protein, several transgenic models, which express the full-length human mHTT (in addition to endogenous mouse Htt), were also developed, namely yeast artificial chromosome (YAC) 128 (Slow et al., 2003) and the bacterial artificial chromosome HD (BACHD) models (Gray et al., 2008). In the R6/2 mice, hallmarks of HD progression appear from a very early age. A loss of brain weight can be observed from the age of 4 weeks, motor functions are impaired from 6 weeks, and the typical phenotype is visible from 8 weeks (Zabel et al., 2009). Pronounced symptoms in the advanced stage are present from 12 weeks, with the disease endpoint at 14–15 weeks. Early changes manifest in the full-length mHTT transgenic mice from the age of 2 months (subtle motor deficit in BACHD mice) (Slow et al., 2003), and 3 months (hyperkinetic phenotype in YAC128 mice) (Gray et al., 2008). The YAC128 mice have a normal lifespan, with a progressive motor deficit that highly correlates with neuronal loss at 6 months; late-onset striatal atrophy is reached by 9 months of age, which is followed by cortical atrophy at 12 months (Slow et al., 2003). Similarly, in the BACHD mice, region-specific brain atrophy also develops by the age of 12 months (Gray et al., 2008).

The brain proteome of the R6/2 mice has been subjected to several proteomics analyses that allowed researchers to identify key players in HD progression. Most studies employed 2-DE followed by MS-based protein identification. In more recent workflows, differential proteins were subjected to independent verification, and pathway analyses were often used to establish biological context. To this end, it was revealed that in the R6/2 model, energy metabolism was upregulated in early to middle HD stages (Perluigi et al., 2005; Zabel et al., 2009, 2010), while ubiquitin-proteasome functions were critical in the terminal stages (Zabel et al., 2009, 2010), which protects cells with aggregates in HD, and also in other neurodegenerative disorders (Zabel et al., 2010). Particular proteins emerging from 2-DE analyses were inspected to recapitulate the process of HD pathogenesis; during early HD, α-enolase (Eno1) was found in increased abundance in the R6/2 mice (Perluigi et al., 2005). Enolase is essential for glucose metabolic pathways that are vital for cellular functions, including adaptation. Eno1, however, was decreased in YAC128 mice, suggesting that the adaptation is specific to the R6/2 model (Wegrzynowicz et al., 2012). Down-regulation of this ubiquitously expressed enzyme was further demonstrated in YAC128 mice exposed to manganese toxicity, which was investigated as an environmental factor affecting the HD phenotype. NfL was also decreased in the late-onset YAC128 model (Wegrzynowicz et al., 2012).

In late to terminal stages, mitochondrial dihydrolipoamide S-succinyltransferase (Dlst) and aspartate aminotransferase (Got2) appeared to increase in symptomatic R6/2 mice, which was followed by a decrease in pyruvate dehydrogenase (Perluigi et al., 2005). Another study focused on the mitochondrial-enriched proteome of the cortices of R6/2 and YAC128 mice with different progression rates (Agrawal & Fox, 2019). This study revealed decreased levels of peroxiredoxin-3 (Prdx3, hydrogen peroxide scavenging enzyme), and stress-70 protein (Hspa9), as well as an increased level of Parkinson disease protein 7 homolog (Park7, chaperone protein). Decreased levels of Hspa9 have also previously been described in the human PD brain (Shi et al., 2008), and in the AD brain (Park et al., 2014). Park7 mutations, on the other side, lead to early-onset recessive PD (Shendelman et al., 2004), and an increased abundance of this chaperone was recorded in the advanced HD (Sajjad et al., 2014). Finally, the alpha-crystallin B chain (Cryab) chaperone, which prevents protein aggregation and promotes cell survival, was decreased in mouse models of neurodegeneration, including R6/2 mice (Zabel et al., 2002, 2006).

Large-scale MS-based quantitative proteomic studies were conducted to reveal neuropathological mechanisms of advanced HD using the R6/2 mice (Benraiss et al., 2022; Burgold et al., 2019; Chiang et al., 2007; Hosp et al., 2017; Skotte et al., 2018; Wishart et al., 2012). Early HD stages were also investigated in these mice (Hosp et al., 2017; Liu et al., 2007), and in the full-length mHTT transgenic models, BACHD (Shirasaki et al., 2012) and YAC128 (Wan et al., 2013). Proteome changes were reported in the whole brain extract of mHTT-affected animals (Karayel-Basar et al., 2022), in the striatum (Burgold et al., 2019; Hosp et al., 2017; Liu et al., 2007; Skotte et al., 2018), cortex (Burgold et al., 2019; Hosp et al., 2017; Skotte et al., 2018), hippocampus and midbrain (Skotte et al., 2018), corpus callosum (Benraiss et al., 2022), as well as in striatal nuclear-fractionated samples (Chiang et al., 2007), and synapse-enriched samples (Wishart et al., 2012). Sample purification was particularly beneficial for the study of the HTT interactome (Shirasaki et al., 2012), as well as an in-depth analysis of inclusion bodies (Hosp et al., 2017), and the characterisation of palmitoyl-proteins (Wan et al., 2013). Symptomatic stages of HD were accompanied by compromised glutamine homeostasis (Liu et al., 2007; Skotte et al., 2018). Glial-specific glutamine synthetase (Glul) was reduced in the R6/2 brains (Hosp et al., 2017; Skotte et al., 2018). Isolated brain slices from the striatum and cortex were analysed using *ex vivo* ^{13}C labelling with substrates that distinguish between the astrocytic and neuronal compartments, with results indicating that limited amounts of glutamine reach the neuronal compartment of the R6/2 mice (Skotte et al., 2018).

A detailed interactome analysis in BACHD mice, complemented by validation in a Drosophila model of HD, provided a system-level view of the spreading pathology typically seen in HD (Shirasaki et al., 2012). Another spatiotemporal system-wide study enabled researchers to describe remodelling of the brain proteome and molecular signatures of HD neurodegeneration (Hosp et al., 2017). Inclusion bodies with HTT aggregates were isolated from the R6/2 mice and quantitative MS analysis revealed extensive protein sequestration to the insoluble fraction. R6/2 striata were significantly enriched for aggregation-prone proteins, and proteins with higher molecular weight, including HTT (Hosp et al., 2017).

Knock-in Mice

In contrast to transgenic models where the fragment of the *mHTT* is inserted at a random site of the mouse genome, in knock-in mice, the *HTT* mutation is inserted at the appropriate locus under the endogenous mouse promoter (*Hdh*) (Menalled, 2005). Knock-in mice represent a genetically more precise paradigm of human HD than transgenic models. Disease progression is slower, but with an earlier emergence of the HD phenotype. Available knock-in mouse models have polyQs from 50 to 200 (Pouladi et al., 2013). HD knock-in mice Q140 and zQ175 show both behavioural symptoms and molecular alterations, while models with shorter polyQs (e.g., Q80, Q92 and Q111) exhibit only molecular changes (Langfelder et al., 2016; Menalled et al., 2003, 2012). The zQ175 mouse model is the first knock-in model that shows HD symptoms as a heterozygote (Menalled et al., 2012). Comparative proteomic studies in HD knock-in models used LC-MS/MS, except for one study (Deschepper et al., 2012) that used both gel-free and 2-D DIGE approaches to characterise whole brain lysates (Deschepper et al., 2012; Sap et al., 2019), striata (Federspiel et al., 2019; Langfelder et al., 2016), cortical, and striatal synaptosomes (Sapp et al., 2020).

As a polyQ protein, HTT is involved in transcriptional regulation (Gerber et al., 1994). Global reductions in gene expression are documented in HD studies, implicating various processes through transcriptional dysregulation including histone acetylation (Hyeon et al., 2021; Sugars & Rubinsztein, 2003). Interactions of Hdac4, a lysine deacetylase associated with transcriptional repression, were analysed in dissected striata from Q140 mice (Federspiel et al., 2019). Hdac4 interactome, as well as striatal proteome and transcriptome datasets were integrated, revealing an overlap in enriched functional classes, particularly in vesicular trafficking and synaptic functions. In addition, proteins and transcripts were significantly different in the striatal tissue of Q140 mice, compared to Q20 mice, and these changes were concerted in association with Hdac4. Another integrated study used RNA-seq and label-free proteomics in the striatal tissue of knock-in mice of six different CAG lengths, and hundreds of differentially expressed genes were detected both at the mRNA and protein levels. These changes were continuous and dependent on the CAG length (Langfelder et al., 2016).

The brain proteomes and their ubiquitination patterns were analysed in the zQ175 and Q20 HD mice (Sap et al., 2019). In insoluble fractions of HD brains, Sap and colleagues observed a notable increase in vimentin (Vim). Glial fibrillary acidic protein (Gfap) and NfL were increased both in abundance and ubiquitination, but these findings require validation. The role of synaptic protein dysfunction was investigated in the synaptosomes from cortices and striata of heterozygous Q140 mice using LC-MS/MS. The proteome changes were age- and brain region-dependent and the decreased level of three proteins was apparent in HD mice, namely sodium channel subunit 4 (Scn4b), phosphodiesterase 10a (Pde10a) and cannabinoid receptor 1 (Cb1) (Sapp et al., 2020).

Transgenic Rat

Compared to mice, the larger body size and weight of rats with consequently bigger inner organs allows for easier surgical intervention (e.g., neurosurgery), better resolution of brain imaging techniques, and easier handling in general. Metabolic alterations and hormonal changes have been documented in HD and so far, the only study of hypothalamic proteome and hormone levels was conducted using a transgenic rat model. The transgenic animals were at different stages of HD progression and the analysis of euglycemic and metabolic hormones in blood plasma revealed a lower concentration of leptin (Lep), amylin (Iapp) and peptide Y (Pyy) in pre-symptomatic HD rats. Proteomic analysis based on LC-MS/MS with iTRAQ labelling measurements of hypothalami from symptomatic animals resulted in the identification of increased phosphatidylethanolamine-binding protein 1 (Pebp1), pyruvate kinase PKM (Pkm2), G1 to S phase transition 2 (Gspt2), and serine/threonine-protein kinase mTOR (Mtor). Gfap, flotillin-2 (Flot2), and prelamin-A/C (Lmna) were decreased in symptomatic animals. Insulin (Ins1/2) was decreased in both pre-symptomatic and symptomatic animals (Cong et al., 2012).

Transgenic Sheep

Rodent models provided invaluable insights into HTT and mHTT functions, but there is a need for large animal models that would enable translational research. Recently, Mears et al. analysed the transcriptome, proteome (LC-MS/MS with iTRAQ labelling), and metabolome of the brain (cerebellum, hippocampus, motor cortex and striatum) and other tissues (e.g., blood plasma) using a pre-symptomatic transgenic HD sheep model to create a publicly available database containing data from all "omic" analyses. Such a database may be very useful for further studies focused on the interplay between metabolites, proteins, and transcripts in response to mHTT, and for data mining in search for novel candidate biomarkers of HD (Mears et al., 2021).

Peripheral Tissues

Besides neurodegeneration, HD patients and animal models develop non-neurological symptoms, such as weight loss, skeletal muscle wasting, diabetes, and reduced fertility. These defects are directly related to mHTT and associated with disease progression and fatality, thereby, monitoring the peripheral pathological changes is clinically important (van der Burg et al., 2009; Sassone et al., 2009).

Nuclear inclusions of mHTT, the pathological hallmark of HD (DiFiglia et al., 1997), are prominent in muscle, but not in the skin or adipose tissue of R6/2 mice (van der Burg et al., 2009; Sassone et al., 2009). The Multiple Tissue Monitoring in Huntington's Disease study (Neueder et al., 2022) was recently conducted to explore the human peripheral molecular signatures of HD and revealed potential biomarkers. Skeletal muscle, adipose tissue and skin tissues were separated from collected biopsies from HD patients and quantitative proteomics revealed an increasing number of differentially expressed proteins associated with disease advancement. The authors reported two proteins consistently dysregulated in pre-manifest and early manifest HD patients, syntaxin-binding protein 3 (STXBP3) and MOB family member 4 (MOB4) in muscle and skin tissues, respectively. Although none of the differentially abundant proteins coincided with differences in levels of mRNA, this is the first omics study of peripheral tissues from HD patients and validation of the reported proteins merits further effort.

In Vitro Models

Comparative proteomics studies focused on human and mouse cells expressing mHTT have provided valuable understanding of affected proteins at a cellular level. Here, we describe the findings that have emerged from these studies through the use of a variety of HD *in vitro* models.

Mouse-Derived Cell Models

Q7Q7, Q7Q111 and Q111Q111 mouse striatal neuronal cells were analysed for proteasome activity (Dasgupta et al., 2015). Using these models, the authors investigated the levels of endogenous peptides (peptidome) resulting from proteasome cleavage, and reported a significant reduction in almost all peptides detected, when comparing Q7Q111 or Q111Q111 to Q7Q7, further supporting the hypothesis of reduced activity of the proteasome in HD. Nevertheless, missing information on the total level of proteins, whose endogenous peptides were depleted because of reduced proteasome activity, makes proper interpretation challenging.

In striatal neuronal and neuroblast (Neuro2A) cells from homozygous Q111 knock-in mice, decreased huntingtin interacting protein K (Hypk) and Vim, together with increased GTP-binding nuclear protein Ran (Ran) and superoxide dismutase [Mn], mitochondrial (Sod2) proteins were reported (Choudhury et al., 2016). Differentially abundant proteins, associated mainly with the cytoskeleton, were identified in Neuro2A cells transiently transfected to express HTTex1 fused with mCherry (Sui et al., 2020). It was found that mHTT affects proteome solubility by altering the localisation of stress granule proteins and by sequestering proteins involved in protein quality control systems (i.e., chaperones) into mHTT aggregates (Sui et al., 2020).

Chou et al. analysed primary cortical astrocytes derived from R6/2 mice and employed an antibody microarray to determine the abundance of chemokines and cytokines secreted into the culture media (Chou et al., 2008). Chemokine (C-C motif) ligand 5/regulated on activation normal T cell expressed and secreted (Ccl5) was found decreased and its lower secretion was confirmed in cortical and striatal, but not cerebellum-derived astrocytes. Conditioned medium from HD astrocytes resulted in limited neurite outgrowth in primary rat cortical neurons, as well as reduced migration and impaired calcium influx after membrane depolarization. Further analyses revealed that secretion of Ccl5 is suppressed by its intracellular accumulation in astrocytes, which was confirmed in transgenic R6/2 mice, as well as in knock-in mice and brain tissues (caudate nucleus, frontal cortex and substantia nigra) of HD patients (Chou et al., 2008).

Human-Derived Cell Models

Studies using human-derived *in vitro* HD models utilised induced pluripotent stem cells (iPSCs) from HD patients (Akimov et al., 2021; The HD iPSC Consortium, 2020), and neural cells or neural progenitor differentiated cells derived from human embryonic stem cells (hESCs) (McQuade et al., 2014; Ooi et al., 2019). Both ESCs and iPSCs can differentiate and mimic a HD phenotype. These studies revealed dysregulated proteins associated with bioenergetic deficits (Akimov et al., 2021; The HD iPSC Consortium, 2020; McQuade et al., 2014; Ooi et al., 2019), alterations in endocytosis, GABA and glutamate signalling, sphingomyelin metabolism and ethanol degradation, cell-to-cell signalling and interaction (Akimov et al., 2021).

Analysis of striatal neural cells differentiated from patients-derived iPSCs and inspected by LC-MS/MS with iTRAQ labelling revealed downregulated glycolysis and upregulated oxidative phosphorylation (McQuade et al., 2014). Moreover, supplementation of cells with late-stage glycolytic metabolites (e.g., pyruvate) restored the normal level of ATP and increased cell viability. These findings indicate that these metabolites might be beneficial for HD therapy (The HD iPSC Consortium, 2020).

Recently, Ooi et al. created an allelic panel of isogenic HD (IsoHD) ESCs with 30, 45, 65 and 81 CAG repeats from human H9 ESCs. The IsoHD ESCs were able to differentiate into neural progenitor cells (NPCs) and neurons. Analysis of HD NPCs recapitulated known HD phenotypes that were more associated with CAG repeat length than cell type. Comparison of proteomes by LC-MS/MS with TMT labelling showed much higher variability at the level of NPCs compared to ESCs, and only a minor overlap of proteins changed in both cell types. A combination of data from RNA-seq and proteomic analyses confirmed the repeatedly discussed limited correlation of mRNA and protein levels, thus supporting the employment of both techniques for complex HD studies (Ooi et al., 2019).

4 Analysis of Overlaps in Identified Dysregulated Proteins Between Multiple Independent Studies

Attempting to summarise a complete picture drawn from the HD proteome studies of patient samples and experimental animal and cell models, we examined their overlap (Fig. 3, Supplementary File 2). In this part, we will discuss promising proteins extracted from proteomics studies based on their consistent detection across multiple independent studies. While not yet verified using an independent method, such proteins certainly deserve attention in future targeted measurements.

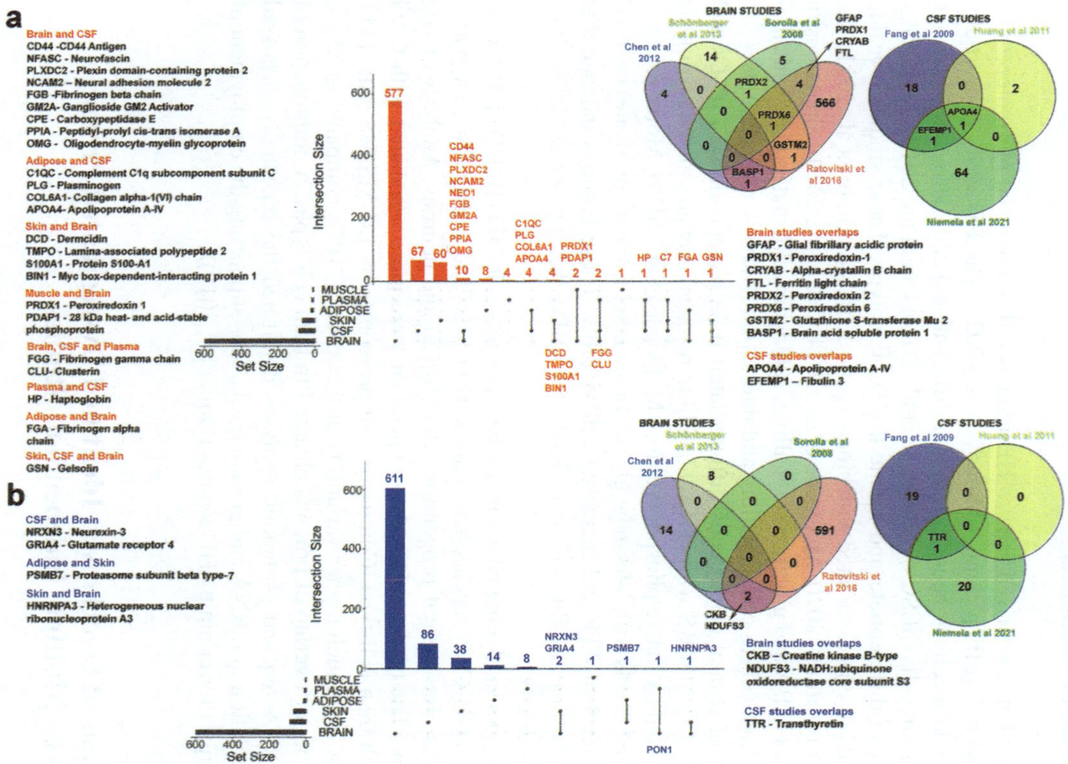

Fig. 3 Graphical summaries of the overlaps from all significantly dysregulated proteins reported in mass spectrometry (MS) studies analysing the global proteome in samples from Huntington's disease patients. The accepted significance was $p < 0.05$ for gel-based and p adjusted<0.05 for gel-free studies. Upregulated proteins are shown in (**a**) and downregulated in (**b**). The UpSet graphs plot the convergence of gene products across tissues. The Venn diagrams show the differentially abundant proteins and their overlaps per study for brain tissue and cerebrospinal fluid (CSF). Note the increase in number of significantly altered proteins reported by more recent studies using tandem MS such as Ratovitski et al. (2016) or Niemela et al. (2021). The shared proteins are indicated as gene names at each intersection and their full names are explained on the left side for Upset plots and the right side for Venn diagrams

Tissues Overlaps

Global proteomes from different parts of post-mortem brains of HD patients were analysed by four studies, indicating 10 proteins dysregulated in the cortex, striatum or substantia nigra. These proteins are listed on Fig. 3a, b (brain studies overlaps), namely GFAP, peroxiredoxins 1, 2 and 6 (PRDX1, PRDX2, PRDX6), CRYAB, ferritin light polypeptide (FTL), glutathione S-transferase Mu-2 (GSTM2), brain abundant membrane attached signal protein 1 (BASP1), creatine kinase (CKB) and NADH dehydrogenase [ubiquinone] iron-sulfur protein 3, mitochondrial (NDUFS3). Differences in the output datasets can be attributed to biological variability and different efficacy of the proteomic MS platform (gel-based vs gel-free).

Brain increases of GFAP were reported in the striatum and cortex from HD patient samples (Ratovitski et al., 2016; Sorolla et al., 2008), and HD mouse models (Sap et al., 2019; Zabel et al., 2009). GFAP is a neuroinflammatory marker of astroglia activation and gliosis related to HD progression and striatal degeneration (Vonsattel et al., 1985). An increase in blood GFAP was correlated with blood NfL levels, disease burden and functional and motor impairment in a cohort of HD mutation carriers, suggesting GFAP as a potential biomarker for evaluating HD progression (You et al., 2021).

Other proteins with a neuroprotective role were increased in HD patients in the striatum (Sorolla et al., 2008) and cortex (Ratovitski et al., 2016) compared to controls, in contrast to proteomic findings from HD animal models. One example is CRYAB, one of the small heat shock proteins, expressed in various neuronal cell types, especially in astrocytes (Zhu & Reiser, 2018). CRYAB attenuates mitochondrial dysfunctions and oxidative stress, reduces the accumulation of misfolded proteins, prevents neuroinflammation and apoptosis (Zhu & Reiser, 2018) and, in BACHD mice, astrocytic overexpression of *Cryab* alleviated the motor impairment *in vivo* (Oliveira et al., 2016). However, in transgenic (Skotte et al., 2018; Zabel et al., 2006, 2009) and knock-in (Sap et al., 2019) models of HD, the gene product of *Cryab* was decreased compared to control mice. Similarly, BASP1 was upregulated in the human striatum (Sorolla et al., 2008) and cortex (Ratovitski et al., 2016), but downregulated in the brains of R6/2 mice (Burgold et al., 2019). BASP1 is a cytoskeleton-related, calmodulin-binding protein with expression induced following nerve regeneration during development or injury. During neuroinflammation, increased levels of cytokines and neurotrophins activate phosphorylation and increase in BASP1 levels which promotes synaptogenesis and neurite outgrowth (Chung et al., 2020). BASP1 interacts with HTT but its role in HD has not been explored (Kaltenbach et al., 2007).

In most proteomic studies, oxidative stress was the major dysregulated pathway, and proteins related to antioxidant defence, redox regulation, and mitochondrial function were commonly altered across samples from HD patients, as well as *in vivo* and *in vitro* HD models. Mammalian cells express six isoforms of peroxiredoxins (Rhee et al., 2005) and in the brain, different peroxiredoxins are cell-specific (Szeliga, 2020). PRDX1, PRDX2 and PRDX6 have been reported as increased in at least two brain studies of HD patients (Ratovitski et al., 2016; Schönberger et al., 2013; Sorolla et al., 2008). Sorolla and colleagues verified the upregulation of PRDX1, PRDX2 and PRDX6, but not of 3, 4 and 5 in the human HD striatum (Sorolla et al., 2008). In addition, PRDX1 was increased in iPSC lines from HD patients compared to control iPSCs and ESCs (Chae et al., 2012). Dysregulated levels of peroxiredoxins were evident in knock-in (Langfelder et al., 2016) and transgenic (Agrawal & Fox, 2019; Skotte et al., 2018; Zabel et al., 2009) mouse models. Another enzyme implicated in antioxidant defence, a cytosolic isoform of glutathione transferase, GSTM2 was increased in the cortex in two studies (Ratovitski et al., 2016; Schönberger et al., 2013). NDUFS3 was downregulated in substantia nigra (Chen et al., 2012) and motor cortex (Ratovitski et al., 2016) in HD patients and mouse models (Federspiel et al., 2019; Zabel et al., 2009).

CKB was downregulated in human HD brains (Chen et al., 2012; Ratovitski et al., 2016), consistent with alterations shown by proteomic findings in R6/2 (Skotte et al., 2018; Zabel et al., 2009) and knock-in mice (Deschepper et al., 2012; Langfelder et al., 2016). Substantial evidence exists that suppression of CKB occurs in the presence of mHTT (Ju et al., 2012), and reduced brain and plasma levels were found in R6/2 mice and HD patients using Western blot and immunohistochemistry (Kim et al., 2010). Clinical trials of creatine supplementation showed that daily administration of creatine significantly reduced oxidative DNA damage in early HD patients (Hersch et al., 2006). It slowed brain atrophy and was well tolerated by pre-manifest HD patients (Rosas et al., 2014), but high-dose creatine failed to slow the functional decline in early manifest HD participants (Hersch et al., 2017).

An increase in FTL was shown in two studies in post-mortem HD brains (Ratovitski et al., 2016; Sorolla et al., 2008) and striatum and midbrain from the R6/2 transgenic mouse (Skotte et al., 2018). FTL is one of the two subunits of ferritin implicated in the intracellular storage and detoxification of iron (Harrison & Arosio, 1996). Alteration of FTL levels is consistent with the role of HTT in iron homeostasis (Hilditch-Maguire et al., 2000), and the dysregulation of metal ions deposition due to the presence of mHTT (Bartzokis et al., 1999; Muller & Leavitt, 2014).

Of note, a number of protein changes consistently reported across tissues in human patients were altered in opposite directions in animal and cell models of HD. For example, CLU was increased in HD patients (Dalrymple et al., 2007; Niemela et al., 2021) but several studies in HD mice models found decreased levels (Burgold et al., 2019; Langfelder et al., 2016; Skotte et al., 2018). Carboxypeptidase E (CPE), neural cell adhesion molecule 2 (NCAM2) and neurofascin (NFASC) were upregulated in HD patient brains (Niemela et al., 2021), but downregulated in iPSCs derived from HD patients (The HD iPSC Consortium, 2020), or in transgenic mice (Burgold et al., 2019) and rats (Cong et al., 2012).

Cerebrospinal Fluid Overlaps

Present in such proximity to the brain and sharing a fraction of its excreted proteins, CSF offers the opportunity to discover HD-specific biomarkers. Two precursors of endogenous opioids, PENK and PDYN stand out as two of the most promising HD-specific biomarkers identified in CSF (Supp. Table 1).

Brain and CSF had the highest convergence in our overlap analysis. Among the CNS-related protein changes, CD44, NFASC, NCAM2, oligodendrocyte-myelin glycoprotein (OMGP), or ganglioside GM2 activator (GM2A) were upregulated in the human CSF and brains (Fig. 3a). GM2A is a lipid transport protein with overexpression linked to the impairment of neurite integrity and reduction in the mean firing rate of human neurons *in vitro* (Hsieh et al., 2022). The upregulation of GM2A was also found in the striatum of symptomatic R6/2 mice (Skotte et al., 2018). Other CNS proteins related to synaptic organisation were downregulated in the brain and CSF (Fig. 3b), namely glutamate receptor 4 (GRIA4) and neurexin 3 (NRXN3). NRXN3 was also downregulated in knock-in mice (Sap et al., 2019). Increased levels of immune-related proteins HP, CLU, and fibrinogen gamma chain (FGG) coincided in plasma and CSF.

CSF candidate protein overlaps across studies included APOA4, Fibulin-3 and T_4-binding protein (TTR) (Fig. 3a, b). Since APOA4 levels are controlled by leptin in the hypothalamus to allow inhibition of food intake, regulation of body weight, cholesterol transport and energy homeostasis (Shen et al., 2007, 2008; Tso et al., 2001), this protein marker could be correlated to the weight loss seen in HD patients. Fibulin-3 was upregulated in CSF in HD patients (Fang et al., 2009; Niemela et al., 2021), and we propose this glycoprotein, which has been implicated in ECM remodelling tissue regeneration and organogenesis (Timpl et al., 2003), as a target for validation studies. Physiologically, fibulin-3 has very low expression levels in rodent brains (Kobayashi et al., 2007), and its upregulation has been reported in gliomas (Hu et al., 2009), but its relationship with HD has not been studied so far. Although CSF levels of TTR were found unaltered between disease stages in the PRM study of Caron et al., TTR did show an association with predicted years to disease onset in premanifest HD mutation carriers (Caron et al., 2022). Also, Cong et al. reported significantly decreased levels of TTR in transgenic rat hypothalamus (Cong et al., 2012). Hypothalamic neurodegenerative changes are well documented in postmortem human brains (Kremer et al., 1990) and neuroendocrine abnormalities occur in HD patients.

Functional Analysis

Inflammation and immune hyperactivity were among the most prominent upregulated biological processes in CSF and plasma in HD patients (Dalrymple et al., 2007; Fang et al., 2009; Huang et al., 2011; Niemela et al., 2021) and the R6/2 mouse model (Hosp et al., 2017). We completed a functional analysis of the dysregulated proteins previously reported in studies of HD patients, and further validated that immune response-related pathways were the most significantly enriched in CSF (Fig. 4a). Although some of the reported immune-related proteins were not verified (Vinther-Jensen et al., 2014, 2016), an early inflammatory response is among the first changes in HD animal models and patients (Jia et al., 2022; Valekova et al., 2016). Microglia expressing mHTT exhibit a pro-inflammatory transcriptional activation and enhanced reactivity (Crotti et al., 2014), and early changes in cytokines IL6 and tumour necrosis factor α appear in pre-symptomatic and early-stage HD patients (Björkqvist et al., 2008). Other biological processes included mitochondrial dysfunction and cellular respiration, synaptic regulation, and cytoskeletal changes (Fig. 4).

5 Overview of Proteomics Findings and Limitations

HD is a hereditary monogenic disorder, which can be confirmed by a simple genetic test. However, there is an urgent need for biomarkers of HD progression in CNS and other tissues, as well as for an assessment of the effectiveness of possible HD therapies. Omics technologies have contributed to numerous studies proposing thousands of candidate biomarkers. In this review, we described all differential proteins that were identified and verified in proteome studies focused on HD pathology (Supp. Table 1). Several verified proteins were identified in parallel proteomics studies, such as GFAP, Glul, Ca2, and Vim. To ensure that relevant non-verified markers were not omitted, we additionally extracted the proteins that appeared differentially abundant in more than one HD proteome study focused on the human brain and biofluids (Fig. 3). These overlapping proteins also include verified proteins, such as PENK, APOA4, CRYAB, PRDX1, PRDX2, PRDX6, CKB, NRXN3, HP, and CLU. Animal models allow us to delve into the molecular profile of asymptomatic HD stages that would not be possible by studies of post-mortem human brains. A few studies compared the proteomic profiling of brains from transgenic (Burgold et al., 2019; Hosp et al., 2017; Perluigi et al., 2005; Zabel et al., 2009) or knock-in (Deschepper et al., 2012; Federspiel et al., 2019) mice in different ages and showed that widespread molecular changes occur mostly at later stages in the brain of diseased animals. Additional candidates for early HD biomarkers coming from both human iPSCs (Akimov et al., 2021; The HD iPSC Consortium, 2020), or ESCs (McQuade et al., 2014; Ooi et al., 2019), and post-mortem HD brains (Ratovitski et al., 2016) included upregulated proteins implicated in bioenergetic dysfunction (e.g., cytosol aminopeptidase, LAP3; mitochondrial aldehyde dehydrogenase, ALDH2) and downregulated proteins implicated in synapse organisation (e.g., cytoplasmic dynein 1 light intermediate chain 1, DYNC1LI1; Liprin-alpha-3, PPFIA3). Finally, all datasets explored in this review showed a substantial similarity

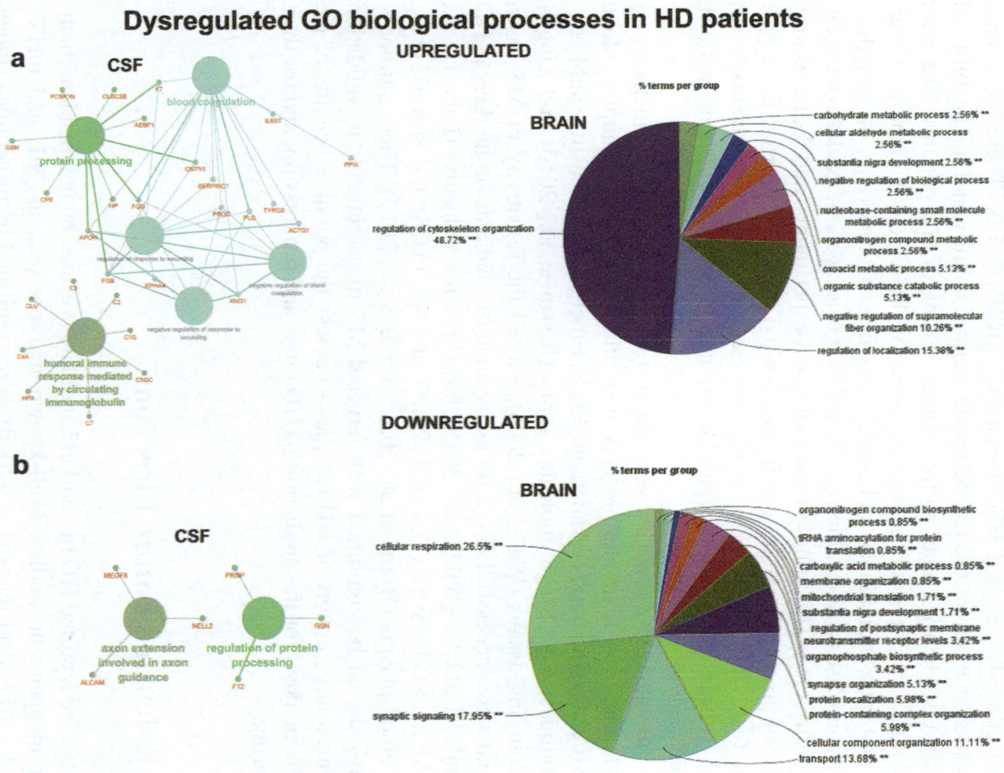

Fig. 4 Results of functional analysis from all significantly dysregulated proteins that were reported in studies analysing the global proteome in cerebrospinal fluid (CSF) and post-mortem brain from Huntington's disease patients. Gene ontology (GO) Biological processes enriched pathways are shown as a network for CSF, and a pie chart for the brain for (**a**) upregulated proteins and (**b**) downregulated proteins. Only non-redundant significant pathways are shown (false discovery rate < 0.001)

at the level of impacted pathways, confirming that dysregulated proteins are implicated in well-documented processes of HD neurodegeneration.

Most of the HD proteomics studies focused on the elucidation of HTT and mHTT (patho)physiological roles, rather than on biomarker discovery. Although many changes in protein abundances were described in various tissues of HD patients and models, and some of them were verified by other methods and/or in independent samples, no clinically applied HD biomarker currently exists. There are several obstacles which may affect the reproducibility, translatability and applicability of these biomarkers when it comes to the clinical testing of HD patients. The most limiting factor for reproducible measurements is biological variability (inter-individual and intra-individual), and model-associated factors (e.g., the length of CAG repeats and disease stage, biology related to species, genetic background, gender, comorbidity, etc.). This applies to all biological experiments, and the translation of results from proteomics studies is further limited by differences in: (i) sample preparation (e.g., homogenisation, protein extraction, or pre-fractionation); (ii) LC-MS/MS settings (e.g., protein/peptide separation, ionisation/fragmentation); and (iii) data analysis workflows.

Another limitation discussed above is the current sensitivity of MS methods for small or extremely low abundant proteins, particularly in complex samples (blood plasma, CSF, etc.) (Anderson & Anderson, 2002). Emerging ultrasensitive antibody-based techniques, e.g., Single-molecule array (Rissin et al., 2010) and Single-molecule counting assays (Wu et al., 2006), exceed the sensitivity of MS-based methods, and were successfully used for assessment of candidate biomarkers in HD patients – mHTT in CSF (Wild et al., 2015), blood plasma and saliva (Parkin et al., 2023), GFAP in blood plasma (You et al., 2021), and NfL in CSF (Al Shweiki et al., 2021) and blood plasma (Byrne et al., 2022; Byrne et al., 2017). These antibody-based assays should be combined with targeted MS measurements in validation studies using a large cohort of HD samples to assess the possible benefits of the candidate biomarkers for the prediction of HD onset and progression, and for therapy monitoring.

6 Conclusion and Future Directions

In this review, we reported HD biomarker candidates verified by independent methods and/or independent studies. Well-known markers such as NfL and mHTT (Byrne et al., 2017; Rodrigues et al., 2020) were confirmed by proteomics studies and novel candidates were also proposed. Validation in large-scale trials is anticipated for HD-specific candidate biomarkers (e.g., PENK or PDYN), CNS-specific proteins presumably associated with neurodegeneration (e.g., GFAP), but also for proteins with limited specificity to neurodegeneration (e.g., proteins associated with homeostasis, immune response or metabolism).

Numerous studies have already contributed to deciphering HD pathology and identifying candidate biomarkers; however, there remains a significant aspect of HD models that have not been analysed by proteomics methods. For instance, iPSC-derived 3D *in vitro* models with cerebral architecture and greater physiological relevance will also require detailed analysis. On the other hand, large animal models (e.g., non-human primate rhesus macaque or minipig HD models), have so far been poorly characterised at a system-wide proteome level. These models share similarities with the human in anatomy (especially body and organ size and weight), physiology (e.g., longer life), or slow HD progression. Such models are more suitable for translational research of HD development and progression, as well as for pre-clinical testing of potential therapy.

In light of the rapid development of separation techniques, MS instruments, sample preparation (and its automatization), and data analyses, we believe that MS-based proteomics with the DIA approach will be the method of choice in the near future. Single-cell MS will deliver unique data in (patho)physiology (Tajik et al., 2022). Comprehensive analyses of PTMs and proteoforms will provide novel insights into the regulation of fundamental biological processes, identifying new targets for disease therapies (Smith & Kelleher, 2018) during ultrafast analyses (up to 5 minutes) with unprecedented high proteome coverage and throughput (Messner et al., 2021). Moreover, the recently announced combination of RNA-seq and MS proteomics at a single-cell level will enable researchers to perform highly complex systems biology studies (Fulcher et al., 2022). However, not all proteins will be accessible by MS (inability to ionisation, very low abundance etc.), and the development of novel approaches that would allow analysis of a single copy of a protein, deciphering its sequence and PTMs is at the forefront of interest, with some candidates already on the horizon (Timp & Timp, 2020).

Acknowledgments This work was supported by the Czech Science Foundation project 22-24983S and by the Operational Programme Research, Development and Education (reg. No. CZ.02.1.01/0.0/0.0/16_019/0000785).

References

Aebersold, R., & Mann, M. (2016). Mass-spectrometric exploration of proteome structure and function. *Nature. Nature Publishing Group, 537*(7620), 347–355.

Aebersold, R., Burlingame, A. L., & Bradshaw, R. A. (2013). Western blots versus selected reaction monitoring assays: time to turn the tables? *Molecular & Cellular Proteomics, 12*(9), 2381–2382.

Agrawal, S., & Fox, J. H. (2019). Novel proteomic changes in brain mitochondria provide insights into mitochondrial dysfunction in mouse models of Huntington's disease. *Mitochondrion, 47*, 318–329.

Akimov, S. S., Jiang, M., Kedaigle, A. J., Arbez, N., Marque, L. O., Eddings, C. R., et al. (2021). Immortalized striatal precursor neurons from Huntington's disease patient-derived iPS cells as a platform for target identification and screening for experimental therapeutics. *Human Molecular Genetics, 30*(24), 2469–2487.

Al Shweiki, M. R., Oeckl, P., Pachollek, A., Steinacker, P., Barschke, P., Halbgebauer, S., et al. (2021). Cerebrospinal fluid levels of Prodynorphin-derived peptides are decreased in Huntington's disease. *Movement Disorders, 36*(2), 492–497.

Anderson, N. L., & Anderson, N. G. (2002). The human plasma proteome: history, character, and diagnostic prospects. *Molecular & Cellular Proteomics, 1*(11), 845–867.

Andre, R., Carty, L., & Tabrizi, S. J. (2016). Disruption of immune cell function by mutant huntingtin in Huntington's disease pathogenesis. *Current Opinion in Pharmacology, 26*, 33–38.

Aylward, E. H., Nopoulos, P. C., Ross, C. A., Langbehn, D. R., Pierson, R. K., Mills, J. A., et al. (2011). Longitudinal change in regional brain volumes in prodromal Huntington disease. *Journal of Neurology, Neurosurgery, and Psychiatry, 82*(4), 405–410.

Bachoud-Lévi, A.-C., Ferreira, J., Massart, R., Youssov, K., Rosser, A., Busse, M., et al. (2019). International guidelines for the treatment of Huntington's disease. *Frontiers in Neurology* [Internet]. [cited 2023 Feb 2], *10*. Available from: https://www.frontiersin.org/articles/10.3389/fneur.2019.00710

Barschke, P., Abu-Rumeileh, S., Al Shweiki, M. H. D. R., Barba, L., Paolini Paoletti, F., Oeckl, P., et al. (2022). Cerebrospinal fluid levels of proenkephalin and prodynorphin are differentially altered in Huntington's and Parkinson's disease. *Journal of Neurology, 269*(9), 5136–5143.

Bartzokis, G., Cummings, J., Perlman, S., Hance, D. B., & Mintz, J. (1999). Increased basal ganglia iron levels in Huntington disease. *Archives of Neurology, 56*(5), 569–574.

Bates, G. P., Dorsey, R., Gusella, J. F., Hayden, M. R., Kay, C., Leavitt, B. R., et al. (2015). Huntington disease. *Nature Reviews. Disease Primers, 1*, 15005.

Bennett, E. J., Shaler, T. A., Woodman, B., Ryu, K.-Y., Zaitseva, T. S., Becker, C. H., et al. (2007). Global changes to the ubiquitin system in Huntington's disease. *Nature. Nature Publishing Group, 448*(7154), 704–708.

Benraiss, A., Mariani, J. N., Tate, A., Madsen, P. M., Clark, K. M., Welle, K. A., et al. (2022). A TCF7L2-responsive suppression of both homeostatic and compensatory remyelination in Huntington disease mice. *Cell Reports, 40*(9), 111291.

Biomarkers Definitions Working Group. (2001). Biomarkers and surrogate endpoints: Preferred definitions and conceptual framework. *Clinical Pharmacology & Therapeutics, 69*(3), 89–95.

Björkqvist, M., Wild, E. J., Thiele, J., Silvestroni, A., Andre, R., Lahiri, N., et al. (2008). A novel pathogenic pathway of immune activation detectable before clinical onset in Huntington's disease. *The Journal of Experimental Medicine, 205*(8), 1869–1877.

Buccitelli, C., & Selbach, M. (2020). mRNAs, proteins and the emerging principles of gene expression control. *Nature Reviews Genetics. Nature Publishing Group, 21*(10), 630–644.

Burgold, J., Schulz-Trieglaff, E. K., Voelkl, K., Gutiérrez-Ángel, S., Bader, J. M., Hosp, F., et al. (2019). Cortical circuit alterations precede motor impairments in Huntington's disease mice. *Scientific Reports. Nature Publishing Group, 9*(1), 6634.

Byrne, L. M., & Wild, E. J. (2016). Cerebrospinal fluid biomarkers for Huntington's disease. *Journal of Huntington's Disease. IOS Press, 5*(1), 1–13.

Byrne, L. M., Rodrigues, F. B., Blennow, K., Durr, A., Leavitt, B. R., Roos, R. A. C., et al. (2017). Neurofilament light protein in blood as a potential biomarker of neurodegeneration in Huntington's disease: a retrospective cohort analysis. *Lancet Neurology, 16*(8), 601–609.

Byrne, L. M., Rodrigues, F. B., Johnson, E. B., Wijeratne, P. A., Vita, E. D., Alexander, D. C., et al. (2018). Evaluation of mutant huntingtin and neurofilament proteins as potential markers in Huntington's disease. *Science Translational Medicine* [Internet]. American Association for the Advancement of Science; [cited 2021 Mar 9], *10*(458). Available from: https://stm.sciencemag.org/content/10/458/eaat7108

Byrne, L. M., Schultz, J. L., Rodrigues, F. B., van der Plas, E., Langbehn, D., Nopoulos, P. C., et al. (2022). Neurofilament light protein as a potential blood biomarker for Huntington's disease in children. *Movement Disorders, 37*(7), 1526–1531.

Caron, N. S., Haqqani, A. S., Sandhu, A., Aly, A. E., Findlay Black, H., Bone, J. N., et al. (2022). Cerebrospinal fluid biomarkers for assessing Huntington disease onset and severity. *Brain Communications, 4*(6), fcac309.

Carroll, J. B., Bates, G. P., Steffan, J., Saft, C., & Tabrizi, S. J. (2015). Treating the whole body in Huntington's disease. *Lancet Neurology, 14*(11), 1135–1142.

Chae, J.-I., Kim, D.-W., Lee, N., Jeon, Y.-J., Jeon, I., Kwon, J., et al. (2012). Quantitative proteomic analysis of induced pluripotent stem cells derived from a human Huntington's disease patient. *The Biochemical Journal, 446*(3), 359–371.

Chen, S., Lu, F. F., Seeman, P., & Liu, F. (2012). Quantitative proteomic analysis of human substantia Nigra in Alzheimer's disease, Huntington's disease and multiple sclerosis. *Neurochemical Research, 37*(12), 2805–2813.

Chiang, M.-C., Juo, C.-G., Chang, H.-H., Chen, H.-M., Yi, E. C., & Chern, Y. (2007). Systematic uncovering of multiple pathways underlying the pathology of Huntington disease by an acid-cleavable isotope-coded affinity tag approach. *Molecular & Cellular Proteomics. United States, 6*(5), 781–797.

Chou, S.-Y., Weng, J.-Y., Lai, H.-L., Liao, F., Sun, S. H., Tu, P.-H., et al. (2008). Expanded-polyglutamine huntingtin protein suppresses the secretion and production of a chemokine (CCL5/RANTES) by astrocytes. *The Journal of Neuroscience, 28*(13), 3277–3290.

Choudhury, K. R., Das, S., & Bhattacharyya, N. P. (2016). Differential proteomic and genomic profiling of mouse striatal cell model of Huntington's disease and control; probable implications to the disease biology. *Journal of Proteomics, 132*, 155–166.

Chung, D., Shum, A., & Caraveo, G. (2020). GAP-43 and BASP1 in axon regeneration: implications for the treatment of neurodegenerative diseases. *Frontiers in Cell and Development Biology, 8*, 567537.

Cong, W., Cai, H., Wang, R., Daimon, C. M., Maudsley, S., Raber, K., et al. (2012). Altered hypothalamic protein expression in a rat model of Huntington's disease. *PLOS ONE. Public Library of Science, 7*(10), e47240.

Corvol, J.-C., Muriel, M.-P., Valjent, E., Féger, J., Hanoun, N., Girault, J.-A., et al. (2004). Persistent increase in olfactory type G-protein α subunit levels may underlie D1 receptor functional hypersensitivity in Parkinson disease. *Journal of Neuroscience. Soc Neurosci, 24*(31), 7007–7014.

Crotti, A., Benner, C., Kerman, B., Gosselin, D., Lagier-Tourenne, C., Zuccato, C., et al. (2014). Mutant huntingtin promotes autonomous microglia activation via myeloid lineage-determining factors. *Nature Neuroscience, 17*(4), 513–521.

Dalrymple, A., Wild, E. J., Joubert, R., Sathasivam, K., Björkqvist, M., Petersén, A., et al. (2007). Proteomic profiling of plasma in Huntington's disease reveals neuroinflammatory activation and biomarker candidates. *Journal of Proteome Research. United States, 6*(7), 2833–2840.

Dasgupta, S., Fishman, M. A., Mahallati, H., Castro, L. M., Tashima, A. K., Ferro, E. S., et al. (2015). Reduced levels of proteasome products in a mouse striatal cell model of Huntington's disease. *PLOS ONE. Public Library of Science, 10*(12), e0145333.

Deschepper, M., Hoogendoorn, B., Brooks, S., Dunnett, S. B., & Jones, L. (2012). Proteomic changes in the brains of Huntington's disease mouse models reflect pathology and implicate mitochondrial changes. *Brain Research Bulletin. United States, 88*(2–3), 210–222.

Dickey, A. S., & La Spada, A. R. (2018). Therapy development in Huntington disease: from current strategies to emerging opportunities. *American Journal of Medical Genetics. Part A, 176*(4), 842–861.

DiFiglia, M., Sapp, E., Chase, K. O., Davies, S. W., Bates, G. P., Vonsattel, J. P., et al. (1997). Aggregation of huntingtin in neuronal intranuclear inclusions and dystrophic neurites in brain. *Science, 277*(5334), 1990–1993.

El-Daher, M.-T., Hangen, E., Bruyère, J., Poizat, G., Al-Ramahi, I., Pardo, R., et al. (2015). Huntingtin proteolysis releases non-polyQ fragments that cause toxicity through dynamin 1 dysregulation. *The EMBO Journal, 34*(17), 2255–2271.

Estevez-Fraga, C., Tabrizi, S. J., & Wild, E. J. (2022). Huntington's disease clinical trials corner: November 2022. *Journal of Huntingtons Disease, 11*(4), 351–367.

Fang, Q., Strand, A., Law, W., Faca, V. M., Fitzgibbon, M. P., Hamel, N., et al. (2009). Brain-specific proteins decline in the cerebrospinal fluid of humans with Huntington disease. *Molecular & Cellular Proteomics, 8*(3), 451–466.

Federspiel, J. D., Greco, T. M., Lum, K. K., & Cristea, I. M. (2019). Hdac4 interactions in Huntington's disease viewed through the prism of multiomics*. *Molecular & Cellular Proteomics, 18*(8, Supplement 1), S92–S113.

Ferrante, R. J., Kowall, N. W., Beal, M. F., Martin, J. B., Bird, E. D., & Richardson, E. P., Jr. (1987). Morphologic and histochemical characteristics of a spared subset of striatal neurons in Huntington's disease. *Journal of Neuropathology and Experimental Neurology, 46*(1), 12–27.

Freitas, A., Aroso, M., Rocha, S., Ferreira, R., Vitorino, R., & Gomez-Lazaro, M. (2021). Bioinformatic analysis of the human brain extracellular matrix proteome in neurodegenerative disorders. *The European Journal of Neuroscience, 53*(12), 4016–4033.

Frohner, I. E., Mudrak, I., Schüchner, S., Anrather, D., Hartl, M., Sontag, J.-M., et al. (2020). PP2AC phospho-Tyr307 antibodies are not specific for this modification but are sensitive to other PP2AC modifications including Leu309 methylation. *Cell Reports, 30*(9), 3171–3182.e6.

Fulcher, J. M., Markillie, L. M., Mitchell, H. D., Williams, S. M., Engbrecht, K. M., Moore, R. J., et al. (2022). Parallel measurement of transcriptomes and proteomes from same single cells using nanodroplet splitting [Internet]. *bioRxiv* [cited 2023 Mar 8], 2022.05.17.492137. Available from: https://www.biorxiv.org/content/10.1101/2022.05.17.492137v1

Fusilli, C., Migliore, S., Mazza, T., Consoli, F., De Luca, A., Barbagallo, G., et al. (2018). Biological and clinical manifestations of juvenile Huntington's disease: a retrospective analysis. *Lancet Neurology, 17*(11), 986–993.

Gadalla, R., Noamani, B., MacLeod, B. L., Dickson, R. J., Guo, M., Xu, W., et al. (2019). Validation of CyTOF against flow cytometry for immunological studies and monitoring of human cancer clinical trials. *Frontiers in Oncology* [Internet]. [cited 2023 Mar 8], *9*. Available from: https://www.frontiersin.org/articles/10.3389/fonc.2019.00415

García-Huerta, P., Troncoso-Escudero, P., Wu, D., Thiruvalluvan, A., Cisternas-Olmedo, M., Henríquez, D. R., et al. (2020). Insulin-like growth factor 2 (IGF2) protects against Huntington's disease through the extracellular disposal of protein aggregates. *Acta Neuropathologica, 140*(5), 737–764.

Gerber, H.-P., Seipel, K., Georgiev, O., Höfferer, M., Hug, M., Rusconi, S., et al. (1994). Transcriptional activation modulated by homopolymeric glutamine and proline stretches. *Science. American Association for the Advancement of Science, 263*(5148), 808–811.

Gerber, S. A., Rush, J., Stemman, O., Kirschner, M. W., & Gygi, S. P. (2003). Absolute quantification of proteins and phosphoproteins from cell lysates by tandem MS. *Proceedings of the National Academy of Sciences, 100*(12), 6940–6945.

Gil, J. M., & Rego, A. C. (2008). Mechanisms of neurodegeneration in Huntington's disease. *The European Journal of Neuroscience, 27*(11), 2803–2820.

Graveland, G. A., Williams, R. S., & DiFiglia, M. (1985). Evidence for degenerative and regenerative changes in neostriatal spiny neurons in Huntington's disease. *Science. American Association for the Advancement of Science, 227*(4688), 770–773.

Gray, M., Shirasaki, D. I., Cepeda, C., André, V. M., Wilburn, B., Lu, X.-H., et al. (2008). Full-length human mutant huntingtin with a stable polyglutamine repeat can elicit progressive and selective neuropathogenesis in BACHD mice. *The Journal of Neuroscience, 28*(24), 6182–6195.

Guo, Q., Huang, B., Cheng, J., Seefelder, M., Engler, T., Pfeifer, G., et al. (2018). The cryo-electron microscopy structure of huntingtin. *Nature, 555*(7694), 117–120.

Haag, A. M. (2016). Mass analyzers and mass spectrometers. In: H. Mirzaei, M. Carrasco (Eds.), *Modern proteomics – Sample preparation, analysis and practical applications* [Internet]. Springer International Publishing; cited 2023 Feb 18]. p. 157–69. Available from: https://doi.org/10.1007/978-3-319-41448-5_7

Hafko, R., Villapol, S., Nostramo, R., Symes, A., Sabban, E. L., Inagami, T., et al. (2013). Commercially available angiotensin II At2 receptor antibodies are nonspecific. *PLOS ONE. Public Library of Science, 8*(7), e69234.

Halliday, G. M., McRitchie, D. A., Macdonald, V., Double, K. L., Trent, R. J., & McCusker, E. (1998). Regional specificity of brain atrophy in Huntington's disease. *Experimental Neurology, 154*(2), 663–672.

Harrison, P. M., & Arosio, P. (1996). The ferritins: molecular properties, iron storage function and cellular regulation. *Biochimica et Biophysica Acta (BBA), 1275*(3), 161–203.

Heemskerk, A.-W., & Roos, R. A. C. (2012). Aspiration pneumonia and death in Huntington's disease. *PLoS Current, 4*, RRN1293.

Hersch, S. M., Gevorkian, S., Marder, K., Moskowitz, C., Feigin, A., Cox, M., et al. (2006). Creatine in Huntington disease is safe, tolerable, bioavailable in brain and reduces serum 8OH2′dG. *Neurology. Wolters Kluwer Health, Inc. on behalf of the American Academy of Neurology, 66*(2), 250–252.

Hersch, S. M., Schifitto, G., Oakes, D., Bredlau, A.-L., Meyers, C. M., Nahin, R., et al. (2017). The CREST-E study of creatine for Huntington disease: a randomized controlled trial. *Neurology, 89*(6), 594–601.

Hilditch-Maguire, P., Trettel, F., Passani, L. A., Auerbach, A., Persichetti, F., & MacDonald, M. E. (2000). Huntingtin: an iron-regulated protein essential for normal nuclear and perinuclear organelles. *Human Molecular Genetics, 9*(19), 2789–2797.

Hodges, A., Strand, A. D., Aragaki, A. K., Kuhn, A., Sengstag, T., Hughes, G., et al. (2006). Regional and cellular gene expression changes in human Huntington's disease brain. *Human Molecular Genetics, 15*(6), 965–977.

Hosp, F., Gutiérrez-Ángel, S., Schaefer, M. H., Cox, J., Meissner, F., Hipp, M. S., et al. (2017). Spatiotemporal proteomic profiling of Huntington's disease inclusions reveals widespread loss of protein function. *Cell Reports, 21*(8), 2291–2303.

Hsieh, Y.-C., Negri, J., He, A., Pearse, R. V., Liu, L., Duong, D. M., et al. (2022). Elevated ganglioside GM2 activator (GM2A) in human brain tissue reduces neurite integrity and spontaneous neuronal activity. *Molecular Neurodegeneration, 17*, 61.

Hsu, J.-L., Huang, S.-Y., Chow, N.-H., & Chen, S.-H. (2003). Stable-isotope dimethyl labeling for quantitative proteomics. *Analytical Chemistry. American Chemical Society, 75*(24), 6843–6852.

Hu, B., Thirtamara-Rajamani, K. K., Sim, H., & Viapiano, M. S. (2009 Nov). FIBULIN-3 is uniquely upregulated in malignant gliomas and promotes tumor cell motility and invasion. *Molecular Cancer Research, 7*(11), 1756–1770.

Huang, Y.-C., Wu, Y.-R., Tseng, M.-Y., Chen, Y.-C., Hsieh, S.-Y., & Chen, C.-M. (2011). Increased prothrombin, apolipoprotein A-IV, and Haptoglobin in the cerebrospinal fluid of patients with Huntington's disease. *PLOS ONE. Public Library of Science, 6*(1), e15809.

Hyeon, J. W., Kim, A. H., & Yano, H. (2021). Epigenetic regulation in Huntington's disease. *Neurochemistry International, 148*, 105074.

Jia, Q., Li, S., Li, X.-J., & Yin, P. (2022). Neuroinflammation in Huntington's disease: from animal models to clinical therapeutics. *Frontiers in Immunology, 13*, 1088124.

Ju, T.-C., Lin, Y.-S., & Chern, Y. (2012). Energy dysfunction in Huntington's disease: insights from PGC-1α, AMPK, and CKB. *Cellular and Molecular Life Sciences, 69*(24), 4107–4120.

Kaltenbach, L. S., Romero, E., Becklin, R. R., Chettier, R., Bell, R., Phansalkar, A., et al. (2007). Huntingtin interacting proteins are genetic modifiers of neurodegeneration. *PLoS Genetics. Orr H, editor, 3*(5), e82.

Karayel-Basar, M., Uras, I., Kiris, I., Sahin, B., Akgun, E., & Baykal, A. T. (2022). Spatial proteomic alterations detected via MALDI-MS imaging implicate neuronal loss in a Huntington's disease mouse (YAC128) brain. *Molecular Omics. The Royal Society of Chemistry, 18*(4), 336–347.

Kim, J., Amante, D. J., Moody, J. P., Edgerly, C. K., Bordiuk, O. L., Smith, K., et al. (2010). Reduced creatine kinase as a central and peripheral biomarker in Huntington's disease. *Biochimica et Biophysica Acta, 1802*(7–8), 673–681.

Kitata, R. B., Yang, J.-C., & Chen, Y.-J. (2022). Advances in data-independent acquisition mass spectrometry towards comprehensive digital proteome landscape. *Mass Spectrom Reviews*, e21781. https://analyticalsciencejournals.onlinelibrary.wiley.com/doi/full/10.1002/mas.21781

Kobayashi, N., Kostka, G., Garbe, J. H. O., Keene, D. R., Bächinger, H. P., Hanisch, F.-G., et al. (2007). A comparative analysis of the Fibulin protein family: biochemical characterization, binding interactions, and tissue localization*. *The Journal of Biological Chemistry, 282*(16), 11805–11816.

Kremer, H. P. H., Roos, R. A. C., Dingjan, G., Marani, E., & Bots, G. T. A. M. (1990). Atrophy of the hypothalamic lateral tuberal nucleus in Huntington's disease. *Journal of Neuropathology and Experimental Neurology, 49*(4), 371–382.

Kremer, B., Goldberg, P., Andrew, S. E., Theilmann, J., Telenius, H., Zeisler, J., et al. (1994). A worldwide study of the Huntington's disease mutation: the sensitivity and specificity of measuring CAG repeats. *The New England Journal of Medicine. Massachusetts Medical Society, 330*(20), 1401–1406.

van der Burg, J. M., Björkqvist, M., & Brundin, P. (2009). Beyond the brain: widespread pathology in Huntington's disease. *Lancet Neurology, 8*(8), 765–774.

Kroksveen, A. C., Opsahl, J. A., Aye, T. T., Ulvik, R. J., & Berven, F. S. (2011). Proteomics of human cerebrospinal fluid: discovery and verification of biomarker candidates in neurodegenerative diseases using quantitative proteomics. *Journal of Proteomics, 74*(4), 371–388.

Kuemmerle, S., Gutekunst, C.-A., Klein, A. M., Li, X.-J., Li, S.-H., Beal, M. F., et al. (1999). Huntingtin aggregates may not predict neuronal death in Huntington's disease. *Annals of Neurology, 46*(6), 842–849.

Kupcova Skalnikova, H., Cizkova, J., Cervenka, J., & Vodicka, P. (2017). Advances in proteomic techniques for cytokine analysis: focus on melanoma research. *International Journal of Molecular Sciences, 18*(12), E2697.

Lange, V., Picotti, P., Domon, B., & Aebersold, R. (2008). Selected reaction monitoring for quantitative proteomics: a tutorial. *Molecular Systems Biology. John Wiley & Sons, Ltd., 4*(1), 222.

Langfelder, P., Cantle, J. P., Chatzopoulou, D., Wang, N., Gao, F., Al-Ramahi, I., et al. (2016). Integrated genomics and proteomics define huntingtin CAG length-dependent networks in mice. *Nature Neuroscience, 19*(4), 623–633.

Liu, X., Miller, B. R., Rebec, G. V., & Clemmer, D. E. (2007). Protein expression in the striatum and cortex regions of the brain for a mouse model of Huntington's disease. *Journal of Proteome Research. American Chemical Society, 6*(8), 3134–3142.

Luthi-Carter, R., Strand, A., Peters, N. L., Solano, S. M., Hollingsworth, Z. R., Menon, A. S., et al. (2000). Decreased expression of striatal signaling genes in a mouse model of Huntington's disease. *Human Molecular Genetics, 9*(9), 1259–1271.

Mangiarini, L., Sathasivam, K., Seller, M., Cozens, B., Harper, A., Hetherington, C., et al. (1996). Exon 1 of the HD gene with an expanded CAG repeat is sufficient to cause a progressive neurological phenotype in transgenic mice. *Cell, 87*(3), 493–506.

Marques Sousa, C., & Humbert, S. (2013). Huntingtin: here, there, everywhere! *Journal of Huntington's Disease. IOS Press, 2*(4), 395–403.

Mazarei, G., Budac, D. P., Lu, G., Adomat, H., Tomlinson Guns, E. S., Möller, T., et al. (2013). Age-dependent alterations of the kynurenine pathway in the YAC128 mouse model of Huntington disease. *Journal of Neurochemistry, 127*(6), 852–867.

Mazarei, G., Neal, S. J., Becanovic, K., Luthi-Carter, R., Simpson, E. M., & Leavitt, B. R. (2010). Expression analysis of novel striatal-enriched genes in Huntington disease. *Human Molecular Genetics, 19*(4), 609–622.

McQuade, L. R., Balachandran, A., Scott, H. A., Khaira, S., Baker, M. S., & Schmidt, U. (2014). Proteomics of Huntington's disease-affected human embryonic stem cells reveals an evolv-

ing pathology involving mitochondrial dysfunction and metabolic disturbances. *Journal of Proteome Research. American Chemical Society, 13*(12), 5648–5659.
Mears, E. R., Handley, R. R., Grant, M. J., Reid, S. J., Day, B. T., Rudiger, S. R., et al. (2021). A multi-Omic Huntington's disease transgenic sheep-model database for investigating disease pathogenesis. *Journal of Huntingtons Disease, 10*(4), 423–434.
Medina, A., Mahjoub, Y., Shaver, L., & Pringsheim, T. (2022). Prevalence and incidence of Huntington's disease: an updated systematic review and meta-analysis. *Movement Disorders, 37*(12), 2327–2335.
Mehta, D., Ahkami, A. H., Walley, J., Xu, S.-L., & Uhrig, R. G. (2022). The incongruity of validating quantitative proteomics using western blots. *Nature Plants. Nature Publishing Group, 8*(12), 1320–1321.
Meier, F., Geyer, P. E., Virreira Winter, S., Cox, J., & Mann, M. (2018). BoxCar acquisition method enables single-shot proteomics at a depth of 10,000 proteins in 100 minutes. *Nature Methods. Nature Publishing Group, 15*(6), 440–448.
Menalled, L. B. (2005). Knock-in mouse models of Huntington's disease. *NeuroRx, 2*(3), 465–470.
Menalled, L. B., Kudwa, A. E., Miller, S., Fitzpatrick, J., Watson-Johnson, J., Keating, N., et al. (2012). Comprehensive behavioral and molecular characterization of a new knock-in mouse model of Huntington's disease: zQ175. *PLoS One, 7*(12), e49838.
Menalled, L. B., Sison, J. D., Dragatsis, I., Zeitlin, S., & Chesselet, M.-F. (2003). Time course of early motor and neuropathological anomalies in a knock-in mouse model of Huntington's disease with 140 CAG repeats. *The Journal of Comparative Neurology, 465*(1), 11–26.
Messner, C. B., Demichev, V., Bloomfield, N., Yu, J. S. L., White, M., Kreidl, M., et al. (2021). Ultra-fast proteomics with scanning SWATH. *Nature Biotechnology. Nature Publishing Group, 39*(7), 846–854.
Michalik, A., & Van Broeckhoven, C. (2003). Pathogenesis of polyglutamine disorders: aggregation revisited. *Human Molecular Genetics, 12*(suppl_2), R173–R186.
Muller, M., & Leavitt, B. R. (2014). Iron dysregulation in Huntington's disease. *Journal of Neurochemistry, 130*(3), 328–350.
Muntel, J., Gandhi, T., Verbeke, L., Bernhardt, O. M., Treiber, T., Bruderer, R., et al. (2019). Surpassing 10 000 identified and quantified proteins in a single run by optimizing current LC-MS instrumentation and data analysis strategy. *Molecular Omics. The Royal Society of Chemistry, 15*(5), 348–360.
Neely, B. A., Dorfer, V., Martens, L., Bludau, I., Bouwmeester, R., Degroeve, S., et al. (2023). Toward an integrated machine learning model of a proteomics experiment. *Journal of Proteome Research* [Internet]. American Chemical Society [cited 2023 Feb 18]; Available from: https://doi.org/10.1021/acs.jproteome.2c00711.
Neueder, A., Kojer, K., Hering, T., Lavery, D. J., Chen, J., Birth, N., et al. (2022). Abnormal molecular signatures of inflammation, energy metabolism, and vesicle biology in human Huntington disease peripheral tissues. *Genome Biology, 23*(1), 189.
Niemela, V., Landtblom, A.-M., Nyholm, D., Kneider, M., Constantinescu, R., Paucar, M., et al. (2021). Proenkephalin decreases in cerebrospinal fluid with symptom progression of Huntington's disease. *Movement Disorders, 36*(2), 481–491.
Oliveira, A. O., Osmand, A., Outeiro, T. F., Muchowski, P. J., & Finkbeiner, S. (2016). αB-Crystallin overexpression in astrocytes modulates the phenotype of the BACHD mouse model of Huntington's disease. *Human Molecular Genetics, 25*(9), 1677–1689.
Ong, S.-E., Blagoev, B., Kratchmarova, I., Kristensen, D. B., Steen, H., Pandey, A., et al. (2002). Stable isotope labeling by amino acids in cell culture, SILAC, as a simple and accurate approach to expression proteomics*. *Molecular & Cellular Proteomics, 1*(5), 376–386.
Ooi, J., Langley, S. R., Xu, X., Utami, K. H., Sim, B., Huang, Y., et al. (2019). Unbiased profiling of isogenic Huntington disease hPSC-derived CNS and peripheral cells reveals strong cell-type specificity of CAG length effects. *Cell Reports, 26*(9), 2494–2508.e7.

Park, S. J., Shin, J. H., Jeong, J. I., Song, J. H., Jo, Y. K., Kim, E. S., et al. (2014). Down-regulation of mortalin exacerbates Aβ-mediated mitochondrial fragmentation and dysfunction. *The Journal of Biological Chemistry, 289*(4), 2195–2204.

Parkin, G. M., Corey-Bloom, J., Snell, C., Smith, H., Laurenza, A., Daldin, M., et al. (2023). Salivary Huntingtin protein is uniquely associated with clinical features of Huntington's disease. *Scientific Reports. Nature Publishing Group, 13*(1), 1034.

Perluigi, M., Poon, H. F., Maragos, W., Pierce, W. M., Klein, J. B., Calabrese, V., et al. (2005). Proteomic analysis of protein expression and oxidative modification in r6/2 transgenic mice: a model of Huntington disease. *Molecular & Cellular Proteomics. United States, 4*(12), 1849–1861.

Peterson, A. C., Russell, J. D., Bailey, D. J., Westphall, M. S., & Coon, J. J. (2012). Parallel reaction monitoring for high resolution and high mass accuracy quantitative, targeted proteomics*. *Molecular & Cellular Proteomics. Elsevier, 11*(11), 1475–1488.

Poon, S., Easterbrook-Smith, S. B., Rybchyn, M. S., Carver, J. A., & Wilson, M. R. (2000). Clusterin is an ATP-independent chaperone with very broad substrate specificity that stabilizes stressed proteins in a folding-competent state. *Biochemistry, 39*(51), 15953–15960.

Pouladi, M. A., Morton, A. J., & Hayden, M. R. (2013). Choosing an animal model for the study of Huntington's disease. *Nature Reviews Neuroscience. Nature Publishing Group, 14*(10), 708–721.

Przybyl, L., Wozna-Wysocka, M., Kozlowska, E., & Fiszer, A. (2021). What, when and how to measure—peripheral biomarkers in therapy of Huntington's disease. *International Journal of Molecular Sciences, 22*(4), 1561.

Ratovitski, T., Chaerkady, R., Kammers, K., Stewart, J. C., Zavala, A., Pletnikova, O., et al. (2016). Quantitative proteomic analysis reveals similarities between Huntington's disease (HD) and Huntington's disease-like 2 (HDL2) human brains. *Journal of Proteome Research. American Chemical Society, 15*(9), 3266–3283.

Ravkov, E. V., Charlton, C. M., Barker, A. P., Hill, H., Peterson, L. K., Slev, P., et al. (2019). Evaluation of mass cytometry in the clinical laboratory. *Cytometry. Part B, Clinical Cytometry, 96*(4), 266–274.

Rezeli, M., Végvári, Á., Silajdžić, E., Björkqvist, M., Tabrizi, S. J., Laurell, T., et al. (2014). Inflammatory markers in Huntington's disease plasma—a robust nanoLC–MRM-MS assay development. *EuPA Open Proteomics, 3*, 68–75.

Rhee, S. G., Chae, H. Z., & Kim, K. (2005). Peroxiredoxins: a historical overview and speculative preview of novel mechanisms and emerging concepts in cell signaling. *Free Radical Biology & Medicine, 38*(12), 1543–1552.

Rifai, N., Gillette, M. A., & Carr, S. A. (2006). Protein biomarker discovery and validation: the long and uncertain path to clinical utility. *Nature Biotechnology. Nature Publishing Group, 24*(8), 971–983.

Riguet, N., Mahul-Mellier, A.-L., Maharjan, N., Burtscher, J., Croisier, M., Knott, G., et al. (2021). Nuclear and cytoplasmic huntingtin inclusions exhibit distinct biochemical composition, interactome and ultrastructural properties. *Nature Communications. Nature Publishing Group, 12*(1), 6579.

Rissin, D. M., Kan, C. W., Campbell, T. G., Howes, S. C., Fournier, D. R., Song, L., et al. (2010). Single-molecule enzyme-linked immunosorbent assay detects serum proteins at subfemtomolar concentrations. *Nature Biotechnology. Nature Publishing Group, 28*(6), 595–599.

Rodrigues, F. B., Byrne, L. M., McColgan, P., Robertson, N., Tabrizi, S. J., Zetterberg, H., et al. (2016). Cerebrospinal fluid inflammatory biomarkers reflect clinical severity in Huntington's disease. *PLoS One, 11*(9), e0163479.

Rodrigues, F. B., Byrne, L. M., Tortelli, R., Johnson, E. B., Wijeratne, P. A., Arridge, M., et al. (2020). Mutant huntingtin and neurofilament light have distinct longitudinal dynamics in Huntington's disease. *Science Translational Medicine, 12*(574), eabc2888.

Rodrigues, F. B., Byrne, L. M., & Wild, E. J. (2018). Biofluid biomarkers in Huntington's disease. *Methods in Molecular Biology. United States, 1780*, 329–396.

Rosas, H. D., Doros, G., Gevorkian, S., Malarick, K., Reuter, M., Coutu, J.-P., et al. (2014). PRECREST: a phase II prevention and biomarker trial of creatine in at-risk Huntington disease. *Neurology, 82*(10), 850–857.

Rosas, H. D., Salat, D. H., Lee, S. Y., Zaleta, A. K., Pappu, V., Fischl, B., et al. (2008). Cerebral cortex and the clinical expression of Huntington's disease: complexity and heterogeneity. *Brain, 131*(Pt 4), 1057–1068.

Rosenblatt, A., Abbott, M. H., Gourley, L. M., Troncoso, J. C., Margolis, R. L., Brandt, J., et al. (2003). Predictors of neuropathological severity in 100 patients with Huntington's disease. *Annals of Neurology, 54*(4), 488–493.

Rosenblatt, A., Kumar, B. V., Mo, A., Welsh, C. S., Margolis, R. L., & Ross, C. A. (2012). Age, CAG repeat length, and clinical progression in Huntington's disease. *Movement Disorders, 27*(2), 272–276.

Rosenblatt, A., Liang, K.-Y., Zhou, H., Abbott, M. H., Gourley, L. M., Margolis, R. L., et al. (2006). The association of CAG repeat length with clinical progression in Huntington disease. *Neurology. Wolters Kluwer Health, Inc. on behalf of the American Academy of Neurology, 66*(7), 1016–1020.

Ross, C. A., Aylward, E. H., Wild, E. J., Langbehn, D. R., Long, J. D., Warner, J. H., et al. (2014). Huntington disease: natural history, biomarkers and prospects for therapeutics. *Nature Reviews. Neurology, 10*(4), 204–216.

Ross, P. L., Huang, Y. N., Marchese, J. N., Williamson, B., Parker, K., Hattan, S., et al. (2004). Multiplexed protein quantitation in Saccharomyces cerevisiae using amine-reactive isobaric tagging reagents*. *Molecular Cell Proteom. Elsevier, 3*(12), 1154–1169.

Ryskamp, D., Wu, J., Geva, M., Kusko, R., Grossman, I., Hayden, M., et al. (2017). The sigma-1 receptor mediates the beneficial effects of pridopidine in a mouse model of Huntington disease. *Neurobiology of Disease, 97*, 46–59.

Sajjad, M. U., Green, E. W., Miller-Fleming, L., Hands, S., Herrera, F., Campesan, S., et al. (2014). DJ-1 modulates aggregation and pathogenesis in models of Huntington's disease. *Human Molecular Genetics, 23*(3), 755–766.

Sap, K. A., Guler, A. T., Bezstarosti, K., Bury, A. E., Juenemann, K., Demmers, J. A., et al. (2019). Global proteome and Ubiquitinome changes in the soluble and insoluble fractions of Q175 Huntington mice brains. *Molecular & Cellular Proteomics, 18*(9), 1705–1720.

Sapp, E., Seeley, C., Iuliano, M., Weisman, E., Vodicka, P., DiFiglia, M., et al. (2020). Protein changes in synaptosomes of Huntington's disease knock-in mice are dependent on age and brain region. *Neurobiology of Disease, 141*, 104950.

Sassone, J., Colciago, C., Cislaghi, G., Silani, V., & Ciammola, A. (2009). Huntington's disease: the current state of research with peripheral tissues. *Experimental Neurology, 219*(2), 385–397.

Saudou, F., & Humbert, S. (2016). The biology of huntingtin. *Neuron, 89*(5), 910–926.

Schönberger, S. J., Jezdic, D., Faull, R. L. M., & Cooper, G. J. S. (2013). Proteomic analysis of the human brain in Huntington's disease indicates pathogenesis by molecular processes linked to other neurodegenerative diseases and to type-2 diabetes. *Journal of Huntingtons Disease. Netherlands, 2*(1), 89–99.

Shen, L., Pearson, K. J., Xiong, Y., Lo, C.-M., Tso, P., Woods, S. C., et al. (2008). Characterization of apolipoprotein A-IV in brain areas involved in energy homeostasis. *Physiology & Behavior, 95*(1–2), 161–167.

Shen, L., Tso, P., Woods, S. C., Sakai, R. R., Davidson, W. S., & Liu, M. (2007). Hypothalamic apolipoprotein A-IV is regulated by leptin. *Endocrinology, 148*(6), 2681–2689.

Shendelman, S., Jonason, A., Martinat, C., Leete, T., & Abeliovich, A. (2004). DJ-1 is a redox-dependent molecular chaperone that inhibits alpha-synuclein aggregate formation. *PLoS Biology, 2*(11), e362.

Shi, M., Jin, J., Wang, Y., Beyer, R. P., Kitsou, E., Albin, R. L., et al. (2008). Mortalin: a protein associated with progression of Parkinson disease? *Journal of Neuropathology and Experimental Neurology, 67*(2), 117–124.

Shiio, Y., & Aebersold, R. (2006). Quantitative proteome analysis using isotope-coded affinity tags and mass spectrometry. *Nature Protocol. Nature Publishing Group, 1*(1), 139–145.
Shirasaki, D. I., Greiner, E. R., Al-Ramahi, I., Gray, M., Boontheung, P., Geschwind, D. H., et al. (2012). Network organization of the huntingtin proteomic interactome in mammalian brain. *Neuron, 75*(1), 41–57.
Silajdžić, E., & Björkqvist, M. (2018). A critical evaluation of wet biomarkers for Huntington's disease: current status and ways forward. *Journal of Huntingtons Disease, 7*(2), 109–135.
Skotte, N. H., Andersen, J. V., Santos, A., Aldana, B. I., Willert, C. W., Nørremølle, A., et al. (2018). Integrative characterization of the R6/2 mouse model of Huntington's disease reveals dysfunctional astrocyte metabolism. *Cell Reports, 23*(7), 2211–2224.
Slow, E. J., van Raamsdonk, J., Rogers, D., Coleman, S. H., Graham, R. K., Deng, Y., et al. (2003). Selective striatal neuronal loss in a YAC128 mouse model of Huntington disease. *Human Molecular Genetics, 12*(13), 1555–1567.
Smith, L. M., & Kelleher, N. L. (2018). Proteoforms as the next proteomics currency. *Science. American Association for the Advancement of Science, 359*(6380), 1106–1107.
Sorolla, M. A., Reverter-Branchat, G., Tamarit, J., Ferrer, I., Ros, J., & Cabiscol, E. (2008). Proteomic and oxidative stress analysis in human brain samples of Huntington disease. *Free Radical Biology & Medicine, 45*(5), 667–678.
Sugars, K. L., & Rubinsztein, D. C. (2003). Transcriptional abnormalities in Huntington disease. *Trends in Genetics, 19*(5), 233–238.
Sui, X., Pires, D. E. V., Ormsby, A. R., Cox, D., Nie, S., Vecchi, G., et al. (2020). Widespread remodeling of proteome solubility in response to different protein homeostasis stresses. *Proceedings of the National Academy of Sciences, 117*(5), 2422–2431.
Szeliga, M. (2020). Peroxiredoxins in neurodegenerative diseases. *Antioxidants (Basel), 9*(12), 1203.
Tabrizi, S. J., Estevez-Fraga, C., van Roon-Mom, W. M. C., Flower, M. D., Scahill, R. I., Wild, E. J., et al. (2022). Potential disease-modifying therapies for Huntington's disease: lessons learned and future opportunities. *Lancet Neurology, 21*(7), 645–658.
Tajik, M., Baharfar, M., & Donald, W. A. (2022). Single-cell mass spectrometry. *Trends in Biotechnology, 40*(11), 1374–1392.
The HD iPSC Consortium. (2020). Bioenergetic deficits in Huntington's disease iPSC-derived neural cells and rescue with glycolytic metabolites. *Human Molecular Genetics, 29*(11), 1757–1771.
The Huntington's Disease Collaborative Research Group. (1993). A novel gene containing a trinucleotide repeat that is expanded and unstable on Huntington's disease chromosomes. The Huntington's Disease Collaborative Research Group. *Cell, 72*(6), 971–983.
Thiede, B., Höhenwarter, W., Krah, A., Mattow, J., Schmid, M., Schmidt, F., et al. (2005). Peptide mass fingerprinting. *Methods, 35*(3), 237–247.
Thompson, A., Schäfer, J., Kuhn, K., Kienle, S., Schwarz, J., Schmidt, G., et al. (2003). Tandem mass tags: a novel quantification strategy for comparative analysis of complex protein mixtures by MS/MS. *Analytical Chemistry. American Chemical Society, 75*(8), 1895–1904.
Timp, W., & Timp, G. (2020). Beyond mass spectrometry, the next step in proteomics. *Science Advances, 6*(2), eaax8978.
Timpl, R., Sasaki, T., Kostka, G., & Chu, M.-L. (2003). Fibulins: a versatile family of extracellular matrix proteins. *Nature Reviews Molecular Cell Biology. Nature Publishing Group, 4*(6), 479–489.
Tsiamis, V., Ienasescu, H.-I., Gabrielaitis, D., Palmblad, M., Schwämmle, V., & Ison, J. (2019). One thousand and one software for proteomics: Tales of the toolmakers of science. *Journal of Proteome Research. American Chemical Society, 18*(10), 3580–3585.
Tso, P., Liu, M., Kalogeris, T. J., & Thomson, A. B. (2001). The role of apolipoprotein A-IV in the regulation of food intake. *Annual Review of Nutrition, 21*, 231–254.
Ünlü, M., Morgan, M. E., & Minden, J. S. (1997). Difference gel electrophoresis. A single gel method for detecting changes in protein extracts. *Electrophoresis, 18*(11), 2071–2077.

Valekova, I., Jarkovska, K., Kotrcova, E., Bucci, J., Ellederova, Z., Juhas, S., et al. (2016). Revelation of the IFNα, IL-10, IL-8 and IL-1β as promising biomarkers reflecting immuno-pathological mechanisms in porcine Huntington's disease model. *Journal of Neuroimmunology, 293*, 71–81.

Vidova, V., & Spacil, Z. (2017). A review on mass spectrometry-based quantitative proteomics: targeted and data independent acquisition. *Analytica Chimica Acta, 964*, 7–23.

Vinther-Jensen, T., Börnsen, L., Budtz-Jørgensen, E., Ammitzbøll, C., Larsen, I. U., Hjermind, L. E., et al. (2016). Selected CSF biomarkers indicate no evidence of early neuroinflammation in Huntington disease. *Neurology Neuroimmunology & Neuroinflammation* [Internet], *3*(6). Available from: http://nn.neurology.org/content/3/6/e287.abstract

Vinther-Jensen, T., Budtz-Jørgensen, E., Simonsen, A. H., Nielsen, J. E., & Hjermind, L. E. (2014). YKL-40 in cerebrospinal fluid in Huntington's disease – a role in pathology or a nonspecific response to inflammation? *Parkinsonism & Related Disorders, 20*(11), 1301–1303.

Vinther-Jensen, T., Simonsen, A. H., Budtz-Jørgensen, E., Hjermind, L. E., & Nielsen, J. E. (2015). Ubiquitin: a potential cerebrospinal fluid progression marker in Huntington's disease. *European Journal of Neurology, 22*(10), 1378–1384.

Vonsattel, J. P., Myers, R. H., Stevens, T. J., Ferrante, R. J., Bird, E. D., & Richardson, E. P. (1985). Neuropathological classification of Huntington's disease. *Journal of Neuropathology and Experimental Neurology, 44*(6), 559–577.

Vonsattel, J. P. G. (2008). Huntington disease models and human neuropathology: similarities and differences. *Acta Neuropathologica, 115*(1), 55–69.

Wan, J., Savas, J. N., Roth, A. F., Sanders, S. S., Singaraja, R. R., Hayden, M. R., et al. (2013). Tracking brain palmitoylation change: predominance of glial change in a mouse model of Huntington's disease. *Chemistry & Biology, 20*(11), 1421–1434.

Wanker, E. E., Ast, A., Schindler, F., Trepte, P., & Schnoegl, S. (2019). The pathobiology of perturbed mutant huntingtin protein–protein interactions in Huntington's disease. *Journal of Neurochemistry, 151*(4), 507–519.

Wegrzynowicz, M., Holt, H. K., Friedman, D. B., & Bowman, A. B. (2012). Changes in the striatal proteome of YAC128Q mice exhibit gene–environment interactions between mutant huntingtin and manganese. *Journal of Proteome Research. American Chemical Society, 11*(2), 1118–1132.

Wexler, N. S. (2004). Venezuelan kindreds reveal that genetic and environmental factors modulate Huntington's disease age of onset. *Proceedings of the National Academy of Sciences of the United States of America, 101*(10), 3498–3503.

Wild, E. J., Boggio, R., Langbehn, D., Robertson, N., Haider, S., Miller, J. R. C., et al. (2015). Quantification of mutant huntingtin protein in cerebrospinal fluid from Huntington's disease patients. *Journal of Clinical Investigation. American Society for Clinical Investigation, 125*(5), 1979–1986.

Wishart, T. M., Rooney, T. M., Lamont, D. J., Wright, A. K., Morton, A. J., Jackson, M., et al. (2012). Combining comparative proteomics and molecular genetics uncovers regulators of synaptic and axonal stability and degeneration in vivo. *PLOS Genetics. Public Library of Science, 8*(8), e1002936.

Wu, A. H., Fukushima, N., Puskas, R., Todd, J., & Goix, P. (2006). Development and preliminary clinical validation of a high sensitivity assay for cardiac troponin using a capillary flow (single molecule) fluorescence detector. *Clinical Chemistry, 52*(11), 2157–2159.

Wyatt, A. R., Yerbury, J. J., & Wilson, M. R. (2009). Structural characterization of clusterin-chaperone client protein complexes. *The Journal of Biological Chemistry, 284*(33), 21920–21927.

You, H., Wu, T., Du, G., Huang, Y., Zeng, Y., Lin, L., et al. (2021). Evaluation of blood glial fibrillary acidic protein as a potential marker in Huntington's disease. *Frontiers in Neurology, 12*, 779890.

Zabel, C., Chamrad, D. C., Priller, J., Woodman, B., Meyer, H. E., Bates, G. P., et al. (2002). Alterations in the mouse and human proteome caused by Huntington's disease. *Molecular & Cellular Proteomics. United States, 1*(5), 366–375.

Zabel, C., Mao, L., Woodman, B., Rohe, M., Wacker, M. A., Kläre, Y., et al. (2009). A large number of protein expression changes occur early in life and precede phenotype onset in a mouse model for Huntington disease*S. *Molecular & Cellular Proteomics, 8*(4), 720–734.

Zabel, C., Nguyen, H. P., Hin, S. C., Hartl, D., Mao, L., & Klose, J. (2010). Proteasome and oxidative phoshorylation changes may explain why aging is a risk factor for neurodegenerative disorders. *Journal of Proteomics. Netherlands, 73*(11), 2230–2238.

Zabel, C., Sagi, D., Kaindl, A. M., Steireif, N., Kläre, Y., Mao, L., et al. (2006). Comparative proteomics in neurodegenerative and non-neurodegenerative diseases suggest nodal point proteins in regulatory networking. *Journal of Proteome Research. United States, 5*(8), 1948–1958.

Zhang, Y., Fonslow, B. R., Shan, B., Baek, M.-C., & Yates, J. R. (2013). Protein analysis by shotgun/bottom-up proteomics. *Chemical Reviews, 113*(4), 2343–2394.

Zhou, S., Bailey, M. J., Dunn, M. J., Preedy, V. R., & Emery, P. W. (2005). A quantitative investigation into the losses of proteins at different stages of a two-dimensional gel electrophoresis procedure. *Proteomics, 5*(11), 2739–2747.

Zhu, Z., & Reiser, G. (2018). The small heat shock proteins, especially HspB4 and HspB5 are promising protectants in neurodegenerative diseases. *Neurochemistry International, 115*, 69–79.

Zuccato, C., Ciammola, A., Rigamonti, D., Leavitt, B. R., Goffredo, D., Conti, L., et al. (2001). Loss of huntingtin-mediated BDNF gene transcription in Huntington's disease. *Science. American Association for the Advancement of Science, 293*(5529), 493–498.

Microbiome and Metabolomic Biomarkers for Huntington's Disease

Bethany A. Masson, Wendy Qin, Chloe J. Love, Carolina Gubert, and Anthony J. Hannan

Abbreviations

AD	Alzheimer's disease
ALS	Amyotrophic lateral sclerosis
BDNF	Brain-derived neurotrophic factor
CAG	Cytosine-adenine-guanine
CNS	Central nervous system
ENS	Enteric nervous system
FMT	Faecal microbiota transplant
GALT	Gut-associated lymphatic tissue
HD	Huntington's disease
HDGEC	Huntington's disease gene expansion carrier
HTT	Huntingtin gene
Htt	Huntingtin protein
HPA	Hypothalamus-pituitary-adrenal
IBS	Irritable bowel syndrome
IFN	Interferon
IL	Interleukin
mHtt	Mutant huntingtin protein

B. A. Masson · W. Qin · C. Gubert (✉)
Florey Institute of Neuroscience and Mental Health, University of Melbourne, Parkville, VIC, Australia
e-mail: carolina.gubert@florey.edu.au

C. J. Love
School of Medicine, Faculty of Health, Geelong Warnum Ponds Campus, Deakin University, Geelong, VIC, Australia

A. J. Hannan (✉)
Florey Institute of Neuroscience and Mental Health, University of Melbourne, Parkville, VIC, Australia

Department of Anatomy and Physiology, University of Melbourne, Parkville, VIC, Australia
e-mail: anthony.hannan@florey.edu.au

E. A. Thomas, G. M. Parkin (eds.), *Biomarkers for Huntington's Disease*, Contemporary Clinical Neuroscience, https://doi.org/10.1007/978-3-031-32815-2_10

MS	Mass spectrometry
NMR	Nuclear magnetic resonance
PCR	Polymerase chain reaction
PD	Parkinson's disease
rRNA	Ribosomal ribonucleic acid
SCFA	Short-chain fatty acids
SPF	Specific pathogen free
WT	Wild-type

1 Introduction

The gut microbiome and its role in both health and disease has been a topic of increasing interest, with advancements in analysing and profiling the gut microbiome allowing for increased understanding (Cryan et al., 2019). Subsequently, the gut microbiome and its components have potential be utilised as biomarkers to monitor disease progression and treatment efficacy. Huntington's disease (HD) is a fatal monogenic neurodegenerative disease with limited therapeutic interventions (Stoker et al., 2022). Recent pre-clinical and clinical studies have established alteration in the gut microbiome profile in HD, and a potential avenue for novel therapeutic approach. This chapter summarises current findings of the role of the gut microbiome in HD and explores the potential to utilise the gut microbiota to establish novel biomarkers and therapeutic options.

2 The Gut Microbiome

The human body hosts various microorganisms within the oral and genital cavities, skin, and gastrointestinal tract, with distinctions between each anatomical area (Peterson et al., 2009). The gastrointestinal tract is the most populated region, containing bacteria, fungi, archaea, and viruses, with bacteria being the dominant microbe (Luckey, 1972; Shanahan et al., 2021). The gut microbiome refers to all these microorganisms, their genomes, the metabolites they produce, and the environmental conditions within the gastrointestinal tract (Marchesi & Ravel, 2015).

The composition of the community of microorganisms, or the gut microbiota, is predominantly acquired from exposure to the maternal microbiota during birth, with varying exposure depending on vaginal or caesarean delivery (Dominguez-Bello et al., 2019). The gut microbiota composition is dynamic throughout infancy, before establishing a more stable profile at 2–3 years of age (Koenig et al., 2011). Subsequently, the gut microbiota composition remains relatively stable across an individual's life, with some dynamic response to environmental influences such as diet, stress, medication, and exercise (Cryan et al., 2019). Overall temporal stability suggests that while environmental factors may shift the microbiota composition, each individual has a homeostatic microbiota state (Walker et al., 2003). This homeostatic state likely consists of a 'core' microbiome profile, referring to an individual's specific set of microbial taxa and its functional traits that allows functional redundancy of the microbiome with some flexibility (Neu et al., 2021). The volatility of the gut microbiome and its ability to adapt and change to external influences may be useful in providing rapid insights into an individual's health and potentially assist with early detection and disease intervention. Conversely, this leads to significant individual variability in gut microbiome composition and can make it challenging to establish ubiquitous biomarkers for disease states.

The Gut Microbiome and Neurodegenerative Disease

DNA sequencing can be used to profile the microbiome, assessing the complex commensal, symbiotic, and pathogenic occupation of the host with the bacterial microbiome, the virome (viruses), the mycobiome (fungi), and the archaeal microbiome (archaea) (Stappenbeck & Virgin, 2016). Together, the intricate relationships between the gut microbiome players help maintain bodily homeostasis and are involved in nutrient absorption, immune function, maintaining intestinal barrier integrity, and pathogen resistance (Garrett et al., 2010).

It has been established that distinct microbiota profiles exist between healthy and disease states. Understanding these compositional differences has allowed researchers to use microbiome-based biomarkers in the diagnosis, prognosis, therapy efficacy, and risk prediction of disease (Shanahan et al., 2021). Microbiota disruption, also known as 'dysbiosis', refers to an imbalance of microbial species or reduction in microbiota diversity (Walker, 2017). Disruption to the microbiota has been established in preclinical models of neurodegenerative disease including Parkinson's disease (PD), Alzheimer's disease (AD), amyotrophic lateral sclerosis (ALS), and Huntington's disease (HD) (Wu et al., 2015; Kong et al., 2018; Perez-Pardo et al., 2018; Lee et al., 2019). Additionally, gut microbiota disruption is also seen clinically in PD, AD, ALS, and HD (Scheperjans et al., 2015; Rowin et al., 2017; Vogt et al., 2017; Wasser et al., 2020).

Alongside gut microbiota disruption, gastrointestinal symptoms, and disorders such as constipation, diarrhoea, delayed gastric emptying, and irritable bowel syndrome (IBS), are frequent across neurodegenerative diseases (Fasano et al., 2015; Scheperjans et al., 2015; Vogt et al., 2017). These gastrointestinal symptoms often predate the onset of illness, persisting throughout its course and significantly impacting patients' quality of life. These findings suggest that the gut microbiota may play an important role in the peripheral pathology of these disorders.

While each neurodegenerative disease has unique pathological symptoms, they are all characterised by neuronal (and glial) deterioration and neuroinflammation, which suggest that neurodegenerative diseases, such as PD, may even originate in the gut, however further studies are needed (Liddle, 2018). A deeper understanding of the gut microbiome composition in these disorders may provide novel insights into their origins and reveal therapeutic targets to improve the management and treatment of these neurodegenerative diseases (Chung et al., 2012; Borre et al., 2014).

The Microbiota-Gut-Brain-Axis

Over the past decade, it has become increasingly evident that resident gut microbiota have influence over brain function, cognition, behavior, and disease outcomes (Sekirov et al., 2010; Clemente et al., 2012). Early research in gut-brain communication focused on its impact on digestive function and satiety (Taché et al., 1980; Konturek et al., 2004). However, recent research has seen an increased focus on the bi-directional communication between the gut microbiota, the gut, and the brain via a wide array of communication pathways that include the nervous system, the immune system, and endocrine signalling (Cryan et al., 2019). The gut microbiota preserves homeostasis within the gut through maintenance of gut integrity, nutrient and drug metabolism, aiding the mucous layer to protect from pathogenic microbes, and assisting in immune maturation and modulation (Jandhyala et al., 2015).

To further investigate the role of the microbiota-gut-brain axis, various studies have utilised germ-free mice and have demonstrated that microbes are essential for the development of a healthy brain and nervous system and even influence the development of the 'second brain', the enteric nervous system (ENS) of the gut (Kunze et al., 2009; Heijtz et al., 2011; Sharon et al., 2016). Germ-free mice also show changes in gastrointestinal physiology, with enlarged caecum, reduced small intestine surface area as well as overall reduced body weight (Al-Asmakh & Zadjali, 2015). The introduction of human gut microbiota into germ-free mice has allowed researchers to study specific microbes and host interactions (Al-Asmakh & Zadjali, 2015) identifying specific microbiota that may be associated with disease outcomes. Other studies have investigated the role of the microbiota-gut-brain axis through specifically targeting the gut microbiota in order to modulate it. This includes using antibiotics to deplete the bacterial population, faecal microbiota transplantation (FMT) to introduce a different gut microbiota profile, gastrointestinal infection, and supplementation of prebiotics (substrates utilized by host microorganisms that provide favourable conditions for beneficial bacteria), and probiotics (live microorganisms that promote health benefits to the host) (Luczynski et al., 2016; Harris et al., 2017; Staley et al., 2017; Kazemi et al., 2019). While the concept of the microbiota-gut-brain axis is relatively new, it provides an exciting new research avenue and therapeutic target in neurological diseases.

The ENS is present throughout the gastrointestinal tract and forms a neuronal pathway of communication between the gut and the brain, facilitating digestion and gut motility through peristalsis (Fleming et al., 2020; Geng et al., 2022). The vagus nerve connects the ENS to the central nervous system (CNS) and allows for bi-directional communication through afferent and efferent fibres (Breit et al., 2018). Vagal nerve stimulation has been reported to beneficial in the treatment of the neurological diseases, like epilepsy and psychiatric disorders, such as depression (Milby et al., 2008). There is evidence that the gut microbiota are able to directly stimulate the vagus nerve. Bravo et al. reported that the positive effects seen in depression and anxiety-like behavior after the preclinical administration of the bacteria *Lactobacillus rhamnosus* were not seen in mice which had undergone a vagotomy (Bravo et al., 2011).

The gut microbiota composition influences the production and composition of neurotransmitters and neuroactive metabolites key to host health, and which play a vital role within the microbiota-gut-brain axis (Agus et al., 2021). Neurotransmitters are molecules which facilitate the communication between neurons in the body and there is increasing evidence that the gut microbiota contribute to neurotransmitter synthesis and composition (Strandwitz, 2018). Fermentation of dietary fibre by the gut microbiota produces short-chain fatty acids (SCFAs) which consist primarily of butyrate, propionate, and acetate (Silva et al., 2020). Key aspects of gut health such as inflammation and permeability can be influenced by SCFA diversity and concentration. Altered SCFA composition has been demonstrated in neurodegenerative diseases including AD and PD (Dalile et al., 2019; Gubert et al., 2020b; Silva et al., 2020).

The gut microbiota are essential in the development of gut-associated lymphoid tissue (GALT) which allows the immune system to detect potential pathogens within the gut while not targeting the commensal microbes (Kamada et al., 2013). Regulatory T cells, immunoglobulin-A secreting B cells and a range of innate immune cells have been reported to be modulated by the gut microbiota during homeostasis (Jandhyala et al., 2015). Systemic circulation of cytokines allows for stimulation of appropriate immune responses and can be generated by enterocytes within the gastrointestinal tract (Cryan et al., 2019). The hypothalamic-pituitary-adrenal (HPA) axis is influenced by the gut microbiota during development and adult function, with germ-free mice having a hyper-responsive HPA axis which is able to be modulated by introduction of gut microbiota from specific pathogen free (SPF) mice (Sudo et al., 2004; Sudo, 2014). Vagal nerve stimulation was shown to preclinically increase corticotropin-releasing factor mRNA levels in the hypothalamus (Hosoi et al., 2000). The gut microbiota is a vital component in various pathways of communication within the microbiota-gut-brain axis, making gut microbiota disruption a valuable target for investigation.

Assessing the Gut Microbiome

The progression of DNA sequencing technology has improved our ability to quantify abundances of complex microbial communities in comparison to culture-based methods (Bonk et al., 2018). 16S metagenomic sequencing and polymerase chain reaction (PCR), use the genetic material of the microbiome to enable inferences regarding the composition, diversity, richness and functional capability of the gut microbiota (Morgan & Huttenhower, 2012). Alongside the use of germ-free mice, raised with no exposure to microorganisms (Luczynski et al., 2016), these tools have allowed researchers to measure changes in the gut microbiome and identify gut markers associated with early disease stages.

16S sequencing characterises gut microbial composition from genetic material, providing relative taxa-specific gene abundances (Bonk et al., 2018). The 16S ribosomal RNA (rRNA) gene is absent in eukaryotes, with conserved regions in bacteria and hypervariable regions that are genera specific (Gao et al., 2021). 16S rRNA can be amplified by PCR and sequenced to generate reads which reflect bacterial abundance or the occurrence of DNA in a sample. Reads are then compared to large databases to infer the bacteria present using taxonomic assignment.

16S sequencing generates phylogenetic information but relies on small genetic targets with hypervariable regions highly prone to mutation, which can lead to sequencing errors (Morgan & Huttenhower, 2012). An increased degree of conservation of the hypervariable regions correlates with a higher-level taxonomy. Some studies suggest that sequencing of the full-length 16S sequence can provide species level resolution, however this can be more expensive, therefore up to genus level information is more commonly characterised (Yang et al., 2016). 16S sequencing is also limited by the primers selected for PCR in characterising relative abundances of 16S amplicons, with no primer pair able to detect all microbial structures (Soergel et al., 2012). Additionally, 16S sequencing cannot assess the functional capacity of a bacterial community. Although phylogeny correlates with functional composition, predictive software such as Phylogenetic Investigation of Communities by Reconstruction of Unobserved States (PICRUSt) is limited by its ability to only differentiate taxa by the degree to which they differ at the amplified marker gene sequence (Langille et al., 2013). While 16S sequencing methods commonly only use a specific region of a single gene, metagenomic sequencing uses all of the genes from a community to provide information about the gene composition as well as the functional capacity of the microbiome (Wang et al., 2015).

Metagenomics usually involves 'shotgun sequencing' the entire genomes of microbial populations without specific PCR primers, eliminating bias due to primer selection found in 16S sequencing (Soergel et al., 2012). Metagenomic sequencing splits bacterial genetic material into short fragments rather than relying on a single rRNA. Fragments are then sequenced to assess species profiles based on phylogenetic markers (Wang et al., 2015). Sequencing approaches need to consider read length, balancing between short, high-quality reads, which are challenging to assemble, and long but low-quality reads, which are easy to assemble (Ghosh et al., 2021). Due to large DNA quantities used in metagenomic analysis, high-powered computational methods are required to effectively sequence, assemble and analyse reads (Thomas et al., 2012). While metagenomic sequencing is a more in-depth assessment of the structure and function of microbial communities, it uses substantially more resources and time than 16S sequencing.

A significant bottleneck of 16S and metagenomic sequencing is the cost and time involved in analysing numerous samples. PCR amplicon-based sequencing sidesteps the DNA extraction process making it a feasible option to rapidly quantify bacterial taxon without high-cost implications (Videvall et al., 2017). Studies by Flores et al. found that a direct PCR approach successfully replicated the richness and phylogenetic diversity of traditional 16S DNA extraction methods across five body microbiota habitats (Flores et al., 2012). No difference was seen in bacterial community composition and relative abundance estimates (Flores et al., 2012). These findings suggest that direct PCR approaches to assessing the microbiome profile may benefit studies with large numbers of samples by increasing speed and efficiency.

Both 16S and metagenomic sequencing use taxonomic profiling to describe the diversity, richness, abundance, and evenness of microbiota communities. α-diversity is composed of richness and evenness measures. Richness refers to the number of distinct taxa observed or inferred within an individual and evenness determines how similar taxa abundance is within an individual (Thukral, 2017; Ghosh et al., 2021). α-diversity can be measured using different metrics, including Observed, Shannon and Inverse Simpson indices which do not take into consideration the identity of species, or using Faith's phylogenetic diversity which does consider species identity (Thukral, 2017). β-diversity measures the dissimilarity between microbial communities (Koleff et al., 2003). The higher the β-diversity, the less similar are the communities. β-diversity can be measured using the Bray-Curtis dissimilarity index and weighted or unweighted UniFrac distances. Weighted UniFrac takes into account the relative abundance of communities between samples, whereas unweighted UniFrac only assesses the presence or absence of taxa (Ghosh et al., 2021). Understanding α- and β-diversity profiles of the gut microbiome can provide insight into characteristics of the microbiota in disease states.

Metabolomics is a process of identifying all the small molecules, or metabolites, existing within a given organism or sample. In the context of microbiome research, metabolomics aims to identify the metabolite production of the gut microbiota, providing insight into its functional capacity. Gut-derived metabolites can modulate several processes within the gastrointestinal tract and move externally into systemic circulation to influence other body systems (Idle & Gonzalez, 2007). Metabolites are measured using nuclear magnetic resonance (NMR) spectroscopy and mass spectrometry (MS) (Zhou et al., 2021). NMR has the capacity to detect amino-acids, short and medium chain fatty acids, sugars and bile acids (Nicholson et al., 2012), however, is significantly less sensitive compared to MS (Zhou et al., 2021). Multi-omic approaches can be useful to identify relationships between metagenomic and metabolomic datasets and establish gut-derived functional profiles (Rohart et al., 2017).

Strengths & Limitations of Gut Microbiome Assessment

Microbiota populate the interfaces between the human body and the external environment, providing a unique opportunity to understand interactions with our environment (Dominguez-Bello et al., 2019). Advances in sequencing methods have increased the accuracy, sensitivity and speed with which we can assess large microbial populations such as the gut microbiota. Current challenges in the field include a lack of a comprehensive understanding of the effects of coprophagic nature of mice and rats (the most commonly used preclinical animal species) on the structural and functional capacity of the gut microbiome (Bogatyrev et al., 2020). Coprophagy can lead to cage effects, with mice within the same cage sharing a distinct microbiota profile with the potential to confound findings if not carefully considered, particularly when mixing genotypes (Singh et al., 2021). Cage effects can contribute to up to 32% of the variance in the microbiome composition between studies (Hildebrand et al., 2013). Batch effects from sample processing can also introduce confounding factors that are not related to any factors of interest (Wang & Lê Cao, 2023). While analytical methods can account for these batch effects, the complex nature of the microbiome and the removal or accounting for batch effects can make findings hard to reproduce (Wang & Lê Cao, 2023).

Additional limitations in microbiome research include incomplete reference databases and technical variation across sample collection, storage, DNA extraction and bioinformatic analysis (Schmidt et al., 2018). Determining the characteristics of a microbiome that can predict health would provide a target for microbial modulations; however, due to the volatile and individualistic nature of the microbiome as well as needing to consider temporal changes, progress in this area is slow. Additionally, there are still challenges in establishing whether disruption to the microbiota profile plays a casual or consequential role in disease states (McBurney et al., 2019).

3 Biomarkers

Functional biomarkers provide accurate and reproducible measurements of a characteristic that correlate with progression of disease (Strimbu & Tavel, 2010). Biomarkers include invasive biopsies, which can provide biochemical measures or be histologically evaluated (Przybyl et al., 2021). Wet biomarkers are identified from blood or other bodily fluids. Additionally, non-invasive imaging can provide valuable information via associated imaging biomarkers (Przybyl et al., 2021).

Useful biomarkers provide reliable and quick diagnostic tests which provide disease-specific information (Przybyl et al., 2021). This enables an initial determination of disease progression as well as continued monitoring of disease advancement. Biomarker tests are clinically significant, as they can be used to assess the effectiveness of applied therapeutics. Additionally, biomarkers related to the premanifest stage of disease allow for early detection and preventative treatment, providing opportunity to delay disease onset and potential disease prevention. Early detection is particularly relevant for HD patients as there are limited effective therapeutic interventions (and no disease-modifying therapies), with therapeutic care focusing on improving quality of life (Paulsen et al., 2014). It is hoped that, when we do have our first disease-modifying therapy available for clinical HD, it would have greater therapeutic benefits when initiated at the earlier stages of disease (as appears to be the case for Alzheimer's disease).

Gut microbiome biomarkers were first identified for their potential in assessing metabolic outcomes in diseases such as metabolic syndrome and obesity due to the vital role of the gut microbiome in the metabolism of food (Finucane et al., 2014). Moreover, the metabolism of nutrients helps regulate host health via key organs including brain function and development (Selkrig et al., 2014). Gut microbiota disruption has been reported in both immune and neurological diseases (Chung et al., 2012; Borre et al., 2014), with specific microbiome profiles able to be utilised as a biomarker to monitor disease progression.

Clinical HD biomarkers currently being used are predominately applicable to symptomatic HD patients, such as brain imaging with striatal atrophy (Montoya et al., 2006). However, therapeutic approaches targeting pre-symptomatic disease require more sensitive biomarkers which can provide valuable information about disease onset and progression in an individual.

4 Huntington's Disease

Huntington's disease is the most prevalent monogenic neurodegenerative disease, resulting in a triad of motor, cognitive, and psychiatric symptoms (Stoker et al., 2022). The clinical onset of HD is determined when an individual develops degenerative motor symptoms, typically chorea, which causes involuntary movements (McColgan & Tabrizi, 2018). Other motor features of HD include impersistence, impaired gait and impaired postural stability (Pandey & Rajamma, 2018; Stoker et al., 2022). Combined with cognitive deficits, the motor symptoms increase the chance of the patient falling and impair their ability to perform everyday tasks, leading to a loss in quality of life.

However, cognitive deficits in HD can occur well before motor onset (Ghosh & Tabrizi, 2018). These are usually observed as issues with judgement, learning, planning, and the ability to switch attention between tasks. Shifts in an individual's mood and personality, typically irritability, are also observed, along with other psychiatric symptoms. In addition, the decline in social cognition can lead to inappropriate responses in social situations, resulting in isolation and stigmatisation. Especially when a patient is unaware that these changes are caused by prodromal HD symptoms, it can lead to difficulties with relationships at home and employment (Ross et al., 2014).

Depression and anxiety are far more prevalent in HD patients than in the general population (Gubert et al., 2020a). This may result from psychosomatic effects of living with the disease as well as changes in the brain causing a predisposition to depression and anxiety well before clinical (motor) onset of HD (Stoker et al., 2022). Suicidal ideation is more common in HD individuals compared to the general population even before motor onset (McColgan & Tabrizi, 2018). Depression and irritability are the most common behavioral changes throughout the course of disease. Other psychiatric manifestations such as delusions and hallucinations can occur but are quite uncommon (Stoker et al., 2022). Most HD patients also experience sleep issues such as insomnia, which can compound cognitive deficits and irritability (Pandey&Rajamma, 2018).

Huntington's Disease Pathology

HD results in neuropathological changes with significant cell loss in the caudate and putamen (Pandey & Rajamma, 2018). Interneurons are generally spared, but imaging shows a loss of striatal volume in moderate-severe HD. However, prior to HD motor onset, no loss in striatal volume can be seen. Other areas of the brain that are affected are the hippocampus, parietal lobe, cerebellum, hypothalamus, thalamus, substantia nigra, and cortical layers 3, 5, and 6.

Since HD is considered a neurodegenerative disease, the extent of peripheral pathology has been understudied, and thus remains relatively unclear. However, increasing interest in the gut microbiota and findings of gut microbiota disruption in other neurodegenerative diseases have resulted in new research on the microbiota-gut-brain axis and its potential role within HD. Individuals with HD experience a wide range of gastrointestinal disturbances; most notably, unintended weight loss (which could also involve changes outside the gut) as well as diarrhoea, constipation, gastritis, and nutrient deficiencies (Djoussé et al., 2002; van der Burg et al., 2017). Some evidence suggests that weight loss could be due to high caloric demands that result from constant choreiform movements and may be further aggravated by depression and insomnia (Underwood et al., 2006). However, these gastrointestinal disturbances could also result from HD-manifested gastrointestinal tract dysfunction (van der Burg et al., 2011; Kong et al., 2018; Gubert et al., 2022b).

5 Huntington's Disease Aetiology

HD is caused by an autosomal dominant monogenic mutation; the pathogenic expansion of the cytosine-adenine-guanine (CAG) trinucleotide repeat in the *huntingtin* (*HTT*) gene (Nance, 2017; Bashir, 2019). This repeat, and the *HHT* gene where it resides, is located in the short arm of chromosome 4 (4p16.3) (Kim et al., 2021). Subsequently to identification of clinical symptoms, diagnosis of HD can easily be confirmed by genetic testing due to its monogenic nature (Walker, 2007). CAG repeat lengths of less than 27 are considered normal and are not associated with HD (Rosenblatt et al., 2012). Typically, the gene carries 10–35 CAG repeats with a mean value of 18 CAG repeats in the population. However, CAG repeats greater than 35 are pathogenic with reduced penetrance observed with alleles of 36–39 repeats and those with greater than 39 repeats are certain to develop HD.

The huntingtin protein (Htt) is expressed in humans and rodents (and most other animal species), with highest levels of expression in neurons in the CNS (DiFiglia et al., 1997). Htt plays a critical role for early embryonic development and in neurogenesis (Nguyen et al., 2013; Jimenez-Sanchez et al., 2017). Htt has also been well characterised as a scaffolding protein, orientating spindle poles during mitosis (Godin et al., 2010). Brain-derived neurotrophic factor (BDNF) has potential to be used as a biomarker because Htt is a transcriptional regulator of the gene that expresses BDNF (Kegel et al., 2002; Zuccato et al., 2003). Since the expression of Htt is necessary for function, the expression of mutant Htt (mHtt) has deleterious effects. In its mutated form, mHtt has an expanded pathogenic polyQ tract caused by the expansion of CAG repeats (Marxreiter et al., 2020). This significantly alters the protein-protein interactions that Htt is involved in, leading to changes in key signaling pathways and biological processes (Wanker et al., 2019).

Models of Huntington's Disease

The R6 transgenic mouse models are common preclinical *in vivo* models used to investigate the pathogenesis of HD and potential therapeutic targets (Farshim & Bates, 2018). The R6 mouse models express the human genomic fragment, exon 1 *HTT* and 200 bp of intron 1 (Mangiarini et al., 1996). This fragment incorporates an *HTT* promoter sequence which promotes transcription of the HTT transgene, including the HD-causing repetition of the CAG trinucleotide. Within the R6 mouse models, the R6/1 and R6/2 lines are the most widely used HD animal models (Farshim & Bates, 2018). The R6/1 line was generated with approximately 115 CAG repeats in the transgene, while the R6/2 line was generated with approximately 150 CAG repeats (Mangiarini et al., 1996). These CAG repeats exhibit some germline instability (in both animal models and HD families) leading to varying repeat lengths in different colonies of the same transgenic lines.

6 The Gut Microbiome in Huntington's Disease

Gastrointestinal disturbances and disruptions to the gut microbiota have been reported preclinically and clinically in HD, with findings indicating changes in the gut microbiome may occur prior to clinical HD diagnosis (Kong et al., 2018; Wasser et al., 2020). While primarily HD research has focused on pathology in the CNS, changes in the plasma levels of gut microbiota-derived metabolites and reports of constipation and diarrhea in HD patients have prompted exploration of the peripheral pathology (Kong et al., 2018; Vicars et al., 2021). HD differs from most neurodegenerative diseases which have reported gut microbiota disturbances, as it is a monogenic disease. Pre-clinical models of HD have reported alterations in the ENS due to the expression of mHtt, however the relationship between mHtt and the gastrointestinal changes is yet to be fully understood (van der Burg et al., 2011). Further investigation into the gut microbiota's potential influence on clinical onset and progression of the disease may be beneficial in developing relevant biomarkers and understanding changes in individuals prior to cognitive, motor, and psychiatric decline (Fig. 1).

Kong et al. (2018) presented the first findings of HD gut microbiota disruption, either preclinically or clinically. This was first demonstrated in the R6/1 mouse

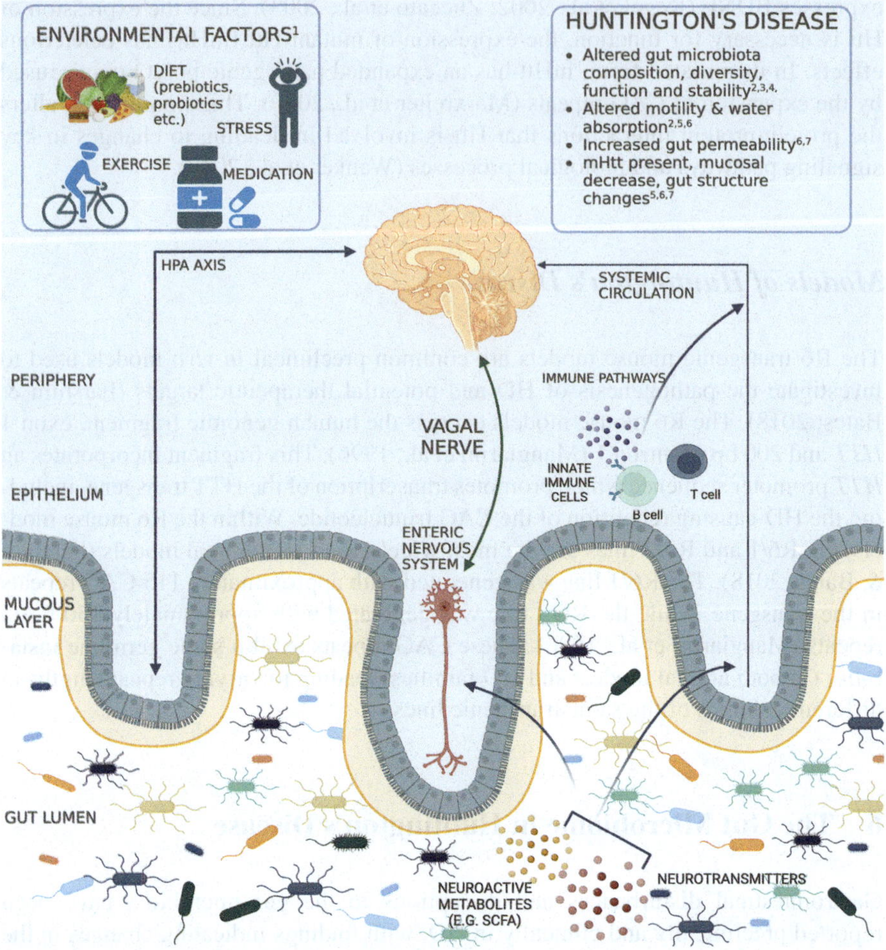

Fig. 1 Schematised visualisation of the bi-directional communication pathways within the microbiota-gut-brain axis, known influential environmental factors, and pathogenic impacts of Huntington's disease. HPA, hypothalamus-pituitary-adrenal axis; mHtt, mutant huntingtin protein; SCFA, short-chain fatty acids. (Created with Biorender.com. [1]Gubert et al. (2020b); [2]Kong et al. (2018); [3]Wasser et al. (2020); [4]Kong et al. (2020); [5]van der Burg (2011); [6]Gubert et al. (2022b); [7]Stan et al. (2020))

model at 12 weeks of age, prior to the onset of motor symptoms (Kong et al., 2018). 16S rRNA analysis revealed differences in the gut microbiota composition of HD mice compared to wild-type (WT) littermates, with male HD mice displaying more distinct differences (Kong et al., 2018). The two dominant phyla in both HD and WT mice were *Bacteroidetes* and *Firmicutes*, however the *Bacteroidetes/Firmicutes* ratio between HD and WT were different with HD mice reporting a higher abundance of *Bacteroidetes* (Kong et al., 2018). Male HD mice had increased α-diversity, while female HD mice did not differ in α-diversity measures (Kong et al., 2018). In β-diversity both male and female HD mice showed significant differences to their WT littermates in the Bray-Curtis measure but not in the unweighted UniFrac measure, indicating that the bacterial composition is phylogenetically similar but has different abundances (Kong et al., 2018). A follow up study analysing the gut microbiota of R6/1 mice investigated the onset of gut microbiota disruption in HD, confirming disruption does not occur before 12 weeks of age (Kong et al., 2020).

Findings of HD gut microbiota disruption were clinically confirmed by Wasser et al., who reported gut microbiota changes between Huntington's disease gene expansion carriers (HDGEC) and age and sex matched healthy controls (HC) using 16S rRNA analysis (Wasser et al., 2020). The HDGEC group was composed of both premanifest and manifest individuals with confirmed expansion of the HD gene; although the study had a limited number for each sub-group (23 and 19 respectively), there was no significant differences in their gut microbiota profiles (Wasser et al., 2020). The premanifest group consists of individuals who have yet to display the overt motor symptoms required for clinical HD diagnosis, supporting the preclinical findings of gut microbiota disturbances prior to motor symptoms in R6/1 mice (Kong et al., 2018; Wasser et al., 2020). The altered *Bacteroidetes/Firmicutes* ratio reported preclinically was not seen in the HDGECs, but changes were observed in both α- and β-diversity measures (Wasser et al., 2020). HDGECs had lower diversity in both Observed and Fisher indices of α-diversity and significantly differed in β-diversity based on unweighted UniFrac distances, indicating gut microbiota structure differed in HDGECs (Wasser et al., 2020). Following on from these initial studies, gut microbiota disruption has been reported preclinically in another mouse model of HD, R6/2, and in another clinical cohort of HD patients (Stan et al., 2020; Du et al., 2021).

Further preclinical exploration of the nature of HD gut microbiota disruption was conducted by Kong et al. (2020) using shotgun sequencing which was used to further clarify the functional capacity of the gut microbiota at a metagenomic level (Kong et al., 2020). This study investigated the stability of the HD gut microbiota over a range of time points from 4 to 12 weeks of age and found increased volatility compared to WT littermates, a feature also seen in inflammatory gastrointestinal tract diseases (Kong et al., 2020; Clooney et al., 2021). Analysis of the altered pathways between HD and WT gut microbiota revealed that those stable in the WT were predominately volatile in the HD profile from adolescence to adulthood (Kong et al., 2020). It has been proposed that mHtt protein present in the gut causes constant pressure on the gut microbiota which may cause instability in both structure and function (van der Burg et al., 2011; Kong et al., 2020). A range of pathways were unaltered and remained stable in both the HD and WT profiles, suggesting there are pathways unaltered by HD pathogenesis, however further studies are required to fully understand this relationship (Kong et al., 2020).

Gut Function in Huntington's Disease

The *HTT* gene is expressed peripherally outside of the CNS. This has been found preclinically where huntingtin inclusions have been reported in a range of cell types in the stomach wall, duodenum, and rectum (Moffitt et al., 2009; Lakra et al., 2019). Weight loss is seen in premanifest HD patients and is a significant predicting factor for quality of life in patients with HD (Lakra et al., 2019). Gastrointestinal disturbances seen in R6/2 mice include huntingtin inclusions in enteric neurons, loss of gut neuropeptides, decreased mucosal thickness and decreased villi length (van der Burg et al., 2011). There was an increase in faecal output to food intake ratio at 12 weeks of age in male R6/2 mice, indicating that there is malabsorption of food content (van der Burg et al., 2011). In this HD model, mice begin to significantly lose weight at 10 weeks of age, indicating that weight loss is not caused by malabsorption, but rather it may be a contributing factor (van der Burg et al., 2011). This is supported by findings in the R6/1 model, where male HD mice have a significantly reduced body weight despite increased food intake (Kong et al., 2018).

Gut motility and water absorption is altered in both the R6/1 and R6/2 models of HD compared with their WT counterparts. In the R6/2 model, HD mice have increased whole gut transit time, increased faecal output despite similar food intake, and increased faecal water content (van der Burg et al., 2011). Studies have reported differing findings in the R6/1 model, with Kong et al. (2018) finding an increase in faecal output in male HD mice and an increase in faecal water content in female HD mice. However, Gubert et al. found a decrease in faecal output in HD mice compared to their WT littermates despite higher food intake and a decrease in faecal water content in female HD mice (Kong et al., 2018; Gubert et al., 2022b). These findings indicate that HD pathogenesis impacts gastrointestinal motility and water absorption, which is reflective of the differing gastrointestinal disturbances reported clinically in HD patients, and may be influenced by environmental factors resulting in different presentations (Vicars et al., 2021).

Gut permeability is increased in both R6/1 and R6/2 mice, despite a study in R6/2 mice finding no significant changes to tight junction protein levels in the barrier of the gastrointestinal tract (Stan et al., 2020; Gubert et al., 2022b). It has been reported that changes to the gut microbiota can transiently affect the tight junction proteins, however further research is required to understand the role of the gut microbiota and its effect on gut permeability in HD (Fasano et al., 1991; Asmar et al., 2002). Preclinically, gastrointestinal structural changes have been reported, with both R6/1 and R6/2 male mice found to have shortened colon length, which is indicative of inflammation and is present in preclinical models of ulcerative colitis (Stan et al., 2020; Xu et al., 2020; Gubert et al., 2022b). Faecal microbiota transplant (FMT) is a tool utilised to establish new gut microbiota, either preclinically or clinically. When used preclinically in R6/1 mice the wild-type (WT) FMT was unable to engraft as successfully in HD mice compared to their WT littermates, particularly in males (Gubert et al., 2022b). This suggests that changes in the gut micro-environment and gut microbiota of the R6/1 model may prevent a stable microbiota from being established.

Gut Microbiome and Inflammation in Huntington's Disease

Inflammation plays an important role in neurodegenerative diseases, however the role of the gut microbiota in this process is not fully understood. It has been hypothesised that gut microbiota disruption may play a role in pathogenesis in PD (Liddle, 2018). Du et al. investigated the link between the gut microbiota, circulating cytokines and clinical outcomes in HD patients (Du et al., 2021). This study included 33 HD gene expansion carriers, both premanifest and manifest, and 33 sex and age matched controls who were relatives of the HD gene expansion carriers (Du et al., 2021). Environment plays a key role in the structure and functioning of the gut microbiome, with co-habitants sharing similar profiles, the use of relatives, including spouses, in this study allows for greater control of individuals differing environmental factors (Valles-Colomer et al., 2023). The gut microbiome was analysed using 16S DNA and it was found that HD gene expansion carriers had increased α- and β-diversity, similar to preclinical findings (Kong et al., 2018; Du et al., 2021). The genus *Intestinimonas* was found to have a higher abundance in HDGECs and positively correlated with total functional capacity score, which looks at and individuals' ability in day-to-day life and level of care required (Du et al., 2021). *Intestinimonas* was also positively correlated with plasma levels of the cytokine interleukin (IL)-4, an anti-inflammatory cytokine, suggesting the gut microbiota and systemic inflammation may contribute to functioning in HD gene expansion carriers (Du et al., 2021). *Bilophila* was found to be negatively correlated with the pro-inflammatory cytokine IL-6, however further studies are required to better understand the relationship of the gut microbiota with systemic inflammation in HD (Du et al., 2021). Preclinical investigation into the cytokine profile of the proximal colon in the R6/1 model at 14 weeks of age, after the onset of gut microbiota disruption, revealed a decrease in interferon (IFN)-γ in female HD mice and an increase in IL-7R in HD males (Gubert et al., 2022b). IFN-γ plays a key role in protection from pathogenic bacteria, while IL-7R has increased signalling in irritable bowel disease (IBD) (Boxx & Cheng, 2016; Belarif et al., 2019). These findings suggest a pro-inflammatory environment is present in the gastrointestinal tract in HD and indicates a need for further exploration of inflammation in the HD gut and its potential role in pathogenesis and how this is reflective in plasma cytokines as a potential biomarker.

Although these recent findings have established disruption to the gut microbiota in HD, the relationship between the disruption and disease progression and symptom severity have yet to be elucidated. *Eubacterium hallii* was found to be negatively correlated with disease onset in premanifest HDGECs, and a lower abundance in the gut microbiota correlated with more severe motor symptoms (Wasser et al., 2020). *E. hallii* was also noted to have associations with memory tests (Hopkins verbal learning test-revived and symbol digit modalities test) but was not sufficient to draw conclusions on the gut microbiota's impact on cognition (Wasser et al., 2020). As previously mentioned, the genus *Intestinimonas* is positively correlated with total functional capacity score, suggesting certain bacteria may be beneficial in the HD gut microbiota (Du et al., 2021). *Lactobacillus* was also reported to have increased abundance in HD patients and found to be negatively correlated with outcomes in the mini-mental state exam indicating that cognitive function may be negatively impacted by the gut microbiota (Du et al., 2021). Alterations in the gut microbiota can impact cognition through the microbiota-gut-brain axis, therapeutic measures have been proposed to target cognitive parameters using the gut microbiota (Cryan et al., 2019; Gubert et al., 2022b). The HD gut microbiota were modulated preclinically in the R6/1 model using FMT from WT littermates, with gut microbiota as an intervention to establish a WT gut microbiota (Gubert et al., 2022b). Cognitive deficits seen in the female HD mice were ameliorated in the group which received WT FMT, specifically in testing fear conditioning and extinction (Gubert et al., 2022b). These initial studies provide insight for further exploration in terms of potential biomarkers in the HD gut microbiota as well as in understanding the role of the gut microbiota in HD pathogenesis and as a possible therapeutic target.

Recent years have seen increased research into the gut microbiota in HD, however there are inconsistent attributes in the gut microbiota disruption seen throughout these studies. Gubert et al. explored the gene-environment interactions in R6/1 mice to investigate the HD gut microbiota as a potential therapeutic avenue (Gubert et al., 2022a). Increased α-diversity was seen in the HD mice, and the gut microbiota was altered in β-diversity between genotype, sex and introduction of exercise equipment or environmental enrichment to the home cage (Gubert et al., 2022a). Environmental enrichment and exercise have been demonstrated to affect the gut microbiota and have reported potential as therapeutic interventions (Gubert et al., 2022a). The gut microbiota profiles differed in both abundance of bacteria and phylogeny, indicating that the environment of the home cage impacted the gut microbiota of both HD and WT mice (Gubert et al., 2022a). The preclinical environment is tightly controlled, lowering the impact of differing environmental factors on mice of the same strain, raised in the same facility in these studies (Gubert et al., 2022a). However, studies in HD patients rely on individuals with varying environmental factors and poses a potential reason why there are inconsistent findings in changes to the gut and gut microbiota in both preclinical and clinical HD studies (Wasser et al., 2020; Du et al., 2021; Gubert et al., 2022a; Valles-Colomer et al., 2023). While 16S studies provide valuable insight into the changes to the gut microbiota composition in HD, more sophisticated techniques such as shotgun-sequencing metabolomics allow researchers to further elucidate the functional capacity of the gut microbiota and its potential role in pathogenesis as well as offer potential biomarkers for HD.

7 Metagenomics & Metabolomics in HD

Blood metabolites may provide insight into processes that contribute to neurodegeneration in HD. Preclinically, metabolomic analysis in the serum of presymptomatic HD rats found a decrease in N-acetylaspartate, as well as increased glutamine, succinic acid, glucose and lactate (Verwaest et al., 2011). Increased glucose and lactate were also found in the cerebrospinal fluid, suggesting an alteration in energy metabolism (Verwaest et al., 2011). Serum metabolic profiling of 10 presymptomatic people with HD, 20 early symptomatic people with HD and 20 control participants found amino acid dysregulation associated with a pro-catabolic state (Underwood et al., 2006), which may contribute to cachexia seen in people with HD (Djoussé et al., 2002). This study design was then mimicked in three groups of mice, with two groups of N171-82Q HD mice and one group of wild-type controls (Underwood et al., 2006). Similar to the human findings, there were notable differences in the metabolic profiles of the N171-82Q HD mice compared to wild-type (Underwood et al., 2006). Over time there was an observed increase in amino acid accumulation from asymptomatic to symptomatic people with HD, consistent with disease progression, however, reduced valine in people with HD suggests a complex dysregulation of amino acids, requiring validation from further studies (Underwood et al., 2006). Another study found changes in tryptophan and antioxidant pathways in 52 premanifest and 102 early symptomatic HD, compared to 140 healthy controls, which may be related to gut-derived oxidative stress or toxicity of the mHtt present in the gut (Rosas et al., 2015).

Few studies have combined metagenomic and metabolomic approaches to explore gut-derived metabolite production in HD, with findings predominantly in preclinical models. Kong et al. (2020) were the first to integrate both approaches in HD (Kong et al., 2020). Using metagenomic techniques, no difference was found in relative abundances of bacterial families between WT and R6/1 HD mice across five time points prior to overt motor symptoms at 12 weeks of age (Kong et al., 2020). However, while no difference was observed in bacterial species, when assessing overall stability of the gut microbiota profile the R6/1 HD mice showed an unstable bacterial temporal profile throughout adolescence and into adulthood, compared to a stable profile see in WT mice (Kong et al., 2020). At 12 weeks of age (prior to onset of overt motor deficits) there were significant differences in β-diversity between R6/1 and WT littermates (Kong et al., 2020). Functional profiling demonstrated differences in 15 plasma metabolites create a distinct signature between WT and R6/1 genotypes with butanoate metabolism increased in the R6/1 HD mice at 12 weeks of age (Kong et al., 2020). Plasma SCFA analysis found decreased propionate and butyrate in the R6/1 HD mice compared to WT littermate controls. Integration of metabolomics and metagenomics was able to identify several *Bacteroides* species in R6/1 HD mice that positively correlated with butyrate levels while negatively associated with plasma ATP and pipecolic acid (Kong et al., 2020). Butyrate, ATP and pipecolic acid can all be microbially derived and play an important role in CNS function (Kong et al., 2020). Increased butanoate metabolism in the gut alongside decreased butyrate concentration in the plasma suggests a potential dysfunction of butyrate uptake (Kong et al., 2020). In-depth characterisation of butyrate metabolism may provide further insights into butyrate as a HD-related marker of gut dysfunction. Butyrate regulates the expression of junction proteins that control the permeability of epithelial cells lining the colon (Siddiqui and Cresci, 2021). Butyrate uptake dysregulation may increase colon permeability and lead to 'leaky gut', increasing the translocation of bacteria and their by-products into systemic circulation and potentially triggering immune cascades, contributing to systemic inflammation (Siddiqui & Cresci, 2021). While Kong et al. (2020) only investigated up to 12 weeks of age, changes in the gut microbiome and metabolome may become more apparent at later disease stages (Kong et al., 2020). However, pre-symptomatic changes remain crucial markers of disease progression prior to clinical diagnosis.

Kong et al. (2022) used metagenomic techniques to characterize the gut mycobiome profile in R6/1 HD mice, finding significant differences in gut mycobiome composition between R6/1 and WT mice at 12 weeks of age, with a signature of 15 fungal species discriminating between the two genotypes (Kong et al., 2022). Greater mycobiome α-diversity was also seen in R6/1 HD mice at 12 weeks (Kong et al., 2022). These findings suggest a disrupted mycobiome profile may be occurring prior to the onset of HD motor symptoms. This could be due to an observed increase in R6/1 HD mice of the species *C. glabrata*, a fungal pathobiont which has previously been found to contribute to gut inflammation and increased intestinal permeability (Charlet et al., 2018; Kong et al., 2022).

We are only beginning to understand the role of fungi in the gut in HD. While the *Bacteroidetes / Firmicutes* ratio is an indicator of gut bacterial health, the mycobiome *Ascomycota/Basidiomycota* ratio may be an indicator of healthy gut fungal composition (Sokol et al., 2017). While no difference was observed between R6/1 HD and wild-type *Ascomycota/Basidiomycota* ratio, this study should be replicated independently with larger numbers to corroborate these findings (Kong et al., 2022). Despite fungi being relatively low abundance withing the gastrointestinal tract, understanding the mycobiome profile and integrating this knowledge with bacterial microbiota studies is essential to providing a deeper understanding of early gut dysfunction in HD.

8 Potential Therapeutic Uses

Progress towards developing and implementing an accurate HD diagnostic system using microbiome biomarkers is still in its early stages. Currently, there are limited treatment options for HD patients, with the gut microbiota being a valuable therapeutic target beginning to be explored. The development of relevant biomarkers is imperative to developing effective treatments and monitoring therapeutic efficacy, especially in a disease where individuals have variable clinical onset and differing rates of degeneration (Fig. 2) (Janse van Rensburg et al., 2022; Stoker et al., 2022).

The gut microbiota can be targeted through alteration of the diet, specifically through supplementation with prebiotics and probiotics. Prebiotics are utilised by the gut microbiota to the benefit of the host (Gibson et al., 2017). They are sourced from the diet from food sources through fruits, vegetables, and grains and have been shown to produce a range of benefits including increased BDNF expression and alter cognitive outcomes, both of which are relevant in HD treatment (Savignac et al.,

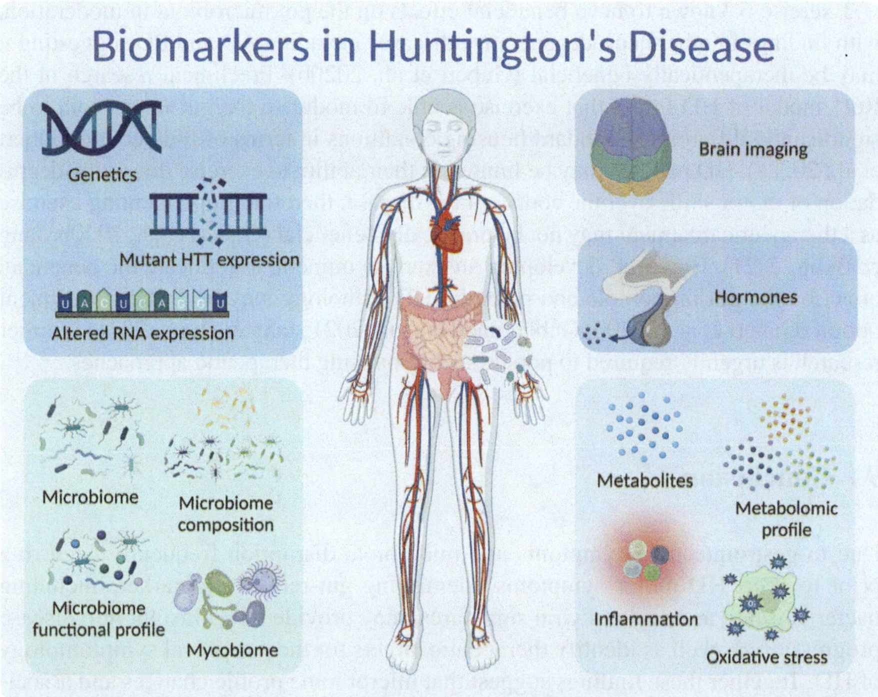

Fig. 2 Biomarkers in Huntington's disease. Diagrammatic representation of current and potential biomarkers used in diagnosis, disease progression, monitoring, and investigation of potential therapeutic options in Huntington's disease. (Created with Biorender.com)

2013; Vázquez et al., 2015; Burokas et al., 2017; Cryan et al., 2019). Alternatively, probiotics are beneficial microbes delivered to the host's gastrointestinal tract (Mitra et al., 2022). Research into the potential therapeutic benefits of probiotic administration in AD found benefits both preclinically and clinically, suggesting a potential therapeutic avenue in neurodegenerative diseases (Yeon et al., 2010; Akbari et al., 2016). Synbiotics, the combination of prebiotics and probiotics, improves the efficacy and survival of the administered probiotics (Gyawali et al., 2019). Future studies in HD investigating the potential therapeutic benefits of prebiotics and probiotics should consider the benefits of synbiotics, especially after preclinical reports of gut microbiota volatility in HD (Kong et al., 2020; Gubert et al., 2022b).

FMT aims to replace the disrupted gut microbiota with that of a 'healthy' individual in order to provide therapeutic benefits (Matheson and Holsinger, 2023). Although FMT has been successfully used to target the gut microbiota in diseases impacting the gastrointestinal tract, their potential to treat neurodegenerative diseases is just beginning to be explored (Matheson and Holsinger, 2023). Preclinically FMT in R6/1 mice was able to positively affect cognitive parameters, however, indicated issues may arise trying to engraft the FMT due to the volatile nature of the HD gut (Gubert et al., 2022b).

Exercise is known to have beneficial effects on the gut microbiota in moderation, with findings in other neurodegenerative diseases including AD and PD suggesting it may be therapeutically beneficial (Gubert et al., 2020b). Preclinical research in the R6/1 model of HD found that exercise is able to modulate the gut microbiota to be significantly different to standard housing conditions in terms of β-diversity (Gubert et al., 2022a). HD patients may be limited in their ability to exercise due to the degradation of motor skills over the course of the disease, therefore implementing exercise as a therapeutic treatment may not be practically beneficial (Gubert et al., 2020b; Jang & Byun, 2021). However, developing an exercise mimetic to replicate the beneficial effect on the gut microbiota and potential HD pathology may be a viable treatment option (Gubert et al., 2020b; Gubert and Hannan, 2021; Jang & Byun, 2021). Further research is urgently required to pursue such promising therapeutic approaches.

9 Conclusion

Due to gastrointestinal symptoms and microbiota disruption frequently occurring prior to overt HD motor symptoms, identifying gut-related biomarkers including bacteria, fungi, archaea and viral signatures, may provide novel insight into disease progression, as well as identify therapeutic targets for the peripheral symptomology of HD. Together these findings suggest that microbiome profile changes and associated metabolomic outputs may provide insight into the role of the gut microbiome in HD as well as potential therapeutic targets. While foundational studies identifying disruption in the HD gut microbiota have been established, well characterised clinical studies of HD are needed to identify microbiome biomarkers and assess the features microbial diversity, taxa, metabolite production that may be used to predict HD onset and severity. This understanding will assist microbiome modulating strategies using probiotics, prebiotics, synbiotics, high-fibre diets and FMT, to test the relationship between the gut microbiome and the host health. The rapidly developing field of multi-omics and the integration of multiple sources into modelled networks is significantly progressing our understanding of microorganisms living within the gut and their roles in neurodegenerative diseases. We are only at the beginning of using these techniques to characterise the gut profile in HD and understand its influence on pathogenic mechanisms, disease onset and progression, peripheral pathology, and phenotypic variability of HD.

References

Agus, A., Clément, K., & Sokol, H. (2021). Gut microbiota-derived metabolites as central regulators in metabolic disorders. *Gut, 70*, 1174–1182.

Akbari, E., Asemi, Z., Daneshvar Kakhaki, R., Bahmani, F., Kouchaki, E., Tamtaji, O. R., Hamidi, G. A., & Salami, M. (2016). Effect of probiotic supplementation on cognitive function and metabolic status in Alzheimer's sisease: A randomized, double-blind and controlled trial. *Frontiers in Aging Neuroscience, 8*, 256.

Al-Asmakh, M., & Zadjali, F. (2015). Use of germ-free animal models in microbiota-related research. *Journal of Microbiology and Biotechnology, 25*, 1583–1588.

Asmar, R. E., Panigrahi, P., Bamford, P., Berti, I., Not, T., Coppa, G. V., Catassi, C., & Fasano, A. (2002). Host-dependent zonulin secretion causes the impairment of the small intestine barrier function after bacterial exposure. *Gastroenterology, 123*, 1607–1615.

Bashir, H. (2019). Emerging therapies in Huntington's disease. *Expert Review of Neurotherapeutics, 19*, 983–995.

Belarif, L., et al. (2019). IL-7 receptor influences anti-TNF responsiveness and T cell gut homing in inflammatory bowel disease. *The Journal of Clinical Investigation, 129*, 1910–1925.

Bogatyrev, S. R., Rolando, J. C., & Ismagilov, R. F. (2020). Self-reinoculation with fecal flora changes microbiota density and composition leading to an altered bile-acid profile in the mouse small intestine. *Microbiome, 8*, 19.

Bonk, F., Popp, D., Harms, H., & Centler, F. (2018). PCR-based quantification of taxa-specific abundances in microbial communities: Quantifying and avoiding common pitfalls. *Journal of Microbiological Methods, 153*, 139–147.

Borre, Y. E., O'Keeffe, G. W., Clarke, G., Stanton, C., Dinan, T. G., & Cryan, J. F. (2014). Microbiota and neurodevelopmental windows: Implications for brain disorders. *Trends in Molecular Medicine, 20*, 509–518.

Boxx, G. M., & Cheng, G. (2016). The roles of type I interferon in bacterial infection. *Cell Host & Microbe, 19*, 760–769.

Bravo, J. A., Forsythe, P., Chew, M. V., Escaravage, E., Savignac, H. M., Dinan, T. G., Bienenstock, J., & Cryan, J. F. (2011). Ingestion of Lactobacillus strain regulates emotional behavior and central GABA receptor expression in a mouse via the vagus nerve. *Proceedings of the National Academy of Sciences of the United States of America, 108*, 16050–16055.

Breit, S., Kupferberg, A., Rogler, G., & Hasler, G. (2018). Vagus nerve as modulator of the brain-gut axis in psychiatric and inflammatory disorders. *Frontiers in Psychiatry, 9*, 44.

Burokas, A., Arboleya, S., Moloney, R. D., Peterson, V. L., Murphy, K., Clarke, G., Stanton, C., Dinan, T. G., & Cryan, J. F. (2017). Targeting the microbiota-gut-brain axis: Prebiotics have anxiolytic and antidepressant-like effects and reverse the impact of chronic stress in mice. *Biological Psychiatry, 82*, 472–487.

Charlet, R., Bortolus, C., Barbet, M., Sendid, B., & Jawhara, S. (2018). A decrease in anaerobic bacteria promotes Candida glabrata overgrowth while β-glucan treatment restores the gut microbiota and attenuates colitis. *Gut Pathogens, 10*, 50.

Chung, H., Pamp, S. J., Hill, J. A., Surana, N. K., Edelman, S. M., Troy, E. B., Reading, N. C., Villablanca, E. J., Wang, S., Mora, J. R., Umesaki, Y., Mathis, D., Benoist, C., Relman, D. A., & Kasper, D. L. (2012). Gut immune maturation depends on colonization with a host-specific microbiota. *Cell, 149*, 1578–1593.

Clemente Jose, C., Ursell Luke, K., Parfrey Laura, W., & Knight, R. (2012). The impact of the gut microbiota on human health: An integrative view. *Cell, 148*, 1258–1270.

Clooney, A. G., Eckenberger, J., Laserna-Mendieta, E., Sexton, K. A., Bernstein, M. T., Vagianos, K., Sargent, M., Ryan, F. J., Moran, C., Sheehan, D., Sleator, R. D., Targownik, L. E., Bernstein, C. N., Shanahan, F., & Claesson, M. J. (2021). Ranking microbiome variance in inflammatory bowel disease: A large longitudinal intercontinental study. *Gut, 70*, 499–510.

Cryan, J. F., et al. (2019). The microbiota-gut-brain axis. *Physiological Reviews, 99*, 1877–2013.

Dalile, B., Van Oudenhove, L., Vervliet, B., & Verbeke, K. (2019). The role of short-chain fatty acids in microbiota-gut-brain communication. *Nature Reviews. Gastroenterology & Hepatology, 16*, 461–478.

DiFiglia, M., Sapp, E., Chase, K. O., Davies, S. W., Bates, G. P., Vonsattel, J. P., & Aronin, N. (1997). Aggregation of huntingtin in neuronal intranuclear inclusions and dystrophic neurites in brain. *Science, 277*, 1990–1993.

Djoussé, L., Knowlton, B., Cupples, L. A., Marder, K., Shoulson, I., & Myers, R. H. (2002). Weight loss in early stage of Huntington's disease. *Neurology, 59*, 1325–1330.

Dominguez-Bello, M. G., Godoy-Vitorino, F., Knight, R., & Blaser, M. J. (2019). Role of the microbiome in human development. *Gut, 68*, 1108–1114.

Du, G., Dong, W., Yang, Q., Yu, X., Ma, J., Gu, W., & Huang, Y. (2021). Altered gut microbiota related to inflammatory responses in patients with Huntington's disease. *Frontiers in Immunology, 11*.

Farshim, P. P., & Bates, G. P. (2018). Mouse models of Huntington's disease. *Methods in Molecular Biology, 1780*, 97–120.

Fasano, A., Baudry, B., Pumplin, D. W., Wasserman, S. S., Tall, B. D., Ketley, J. M., & Kaper, J. B. (1991). Vibrio cholerae produces a second enterotoxin, which affects intestinal tight junctions. *Proceedings of the National Academy of Sciences, 88*, 5242–5246.

Fasano, A., Visanji, N. P., Liu, L. W., Lang, A. E., & Pfeiffer, R. F. (2015). Gastrointestinal dysfunction in Parkinson's disease. *Lancet Neurology, 14*, 625–639.

Finucane, M. M., Sharpton, T. J., Laurent, T. J., & Pollard, K. S. (2014). A taxonomic signature of obesity in the microbiome? Getting to the guts of the matter. *PLoS One, 9*, e84689.

Fleming, M. A., 2nd, Ehsan, L., Moore, S. R., & Levin, D. E. (2020). The enteric nervous system and its emerging role as a therapeutic target. *Gastroenterology Research and Practice, 2020*, 8024171.

Flores, G. E., Henley, J. B., & Fierer, N. (2012). A direct PCR approach to accelerate analyses of human-associated microbial communities. *PLoS One, 7*, e44563.

Gao, B., Chi, L., Zhu, Y., Shi, X., Tu, P., Li, B., Yin, J., Gao, N., Shen, W., & Schnabl, B. (2021). An introduction to next generation sequencing bioinformatic analysis in gut microbiome studies. *Biomolecules & Therapeutics, 11*.

Garrett, W. S., Gordon, J. I., & Glimcher, L. H. (2010). Homeostasis and inflammation in the intestine. *Cell, 140*, 859–870.

Geng, Z. H., Zhu, Y., Li, Q. L., Zhao, C., & Zhou, P. H. (2022). Enteric nervous system: The bridge between the gut microbiota and neurological disorders. *Frontiers in Aging Neuroscience, 14*, 810483.

Ghosh, R., & Tabrizi, S. J. (2018). Huntington disease. *Handbook of Clinical Neurology, 147*, 255–278.

Ghosh, A., Firdous, S., & Saha, S. (2021). Bioinformatics for human microbiome. pp. 333–350.

Gibson, G. R., Hutkins, R., Sanders, M. E., Prescott, S. L., Reimer, R. A., Salminen, S. J., Scott, K., Stanton, C., Swanson, K. S., Cani, P. D., Verbeke, K., & Reid, G. (2017). Expert consensus document: The international scientific Association for Probiotics and Prebiotics (ISAPP) consensus statement on the definition and scope of prebiotics. *Nature Reviews. Gastroenterology & Hepatology, 14*, 491–502.

Godin, J. D., Poizat, G., Hickey, M. A., Maschat, F., & Humbert, S. (2010). Mutant huntingtin-impaired degradation of beta-catenin causes neurotoxicity in Huntington's disease. *The EMBO Journal, 29*, 2433–2445.

Gubert, C., & Hannan, A. J. (2021). Exercise mimetics: Harnessing the therapeutic effects of physical activity. *Nature Reviews. Drug Discovery, 20*, 862–879.

Gubert, C., Renoir, T., & Hannan, A. J. (2020a). Why Woody got the blues: The neurobiology of depression in Huntington's disease. *Neurobiology of Disease, 142*, 104958.

Gubert, C., Kong, G., Renoir, T., & Hannan, A. J. (2020b). Exercise, diet and stress as modulators of gut microbiota: Implications for neurodegenerative diseases. *Neurobiology of Disease, 134*, 104621.

Gubert, C., Love, C. J., Kodikara, S., Mei Liew, J. J., Renoir, T., Lê Cao, K.-A., & Hannan, A. J. (2022a). Gene-environment-gut interactions in Huntington's disease mice are associated with environmental modulation of the gut microbiome. *iScience, 25*.

Gubert, C., Choo, J. M., Love, C. J., Kodikara, S., Masson, B. A., Liew, J. J. M., Wang, Y., Kong, G., Narayana, V. K., Renoir, T., Lê Cao, K. A., Rogers, G. B., & Hannan, A. J. (2022b). Faecal microbiota transplant ameliorates gut dysbiosis and cognitive deficits in Huntington's disease mice. *Brain Communiations, 4*.

Gyawali, R., Nwamaioha, N., Fiagbor, R., Zimmerman, T., Newman, R. H., & Ibrahim, S. A. (2019). Chapter 12 - The role of prebiotics in disease prevention and health promotion. In R. R. Watson & V. R. Preedy (Eds.), *Dietary interventions in gastrointestinal diseases* (pp. 151–167). Academic Press.

Harris, V. C., Haak, B. W., Boele van Hensbroek, M., & Wiersinga, W. J. (2017). The intestinal microbiome in infectious diseases: The clinical relevance of a rapidly emerging field. Open forum. *Infectious Diseases, 4*.

Heijtz, R. D., Wang, S., Anuar, F., Qian, Y., Björkholm, B., Samuelsson, A., Hibberd, M. L., Forssberg, H., & Pettersson, S. (2011). Normal gut microbiota modulates brain development and behavior. *Proceedings of the National Academy of Sciences, 108*, 3047–3052.

Hildebrand, F., Nguyen, T. L., Brinkman, B., Yunta, R. G., Cauwe, B., Vandenabeele, P., Liston, A., & Raes, J. (2013). Inflammation-associated enterotypes, host genotype, cage and inter-individual effects drive gut microbiota variation in common laboratory mice. *Genome Biology, 14*, R4.

Hosoi, T., Okuma, Y., & Nomura, Y. (2000). Electrical stimulation of afferent vagus nerve induces IL-1β expression in the brain and activates HPA axis. *American Journal of Physiology—Regulatory, Integrative and Comparative Physiology, 279*, R141–R147.

Idle, J. R., & Gonzalez, F. J. (2007). Metabolomics. *Cell Metabolism, 6*, 348–351.

Jandhyala, S. M., Talukdar, R., Subramanyam, C., Vuyyuru, H., Sasikala, M., & Nageshwar Reddy, D. (2015). Role of the normal gut microbiota. *World Journal of Gastroenterology, 21*, 8787–8803.

Jang, Y. J., & Byun, S. (2021). Molecular targets of exercise mimetics and their natural activators. *BMB Reports, 54*, 581–591.

Janse van Rensburg, H. J., Spiliopoulou, P., & Siu, L. L. (2022). Circulating biomarkers for therapeutic monitoring of anti-cancer agents. *The Oncologist, 27*, 352–362.

Jimenez-Sanchez, M., Licitra, F., Underwood, B. R., & Rubinsztein, D. C. (2017). Huntington's disease: Mechanisms of pathogenesis and therapeutic strategies. *Cold Spring Harbor Perspectives in Medicine, 7*.

Kamada, N., Seo, S. U., Chen, G. Y., & Núñez, G. (2013). Role of the gut microbiota in immunity and inflammatory disease. *Nature Reviews. Immunology, 13*, 321–335.

Kazemi, A., Noorbala, A. A., Azam, K., Eskandari, M. H., & Djafarian, K. (2019). Effect of probiotic and prebiotic vs placebo on psychological outcomes in patients with major depressive disorder: A randomized clinical trial. *Clinical Nutrition, 38*, 522–528.

Kegel, K. B., Meloni, A. R., Yi, Y., Kim, Y. J., Doyle, E., Cuiffo, B. G., Sapp, E., Wang, Y., Qin, Z. H., Chen, J. D., Nevins, J. R., Aronin, N., & DiFiglia, M. (2002). Huntingtin is present in the nucleus, interacts with the transcriptional corepressor C-terminal binding protein, and represses transcription. *The Journal of Biological Chemistry, 277*, 7466–7476.

Kim, A., Lalonde, K., Truesdell, A., Gomes Welter, P., Brocardo, P. S., Rosenstock, T. R., & Gil-Mohapel, J. (2021). New avenues for the treatment of Huntington's disease. *International Journal of Molecular Sciences, 22*.

Koenig, J. E., Spor, A., Scalfone, N., Fricker, A. D., Stombaugh, J., Knight, R., Angenent, L. T., & Ley, R. E. (2011). Succession of microbial consortia in the developing infant gut microbiome. *Proceedings of the National Academy of Sciences of the United States of America, 108*(Suppl 1), 4578–4585.

Koleff, P., Gaston, K. J., & Lennon, J. J. (2003). Measuring beta diversity for presence–absence data. *The Journal of Animal Ecology, 72*, 367–382.

Kong, G., Cao, K. L., Judd, L. M., Li, S., Renoir, T., & Hannan, A. J. (2018). Microbiome profiling reveals gut dysbiosis in a transgenic mouse model of Huntington's disease. *Neurobiology of Disease, 135*, 104268.

Kong, G., Ellul, S., Narayana, V. K., Kanojia, K., Ha, H. T. T., Li, S., Renoir, T., Cao, K. L., & Hannan, A. J. (2020). An integrated metagenomics and metabolomics approach implicates the microbiota-gut-brain axis in the pathogenesis of Huntington's disease. *Neurobiology of Disease, 148*, 105199.

Kong, G., Lê Cao, K. A., & Hannan, A. J. (2022). Alterations in the gut fungal community in a Mouse model of Huntington's disease. *Microbiology Spectrum, 10*, e0219221.

Konturek, S. J., Konturek, J. W., Pawlik, T., & Brzozowski, T. (2004). Brain-gut axis and its role in the control of food intake. *Journal of Physiology and Pharmacology, 55*, 137–154.

Kunze, W. A., Mao, Y. K., Wang, B., Huizinga, J. D., Ma, X., Forsythe, P., & Bienenstock, J. (2009). Lactobacillus reuteri enhances excitability of colonic AH neurons by inhibiting calcium-dependent potassium channel opening. *Journal of Cellular and Molecular Medicine, 13*, 2261–2270.

Lakra, P., Aditi, K., & Agrawal, N. (2019). Peripheral expression of mutant huntingtin is a critical determinant of weight loss and metabolic disturbances in Huntington's disease. *Scientific Reports, 9*, 10127.

Langille, M. G., Zaneveld, J., Caporaso, J. G., McDonald, D., Knights, D., Reyes, J. A., Clemente, J. C., Burkepile, D. E., Vega Thurber, R. L., Knight, R., Beiko, R. G., & Huttenhower, C. (2013). Predictive functional profiling of microbial communities using 16S rRNA marker gene sequences. *Nature Biotechnology, 31*, 814–821.

Lee, H.-J., Lee, K.-E., Kim, J.-K., & Kim, D.-H. (2019). Suppression of gut dysbiosis by Bifidobacterium longum alleviates cognitive decline in 5XFAD transgenic and aged mice. *Scientific Reports, 9*, 11814.

Liddle, R. A. (2018). Parkinson's disease from the gut. *Brain Research, 1693*, 201–206.

Luckey, T. D. (1972). Introduction to intestinal microecology. *The American Journal of Clinical Nutrition, 25*, 1292–1294.

Luczynski, P., McVey Neufeld, K. A., Oriach, C. S., Clarke, G., Dinan, T. G., & Cryan, J. F. (2016). Growing up in a bubble: Using germ-free animals to assess the influence of the gut microbiota on brain and behavior. *The International Journal of Neuropsychopharmacology, 19*, pyw020.

Mangiarini, L., Sathasivam, K., Seller, M., Cozens, B., Harper, A., Hetherington, C., Lawton, M., Trottier, Y., Lehrach, H., Davies, S. W., & Bates, G. P. (1996). Exon 1 of the HD gene with an expanded CAG repeat is sufficient to cause a progressive neurological phenotype in transgenic mice. *Cell, 87*, 493–506.

Marchesi, J. R., & Ravel, J. (2015). The vocabulary of microbiome research: A proposal. *Microbiome, 3*, 31.

Marxreiter, F., Stemick, J., & Kohl, Z. (2020). Huntingtin lowering strategies. *International Journal of Molecular Sciences, 21*, 2146.

Matheson, J. T., & Holsinger, R. M. D. (2023). The role of fecal microbiota transplantation in the treatment of neurodegenerative diseases: A review. *International Journal of Molecular Sciences, 24*.

McBurney, M. I., Davis, C., Fraser, C. M., Schneeman, B. O., Huttenhower, C., Verbeke, K., Walter, J., & Latulippe, M. E. (2019). Establishing what constitutes a healthy human gut microbiome: State of the science, regulatory considerations, and future directions. *The Journal of Nutrition, 149*, 1882–1895.

McColgan, P., & Tabrizi, S. J. (2018). Huntington's disease: A clinical review. *European Journal of Neurology, 25*, 24–34.

Milby, A. H., Halpern, C. H., & Baltuch, G. H. (2008). Vagus nerve stimulation for epilepsy and depression. *Neurotherapeutics, 5*, 75–85.

Mitra, S., Dash, R., Nishan, A. A., Habiba, S. U., & Moon, I. S. (2022). Brain modulation by the gut microbiota: From disease to therapy. *Journal of Advanced Research*.

Moffitt, H., McPhail, G. D., Woodman, B., Hobbs, C., & Bates, G. P. (2009). Formation of polyglutamine inclusions in a wide range of non-CNS tissues in the HdhQ150 knock-in mouse model of Huntington's disease. *PLoS One, 4*, e8025.

Montoya, A., Price, B. H., Menear, M., & Lepage, M. (2006). Brain imaging and cognitive dysfunctions in Huntington's disease. *Journal of Psychiatry & Neuroscience, 31*, 21–29.

Morgan, X. C., & Huttenhower, C. (2012). Chapter 12: Human microbiome analysis. *PLoS Computational Biology, 8*, e1002808.

Nance, M. A. (2017). Genetics of Huntington disease. *Handbook of Clinical Neurology, 144*, 3–14.

Neu, A. T., Allen, E. E., & Roy, K. (2021). Defining and quantifying the core microbiome: Challenges and prospects. *Proceedings of the National Academy of Sciences, 118*, e2104429118.

Nguyen, G. D., Molero, A. E., Gokhan, S., & Mehler, M. F. (2013). Functions of huntingtin in germ layer specification and organogenesis. *PLoS One, 8*, e72698.

Nicholson, J. K., Holmes, E., Kinross, J., Burcelin, R., Gibson, G., Jia, W., & Pettersson, S. (2012). Host-gut microbiota metabolic interactions. *Science, 336*, 1262–1267.

Pandey, M., & Rajamma, U. (2018). Huntington's disease: The coming of age. *Journal of Genetics, 97*, 649–664.

Paulsen, J., Long, J., Johnson, H., Aylward, E., Ross, C., Williams, J., Nance, M., Erwin, C., Westervelt, H., Harrington, D., Bockholt, H., Zhang, Y., McCusker, E., Chiu, E., & Panegyres, P. (2014). Clinical and biomarker changes in premanifest Huntington disease show trial feasibility: A decade of the PREDICT-HD study. *Frontiers in Aging Neuroscience, 6*.

Perez-Pardo, P., Dodiya, H. B., Engen, P. A., Naqib, A., Forsyth, C. B., Green, S. J., Garssen, J., Keshavarzian, A., & Kraneveld, A. D. (2018). Gut bacterial composition in a mouse model of Parkinson's disease. *Beneficial Microbes, 9*, 799–814.

Peterson, J., et al. (2009). The NIH human microbiome project. *Genome Research, 19*, 2317–2323.

Przybyl, L., Wozna-Wysocka, M., Kozlowska, E., & Fiszer, A. (2021). What, when and how to measure—Peripheral biomarkers in therapy of Huntington's disease. *International Journal of Molecular Sciences, 22*, 1561.

Rohart, F., Gautier, B., Singh, A., & Lê Cao, K. A. (2017). mixOmics: An R package for 'omics feature selection and multiple data integration. *PLoS Computational Biology, 13*, e1005752.

Rosas, H. D., Doros, G., Bhasin, S., Thomas, B., Gevorkian, S., Malarick, K., Matson, W., & Hersch, S. M. (2015). A systems-level "misunderstanding": The plasma metabolome in Huntington's disease. *Annals of Clinical Translational Neurology, 2*, 756–768.

Rosenblatt, A., Kumar, B. V., Mo, A., Welsh, C. S., Margolis, R. L., & Ross, C. A. (2012). Age, CAG repeat length, and clinical progression in Huntington's disease. *Movement Disorders, 27*, 272–276.

Ross, C. A., Aylward, E. H., Wild, E. J., Langbehn, D. R., Long, J. D., Warner, J. H., Scahill, R. I., Leavitt, B. R., Stout, J. C., Paulsen, J. S., Reilmann, R., Unschuld, P. G., Wexler, A., Margolis, R. L., & Tabrizi, S. J. (2014). Huntington disease: Natural history, biomarkers and prospects for therapeutics. *Nature Reviews. Neurology, 10*, 204–216.

Rowin, J., Xia, Y., Jung, B., & Sun, J. (2017). Gut inflammation and dysbiosis in human motor neuron disease. *Physiological Reports, 5*, e13443.

Savignac, H. M., Corona, G., Mills, H., Chen, L., Spencer, J. P. E., Tzortzis, G., & Burnet, P. W. J. (2013). Prebiotic feeding elevates central brain derived neurotrophic factor, N-methyl-d-aspartate receptor subunits and d-serine. *Neurochemistry International, 63*, 756–764.

Scheperjans, F., Aho, V., Pereira, P. A., Koskinen, K., Paulin, L., Pekkonen, E., Haapaniemi, E., Kaakkola, S., Eerola-Rautio, J., Pohja, M., Kinnunen, E., Murros, K., & Auvinen, P. (2015). Gut microbiota are related to Parkinson's disease and clinical phenotype. *Movement Disorders, 30*, 350–358.

Schmidt, T. S. B., Raes, J., & Bork, P. (2018). The human gut microbiome: From association to modulation. *Cell, 172*, 1198–1215.

Sekirov, I., Russell, S. L., Antunes, L. C. M., & Finlay, B. B. (2010). Gut microbiota in health and disease. *Physiological Reviews, 90*, 859–904.

Selkrig, J., Wong, P., Zhang, X., & Pettersson, S. (2014). Metabolic tinkering by the gut microbiome: Implications for brain development and function. *Gut Microbes, 5*, 369–380.

Shanahan, F., Ghosh, T. S., & O'Toole, P. W. (2021). The healthy microbiome-what is the definition of a healthy gut microbiome? *Gastroenterology, 160*, 483–494.

Sharon, G., Sampson, T. R., Geschwind, D. H., & Mazmanian, S. K. (2016). The central nervous system and the gut microbiome. *Cell, 167*, 915–932.

Siddiqui, M. T., & Cresci, G. A. M. (2021). The immunomodulatory functions of butyrate. *Journal of Inflammation Research, 14*, 6025–6041.

Silva, Y. P., Bernardi, A., & Frozza, R. L. (2020). The role of short-chain fatty acids from gut microbiota in gut-brain communication. *Frontiers in Endocrinology (Lausanne), 11*, 25.

Singh, G., Brass, A., Cruickshank, S. M., & Knight, C. G. (2021). Cage and maternal effects on the bacterial communities of the murine gut. *Scientific Reports, 11*, 9841.

Soergel, D. A. W., Dey, N., Knight, R., & Brenner, S. E. (2012). Selection of primers for optimal taxonomic classification of environmental 16S rRNA gene sequences. *The ISME Journal, 6*, 1440–1444.

Sokol, H., Leducq, V., Aschard, H., Pham, H. P., Jegou, S., Landman, C., Cohen, D., Liguori, G., Bourrier, A., Nion-Larmurier, I., Cosnes, J., Seksik, P., Langella, P., Skurnik, D., Richard, M. L., & Beaugerie, L. (2017). Fungal microbiota dysbiosis in IBD. *Gut, 66*, 1039–1048.

Staley, C., Kaiser, T., Beura, L. K., Hamilton, M. J., Weingarden, A. R., Bobr, A., Kang, J., Masopust, D., Sadowsky, M. J., & Khoruts, A. (2017). Stable engraftment of human microbiota into mice with a single oral gavage following antibiotic conditioning. *Microbiome, 5*, 87.

Stan, T. L., Soylu-Kucharz, R., Burleigh, S., Prykhodko, O., Cao, L., Franke, N., Sjögren, M., Haikal, C., Hållenius, F., & Björkqvist, M. (2020). Increased intestinal permeability and gut dysbiosis in the R6/2 mouse model of Huntington's disease. *Scientific Reports, 10*, 18270.

Stappenbeck, T. S., & Virgin, H. W. (2016). Accounting for reciprocal host-microbiome interactions in experimental science. *Nature, 534*, 191–199.

Stoker, T. B., Mason, S. L., Greenland, J. C., Holden, S. T., Santini, H., & Barker, R. A. (2022). Huntington's disease: Diagnosis and management. *Practical Neurology, 22*, 32–41.

Strandwitz, P. (2018). Neurotransmitter modulation by the gut microbiota. *Brain Research, 1693*, 128–133.

Strimbu, K., & Tavel, J. A. (2010). What are biomarkers? *Current Opinion in HIV and AIDS, 5*, 463–466.

Sudo, N. (2014). Microbiome, HPA axis and production of endocrine hormones in the gut. *Advances in Experimental Medicine and Biology, 817*, 177–194.

Sudo, N., Chida, Y., Aiba, Y., Sonoda, J., Oyama, N., Yu, X. N., Kubo, C., & Koga, Y. (2004). Postnatal microbial colonization programs the hypothalamic-pituitary-adrenal system for stress response in mice. *The Journal of Physiology, 558*, 263–275.

Taché, Y., Vale, W., Rivier, J., & Brown, M. (1980). Brain regulation of gastric secretion: Influence of neuropeptides. *Proceedings of the National Academy of Sciences, 77*, 5515–5519.

Thomas, T., Gilbert, J., & Meyer, F. (2012). Metagenomics - a guide from sampling to data analysis. *Microbial Informatics and Experimentation, 2*, 3.

Thukral, A. (2017). A review on measurement of Alpha diversity in biology. *Agricultural Research Journal, 54*, 1.

Underwood, B. R., Broadhurst, D., Dunn, W. B., Ellis, D. I., Michell, A. W., Vacher, C., Mosedale, D. E., Kell, D. B., Barker, R. A., Grainger, D. J., & Rubinsztein, D. C. (2006). Huntington disease patients and transgenic mice have similar pro-catabolic serum metabolite profiles. *Brain, 129*, 877–886.

Valles-Colomer, M., et al. (2023). The person-to-person transmission landscape of the gut and oral microbiomes. *Nature, 614*, 125–135.

van der Burg, J. M. M., Winqvist, A., Aziz, N. A., Maat-Schieman, M. L. C., Roos, R. A. C., Bates, G. P., Brundin, P., Björkqvist, M., & Wierup, N. (2011). Gastrointestinal dysfunction contributes to weight loss in Huntington's disease mice. *Neurobiology of Disease, 44*, 1–8.

van der Burg, J. M. M., Gardiner, S. L., Ludolph, A. C., Landwehrmeyer, G. B., Roos, R. A. C., & Aziz, N. A. (2017). Body weight is a robust predictor of clinical progression in Huntington disease. *Annals of Neurology, 82*, 479–483.

Vázquez, E., Barranco, A., Ramírez, M., Gruart, A., Delgado-García, J. M., Martínez-Lara, E., Blanco, S., Martín, M. J., Castanys, E., Buck, R., Prieto, P., & Rueda, R. (2015). Effects of a human milk oligosaccharide, 2′-fucosyllactose, on hippocampal long-term potentiation and learning capabilities in rodents. *The Journal of Nutritional Biochemistry, 26*, 455–465.

Verwaest, K. A., Vu, T. N., Laukens, K., Clemens, L. E., Nguyen, H. P., Van Gasse, B., Martins, J. C., Van Der Linden, A., & Dommisse, R. (2011). 1H NMR based metabolomics of CSF and blood serum: A metabolic profile for a transgenic rat model of Huntington disease. *Biochimica et Biophysica Acta (BBA) - Molecular Basis of Disease, 1812*, 1371–1379.

Vicars, B. G., Liu, A. B., Holt, S., Jayadev, S., Bird, T., & Yang, C. C. (2021). High frequency of concomitant bladder, bowel, and sexual symptoms in huntington's disease: A self-reported questionnaire study. *J Pers Med, 11*.

Videvall, E., Strandh, M., Engelbrecht, A., Cloete, S., & Cornwallis, C. K. (2017). Direct PCR offers a fast and reliable alternative to conventional DNA isolation methods for gut Mmicrobiomes. *mSystems, 2*, e00132.

Vogt, N. M., Kerby, R. L., Dill-McFarland, K. A., Harding, S. J., Merluzzi, A. P., Johnson, S. C., Carlsson, C. M., Asthana, S., Zetterberg, H., Blennow, K., Bendlin, B. B., & Rey, F. E. (2017). Gut microbiome alterations in Alzheimer's disease. *Scientific Reports, 7*, 13537.

Walker, F. O. (2007). Huntington's disease. *Lancet, 369*, 218–228.

Walker, W. A. (2017). Dysbiosis. pp 227–232.

Walker, B., Holling, C. S., Carpenter, S., & Kinzig, A. (2003). Resilience, adaptability and transformability in social-ecological systems. *Ecology and Society, 9*.

Wang, Y., & Lê Cao, K.-A. (2023). PLSDA-batch: A multivariate framework to correct for batch effects in microbiome data. *Briefings in Bioinformatics, 24*.

Wang, W. L., Xu, S. Y., Ren, Z. G., Tao, L., Jiang, J. W., & Zheng, S. S. (2015). Application of metagenomics in the human gut microbiome. *World Journal of Gastroenterology, 21*, 803–814.

Wanker, E. E., Ast, A., Schindler, F., Trepte, P., & Schnoegl, S. (2019). The pathobiology of perturbed mutant huntingtin protein-protein interactions in Huntington's disease. *Journal of Neurochemistry, 151*, 507–519.

Wasser, C. I., Mercieca, E. C., Kong, G., Hannan, A. J., McKeown, S. J., Glikmann-Johnston, Y., & Stout, J. C. (2020). Gut dysbiosis in Huntington's disease: Associations among gut microbiota, cognitive performance and clinical outcomes. *Brain Communications, 2*, fcaa110.

Wu, S., Yi, J., Zhang, Y. G., Zhou, J., & Sun, J. (2015). Leaky intestine and impaired microbiome in an amyotrophic lateral sclerosis mouse model. *Physiological Reports, 3*, e12356.

Xu, X., Lin, S., Yang, Y., Gong, X., Tong, J., Li, K., & Li, Y. (2020). Histological and ultrastructural changes of the colon in dextran sodium sulfate-induced mouse colitis. *Experimental and Therapeutic Medicine, 20*, 1987–1994.

Yang, B., Wang, Y., & Qian, P.-Y. (2016). Sensitivity and correlation of hypervariable regions in 16S rRNA genes in phylogenetic analysis. *BMC Bioinformatics, 17*, 135.

Yeon, S.-W., You, Y. S., Kwon, H.-S., Yang, E. H., Ryu, J.-S., Kang, B. H., & Kang, J.-H. (2010). Fermented milk of lactobacillus helveticus IDCC3801 reduces beta-amyloid and attenuates memory deficit. *Journal of Functional Foods, 2*, 143–152.

Zhou, L., Yu, D., Zheng, S., Ouyang, R., Wang, Y., & Xu, G. (2021). Gut microbiota-related metabolome analysis based on chromatography-mass spectrometry. *Trends in Analytical Chemistry, 143*, 116375.

Zuccato, C., Tartari, M., Crotti, A., Goffredo, D., Valenza, M., Conti, L., Cataudella, T., Leavitt, B. R., Hayden, M. R., Timmusk, T., Rigamonti, D., & Cattaneo, E. (2003). Huntingtin interacts with REST/NRSF to modulate the transcription of NRSE-controlled neuronal genes. *Nature Genetics, 35*, 76–83.

Part V
Biochemical and Genetic Biomarkers in Huntington's Disease

Inflammation Biomarkers in Huntington's Disease

Maria Björkqvist

Abbreviations

AD	Alzheimer's disease
ALS	Amyotrophic lateral sclerosis
APPs	Acute phase proteins
CNS	Central nervous system
CRP	C-reactive protein
CSF	Cerebrospinal fluid
GRO/KC	Mouse homologue IL-8
HD	Huntington's disease
IFNg	Interferon gamma
IgG	Immunoglobulin G
IGHG	Heavy chain constant domain of IgG
IL	Interleukin
JAK	Janus kinase
MCP	Monocyte chemoattractant protein
MIP	Macrophage inflammatory protein
MMP	Matrix metalloprotease
NFKB	Nuclear factor kappa beta
PD	Parkinson's disease
PET	Positron emission tomography
Sil-2r	Soluble interleukin-2-receptor
STAT	Signal transducer and activator of transcription
sTNF-R	55-kDa-type soluble tumour necrosis factor receptor
TGF-1b	Transforming growth factor
TNF-a	Tumour necrosis factor alpha
VEGF	Vascular endothelial growth factor

M. Björkqvist (✉)
Brain Disease Biomarker Unit, Wallenberg Neuroscience Center,
Department of Experimental Medical Science, Lund University, Lund, Sweden
e-mail: maria.bjorkqvist@med.lu.se

E. A. Thomas, G. M. Parkin (eds.), *Biomarkers for Huntington's Disease*,
Contemporary Clinical Neuroscience, https://doi.org/10.1007/978-3-031-32815-2_11

1 Introduction

The Immune System, Immune Cells and Immune Markers in Short

The immune system is a complex and dynamic network, involving multiple interactions between various organs, cells and proteins. The immune system and our immune cells play important roles in healthy tissue physiology, as well as in disease. There are two main parts of the immune system: the innate immune system and the adaptive immune system. The innate immune system is inherited, and provides a set response as the first line of defense. The adaptive immune system, also known as the acquired immune system, yields a specific response to a stimulus by recognition of molecules it has previously encountered.

The monocyte present in blood, can differentiate to a macrophage and a macrophage can be resident in tissues throughout the body. The resident macrophages of the brain are the microglia. Monocytes, macrophages and microglia, are all part of the innate immune system and they also influence the adaptive immune response. Cells involved in an immune response communicate via soluble compounds, such as cytokines. This communication can be autocrine, paracrine or endocrine. After being produced and secreted, a cytokine binds to a receptor and an intracellular signaling pathway is initiated, for example the Nuclear factor kappa beta (NFκB) or the JAK/STAT signaling pathway (for background textbook knowledge see Sharon et al. Kuby Immunology, 2019).

Immune markers are important signaling molecules contributing to brain-periphery cross-talk. Until relatively recently, the central nervous system (CNS), was considered immune privileged; however this view is changing. Brain-periphery cross-talk is important for maintaining homeostasis and occurs through various complex pathways, including hormonal signaling, neural pathways, and immune responses. For example, inflammation in the body can send signals to the brain, which can result in changes in behavior or mood (reviewed in Dantzer et al., 2008).

Markers of Inflammation as Biomarkers

Inflammatory markers circulate in blood and can be detected in cerebrospinal fluid (CSF), as well as saliva. Biofluid inflammatory markers are thereby a rich source of potential biomarkers. A biomarker is a measurable substance or characteristic in biological samples, such as blood, urine, or tissue, that provides information about a normal or abnormal biological process, a disease state, or a response to a therapeutic intervention (Biomarkers Definitions Working Group, 2001). A biomarker can be related to the disease or the treatment effect. A good disease biomarker should be closely linked to pathophysiology, as well as being reliable, sensitive, specific, and reproducible. *What role could then immune markers play in HD research and drug development?* Considering this, immune markers have the potential to be useful both as tools to understand pathology, and thus leading onto new disease modification strategies, as well as being important tools to evaluate potential immune modulation strategies. Immune markers have the potential to monitor disease progression or treatment response, as well as to predict clinical outcomes.

Neuroinflammation, Systemic Inflammation and Circulating Immune Mediators in Neurodegenerative Disease

The immune response plays an important role in the pathogenesis of neurodegenerative diseases, such as Alzheimer's disease (AD), Parkinson's disease (PD), Amyotrophic lateral sclerosis (ALS) and Huntington's disease (HD). In neurodegeneration, the immune system can, for example, respond to abnormal protein aggregates, such as beta-amyloid plaques in AD or Lewy bodies in PD, by an inflammatory response. This inflammation can contribute to damage of neurons and further exacerbate the disease. While the immune response in neurodegeneration is intended to protect the brain from damage, excessive or prolonged inflammation can contribute to further damage and disease progression. Therefore, understanding and modulating the immune response may be a promising approach for developing new treatments for neurodegenerative diseases.

A role for immune response in neurodegenerative disease was suggested as early as the 1940s and circulating immune mediators have repeatedly been suggested as markers for disease progression (see Fig. 1 for an illustration of the time-course of publications on the topic). Neuroinflammation is a shared feature in neurodegenerative disorders and has been shown to be associated with the underlying pathogenesis (reviewed by Glass et al. in Cell Glass et al., 2010; Suescun et al., 2019). In the central nervous system, the major cell-types contributing to inflammatory processes are microglia, oligodendrocytes, and astrocytes. In a variety of neurodegenerative settings activated microglia can exert negative effects, and mutations in microglial genes have been demonstrated to be linked to neurodegenerative diseases (reviewed in Song & Colonna, 2018). Also, neuroinflammation involves alterations in microglial neuronal cross-talk as a consequence of the activation of microglia and astrocytes. In addition to microglia, other immune cells, such as T cells, B cells, and monocytes, can also contribute to the immune response in neurodegeneration.

In addition to neuroinflammation, systemic inflammation has been shown to alter progression of neuronal injury (reviewed in Hickman et al., 2018). A low-grade immune response refers to a mild or a sub-optimal immune response that is insufficient to fully eradicate a pathogen. This type of response may occur when the immune system is compromised or not functioning properly, such as in autoimmune disorders or in chronic disease. Despite the heterogeneity of neurodegenerative disease, systemic immune activation and a low-grade immune response is a common feature (Stephenson et al., 2018). Immune activation can function as a double-edged

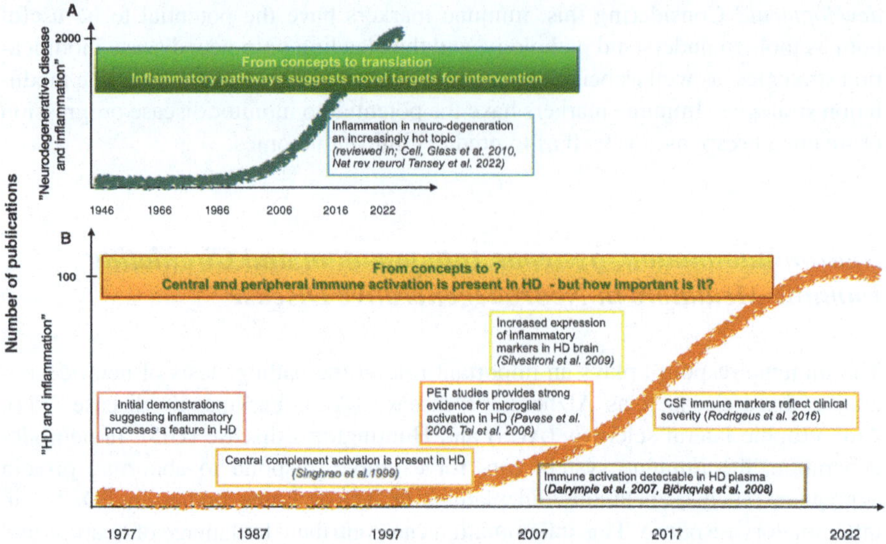

Fig. 1 Timelines illustrating publications linking neurodegenerative disease and Huntington's disease to inflammation. The small line graph illustrates, as a comparison to the HD field; the Pubmed search term "neurodegenerative disease and inflammation" that resulted in 17,732 results. Publications in this field has increased dramatically over the last years, as the research field has recognized the importance of immune involvement in pathogenesis, the potential of immune mediators as disease markers and immune modulation as treatment strategies. The larger line graph summarizes major findings related to inflammation in Huntington's disease. The search term Huntington's disease and inflammation" resulted in 725 publications, with the first publication appearing in 1970s. Both involvement of innate and adaptive immune responses have been identified in HD and several immune markers accessible in biofluids have been shown to be altered in HD. Several markers have been shown to be linked with pathology as well as clinical features. However, there are still remaining questions on how valuable these markers are

sword and, on one hand, aid in cellular repair, and, on the other hand, be detrimental. Many contributing factors add to the complexity behind immune responses, such as aging and metabolism. Increasing age is a known risk factor for neurodegenerative disease. Interestingly, ageing is also linked to an altered immune response (both innate and adaptive immune response) and ageing has been demonstrated to be associated with a low-grade immune response, frequently termed "*inflammaging*" (Franceschi et al., 2000). Obesity has also been demonstrated to be linked to neuroinflammation (Gómez-Apo et al., 2021). In obesity, adipocytes and macrophages of the adipose tissue contribute a chronic low-grade systemic inflammation with the release of cytokines and chemokines that can cross the blood-brain barrier, thereby stimulating microglia (Gómez-Apo et al., 2021).

Immune signaling pathways (such as the JAK/STAT and NFκB signaling pathways) are important under both healthy and neurodegenerative conditions. During homeostasis, a basal and stable activity of these pathways is maintained, whereas a disturbance of these pathways is linked to chronic inflammation. Alongside central activation of microglia and astrocytes and a low-grade systemic immune response, elevated levels of pro-inflammatory cytokines have been observed in AD (Fakhoury, 2018; Villegas-Llerena et al., 2016), PD (Le et al., 2016) and ALS (Kuhle et al., 2009). Accumulating evidence supports a role of the immune response and inflammatory markers in the pathogenesis, diagnosis, and treatment of neurodegenerative diseases. Increased understanding of immune activation and peripheral-central immune cross-talk in neurodegenerative disease may open up potential therapeutic avenues for the future, and studies of inflammation-related markers in peripheral fluids have been suggested to be useful to better characterize disease features.

Lessons from Alzheimer's Disease, Parkinson's Disease and ALS

Alzheimer's Disease

In AD, microglia are well established as key players in neuroinflammation and substantial evidence suggest both spatial and temporal variations in microglial phenotypes under different conditions (reviewed in Leng & Edison, 2021). Furthermore, the identification of AD risk genes that are associated with innate immune functions has provided strong evidence for a role for neuroinflammation in AD pathogenesis (Leng & Edison, 2021). Longitudinal studies show that inflammation and microglial activation occur years before AD onset (Nordengen et al., 2019). In addition, there is an association between neuroinflammation and amyloid and tau accumulation in the AD brain (Nordengen et al., 2019; Edison & Brooks, 2018).

Repeatedly plasma immune markers have been suggested candidate biomarkers for AD and numerous studies have demonstrated inflammatory markers associated with AD pathology. For example, a proteomic study in 2020 demonstrated that proteins involved in interleukin-6 (IL-6) signaling were significantly altered in both plasma and brains of AD patients (Chen & Xia, 2020) and an immune response has been shown to be detected before clinical onset of dementia, in both AD and in vascular dementia (Engelhart et al., 2004).

Even though many studies have suggested proinflammatory immune markers to be altered in AD, follow up studies and longitudinal studies have demonstrated no value as predictors. For example, no associations between plasma levels of the pro-inflammatory markers interleukin (IL)-1 beta, IL-6 and Tumour necrosis factor alpha (TNF-α) and diagnosis nor cognitive decline during a two years of follow-up was shown (Julian et al., 2018).

Even though there are conflicting results on AD immune markers in CSF, a large meta analysis has been conducted that puts emphasis on the usefulness of immune markers in AD. In 2017, Lai and colleagues conducted a meta-analyses comprising 175 studies yielding a review of 51 analytes in 13,344 AD and 12,912 HC patients. The results suggested that a peripheral immune response is present in AD and that IL-6 might be useful marker for AD cognitive impairment (Lai et al., 2017). Another meta-analyses published in 2019 comprising a systematic review of 170 studies, demonstrated convincingly that significantly altered levels of inflammatory markers were verified in comparison between AD, mild cognitive impairment and control subjects (Shen et al., 2019).

Parkinson's Disease

In PD, it is has been shown that dopaminergic neuronal loss is accompanied by inflammatory alterations in microglia, astrocytes, innate immune cells as well as in infiltrating peripheral immune cells (Le et al., 2016). Accumulating evidence suggests that immune system dysfunction has a role in PD (reviewed in Tansey et al., 2022) and neuroinflammation has been suggested as a treatment target (reviewed in Hirsch & Hunot, 2009). Both peripheral blood and CSF, have been shown to display altered levels of markers of inflammation and immune cell populations, potentially playing a role in neuroinflammation and the neurodegenerative process (Tansey et al., 2022). Peripherally accessible and measurable immune markers have been suggested as a tool to monitor the immune dysfunction found in PD (Bartl et al., 2022) and importantly, this pro-inflammatory immune activation is thought to occur in prodromal PD (Bartl et al., 2022).

ALS

ALS is a complex neurodegenerative disease (recently reviewed in Lancet by Feldman et al., 2022). The potential role of neuroinflammation has gained increasing attention in ALS research. Increased levels of inflammatory markers have been observed in both ALS patients and animal models (Jin et al., 2020) and astroglial cells has been shown to contribute to the neuroinflammation-mediated neurodegeneration in *in vitro* models of ALS (Mishra et al., 2016). Interestingly, a link between NFκB activation and ALS pathogenesis has been suggested. NFκB signaling has been shown to be activated in several cell types in ALS models and inhibition of NFκB signaling is suggested as a potential avenue for disease modification (reviewed in Källstig et al., 2021). Upregulation of neuroinflammatory factors have been demonstrated in both ALS patients and animal models of ALS. Peripheral immune markers have been demonstrated to be altered in ALS. For example, IL-1 beta, IL-6 and interferon-gamma (IFN-gamma) have been shown to be increased, alongside a reduction of IL-10 (Jin et al., 2020).

Taken together, there is evidence of immune response alterations in neurodegenerative disorders. Considering the complex nature of these diseases, and complex nature of immune responses, it is challenging to fully understand the role for immune response alterations in the pathogenesis of these diseases, as well as to fully grasp the potential of these inflammatory markers in diagnosis. As this book-chapter now moves on to Huntington's disease (HD) and a role for an immune response, it is important to keep in mind that potentially important lessons can be learned from similarities and differences looking at other neurodegenerative disorders.

2 A Central and Peripheral Immune Response in Huntington's Disease

HD is characterized by progressive neuronal dysfunction and atrophy of the striatum and cortex in advanced disease (Tabrizi et al., 2020). Although HD is characterized by loss of neurons, huntingtin is ubiquitously expressed and both huntingtin and mutant huntingtin exerts effects in many cells and tissues (Carroll et al., 2015). While the underlying cause of HD is the accumulation of the mutant huntingtin protein, there is also evidence that immune dysfunction may contribute to the disease. Inflammation and neuroinflammation have been suggested to be linked to HD pathogenesis, as part of the multiple processes that are found. Studies have shown that there is a chronic inflammatory response in the brain of HD patients and microglial activation has been suggested to play a critical role in neuronal injury in HD (Crotti & Glass, 2015). This inflammatory response may contribute to the neurodegeneration observed in HD by promoting oxidative stress, excitotoxicity, and neuronal damage. In addition, there is evidence that the immune system may play a role in the clearance of mutant huntingtin protein (Jia et al., 2022).

Already in 1999, Singhrao and colleagues demonstrated that central complement activation is present in HD, and suggested that this might be a link to activation of glial cells (Singhrao et al., 1999).Sapp and co-workers showed in 2001 that a microglia accumulation was present in all grades of HD brains (Sapp et al., 2001). However, it was utilizing *in vivo* positron emission tomography (PET) observations that captured the attention of the research community. It was demonstrated that microglia activation is significant in HD brain regions (Pavese et al., 2006), and that microglia activation is present in premanifest HD and throughout disease course (Tai et al., 2007a, b; Politis et al., 2011). (See Fig. 1 for a time-line summarizing the major findings related to inflammation in HD).

The Role of Microglia, Macrophages and Monocytes in HD

Importantly, the presence of proinflammatory microglia has been shown to correlate with HD severity (Pavese et al., 2006; Sapp et al., 2001). Microglia activation is typically characterized by increased numbers of microglia, as well as morphological changes, and these morphological changes are in turn associated with increased production of cytokines, such as IL-6. Actived microglia have been demonstrated in low-grade HD human brains, suggesting an early microglia response in HD (Tai et al., 2007a, b). In addition to microglial activation, altered gene expression in microglia has been shown linked to HD. In 2015 Labardorf and co-workers presented HD brain RNA sequence data demonstrating extensive increases in inflammatory-related gene expression (Labadorf et al., 2015).

HD presents with multiple systemic alterations (van der Burg et al., 2009; Carroll et al., 2015) and in addition to central immune activation, circulating markers of inflammation have been shown to mirror brain inflammatory changes. The activation of microglia that is evident in premanifest HD gene carriers, and detectable as early as 15 years prior to predicted age of onset (Tai et al., 2007b), is mirrored peripherally with activated monocytes and increased levels of cytokines that are detectable in plasma (Björkqvist et al., 2008). Blood levels of several immune proteins have been shown to correlate with disease progression (see Table 1). In 2007, Dalrymple and colleagues presented an unbiased proteomics screen of HD plasma and identified elevated levels of immune markers, including the proinflammatory cytokine IL-6, acute-phase protein alpha-2-macroglobulin, complement factors and a complement inhibitor clusterin, in HD subjects compared to healthy controls (Dalrymple et al., 2007). The authors suggested that the increased levels of IL-6, in turn, led to an acute phase response and subsequent increase in clusterin (Dalrymple et al., 2007). Clusterin has been repeatedly assessed and interestingly has been shown to be either increased or not changed (see Table 1).

In 2008, it was suggested that a peripheral immune response mirrors the central immune response in HD, and that the immune response can be detected n premanifest HD with increased IL-6 levels (Björkqvist et al., 2008). The subtle innate immune response detectable in plasma throughout the course of HD has been demonstrated to be associated with gene expression alterations. Utilizing RNA-seq, Andrade-Navarro and colleauges showed that peripheral blood from HD subjects displayed a gene expression signature illustrating an inflammatory response (Andrade-Navarro et al., 2020). In line with this, levels of neuroinflammatory mRNAs investigated in peripheral blood cells (Runne et al., 2007) also illustrate the presence of transcriptional changes that could be linked to an immune response.

Why Does an Immune Response Occur in HD: Direct or Indirect Effects of Mutant Huntingtin?

Mutant huntingtin has been shown to contribute to the progressive neurodegeneration in HD, through both cell-autonomous and non-cell-autonomous mechanisms (Crotti & Glass, 2015). The non-cell autonomous effects that have been suggested include both an immune response, as a reaction to neuronal damage, and mutant huntingtin aggregation and alterations of the blood-brain barrier, which normally help protect the brain from harmful substances. As the blood-brain barrier becomes more compromised, immune cells and molecules can enter the brain and cause inflammation.

From an immune biomarker perspective, cell-autonomous immune activation mechanisms are possibly more relevant. Huntingtin is expressed, not only in neurons, but in cells and tissue throughout the body and mutant huntingtin inclusion formation is wide-spread (Sathasivam et al., 1999). Mutant huntingtin has been shown to induce immune activation both centrally and peripherally (Björkqvist et al., 2008; Politis et al., 2015) by a direct effect of mutant huntingtin within immune cells, exerting effects at the molecular level affecting intracellular signaling pathways.

Table 1 A summary of immune markers assessed in blood (serum/plasma), CSF and saliva

Biofluid assessed immune markers			
Analyte	Result	Correlation	References
Blood (serum or plasma) in HD compared to control subjects			
sTNF-R	↑	NA	Leblhuber et al. (1998), Widner et al. (1999)
Clusterin	↑	NA	Dalrymple et al. (2007)
	=	NA	Sánchez-López et al. (2012)
CRP	↑	NA	Sánchez-López et al. (2012), Stoy et al. (2005), Silajdžić et al. (2013)
	↓	NA	Silajdžić et al. (2013)
	=	NA	Mochel et al. (2007), Silajdžić et al. (2013), Gruber et al. (2013), Bouwens et al. (2014), Leblhuber et al. (1998)
	=	NA	Corey-Bloom et al. (2020)
IL-6	↑	Independence TFC Cognition	Björkqvist et al. (2008), Dalrymple et al. (2007), Sánchez-López et al. (2012), Chang et al. (2015), Bouwens et al. (2016)
	↑	NA	Corey-Bloom et al. (2020)
	=	Microglial activation	Wang et al. (2014), Politis et al. (2015), Mochel et al. (2007)
Neopterin	↑	Cognition	Stoy et al. (2005), Leblhuber et al. (1998), Widner et al. (1999), Christofides et al. (2006)
sIL-2R	↑	NA	Leblhuber et al. (1998), Widner et al. (1999)
MMP-9, VEGF, TGF-B1	↑	NA	Chang et al. (2015)
IL-18	↓	NA	Chang et al. (2015)
Eotaxin, MCP-1	↑	Motor score, TFC	Wild et al. (2011)
MDC, IP-10, TARC	=	NA	Wild et al. (2011)
Eotaxin-3, MIP-1β, eotaxin, MCP-4	↑	NA	Wild et al. (2011)
Clusterin, C3, C3a, C4, C4a, C5a, CFH, α-2 macroglobulin, α1-antitrypsin, SAP, PEDF, S100A8-A9, S100A12	=	NA	Silajdžić et al. (2013)
Blood (serum or plasma) in premanifest HD compared to control			
IL-1B	↑	Microglial activation	Politis et al. (2015)
TNF-a	=	Microglial activation	Politis et al. (2015)
CSF in HD compared to control			
Clusterin	↑	NA	Dalrymple et al. (2007)
Chitotriosidase	↑	NA	Rodrigues et al. (2016)

(continued)

Table 1 (continued)

Biofluid assessed immune markers			
Analyte	Result	Correlation	References
IL-6	↑	NA	Rodrigues et al. (2016)
IL-8	=	NA	Rodrigues et al. (2016)
sCD27	↑	NA	Niemelä et al. (2018)
C4b, albumin	=	NA	Caron et al. (2022)
C4b	↑	NA	Fang et al. (2009)
IGHG1	↑	NA	Caron et al. (2022)
Saliva in HD compared to control			
CRP	↑	NA	Corey-Bloom et al. (2020)
IL-6	↑	Motor score TFC	Corey-Bloom et al. (2020)
IL-1B	↑	Motor score TFC	Corey-Bloom et al. (2020)

The direction of change is indicated with arrows, or no change (=) and correlation, if demonstrated, is described in brief. References are indicated. This table is based on a Pubmed search employing the following search terms "Huntington's disease" and "inflammation" and "cytokine" and "IL-6" and "saliva" or "urine" or "blood" or "plasma" or "serum" or "cerebrospinal fluid" or "CSF" (and filtering on English language articles). Additional references were added when found to be relevant and not retrieved by the PubMed research. There is a possibility that the table does not represent all published papers within the topic and the table should be viewed as a summary and illustrates the potential to assess immune markers in different biofluids as well as illustrates the possibility that immune markers could correlate with disease progression. Motor score (UHDRS motor score scores) and TFC (total functional capacity)

Strong evidence now illustrates that expression of the mutant huntingtin protein results in a cell-autonomous activation of microglia and, to some extent, astrocytes (reviewed in Crotti & Glass, 2015). Huntingtin mRNA is expressed in microglia (Hickman et al., 2013) and mutant huntingtin expression results in a cell-autonomous increase in proinflammatory expression of *IL-6* and *TNF-α* (Crotti et al., 2014). Human HD post-mortem tissue has been shown to present with a clear inflammatory effector profile, with TNF-α and IL-1β, specifically elevated in the striatum, and matrix metalloprotease (MMP)-9, IL-6, and IL-8 up-regulated in the cortex and cerebellum region (Silvestroni et al., 2009).

Monocytes and macrophages expressing mutant huntingtin have been shown to be hyperactive upon stimulation (Träger et al., 2014). In addition, macrophages from mouse models of HD, as well as monocytes from HD individuals, exhibit defective migratory capacities (Kwan et al., 2012a). Huntingtin expression is higher in neurons compared to other cell types. However, HD subject immune cells, such as monocytes and macrophages, as well as B cells and T cells have been shown to have detectable levels of HTT that correlated with disease status (Pido-Lopez et al., 2018). To further support the direct effect of mutant huntingtin within immune cells, it has been shown that human iPSCs generated microglia successfully, irrespective of the expression of mutant huntingtin. These cells were hyper-reactive upon stimulation and produced more pro-inflammatory cytokines such as IL-6 and TNF-α (O'Regan et al., 2021).

In support of the notion that huntingtin expression within immune cells is linked to HD pathology, a study demonstrated that transgenic mice selectively expressing mutant huntingtin in astrocytes with time displayed signs of HD, demonstrating that mutant huntingtin expression in immune cells is sufficient to cause neurological symptoms (Bradford et al., 2009).

It has been demonstrated that hyperreactivity of HD monocytes and microglia is a result of cell-autonomous effects of mutant huntingtin, and, at least in part due to dysregulation of the NFκB pathway (Träger et al., 2014). NFκB is a central modulator of inflammatory responses, regulating the expression of proinflammatory cytokines including TNF and other genes that are part of the immune response (reviewed in Barnes & Karin, 1997).

A Role for Wild Type Huntingtin in Immune Cells

A role of wild type huntingtin has been demonstrated to be of potential importance in myeloid cells. O'Regan and colleagues demonstrated in 2020 that the wild-type form of huntingtin, has a role in normal human macrophage function (O'Regan et al., 2020). The authors showed that, when lowering wild-type HTT expression in primary monocyte-derived macrophages from healthy subjects, this resulted in a reduced release of cytokines upon stimulation, as well as both elevated phagocytosis and increased stress vulnerability (O'Regan et al., 2020).

The Immune Response in HD: Innate or Adaptive?

As presented above, substantial evidence has now been generated to support the notion that an activation of the innate immune system is present in HD, including activation of microglia with expression of proinflammatory cytokines and impaired migration of macrophages. In addition, studies have also demonstrated involvement of an adaptive immune response in HD with involvement of dendritic cells that prime T-cell responses secreting inflammatory mediators. An adaptive immune response could further increase interest in the peripheral immune system involvement in HD and potentially both innate and adaptive immune response are important for HD pathogenesis (reviewed in Ellrichmann et al., 2013).

3 Biofluid Immune Mediators in Huntington's Disease

The activation of microglia that is evident in premanifest HD gene carriers, and detectable as early as 15 years prior to predicted age of onset (Tai et al., 2007b), is mirrored peripherally with activated monocytes and increased levels of plasma cytokines (Björkqvist et al., 2008). As discussed in above section, the summarized knowledge now suggests that immune markers in CSF samples and peripheral blood samples may reflect the immune pathology associated with HD and that the expression of mutant huntingtin within immune cells could result in accessible and measurable alterations linked to functional and transcriptional changes.

Several studies have been conducted focusing on immune markers in biofluids. Blood, CSF and saliva levels of several immune proteins have been shown to correlate with disease progression (see Table 1). In the below sections groups of markers are highlighted.

Cytokines

Cytokines are a broad group of small proteins that are important signaling molecules within the immune system. They are produced by various immune cells and act as messengers in cell-to-cell communication. Cytokines are involved in a wide range of processes, including the regulation of immune responses. They can act by stimulating or inhibiting the activity of other cells in the immune system and they play an essential role in the coordination of the immune response (Sharon et al. Kuby immunology, 2019). As cytokines play an important part in the immune system and their potential as therapeutic targets, cytokines and their receptors are the subject of ongoing research in many areas, and cytokines are often suggested as potential markers of disease.

Among other cytokines, IL-6 has been assessed in several studies and increased IL-6 levels in HD plasma have been repeatedly demonstrated (Björkqvist et al., 2008; Dalrymple et al., 2007; Sánchez-López et al., 2012; Chang et al., 2015; Bouwens et al., 2016), however no change has also been reported (Wang et al., 2014; Politis et al., 2015; Mochel et al., 2007). Plasma IL-6 levels have been shown to correlate with disease severity and in CSF obtained from HD patients there is evidence of immune activation, with upregulation of IL-6 alongside, IL-8 and TNF-a (Björkqvist et al., 2008). Interestingly, it has been demonstrated that levels of IL-6 in saliva can be assessed and correlates with clinical measures in HD (Corey-Bloom et al., 2020). This indicates that the accessible body fluid saliva can actually provide future possibilities as source of non-invasive markers related to HD symptoms. Also, reduced levels of cytokines have been described. Plasma levels of IL-18 has been shown significantly reduced in HD patients in comparison with control subjects (Chang et al., 2015).

Chemokines

Chemokines are a class of small protein molecules that are involved in many important physiological processes, such as the immune response to infection and inflammation. They help to direct the migration of immune cells, such as T-cells and macrophages, to the sites of infection or injury. Chemokines can also be involved in the development and progression of certain diseases, including cancer, autoimmune disorders and chronic inflammatory diseases (Sharon et al. Kuby immunology, 2019). Because of their important roles in the immune system and their potential as therapeutic targets, chemokines and their receptors are the subjects of ongoing research in many areas and they have often been investigated as potential markers of disease. Chemokine levels may shed light on the status of both the immune system as a whole and the function of specific immune components in HD and chemokine levels have been suggested markers in neurodegenerative disease (Lucin & Wyss-Coray, 2009).

In HD, levels of chemokines have been demonstrated to be altered in biofluids. The chemokines; eotaxin-3, macrophage inflammatory protein (MIP)-1β, eotaxin, monocyte chemoattractant protein (MCP)-1 and MCP-4 have been shown to be significantly altered in HD plasma (Wild et al., 2011) as well as MMP-9, vascular endothelial growth factor (VEGF) and transforming growth factor (TGF)-1β (Chang et al., 2015).

Acute-Phase Proteins

Acute phase proteins (APPs) are proteins synthesized in response to inflammation or infection. They are called "acute phase" proteins because their levels increase rapidly and dramatically during the acute phase of an immune response. Acute phase proteins are produced mainly by the liver in response to cytokines. Acute phase proteins include for example C-reactive protein (CRP) (Sharon et al. 2019). Levels of these proteins in blood can be used as a diagnostic tool to monitor the severity of an immune response or as a marker for treatment response.

CRP has been suggested a potential biomarkers in HD. Studies have reported both increased CRP levels in HD subjects (Sánchez-López et al., 2012; Wang et al., 2014; Gruber et al., 2013; Stoy et al., 2005), as well as decreased levels of CRP in HD or no significant difference (Silajdžić et al., 2013). Increased CRP in HD, could be related to an acute-phase response in HD and indeed studies also suggest other acute-phase proteins such as alpha-1 antitrypsin (Silajdžić et al., 2013) and alpha-2-macroglobulin (Dalrymple et al., 2007; Silajdžić et al., 2013) as potential markers.

However, there is also a possibility that the increased levels of CRP could be related to the use of antipsychotics (Bouwens et al., 2014). Even though a thorough study is needed to conclude the potential usefulness of CRP levels as a marker that increase with disease progression in HD, it is interesting to note that the suggested increase in CRP levels in HD is found in both serum/plasma (Dalrymple et al., 2007), CSF (Dalrymple et al., 2007) and in saliva (Corey-Bloom et al., 2020).

Complement Factors

Complement factors have also been investigated as potential biomarkers. Upregulation of early complement activators and regulators from reactive microglia has been reported in both brain tissue and CSF. C4b, involved in the complement system, has both been shown to be decreased in CSF from HD subjects compared to controls (Fang et al., 2009) and not changed (Caron et al., 2022).

In brain tissue, several complement factors involved in the classical complement pathway have been shown to be expressed at higher levels in HD subjects; components such as C1q C chain, C1r, C3, and C4), as well as complement regulators C1 inhibitor and methyl-accepting chemotaxis protein (MCP) (Singhrao et al., 1999). C1Q (composed of A, B and C polypeptide chains), is a component of the complement C1 recognition complex of the classical pathway that is released from CNS cells in response to inflammatory stimuli. Caron et al. showed that CSF C1QB was increased in early/mid HD compared with preHD and significantly reduced in late HD compared with early/mid HD, illustrating that CSF C1QB levels may reflect early HD-associated complement activation in HD brain and detectable in CSF (Caron et al., 2022).

Albumin is a large molecule that can be measured in CSF and blood to form the albumin index, and used as measure of blood-brain barrier permeability. Both levels of albumin and immuno globulin G (IgG) are routinely used to measure blood–brain barrier permeability as well as intrathecal IgG production. Interestingly, a significant increase of CSF IGHG1 (heavy chain constant domain of IgG) in late-stage HD compared with controls has been demonstrated (Caron et al., 2022). The increased CSF IGHG1 could reflect increased local CNS IgG synthesis as a sign of CNS inflammation.

Neopterin is a biomarker of cellular immune system activation and produced primarily by macrophages. Neopterin release is induced by interferon-gamma, a cytokine that is produced by T cells and natural killer cells in response to an immune response.

Studies have reported increased neopterin levels in HD (Stoy et al., 2005; Leblhuber et al., 1998; Widner et al., 1999; Christofides et al., 2006).

In addition, other adaptive immune cellular mediators of inflammation have been demonstrated altered in HD. sCD27, a marker of adaptive immunity, has been shown to be increased in CSF in premanifest HD compared to control subjects (Niemelä et al., 2018). This indicates early T cell-mediated neuroinflammation that in turn indicates involvement of adaptive immunity in HD.

Immune Markers in Animal Models of HD

Research using animal models and human post-mortem tissue have identified key immune response actors, including microglia, neurons, astrocytes, oligodendrocytes, vessel-associated cells, as well as peripheral immune cells and infiltrating immune cells. HD animal model research have produced substantial evidence on the complex role of an immune dysfunction linked to HD pathology and immune markers have been shown altered across animal models (see Table 2).

Several studies have demonstrated an increase of circulating inflammatory markers in the R6/2 transgenic mouse model (Dalrymple et al., 2007; Björkqvist et al., 2008; Chang et al., 2015; Pido-Lopez et al., 2018; Pido-Lopez et al., 2019; Disatnik et al., 2016; Bondulich et al., 2021). In transgenic R6/2 mice there seem to be a progressive increase in circulating inflammatory markers as it has been demonstrated that at 10 weeks there is no change in cytokine levels (Pido-Lopez et al., 2018), however at 14 weeks, robust data, with an n number of 16–26 mice per group, demonstrate that TNF-a, IL-1β, IL-6, IL-2 and IL-10 are increased in R6/2 mice compared to control mice (Pido-Lopez et al., 2019).

A study following plasma concentration of inflammatory markers over time in the R6/1 transgenic mouse model for HD, revealed altered inflammatory markers all through disease course (Podlacha et al., 2022). The anti-inflammatory marker IL-10 was seen increased throughout disease, indicating a possible compensatory mechanisms and also TNF-a, IL-1B and IL-12 where found increased. In this study, IL-6 was found to be decreased (Podlacha et al., 2022).

In full length HD mouse models, the summarized results are not clear. A study utilizing the YAC128 transgenic mouse model revealed significant increases in IL-6, CXCL1 (the functional homolog of IL-8 in mice), IFN-γ and IL-10 (Kwan et al., 2012b). Also the BAC-225Q mouse have been shown to display increased levels of IL-8 (Wegrzynowicz et al., 2015). However, in serum obtained from BACHD mice (12-month-old) one study has shown no change in cytokines levels, however the expression of inflammatory markers in peripheral tissueswerefound to be altered (Valadão et al., 2019). In addition, the BAC-225Q mouse displayed no altered levels of IFNγ, IL-6 and TNFα (Wegrzynowicz et al., 2015).

In addition to rodent models, transgenic HD rhesus monkeys exhibit increased peripheral levels of cytokines (Raper et al., 2016) and in transgenic HD minipigs an increase in IL-8 has been detected at the pre-symptomatic disease stage, in both serum and CSF (Valekova et al., 2016).

To summarize, there is a consistency in the results across species indicating a subtle immune response present and detectable in blood. Even though this consistency in the presence of an immune response there is however an inconsistency in actual changes and levels. The differences in IL-6 results likely illustrates the problem assessing subtle changes in serum or plasma as markers for disease (as will be further discussed below).

Table 2 A summary of immune markers assessed in blood (serum/plasma) in animal models of HD

Animal model biofluid assessed immune markers in HD compared to wild-type in blood (serum /plasma)			
Analyte	Result	Strain	References
HD mice			
IL-6	↑	R6/2	Dalrymple et al. (2007), Björkqvist et al. (2008), Chang et al. (2015), Pido-Lopez et al. (2018, 2019), Disatnik et al. (2016), Bondulich et al. (2021)
IL-2	↑	R6/2	Pido-Lopez et al. (2018, 2019)
IL-10	↑	R6/2	Pido-Lopez et al. (2018, 2019)
IL-1B	↑	R6/2	Pido-Lopez et al. (2018, 2019), Bondulich et al. (2021)
TNF-a	↑	R6/2	Pido-Lopez et al. (2018, 2019), Disatnik et al. (2016), Bondulich et al. (2021)
IL-4	↑		Bondulich et al. (2021)
IL-5	↓	R6/2	Bondulich et al. (2021)
IL-18	↓	R6/2	Chang et al. (2015)
MMP-9	↑	R6/2	Chang et al. (2015)
TGF-B1	↑	R6/2	Chang et al. (2015)
VEGF	↑	R6/2	Chang et al. (2015)
IFNg, IL-2	=	R6/2	Bondulich et al. (2021)
IL-4, IL-5, IL-6 IL-12p70, IFN-γ TNF-α	=	BACHD	Valadão et al. (2019)
IFNg, IL-6, TNF-a	=	BAC -225Q	Wegrzynowicz et al. (2015)
IL-6	↑	YAC128	Kwan et al. (2012b)
CXCL1 (IL-8)	↑	YAC128	Kwan et al. (2012b)
IFN-g	↑	YAC128	Kwan et al. (2012b)
IL-10	↑	YAC128	Kwan et al. (2012b)
IL-6	↓	R6/1	Podlacha et al. (2022)
IL-10	↓	R6/1	Podlacha et al. (2022)
TNF-a	↑	R6/1	Podlacha et al. (2022)
IL-1B	↑	R6/1	Podlacha et al. (2022)
IL-12	↑	R6/1	Podlacha et al. (2022)
HD rat			
GRO/KC, IL-1β, IL-13 and TNF-α	=	Transgenic HD rat	Nguyen et al. (2010)
HD monkey			
TNF-a	↑	Transgenic HD rhesus monkeys	Raper et al. (2016)
HD pig			
IL-8	↑	Transgenic mini pig	Valekova et al. (2016)

The direction of change is indicated with arrows, or no change (=). References are indicated. This table is based on a Pubmed search employing the following search terms "Huntington's disease" and "inflammation" and "cytokine" and "IL-6" and "species" and filtering on English language articles. Additional references were added when found to be relevant and not retrieved by the PubMed research. There are many studies suing cytokine levels as a readout evaluating different treatment strategies. This table should therefore not be regarded as a full evaluation of published papers, but as a summary

4 The Potential Use of Immune Markers in Huntington's Disease

Out of the immune markers until today tested, none has so far shown the potential to track with HD disease progression robustly and longitudinally. There are however interesting aspects; immune markers can be used as translational markers for HD, immune markers can lead research to new therapeutic interventions and immune markers could be part of panel of marker for disease. Importantly, several biofluids have shown promise as potential source for immune markers and the accessible body fluid saliva can actually provide future possibilities as source of non-invasive markers related to HD symptoms (Corey-Bloom et al., 2020).

The fact that HD immune dysfunction seems to be conserved across human and animal models increases the usefulness of immune markers in HD. Findings in mouse models of HD, mirror findings demonstrating that HD patients have increased plasma levels of inflammatory cytokines (see Tables 1 and 2) and circulating monocytes that are hyper-responsive to immune stimuli. Myeloid cells isolated from three mouse models; R6/2 transgenic mice, HdhQ150 knock-in mice and YAC128 transgenic mice, showed increased cytokine production upon stimulation (Träger et al., 2015), similar to the hyperreactivity found in human HD monocytes and microglia (Träger et al., 2014).

Immune Modulation in HD: A Role for Immune Markers

The growing evidence demonstrating a measurable immune response in HD and the notion that immune dysfunction may contribute to HD pathogenesis has led to studies investigating the possible beneficial disease modification by targeting the immune response (reviewed in Colpo et al., 2017; Tabrizi et al., 2022). For example, the immunomodulator laquinimod, with a demonstrated effect to downregulate both the production of pro-inflammatory cytokines in peripheral blood mononuclear cells and neuroglial activation in the brain, has been shown to be effective in HD mouse models (Ehrnhoefer et al., 2016; Ellrichmann et al., 2017). Even though, the exact mechanism of action of laquinimod is unclear, laquinimod has been shown to significantly reduce the release of cytokines from stimulated monocytes isolated from premanifest and manifest HD subjects (Dobson et al., 2016). Despite these positive preclinical HD data, laquinimod treatment did not present efficacy data in the Phase 2 clinical trial in HD (https://clinicaltrials.gov/ct2/show/NCT02215616).

The increase in IL-6 that is repeatedly demonstrated in HD (both centrally and peripherally) has inspired researchers to test the hypothesis that lack of IL-6 might be beneficial or protective. However, IL-6 deficient R6/2 mice were shown to present with a more severe phenotype compared to R6/2 mice (Wertz et al., 2020).

Along these lines, by using an anti TNF-a strategy, etanercept therapy, it was shown that it was possible to lower the plasma levels of several proinflammatory cytokines in the R6/2 mouse model (Pido-Lopez et al., 2018). This study importantly led to a deceleration of brain atrophy, however without affecting motor or cognitive declines (Pido-Lopez et al., 2018).

In 2016 a study using an inhibitor of pathological mitochondrial fragmentation, P110, to inhibit neurotoxicity in HD, illustrated that the P110 treatment effectively reduced circulating levels of IL-6 and TNF-a (Disatnik et al., 2016). This illustrates the potential of circulating inflammatory markers to assess HD progression and the benefit of intervention for future clinical trials (Disatnik et al., 2016).

Immune Markers: Complexity and Potential Caveats

Based on the knowledge obtained from the literature search of HD inflammatory markers, the conclusion can be made that there is still no biofluid accessible immune marker that have been reliably, and consistently reproduced and longitudinal studies are lacking.

The complexity of the immune response and caveats using immune markers should be considered. Many of the immune markers discussed in this chapter are either non-specific for HD or variable in response to factors not relating to HD, or variable over 24 hours, or related to age or general state (for example illness). It was described back in 1992 by Wei and co-authors that an increase in plasma IL-6 correlates with age in healthy subjects (Wei et al., 1992). The very subtle immune response seen in HD and subsequent low levels of changes in immune markers, contributes to the limitations of using immune markers reliably. Very sensitive assays are needed and in addition, low level of changes are easily masked by natural variability.

Adding to the complexity in studying inflammatory markers and correlations to disease features, studies have shown association between inflammatory markers and diet, sleep duration (as well as sleep quality) and lack of physical activity and adding even more to the complexity, there is a possibility that these associations are more prevalent in elderly (Basta et al., 2022). Levels of immune markers such as neopterin has been demonstrated to follow a circadian rhythm in humans (Seizer et al., 2022). Circadian variation is demonstrated also for cytokines (Knudsen et al., 2008). As alterations in cytokine levels in HD have been shown to be subtle (Björkqvist et al., 2008), time of sampling might therefore affect results. To avoid circadian interference, time of sampling needs to be taken into consideration to allow comparisons between studies.

To further illustrated the complexity, a systemic immune response can be seen linked to the gut microbiome. Emerging evidence illustrates that the microbiome influences CNS function and that disturbances in the microbiota–gut–brain axis may play a role in both neuroinflammation and neurodegeneration (Clemente et al., 2012). In addition to neuroinflammation and systemic inflammation, the gut microbiome has also been shown to alter progression of neuronal injury (reviewed in Hickman et al., 2018). Studies have linked functional alterations in the gut microbiota to several neurodegenerative diseases (Quigley, 2017) and Du and co-workers demonstrated in 2021 that an altered gut microbiota was seen related to the inflammatory response in HD (Du et al., 2021).

Independent validation of results has repeatedly been shown necessary in explorative biomarker research. To exemplify; in 2007 a signature of 18 plasma proteins (immune markers) that could identify AD during pre-dementia and dementia stages was published in Nature Medicine (Ray et al., 2007). In a subsequent study, analyzing the same set of markers, replication of results was however not possible (Björkqvist et al., 2012).

Additional Potential of Immune Markers Useful in HD

Importantly, immune markers and alterations of these, might guide the research field in terms of potential answers to remaining pathological questions. Especially immune marker alterations, observed in premanifest HD should be pursued, as these may directly linked to pathogenesis with the potential to inform both basic understanding of disease, as well as inform further search for disease markers.

A panel of markers has the potential of yielding for example a more robust discrimination between disease states or subject groups, and immune markers could have a potential part in such panels of markers. To illustrated this, it has been demonstrated that the immune marker CB4, although not significantly different in between groups, was beneficial, as part of a panel of markers, for discriminating in between subject groups (Caron et al., 2022).

Possibly, there is also a future role for immune markers as beneficial tools for patient stratification, indicators of a patient group that are more likely to respond to a certain immune modulation strategy. Or possibly there is a role for immune markers as disease activity biomarkers in future evaluation of possible clinical benefit from drug candidates targeting immune alterations.

Future interventions for HD potentially include strategies targeting inflammation (Tabrizi et al., 2022). Immune markers could then aid in the development as well as potentially improve the efficacy of a HD clinical trial focusing on inflammatory processes. An immune marker might then enable more refined stratification and statistical analysis as well as potentially increase trial power.

5 Conclusions

In HD, a central immune activation is present already before onset of clinical features and circulating markers of inflammation has been shown to mirror brain inflammatory changes. This indicates a possibility that circulating immune markers can be used to monitor disease alterations. A literature search of immune markers accessible in body fluids in HD illustrates that there are currently no immune markers that have been reliably reproduced in several cohorts, nor longitudinally. Future studies utilizing structured protocols and large patient cohorts will be needed in to increase understanding of the usefulness of immune-marker alterations is HD.

The importance of accumulated knowledge on markers of inflammation might be in the potential to gain increased understanding of the physiological processes involved in low-grade immune response in HD. In turn this might guide new therapeutic applications. Indeed, the role of inflammation in HD pathogenesis has been suggested repeatedly and translates across species, which further strengthens the notion that anti-inflammatory therapeutics will be important treatment avenues for HD.

References

Andrade-Navarro, M. A., Mühlenberg, K., Spruth, E. J., Mah, N., González-López, A., Andreani, T., Russ, J., Huska, M. R., Muro, E. M., Fontaine, J. F., Amstislavskiy, V., Soldatov, A., Nietfeld, W., Wanker, E. E., & Priller, J. (2020). RNA sequencing of human peripheral blood cells indicates upregulation of immune-related genes in Huntington's disease. *Frontiers in Neurology, 11*, 573560. https://doi.org/10.3389/fneur.2020.573560. PMID: 33329316; PMCID: PMC7731869.

Barnes, P. J., & Karin, M. (1997). Nuclear factor-kappaB: a pivotal transcription factor in chronic inflammatory diseases. *The New England Journal of Medicine, 336*(15), 1066–1071. https://doi.org/10.1056/NEJM199704103361506. PMID: 9091804.

Bartl, M., Xylaki, M., Bähr, M., Weber, S., Trenkwalder, C., & Mollenhauer, B. (2022). Evidence for immune system alterations in peripheral biological fluids in Parkinson's disease. *Neurobiology of Disease, 170*, 105744. https://doi.org/10.1016/j.nbd.2022.105744. Epub 2022 May 2 PMID: 35513230.

Basta, M., Belogianni, C., Yannakoulia, M., Zaganas, I., Panagiotakis, S., Simos, P., & Vgontzas, A. N. (2022). Poor diet, long sleep, and lack of physical activity are associated with inflammation among non-demented community-dwelling elderly. *Healthcare (Basel), 10*(1), 143. https://doi.org/10.3390/healthcare10010143. PMID: 35052306; PMCID: PMC8775386.

Biomarkers Definitions Working Group. (2001). Biomarkers and surrogate endpoints: preferred definitions and conceptual framework. *Clinical Pharmacology and Therapeutics, 69*(3), 89–95. https://doi.org/10.1067/mcp.2001.113989. PMID: 11240971.

Björkqvist, M., Wild, E. J., Thiele, J., Silvestroni, A., Andre, R., Lahiri, N., Raibon, E., Lee, R. V., Benn, C. L., Soulet, D., Magnusson, A., Woodman, B., Landles, C., Pouladi, M. A., Hayden, M. R., Khalili-Shirazi, A., Lowdell, M. W., Brundin, P., Bates, G. P., Leavitt, B. R., Möller, T., & Tabrizi, S. J. (2008). A novel pathogenic pathway of immune activation detectable before clinical onset in Huntington's disease. *The Journal of Experimental Medicine, 205*(8), 1869–1877. https://doi.org/10.1084/jem.20080178. Epub 2008 Jul 14. PMID: 18625748; PMCID: PMC2525598.

Björkqvist, M., Ohlsson, M., Minthon, L., & Hansson, O. (2012). Evaluation of a previously suggested plasma biomarker panel to identify Alzheimer's disease. *PLoS One, 7*(1), e29868. https://doi.org/10.1371/journal.pone.0029868. Epub 2012 Jan 18. PMID: 22279551; PMCID: PMC3261152.

Bondulich, M. K., Fan, Y., Song, Y., Giorgini, F., & Bates, G. P. (2021). Ablation of kynurenine 3-monooxygenase rescues plasma inflammatory cytokine levels in the R6/2 mouse model of Huntington's disease. *Scientific Reports, 11*(1), 5484. https://doi.org/10.1038/s41598-021-84858-7. PMID: 33750843; PMCID: PMC7943810.

Bouwens, J. A., Hubers, A. A., van Duijn, E., Cobbaert, C. M., Roos, R. A., van der Mast, R. C., & Giltay, E. J. (2014). Acute-phase proteins in relation to neuropsychiatric symptoms and use of psychotropic medication in Huntington's disease. *European Neuropsychopharmacology, 24*(8), 1248–1256. https://doi.org/10.1016/j.euroneuro.2014.05.004. Epub 2014 May 20 PMID: 24957738.

Bouwens, J. A., van Duijn, E., Cobbaert, C. M., Roos, R. A., van der Mast, R. C., & Giltay, E. J. (2016). Plasma cytokine levels in relation to neuropsychiatric symptoms and cognitive dysfunction in Huntington's disease. *Journal of Huntingtons Disease, 5*(4), 369–377. https://doi.org/10.3233/JHD-160213. PMID: 27983562.

Bradford, J., Shin, J. Y., Roberts, M., Wang, C. E., Li, X. J., & Li, S. (2009). Expression of mutant huntingtin in mouse brain astrocytes causes age-dependent neurological symptoms. *Proceedings of the National Academy of Sciences of the United States of America, 106*(52), 22480–22485. https://doi.org/10.1073/pnas.0911503106. Epub 2009 Dec 11. PMID: 20018729; PMCID: PMC2799722.

Caron, N. S., Haqqani, A. S., Sandhu, A., Aly, A. E., Findlay Black, H., Bone, J. N., McBride, J. L., Abulrob, A., Stanimirovic, D., Leavitt, B. R., & Hayden, M. R. (2022). Cerebrospinal fluid biomarkers for assessing Huntington disease onset and severity. *Brain Communications, 4*(6), fcac309. https://doi.org/10.1093/braincomms/fcac309. PMID: 36523269; PMCID: PMC9746690.

Carroll, J. B., Bates, G. P., Steffan, J., Saft, C., & Tabrizi, S. J. (2015). Treating the whole body in Huntington's disease. *Lancet Neurology, 14*(11), 1135–1142. https://doi.org/10.1016/S1474-4422(15)00177-5. PMID: 26466780.

Chang, K. H., Wu, Y. R., Chen, Y. C., & Chen, C. M. (2015). Plasma inflammatory biomarkers for Huntington's disease patients and mouse model. *Brain, Behavior, and Immunity, 44*, 121–127. https://doi.org/10.1016/j.bbi.2014.09.011. Epub 2014 Sep 28. PMID: 25266150.

Chen, M., & Xia, W. (2020). Proteomic profiling of plasma and brain tissue from Alzheimer's disease patients reveals candidate network of plasma biomarkers. *Journal of Alzheimer's Disease, 76*(1), 349–368. https://doi.org/10.3233/JAD-200110. PMID: 32474469; PMCID: PMC7457324.

Christofides, J., Bridel, M., Egerton, M., Mackay, G. M., Forrest, C. M., Stoy, N., Darlington, L. G., & Stone, T. W. (2006). Blood 5-hydroxytryptamine, 5-hydroxyindoleacetic acid and melatonin levels in patients with either Huntington's disease or chronic brain injury. *Journal of Neurochemistry, 97*(4), 1078–1088. https://doi.org/10.1111/j.1471-4159.2006.03807.x. Epub 2006 Mar 29. PMID: 16573644.

Clemente, J. C., Ursell, L. K., Parfrey, L. W., & Knight, R. (2012). The impact of the gut microbiota on human health: an integrative view. *Cell, 148*(6), 1258–1270. https://doi.org/10.1016/j.cell.2012.01.035. PMID: 22424233; PMCID: PMC5050011.

Colpo, G. D., Stimming, E. F., Rocha, N. P., & Teixeira, A. L. (2017). Promises and pitfalls of immune-based strategies for Huntington's disease. *Neural Regeneration Research, 12*(9), 1422–1425. https://doi.org/10.4103/1673-5374.215245. PMID: 29089980; PMCID: PMC5649455.

Corey-Bloom, J., Fischer, R. S., Kim, A., Snell, C., Parkin, G. M., Granger, D. A., Granger, S. W., & Thomas, E. A. (2020). Levels of interleukin-6 in saliva, but not plasma, correlate with clinical metrics in Huntington's disease patients and healthy control subjects. *International Journal of Molecular Sciences, 21*(17), 6363. https://doi.org/10.3390/ijms21176363. PMID: 32887270; PMCID: PMC7503233.

Crotti, A., & Glass, C. K. (2015). The choreography of neuroinflammation in Huntington's disease. *Trends in Immunology, 36*(6), 364–373. https://doi.org/10.1016/j.it.2015.04.007. Epub 2015 May 20. PMID: 26001312; PMCID: PMC4786070.

Crotti, A., Benner, C., Kerman, B. E., Gosselin, D., Lagier-Tourenne, C., Zuccato, C., Cattaneo, E., Gage, F. H., Cleveland, D. W., & Glass, C. K. (2014). Mutant Huntingtin promotes autonomous microglia activation via myeloid lineage-determining factors. *Nature Neuroscience, 17*(4), 513–521. https://doi.org/10.1038/nn.3668. Epub 2014 Mar 2. PMID: 24584051; PMCID: PMC4113004.

Dalrymple, A., Wild, E. J., Joubert, R., Sathasivam, K., Björkqvist, M., Petersén, A., Jackson, G. S., Isaacs, J. D., Kristiansen, M., Bates, G. P., Leavitt, B. R., Keir, G., Ward, M., & Tabrizi, S. J. (2007). Proteomic profiling of plasma in Huntington's disease reveals neuroinflammatory activation and biomarker candidates. *Journal of Proteome Research, 6*(7), 2833–2840. https://doi.org/10.1021/pr0700753. Epub 2007 Jun 7. PMID: 17552550.

Dantzer, R., O'Connor, J. C., Freund, G. G., Johnson, R. W., & Kelley, K. W. (2008). From inflammation to sickness and depression: when the immune system subjugates the brain. *Nature Reviews. Neuroscience, 9*(1), 46–56. https://doi.org/10.1038/nrn2297. PMID: 18073775; PMCID: PMC2919277.

Disatnik, M. H., Joshi, A. U., Saw, N. L., Shamloo, M., Leavitt, B. R., Qi, X., & Mochly-Rosen, D. (2016). Potential biomarkers to follow the progression and treatment response of Huntington's disease. *The Journal of Experimental Medicine, 213*(12), 2655–2669. https://doi.org/10.1084/jem.20160776. Epub 2016 Nov 7. PMID: 27821553; PMCID: PMC5110026.

Dobson, L., Träger, U., Farmer, R., Hayardeny, L., Loupe, P., Hayden, M. R., & Tabrizi, S. J. (2016). Laquinimod dampens hyperactive cytokine production in Huntington's disease patient myeloid cells. *Journal of Neurochemistry, 137*(5), 782–794. https://doi.org/10.1111/jnc.13553. Epub 2016 Apr 5. PMID: 26823290; PMCID: PMC4982105.

Du, G., Dong, W., Yang, Q., Yu, X., Ma, J., Gu, W., & Huang, Y. (2021). Altered gut microbiota related to inflammatory responses in patients with Huntington's disease. *Frontiers in Immunology, 11*, 603594. https://doi.org/10.3389/fimmu.2020.603594. PMID: 33679692; PMCID: PMC7933529.

Edison, P., & Brooks, D. J. (2018). Role of Neuroinflammation in the trajectory of Alzheimer's disease and in vivo quantification using PET. *Journal of Alzheimer's Disease, 64*(s1), S339–S351. https://doi.org/10.3233/JAD-179929. PMID: 29865053.

Ehrnhoefer, D. E., Caron, N. S., Deng, Y., Qiu, X., Tsang, M., & Hayden, M. R. (2016). Laquinimod decreases Bax expression and reduces caspase-6 activation in neurons. *Experimental Neurology, 283*(Pt A), 121–128. https://doi.org/10.1016/j.expneurol.2016.06.008. Epub 2016 Jun 11. PMID: 27296315.

Ellrichmann, G., Reick, C., Saft, C., & Linker, R. A. (2013). The role of the immune system in Huntington's disease. *Clinical & Developmental Immunology, 2013*, 541259. https://doi.org/10.1155/2013/541259. Epub 2013 Jul 15. PMID: 23956761; PMCID: PMC3727178.

Ellrichmann, G., Blusch, A., Fatoba, O., Brunner, J., Reick, C., Hayardeny, L., Hayden, M., Sehr, D., Winklhofer, K. F., Saft, C., & Gold, R. (2017). Laquinimod treatment in the R6/2 mouse model. *Scientific Reports, 7*(1), 4947. https://doi.org/10.1038/s41598-017-04990-1. Erratum in: Sci Rep. 2019 Mar 15;9(1):4960. PMID: 28694434; PMCID: PMC5504033.

Engelhart, M. J., Geerlings, M. I., Meijer, J., Kiliaan, A., Ruitenberg, A., van Swieten, J. C., Stijnen, T., Hofman, A., Witteman, J. C., & Breteler, M. M. (2004). Inflammatory proteins in plasma and the risk of dementia: the Rotterdam study. *Archives of Neurology, 61*(5), 668–672. https://doi.org/10.1001/archneur.61.5.668. PMID: 15148142.

Feldman, E. L., Goutman, S. A., Petri, S., Mazzini, L., Savelieff, M. G., Shaw, P. J., & Sobue, G. (2022). Amyotrophic lateral sclerosis. *Lancet, 400*(10360), 1363–1380. https://doi.org/10.1016/S0140-6736(22)01272-7. Epub 2022 Sep 15. PMID: 36116464.

Fakhoury, M. (2018). Microglia and astrocytes in Alzheimer's disease: implications for therapy. *Current Neuropharmacology, 16*(5), 508–518. https://doi.org/10.2174/1570159X15666170720095240. PMID: 28730967; PMCID: PMC5997862.

Fang, Q., Strand, A., Law, W., Faca, V. M., Fitzgibbon, M. P., Hamel, N., Houle, B., Liu, X., May, D. H., Poschmann, G., Roy, L., Stühler, K., Ying, W., Zhang, J., Zheng, Z., Bergeron, J. J., Hanash, S., He, F., Leavitt, B. R., Meyer, H. E., Qian, X., & McIntosh, M. W. (2009). Brain-specific proteins decline in the cerebrospinal fluid of humans with Huntington disease. *Molecular & Cellular Proteomics, 8*(3), 451–466. https://doi.org/10.1074/mcp.M800231-MCP200. Epub 2008 Nov 4. PMID: 18984577; PMCID: PMC2649809.

Franceschi, C., Bonafè, M., Valensin, S., Olivieri, F., De Luca, M., Ottaviani, E., & De Benedictis, G. (2000). Inflamm-aging. An evolutionary perspective on immunosenescence. *Annals of the*

New York Academy of Sciences, 908, 244–254. https://doi.org/10.1111/j.1749-6632.2000.tb06651.x. PMID: 10911963.

Gómez-Apo, E., Mondragón-Maya, A., Ferrari-Díaz, M., & Silva-Pereyra, J. (2021). Structural brain changes associated with overweight and obesity. *Journal of Obesity, 2021*, 6613385. https://doi.org/10.1155/2021/6613385. PMID: 34327017; PMCID: PMC8302366.

Glass, C. K., Saijo, K., Winner, B., Marchetto, M. C., & Gage, F. H. (2010). Mechanisms underlying inflammation in neurodegeneration. *Cell, 140*(6), 918–934. https://doi.org/10.1016/j.cell.2010.02.016. PMID: 20303880; PMCID: PMC2873093.

Gruber, B., Kłaczkow, G., Jaworska, M., Krzysztoń-Russjan, J., Anuszewska, E. L., Zielonka, D., Klimberg, A., & Marcinkowski, J. T. (2013). Huntington' disease--imbalance of amino acid levels in plasma of patients and mutation carriers. *Annals of Agricultural and Environmental Medicine, 20*(4), 779–783. PMID: 24364452.

Hickman, S. E., Kingery, N. D., Ohsumi, T. K., Borowsky, M. L., Wang, L. C., Means, T. K., & El Khoury, J. (2013). The microglial sensome revealed by direct RNA sequencing. *Nature Neuroscience, 16*(12), 1896–1905. https://doi.org/10.1038/nn.3554. Epub 2013 Oct 27. PMID: 24162652; PMCID: PMC3840123.

Hickman, S., Izzy, S., Sen, P., Morsett, L., & El Khoury, J. (2018). Microglia in neurodegeneration. *Nature Neuroscience, 21*(10), 1359–1369. https://doi.org/10.1038/s41593-018-0242-x. Epub 2018 Sep 26. PMID: 30258234; PMCID: PMC6817969.

Hirsch, E. C., & Hunot, S. (2009). Neuroinflammation in Parkinson's disease: a target for neuroprotection? *Lancet Neurology, 8*(4), 382–397. https://doi.org/10.1016/S1474-4422(09)70062-6. PMID: 19296921.

Jia, Q., Li, S., Li, X. J., & Yin, P. (2022). Neuroinflammation in Huntington's disease: From animal models to clinical therapeutics. *Frontiers in Immunology, 13*, 1088124. https://doi.org/10.3389/fimmu.2022.1088124. PMID: 36618375; PMCID: PMC9815700.

Jin, M., Günther, R., Akgün, K., Hermann, A., & Ziemssen, T. (2020). Peripheral proinflammatory Th1/Th17 immune cell shift is linked to disease severity in amyotrophic lateral sclerosis. *Scientific Reports, 10*(1), 5941. https://doi.org/10.1038/s41598-020-62756-8. PMID: 32246039; PMCID: PMC7125229.

Julian, A., Rioux-Bilan, A., Ragot, S., Krolak-Salmon, P., Berrut, G., Dantoine, T., Hommet, C., Hanon, O., Page, G., & Paccalin, M. (2018). Blood inflammatory mediators and cognitive decline in Alzheimer's disease: a two years longitudinal study. *Journal of Alzheimer's Disease, 63*(1), 87–92. https://doi.org/10.3233/JAD-171131. PMID: 29614665.

Knudsen, L. S., Christensen, I. J., Lottenburger, T., Svendsen, M. N., Nielsen, H. J., Nielsen, L., Hørslev-Petersen, K., Jensen, J. E., Kollerup, G., & Johansen, J. S. (2008). Pre-analytical and biological variability in circulating interleukin 6 in healthy subjects and patients with rheumatoid arthritis. *Biomarkers, 13*(1), 59–78. https://doi.org/10.1080/13547500701615017. PMID: 17852075.

Källstig, E., McCabe, B. D., & Schneider, B. L. (2021). The links between ALS and NF-κB. *International Journal of Molecular Sciences, 22*(8), 3875. https://doi.org/10.3390/ijms22083875. PMID: 33918092; PMCID: PMC8070122.

Kuhle, J., Lindberg, R. L., Regeniter, A., Mehling, M., Steck, A. J., Kappos, L., & Czaplinski, A. (2009). Increased levels of inflammatory chemokines in amyotrophic lateral sclerosis. *European Journal of Neurology, 16*(6), 771–774. https://doi.org/10.1111/j.1468-1331.2009.02560.x. Epub 2009 Feb 19. PMID: 19236470.

Kwan, W., Träger, U., Davalos, D., Chou, A., Bouchard, J., Andre, R., Miller, A., Weiss, A., Giorgini, F., Cheah, C., Möller, T., Stella, N., Akassoglou, K., Tabrizi, S. J., & Muchowski, P. J. (2012a). Mutant huntingtin impairs immune cell migration in Huntington disease. *The Journal of Clinical Investigation, 122*(12), 4737–4747. https://doi.org/10.1172/JCI64484. Epub 2012 Nov 19. PMID: 23160193; PMCID: PMC3533551.

Kwan, W., Magnusson, A., Chou, A., Adame, A., Carson, M. J., Kohsaka, S., Masliah, E., Möller, T., Ransohoff, R., Tabrizi, S. J., Björkqvist, M., & Muchowski, P. J. (2012b). Bone marrow transplantation confers modest benefits in mouse models of Huntington's disease. *The Journal of Neuroscience, 32*(1), 133–142. https://doi.org/10.1523/JNEUROSCI.4846-11.2012. PMID: 22219276; PMCID: PMC3571858.

Labadorf, A., Hoss, A. G., Lagomarsino, V., Latourelle, J. C., Hadzi, T. C., Bregu, J., MacDonald, M. E., Gusella, J. F., Chen, J. F., Akbarian, S., Weng, Z., & Myers, R. H. (2015). RNA sequence analysis of human Huntington disease brain reveals an extensive increase in inflammatory and developmental gene expression. *PLoS One, 10*(12), e0143563. https://doi.org/10.1371/journal.pone.0143563. Erratum in: PLoS One. 2016;11(7):e0160295. PMID: 26636579; PMCID: PMC4670106.

Lai, K. S. P., Liu, C. S., Rau, A., Lanctôt, K. L., Köhler, C. A., Pakosh, M., Carvalho, A. F., & Herrmann, N. (2017). Peripheral inflammatory markers in Alzheimer's disease: a systematic review and meta-analysis of 175 studies. *Journal of Neurology, Neurosurgery, and Psychiatry, 88*(10), 876–882. https://doi.org/10.1136/jnnp-2017-316201. Epub 2017 Aug 9. PMID: 28794151.

Le, W., Wu, J., & Tang, Y. (2016). Protective microglia and their regulation in Parkinson's disease. *Frontiers in Molecular Neuroscience, 9*, 89. https://doi.org/10.3389/fnmol.2016.00089. PMID: 27708561; PMCID: PMC5030290.

Leblhuber, F., Walli, J., Jellinger, K., Tilz, G. P., Widner, B., Laccone, F., & Fuchs, D. (1998). Activated immune system in patients with Huntington's disease. *Clinical Chemistry and Laboratory Medicine, 36*(10), 747–750. https://doi.org/10.1515/CCLM.1998.132. PMID: 9853799.

Leng, F., & Edison, P. (2021). Neuroinflammation and microglial activation in Alzheimer disease: where do we go from here? *Nature Reviews. Neurology, 17*(3), 157–172. https://doi.org/10.1038/s41582-020-00435-y. Epub 2020 Dec 14. PMID: 33318676.

Lucin, K. M., & Wyss-Coray, T. (2009). Immune activation in brain aging and neurodegeneration: too much or too little? *Neuron, 64*(1), 110–122. https://doi.org/10.1016/j.neuron.2009.08.039. PMID: 19840553; PMCID: PMC2834890.

Mishra, P. S., Dhull, D. K., Nalini, A., Vijayalakshmi, K., Sathyaprabha, T. N., Alladi, P. A., & Raju, T. R. (2016). Astroglia acquires a toxic neuroinflammatory role in response to the cerebrospinal fluid from amyotrophic lateral sclerosis patients. *Journal of Neuroinflammation, 13*(1), 212. https://doi.org/10.1186/s12974-016-0698-0. PMID: 27578023; PMCID: PMC5006495.

Mochel, F., Charles, P., Seguin, F., Barritault, J., Coussieu, C., Perin, L., Le Bouc, Y., Gervais, C., Carcelain, G., Vassault, A., Feingold, J., Rabier, D., & Durr, A. (2007). Early energy deficit in Huntington disease: identification of a plasma biomarker traceable during disease progression. *PLoS One, 2*(7), e647. https://doi.org/10.1371/journal.pone.0000647. PMID: 17653274; PMCID: PMC1919424.

Niemelä, V., Burman, J., Blennow, K., Zetterberg, H., Larsson, A., & Sundblom, J. (2018). Cerebrospinal fluid sCD27 levels indicate active T cell-mediated inflammation in premanifest Huntington's disease. *PLoS One, 13*(2), e0193492. https://doi.org/10.1371/journal.pone.0193492. PMID: 29474427; PMCID: PMC5825143.

Nordengen, K., Kirsebom, B. E., Henjum, K., Selnes, P., Gísladóttir, B., Wettergreen, M., Torsetnes, S. B., Grøntvedt, G. R., Waterloo, K. K., Aarsland, D., Nilsson, L. N. G., & Fladby, T. (2019). Glial activation and inflammation along the Alzheimer's disease continuum. *Journal of Neuroinflammation, 16*(1), 46. https://doi.org/10.1186/s12974-019-1399-2. PMID: 30791945; PMCID: PMC6383268.

Nguyen, H. P., Björkqvist, M., Bode, F. J., Stephan, M., & von Hörsten, S. (2010). Serum levels of a subset of cytokines show high interindividual variability and are not altered in rats transgenic for Huntington's disease. *PLoS Current, 2*, RRN1190. https://doi.org/10.1371/currents.RRN1190. PMID: 20981129; PMCID: PMC2964049.

O'Regan, G. C., Farag, S. H., Ostroff, G. R., Tabrizi, S. J., & Andre, R. (2020). Wild-type huntingtin regulates human macrophage function. *Scientific Reports, 10*(1), 17269. https://doi.org/10.1038/s41598-020-74042-8. PMID: 33057179; PMCID: PMC7560844.

O'Regan, G. C., Farag, S. H., Casey, C. S., Wood-Kaczmar, A., Pocock, J. M., Tabrizi, S. J., & Andre, R. (2021). Human Huntington's disease pluripotent stem cell-derived microglia develop normally but are abnormally hyper-reactive and release elevated levels of reactive oxygen species. *Journal of Neuroinflammation, 18*(1), 94. https://doi.org/10.1186/s12974-021-02147-6. PMID: 33874957; PMCID: PMC8054367.

Pavese, N., Gerhard, A., Tai, Y. F., Ho, A. K., Turkheimer, F., Barker, R. A., Brooks, D. J., & Piccini, P. (2006). Microglial activation correlates with severity in Huntington disease: a clinical and PET study. *Neurology, 66*(11), 1638–1643. https://doi.org/10.1212/01.wnl.0000222734.56412.17. PMID: 16769933.

Pido-Lopez, J., Andre, R., Benjamin, A. C., Ali, N., Farag, S., Tabrizi, S. J., & Bates, G. P. (2018). In vivo neutralization of the protagonist role of macrophages during the chronic inflammatory stage of Huntington's disease. *Scientific Reports, 8*(1), 11447. https://doi.org/10.1038/s41598-018-29792-x. PMID: 30061661; PMCID: PMC6065433.

Pido-Lopez, J., Tanudjojo, B., Farag, S., Bondulich, M. K., Andre, R., Tabrizi, S. J., & Bates, G. P. (2019). Inhibition of tumour necrosis factor alpha in the R6/2 mouse model of Huntington's disease by etanercept treatment. *Scientific Reports, 9*(1), 7202. https://doi.org/10.1038/s41598-019-43627-3. PMID: 31076648; PMCID: PMC6510744.

Podlacha, M., Pierzynowska, K., Gaffke, L., Jerzemowska, G., Piotrowska, E., & Węgrzyn, G. (2022). Behavioral- and blood-based biomarkers for Huntington's disease: studies on the R6/1 mouse model with prospects for early diagnosis and monitoring of the disease. *Brain Behavior & Immunology Health, 23*, 100482. https://doi.org/10.1016/j.bbih.2022.100482. PMID: 35799674; PMCID: PMC9253406.

Politis, M., Pavese, N., Tai, Y. F., Kiferle, L., Mason, S. L., Brooks, D. J., Tabrizi, S. J., Barker, R. A., & Piccini, P. (2011 Feb). Microglial activation in regions related to cognitive function predicts disease onset in Huntington's disease: a multimodal imaging study. *Human Brain Mapping, 32*(2), 258–270. https://doi.org/10.1002/hbm.21008

Politis, M., Lahiri, N., Niccolini, F., Su, P., Wu, K., Giannetti, P., Scahill, R. I., Turkheimer, F. E., Tabrizi, S. J., & Piccini, P. (2015). Increased central microglial activation associated with peripheral cytokine levels in premanifest Huntington's disease gene carriers. *Neurobiology of Disease, 83*, 115–121. https://doi.org/10.1016/j.nbd.2015.08.011. Epub 2015 Aug 18. PMID: 26297319.

Quigley, E. M. M. (2017). Microbiota-brain-gut axis and neurodegenerative diseases. *Current Neurology and Neuroscience Reports, 17*(12), 94. https://doi.org/10.1007/s11910-017-0802-6. PMID: 29039142.

Ray, S., Britschgi, M., Herbert, C., Takeda-Uchimura, Y., Boxer, A., Blennow, K., Friedman, L. F., Galasko, D. R., Jutel, M., Karydas, A., Kaye, J. A., Leszek, J., Miller, B. L., Minthon, L., Quinn, J. F., Rabinovici, G. D., Robinson, W. H., Sabbagh, M. N., So, Y. T., Sparks, D. L., Tabaton, M., Tinklenberg, J., Yesavage, J. A., Tibshirani, R., & Wyss-Coray, T. (2007). Classification and prediction of clinical Alzheimer's diagnosis based on plasma signaling proteins. *Nature Medicine, 13*(11), 1359–1362. https://doi.org/10.1038/nm1653. Epub 2007 Oct 14. PMID: 17934472.

Raper, J., Bosinger, S., Johnson, Z., Tharp, G., Moran, S. P., & Chan, A. W. S. (2016). Increased irritability, anxiety, and immune reactivity in transgenic Huntington's disease monkeys. *Brain Behavior & Immunology, 58*, 181–190. https://doi.org/10.1016/j.bbi.2016.07.004. Epub 2016 Jul 7. PMID: 27395434; PMCID: PMC5067193.

Rodrigues, F. B., Byrne, L. M., McColgan, P., Robertson, N., Tabrizi, S. J., Zetterberg, H., & Wild, E. J. (2016). Cerebrospinal fluid inflammatory biomarkers reflect clinical severity in Huntington's disease. *PLoS One, 11*(9), e0163479. https://doi.org/10.1371/journal.pone.0163479. PMID: 27657730; PMCID: PMC5033331.

Runne, H., Kuhn, A., Wild, E. J., Pratyaksha, W., Kristiansen, M., Isaacs, J. D., Régulier, E., Delorenzi, M., Tabrizi, S. J., & Luthi-Carter, R. (2007). Analysis of potential transcriptomic biomarkers for Huntington's disease in peripheral blood. *Proceedings of the National Academy of Sciences of the United States of America, 104*(36), 14424–14429. https://doi.org/10.1073/pnas.0703652104. Epub 2007 Aug 27. PMID: 17724341; PMCID: PMC1964868.

Sánchez-López, F., Tasset, I., Agüera, E., Feijóo, M., Fernández-Bolaños, R., Sánchez, F. M., Ruiz, M. C., Cruz, A. H., Gascón, F., & Túnez, I. (2012). Oxidative stress and inflammation biomarkers in the blood of patients with Huntington's disease. *Neurological Research, 34*(7), 721–724. https://doi.org/10.1179/1743132812Y.0000000073. PMID: 22889672.

Sapp E, Kegel KB, Aronin N, Hashikawa T, Uchiyama Y, Tohyama K, Bhide PG, Vonsattel JP, DiFiglia M. Early and progressive accumulation of reactive microglia in the Huntington disease brain. Journal of Neuropathology and Experimental Neurology 2001;60(2):161–172. doi: https://doi.org/10.1093/jnen/60.2.161. PMID: 11273004

Sathasivam, K., Hobbs, C., Turmaine, M., Mangiarini, L., Mahal, A., Bertaux, F., Wanker, E. E., Doherty, P., Davies, S. W., & Bates, G. P. (1999). Formation of polyglutamine inclusions in non-CNS tissue. *Human Molecular Genetics, 8*(5), 813–822. https://doi.org/10.1093/hmg/8.5.813. PMID: 10196370.

Seizer, L., Cornélissen-Guillaume, G., Schiepek, G. K., Chamson, E., Bliem, H. R., & Schubert, C. (2022). About-weekly pattern in the dynamic complexity of a healthy subject's cellular immune activity: a biopsychosocial analysis. *Frontiers in Psychiatry, 13*, 799214. https://doi.org/10.3389/fpsyt.2022.799214

Punt, J., Stranford, S., Jones, P., & Owen, J. A. (2019). *Kuby immunology* (8th ed.). Macmillan Education.

Shen, X. N., Niu, L. D., Wang, Y. J., Cao, X. P., Liu, Q., Tan, L., Zhang, C., & Yu, J. T. (2019). Inflammatory markers in Alzheimer's disease and mild cognitive impairment: a meta-analysis and systematic review of 170 studies. *Journal of Neurology, Neurosurgery, and Psychiatry, 90*(5), 590–598. https://doi.org/10.1136/jnnp-2018-319148. Epub 2019 Jan 10. PMID: 30630955.

Silajdžić, E., Rezeli, M., Végvári, Á., Lahiri, N., Andre, R., Magnusson-Lind, A., Nambron, R., Kalliolia, E., Marko-Varga, G., Warner, T. T., Laurell, T., Tabrizi, S. J., & Björkqvist, M. (2013). A critical evaluation of inflammatory markers in Huntington's disease plasma. *J Huntingtons Dis, 2*(1), 125–134. https://doi.org/10.3233/JHD-130049. PMID: 25063434.

Silvestroni, A., Faull, R. L., Strand, A. D., & Möller, T. (2009). Distinct neuroinflammatory profile in post-mortem human Huntington's disease. *NeuroReport, 20*(12), 1098–1103. https://doi.org/10.1097/WNR.0b013e32832e34ee. PMID: 19590393.

Singhrao, S. K., Neal, J. W., Morgan, B. P., & Gasque, P. (1999). Increased complement biosynthesis by microglia and complement activation on neurons in Huntington's disease. *Experimental Neurology, 159*(2), 362–376. https://doi.org/10.1006/exnr.1999.7170. PMID: 10506508.

Song, W. M., & Colonna, M. (2018). The identity and function of microglia in neurodegeneration. *Nature Immunology, 19*(10), 1048–1058. https://doi.org/10.1038/s41590-018-0212-1. Epub 2018 Sep 24. PMID: 30250185.

Stephenson, J., Nutma, E., van der Valk, P., & Amor, S. (2018). Inflammation in CNS neurodegenerative diseases. *Immunology, 154*(2), 204–219. https://doi.org/10.1111/imm.12922. Epub 2018 Apr 17. PMID: 29513402; PMCID: PMC5980185.

Stoy, N., Mackay, G. M., Forrest, C. M., Christofides, J., Egerton, M., Stone, T. W., & Darlington, L. G. (2005). Tryptophan metabolism and oxidative stress in patients with Huntington's disease. *Journal of Neurochemistry, 93*(3), 611–623. https://doi.org/10.1111/j.1471-4159.2005.03070.x. PMID: 15836620.

Suescun, J., Chandra, S., & Schiess, M. C. (2019). Chapter 13—The role of neuroinflammation in neurodegenerative disorders. In J.K. Actor & K.C. Smith (Eds.), *Translational inflammation* (pp. 241–267). Academic Press. Perspectives in Translational Cell Biology.

Tabrizi, S. J., Flower, M. D., Ross, C. A., & Wild, E. J. (2020). Huntington disease: new insights into molecular pathogenesis and therapeutic opportunities. *Nature Reviews. Neurology, 16*(10), 529–546. https://doi.org/10.1038/s41582-020-0389-4. Epub 2020 Aug 14. PMID: 32796930.

Tabrizi, S. J., Estevez-Fraga, C., van Roon-Mom, W. M. C., Flower, M. D., Scahill, R. I., Wild, E. J., Muñoz-Sanjuan, I., Sampaio, C., Rosser, A. E., & Leavitt, B. R. (2022). Potential disease-modifying therapies for Huntington's disease: lessons learned and future opportunities. *Lancet Neurology, 21*(7), 645–658. https://doi.org/10.1016/S1474-4422(22)00121-1. PMID: 35716694; PMCID: PMC7613206.

Tai, Y. F., Pavese, N., Gerhard, A., Tabrizi, S. J., Barker, R. A., Brooks, D. J., & Piccini, P. (2007a). Imaging microglial activation in Huntington's disease. *Brain Research Bulletin, 72*(2–3), 148–151. https://doi.org/10.1016/j.brainresbull.2006.10.029. Epub 2006 Nov 27. PMID: 17352938.

Tai, Y. F., Pavese, N., Gerhard, A., Tabrizi, S. J., Barker, R. A., Brooks, D. J., & Piccini, P. (2007b). Microglial activation in presymptomatic Huntington's disease gene carriers. *Brain, 130*(Pt 7), 1759–1766. https://doi.org/10.1093/brain/awm044. Epub 2007 Mar 30. PMID: 17400599.

Tansey, M. G., Wallings, R. L., Houser, M. C., Herrick, M. K., Keating, C. E., & Joers, V. (2022). Inflammation and immune dysfunction in Parkinson disease. *Nature Reviews. Immunology, 22*(11), 657–673. https://doi.org/10.1038/s41577-022-00684-6. Epub 2022 Mar 4. PMID: 35246670; PMCID: PMC8895080.

Träger, U., Andre, R., Lahiri, N., Magnusson-Lind, A., Weiss, A., Grueninger, S., McKinnon, C., Sirinathsinghji, E., Kahlon, S., Pfister, E. L., Moser, R., Hummerich, H., Antoniou, M., Bates, G. P., Luthi-Carter, R., Lowdell, M. W., Björkqvist, M., Ostroff, G. R., Aronin, N., & Tabrizi, S. J. (2014). HTT-lowering reverses Huntington's disease immune dysfunction caused by NFκB pathway dysregulation. *Brain, 137*(Pt 3), 819–833. https://doi.org/10.1093/brain/awt355. Epub 2014 Jan 22. PMID: 24459107; PMCID: PMC3983408.

Träger, U., Andre, R., Magnusson-Lind, A., Miller, J. R., Connolly, C., Weiss, A., Grueninger, S., Silajdžić, E., Smith, D. L., Leavitt, B. R., Bates, G. P., Björkqvist, M., & Tabrizi, S. J. (2015). Characterisation of immune cell function in fragment and full-length Huntington's disease mouse models. *Neurobiology of Disease, 73*, 388–398. https://doi.org/10.1016/j.nbd.2014.10.012. Epub 2014 Oct 29. PMID: 25447230; PMCID: PMC4262574.

Valadão, P. A. C., Oliveira, B. D. S., Joviano-Santos, J. V., Vieira, É. L. M., Rocha, N. P., Teixeira, A. L., Guatimosim, C., & de Miranda, A. S. (2019). Inflammatory changes in peripheral organs in the BACHD murine model of Huntington's disease. *Life Sciences, 232*, 116653. https://doi.org/10.1016/j.lfs.2019.116653. Epub 2019 Jul 11. PMID: 31302194.

Valekova, I., Jarkovska, K., Kotrcova, E., Bucci, J., Ellederova, Z., Juhas, S., Motlik, J., Gadher, S. J., & Kovarova, H. (2016). Revelation of the IFNα, IL-10, IL-8 and IL-1β as promising biomarkers reflecting immuno-pathological mechanisms in porcine Huntington's disease model. *Journal of Neuroimmunology, 293*, 71–81. https://doi.org/10.1016/j.jneuroim.2016.02.012. Epub 2016 Feb 20. PMID: 27049565.

van der Burg, J. M., Björkqvist, M., & Brundin, P. (2009). Beyond the brain: widespread pathology in Huntington's disease. *Lancet Neurology, 8*(8), 765–774. https://doi.org/10.1016/S1474-4422(09)70178-4. PMID: 19608102.

Villegas-Llerena, C., Phillips, A., Garcia-Reitboeck, P., Hardy, J., & Pocock, J. M. (2016). Microglial genes regulating neuroinflammation in the progression of Alzheimer's disease. *Current Opinion in Neurobiology, 36*, 74–81. https://doi.org/10.1016/j.conb.2015.10.004. Epub 2015 Oct 24. PMID: 26517285.

Wang, R., Ross, C. A., Cai, H., Cong, W. N., Daimon, C. M., Carlson, O. D., Egan, J. M., Siddiqui, S., Maudsley, S., & Martin, B. (2014). Metabolic and hormonal signatures in pre-manifest and manifest Huntington's disease patients. *Frontiers in Physiology, 5*, 231. https://doi.org/10.3389/fphys.2014.00231. PMID: 25002850; PMCID: PMC4066441.

Wei, J., Xu, H., Davies, J. L., & Hemmings, G. P. (1992). Increase of plasma IL-6 concentration with age in healthy subjects. *Life Sciences, 51*(25), 1953–1956. https://doi.org/10.1016/0024-3205(92)90112-3. PMID: 1453878.

Wegrzynowicz, M., Bichell, T. J., Soares, B. D., Loth, M. K., McGlothan, J. S., Mori, S., Alikhan, F. S., Hua, K., Coughlin, J. M., Holt, H. K., Jetter, C. S., Pomper, M. G., Osmand, A. P., Guilarte, T. R., & Bowman, A. B. (2015). Novel BAC mouse model of Huntington's disease with 225 CAG repeats exhibits an early widespread and stable degenerative phenotype. *Journal of Huntingtons Disease, 4*(1), 17–36. PMID: 26333255; PMCID: PMC4657874.

Wertz, M. H., Pineda, S. S., Lee, H., Kulicke, R., Kellis, M., & Heiman, M. (2020). Interleukin-6 deficiency exacerbates Huntington's disease model phenotypes. *Molecular Neurodegeneration, 15*(1), 29. https://doi.org/10.1186/s13024-020-00379-3. PMID: 32448329; PMCID: PMC7247164.

Widner, B., Leblhuber, F., Walli, J., Tilz, G. P., Demel, U., & Fuchs, D. (1999). Degradation of tryptophan in neurodegenerative disorders. *Advances in Experimental Medicine and Biology, 467*, 133–138. https://doi.org/10.1007/978-1-4615-4709-9_19. PMID: 10721050.

Wild, E., Magnusson, A., Lahiri, N., Krus, U., Orth, M., Tabrizi, S. J., & Björkqvist, M. (2011). Abnormal peripheral chemokine profile in Huntington's disease. *PLoS Current, 3*, RRN1231. https://doi.org/10.1371/currents.RRN1231. PMID: 21826115; PMCID: PMC3082446.

Astrocytes in Huntington's Disease Pathology: Implications for Biomarkers

Huajing You and Zhong Pei

Abbreviations

AA	Ascorbic acid
Aβ-PET	amyloid- positron emission tomography
AD	Alzheimer's disease
AHI	Apnea-hypopnea index
ALS	Amyotrophic lateral sclerosis
ALSFRS-R	ALS Functional Rating Scale
ApoE	Apolipoprotein E
BDNF	Brain-derived neurotrophic factor
CI Aβ+	Amyloid-β-positive cognitively impaired
CI Aβ-	Amyloid-β-negative cognitively impaired
CNS	Central nervous system
CNTF	Ciliary neurotrophic factor
CSF	Cerebrospinal fluid
CU Aβ+	Amyloid-β-positive cognitively unimpaired
CU Aβ-	Amyloid-β-negative cognitively unimpaired
EAAT1 and 2	Excitatory amino acid transporter genes 1 and 2
ECAS	Edinburgh Cognitive and Behavioral ALS Screen
EMA	Eye movement abnormalities
ERK1/2	Extracellular signal-regulated kinase ½
GAT-3	GABA transporter-3
GFAP	Glial Fibrillary Acidic Protein
GS	Glutamine synthetase
HACE1	HECT domain and Ankyrin repeat containing E3 ubiquitin protein ligase 1
HC	Healthy controls
HD	Huntington's disease

H. You · Z. Pei (✉)
Department of Neurology, The First Affiliated Hospital, Sun Yat-sen University, Guangdong Provincial Key Laboratory of Diagnosis and Treatment of Major Neurological Diseases, National Key Clinical Department and Key Discipline of Neurology, Guangzhou, China
e-mail: peizhong@mail.sysu.edu.cn

E. A. Thomas, G. M. Parkin (eds.), *Biomarkers for Huntington's Disease*,
Contemporary Clinical Neuroscience, https://doi.org/10.1007/978-3-031-32815-2_12

IF	Intermediate filament
IFNγ	Interferon gamma
iPSCs	Induced pluripotent stem cells
IL	Interleukins
IKK	Inhibitor of κB kinase
JAK/STAT	Janus Kinase/Signal Transducer and Activator of Transcription
JNK	c-jun N-terminal kinase
Kir4.1	Inward rectifier potassium channel 4.1
LMN	Lower motor neuron
LPS	Lipopolysaccharide
MAPK	Mitogen-Activated Protein Kinase
mHTT	Mutant huntingtin protein
MMSE	Mini-Mental State Examination
MSA	Multiple system atrophy
MSN	Medium spiny neuron
NfL	Neurofilament light
NF-κB	Nuclear Factor of Kappa light polypeptide gene enhancer in B-cells
NGF	Nerve growth factor
ODI	Oxygen desaturation index
PD	Parkinson's disease
PreHD	premanifest Huntington disease
PRR	Pattern recognition receptors
PSP	Progressive supranuclear palsy
ROS	Reactive oxygen species
SCA3	Spinocerebellar ataxia type 3
SER	STAT responsive element
SOCS3	Suppressor of cytokine signaling 3
TFC	Total Functional Capacity
TMS	Total Motor Score
TNFα	Tumor necrosis factor alpha
UHDRS	Unified Huntington's Disease Rating Scale
UMN	Upper motor neuron
UPS	Ubiquitin-proteasome system

1 Introduction

Astrocytes are the housekeepers of the brain. Under normal conditions, they maintain brain function by regulating the maturation of synapses, neurotransmitter homeostasis, water and ion homeostasis, neurovascular coupling, and the formation of the blood-brain barrier (Palpagama et al., 2019). Pathologically, astrocytes can become reactive and lose their normal function, which leads to metabolic disturbances, such as increased release of gliotransmitters, as well as inadequate synthesis of trophic factors and antioxidants (Palpagama et al., 2019). In addition, astrocytes can lose their ability to clear aggregated proteins, which has implications for many neurodegenerative diseases. The concept of astrocyte reactivity emerged with the discovery and immunohistological staining of the intermediate filament (IF) protein Glial fibrillary acidic protein (GFAP) (Eng et al., 1971, 2000). GFAP is one of the best-known markers for the activation of astrocytes in neurodegenerative diseases. However, it is important to note that other IF proteins, such as vimentin and nestin, are also up-regulated by reactive astrocytes (Potokar et al., 2020).

Astrocytes are fundamental for the maintenance of neural circuits, but their role in Huntington's disease (HD) is still unclear. HD is caused by an abnormal expansion of a CAG repeat region in the Huntingtin (*HTT*) gene and is characterized by progressively deteriorative motor, psychiatric, and cognitive function. The huntingtin protein (HTT) in its mutant form (mHTT) is known to aggregate in different cell types in the brain, causing disruption of cortico-striatal-thalamo-cortical neural circuits (Eidelberg & Surmeier, 2011). Interestingly, mHTT protein is also present in cortical and striatal astrocytes (Shin et al., 2005; Palpagama et al., 2019) and previous studies have suggested that dysfunction of astrocytes, possibly via dysregulation of mHTT in these cells, contributes to HD pathology (Stanek et al., 2019; Faideau et al., 2010). For example, in a YAC128 mouse model, *HTT* lowering therapy in both neurons and astrocytes was required to achieve maximal functional benefit (Stanek et al., 2019). Moreover, Faideau et al. found that lentiviral-mediated expression of mHTT, specifically in striatal astrocytes, increased GFAP expression and induced cellular hypertrophy (Faideau et al., 2010). In this chapter, we will provide an overview of astrocytic extracellular signaling and activation of signaling cascades under normal conditions, as well as the metabolic disturbances and other cellular dysfunction of astrocytes that occur during pathological conditions. We also describe the potential for astrocytic markers to serve as biomarkers of HD.

2 Identification of Extracellular Signals by Astrocytes

Under normal conditions, resting astrocytes are exposed to neuronally-secreted cytokines, such as interleukins (IL), tumor necrosis factor alpha (TNFα) and interferon gamma (IFNγ), which can induce transcriptional and functional changes in the astrocytes (Sofroniew & Vinters, 2010). Astrocytes are equipped with many receptors, such as pattern recognition receptors (PRR), and intracellular signaling cascades to respond quickly to alterations in brain homeostasis. Accordingly, evidence suggests that glial cells (both astrocytes and microglia) can detect mild neuronal dysfunction (Buffo et al., 2010). PRR can detect abnormal signals, including bacteria and aggregated proteins in the extracellular space and can also recognize increases in cytokines, chemokines and purines from neighboring cells, even the absence of "normal" signals, including growth factors and neurotransmitters (Buffo et al., 2010; Burda & Sofroniew, 2014). Once activated by the above signals, astrocytes transition into a reactive state, subsequently releasing active molecules, including purines, pro-inflammatory cytokines and growth factors, to set up a reactive micro-environment (Buffo et al., 2010).

Reactive astrocytes display an enlarged cell body (hypertrophy) and prolonged processes, which polarize towards the injured site. Most central nervous system (CNS) astrocytes express GFAP at undetectable levels under physiological conditions. Upon injury or disease, GFAP expression in reactive astrocytes is up-regulated. This increase in GFAPGlial fibrillary acidic protein (GFAP) expression does not appear to be due to an increase in astrocyte proliferation (Ben Haim et al., 2015a). A possible explanation for increased GFAP-positive astrocytes in immunostaining is that cortical atrophy contributes to the perception of increased glia density (Serrano-Pozo et al., 2013).

3 Activated Signaling Cascades in HD

In HD, molecular triggers of reactive astrocyte conversion bind to their cognate receptors at the astrocyte membrane and activate several intracellular signaling cascades. These signaling cascades include the Janus Kinase/Signal Transducer and Activator of Transcription (JAK/STAT) pathway, the Nuclear Factor of Kappa light polypeptide gene enhancer in B-cells (NF-κB) pathway and the Mitogen-Activated Protein Kinase (MAPK) pathway (see Fig. 1) (Ben Haim et al., 2015a, b).

The JAK/STAT3 pathway is a ubiquitous cascade that predominantly mediates cytokine signaling in cells. It regulates the expression of target genes to participate in cell growth, proliferation, differentiation and inflammation. STAT3 has been shown to accumulate in the nucleus of reactive astrocytes in the striatum of murine and primate models of HD (Ben Haim et al., 2015a, b). Ben Haim et al. used lentiviral vectors to over-express suppressor of cytokine signaling 3 (SOCS3), the endogenous inhibitor of the JAK/STAT3 pathway, in two transgenic mouse models of Alzheimer's disease and in a mouse and a nonhuman primate lentiviral vector-based model HD. Selective SOCS3 over-expression in astrocytes prevented the nuclear accumulation of STAT3, as well as GFAP up-regulation. Furthermore, SOCS3-expressing astrocytes displayed a resting morphology, indicating that the JAK/STAT3 pathway is responsible for astrocyte reactivity (Ben Haim et al., 2015a, b).

The NF-κB pathway is another pathway associated with immune responses, inflammation, cell division and apoptosis. This pathway is found to be predominantly active in microglial cells and also active in many cell types in addition to reactive astrocytes (Fig. 1) (Ben Haim et al. 2015a; Frakes et al., 2014). It is activated by several known pro-inflammatory agents (e.g., lipopolysaccharide [LPS], interleukin-1β [IL-1β], TNFα) (Carrero et al., 2012). A previous study has shown that NF-κB accumulates in the nuclei of astrocytes in the R6/2 model of HD (Hsiao et al., 2013). However, another study showed that inhibiting NF-κB signaling selectively in astrocytes only transiently attenuated their reactivity at the onset of amyotrophic lateral sclerosis (ALS) (Crosio et al., 2011), suggesting that, while the

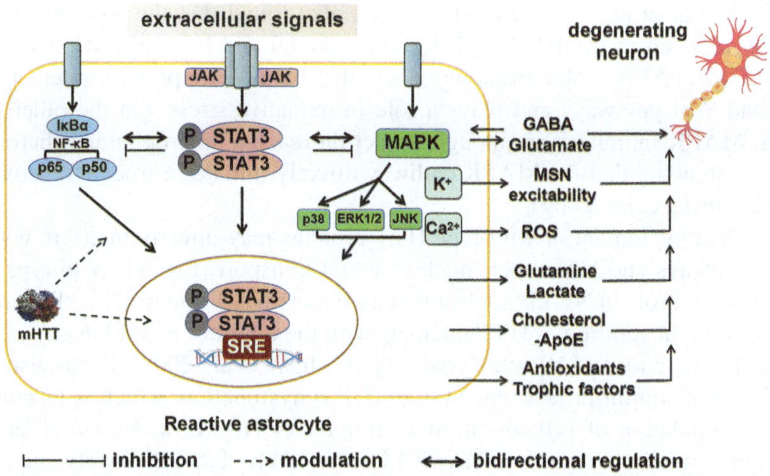

Fig. 1 Intracellular signaling cascades and the secretome of activated astrocytes in HD, which influence neuronal survival and synaptic activity. Many extracellular signals, such as cytokines, chemokines, purines, mHTT and neural dysfunction, can trigger astrocyte reactivity. In the **JAK/STAT3** pathway, cytokine binding results in activation of JAK and subsequent recruitment of STAT3 to the receptor. JAK phosphorylates STAT3, which dimerizes and translocates to the nucleus. Then STAT3 binds to consensus sequences (STAT responsive element, SRE) in the promoter region of its target genes. Activation of the **NF-κB** pathway leads to the phosphorylation and polyubiquitination of IκBα, the master inhibitor of NF-κB. The phosphorylated and ubiquitinated IκBα is then targeted by the proteasome for degradation. The NF-κB subunits p50 and p65 translocate to the nucleus to activate the transcription of target genes. Activation of the **MAPK** pathway results in a phosphorylation cascade. Its downstream effectors ERK1/2, p38 and c-jun regulate gene transcription through transcription factors. These cascades eventually converge on the JAK/STAT3 pathway, triggering a transcriptional program to activate astrocytes. Activated astrocytes also secrete many active molecules to influence neuronal survival and synaptic activity through neuron-astrocyte communication. In HD, reactive astrocytes secrete more antioxidants, fewer trophic factors, cholesterol or glutamine and produce more ROS than resting astrocytes. Dysregulation of glutamate and GABA also occurs. Decreased Kir4.1 functional expression leads to increased extracellular K+ and MSN excitability. Dysregulated intracellular Ca^{2+} level stimulates the release of gliotransmitters such as glutamate and ATP

NF-κB pathway plays a key role in microglial activation, it doesn't seem essential to initiate astrocyte reactivity.

The MAPK pathway is activated by cellular stress, extracellular matrix proteins or cytokines. This pathway is mediated by the activation of small GTP-ase proteins (RAS) and the successive phosphorylation of MAP3K, MAP2K, and MAPK (Jeffrey et al., 2007). There are three main phosphorylation cascades, with p38, extracellular signal-regulated kinase 1/2 (ERK1/2) or c-jun N-terminal kinase (JNK) as downstream effectors. All cascades result in the activation of different transcription factors (Ben Haim et al., 2015a). The ERK pathway has been implicated in HD pathogenesis, with ERK1/2 activation shown to be dysregulated in YAC128 induced pluripotent stem cells (iPSCs) (Maher et al., 2011; Szlachcic et al., 2015). ERK up-regulation was also shown to be beneficial in mouse models of HD (Maher et al., 2011) Szlachcic et al. also found that the level of p53 was strongly reduced in the HD YAC128 mouse model and human iPSCs (Szlachcic et al., 2015). p53 is a pleiotropic molecule that binds HTT protein, interacts with ERK and Wnt pathways and plays a role in oxidative stress. On the other hand, several MAPK inhibitors are thought to act on reactive microglia and there is no evidence showing that the MAPK pathway directly initiates astrocyte reactivity in HD (Kaminska et al., 2009).

Interestingly, mutant or wild-type HTT proteins may directly interfere with the above pathways and affect their nuclear signal transport (Fig. 1). Wild-type HTT protein plays a role in NF-κB nuclear transport and mHTT impairs NF-κB signaling in astrocytes. In addition, mHTT interacts with the inhibitor of κB kinase (IKK) to inhibit IKK activity and NF-κB signaling (Khoshnan et al., 2004). HD is also associated with ubiquitin-proteasome system (UPS) dysfunction, which is responsible for the degradation of IκB, the master inhibitor of NF-κB, and toxic-N-terminal mHTT fragments (Ortega & Lucas, 2014). It is likely that UPS involvement may indirectly influence the activity of signaling cascades in astrocytes.

4 Metabolic Disturbance of Astrocytes in HD Pathogenesis

Astrocytes are responsible for most glutamate uptake at synapses through transporters encoded by excitatory amino acid transporter genes 1 and 2 (EAAT1 and 2, also called GLAST and GLT1 in rodents) (Danbolt, 2001). Inefficient glutamate uptake leads to over-stimulation of glutamate receptors, which can cause excitotoxic cell death in neurons. In one study, astrocytes expressing mHTT were shown to become reactive with remarkably decreased levels of EAAT1 and 2 expression and glutamate uptake (Maragakis & Rothstein, 2004). Another important study showed a significant decrease of EAAT2 mRNA and protein levels in the caudate and putamen of HD patients, which was dependent on disease stage, suggesting that mHTT in astrocytes alters glial glutamate transport capacity in the early stage of the disease and may contribute to HD pathogenesis (Faideau et al., 2010). Once taken up from the synaptic cleft by astrocytes, glutamate is metabolized into glutamine by Glutamine synthetase (GS). Glutamine is then transported back to neurons and used for the production of glutamate and gamma-aminobutyric acid (GABA) (Danbolt, 2001). GS mRNA levels have been shown to be lower in HD mouse models compared to their wild-type littermates (Boussicault et al., 2014). Therefore, alterations of the glutamate-glutamine cycle may directly contribute to neuronal dysfunction in HD.

The pathological state of HD leads to increased neuronal activity. According to the astrocyte-to-neuron lactate shuttle hypothesis, this increase in neuronal activity would require more glucose to be taken up by astrocytes, oxidized through glycolysis and then redistributed to neurons in the form of lactate (Pellerin & Magistretti, 1994). Interestingly, a study of neuron-astrocyte co-cultures showed that mHTT expression in astrocytes didn't affect their own rate of glucose uptake. It impaired glucose uptake in neurons, suggesting that HD astrocytes indirectly regulated neuronal glucose metabolism by diffusible factors (Boussicault et al., 2014). Indeed, the expression of mHTT doesn't directly affect metabolic activity in a cell autonomous manner. Co-culture of neurons with astrocytes from wild-type or BACHD mice indicated that mutant astrocytes affected neuron energy metabolism through adverse non-cell autonomous effects of mHTT (Boussicault et al., 2014).

In addition to glucose uptake, mitochondrial dysfunction also occurs in HD. Impaired respiratory activity and reactive oxygen species (ROS) production is found not only in neurons, but also in reactive astrocytes. In astrocytes, mitochondria undergo more fission events in the reactive compared to the resting state (Motori et al., 2013). Ehrnhoefer et al. found that mHTT expression and loss of HECT domain and Ankyrin repeat containing E3 ubiquitin protein ligase 1 (HACE1) caused mitochondrial dysfunction and impaired astrocyte proliferation in HACE1$^{-/-}$ x YAC128 mice, suggesting that HACE1 is essential for astrocyte mitochondrial function in vivo (Ehrnhoefer et al., 2018).

Astrocytes are also associated with the regulation of cholesterol levels in the brain. Cholesterol is the most common steroid in humans and contributes to synapse formation and neuronal activity in the brain. As part of this process, astrocytes express high levels of apolipoprotein E (ApoE) that carries cholesterol to neurons (Bu, 2009). mRNA levels for cholesterol biosynthesis and efflux were found to be lower in primary astrocytes from HD mice compared to control mice (Valenza et al., 2010). ApoE also formed smaller lipoprotein particles in the cerebrospinal fluid (CSF) of HD mice, suggesting reduced cholesterol transport via these lipoproteins (Valenza et al., 2010). Disruption of astrocyte-neuron cholesterol cross talk affects neuronal function, and the severity of impairment in cholesterol biosynthesis has been shown to correlate with the number of CAG repeats (Leoni & Caccia, 2014).

5 Other Cellular Dysfunction of Astrocytes in HD Pathogenesis

In addition to signaling cascades and metabolic disturbances, other cellular dysfunction, such as increased release of gliotransmitters, inadequate synthesis of trophic factors, altered production of antioxidants and insufficient mHTT clearance, can also occur in astrocytes and play a significant role in HD pathogenesis (Ben Haim et al., 2015a). For example, another function of astrocytes is to buffer potassium (K+) levels by specific channels and transporters, which are enriched in perisynaptic processes and vascular endfeet (Ben Haim et al., 2015a). The maintenance of K+ homeostasis by astrocytes is essential for synaptic transmission. In HD mouse models, decreased inward rectifier potassium channel 4.1 (Kir4.1) functional expression has been shown to result in increased levels of striatal extracellular K+ in vivo and increased medium spiny neuron (MSN) excitability in vitro. Viral delivery of Kir4.1 channels to striatal astrocytes restored Kir4.1 function, prolonged survival and attenuated some motor phenotypes of R6/2 transgenic mice (Tong et al., 2014).

The cellular processes related to the release of gliotransmitters and their downstream physiological function has been studied extensively. Glutamate is one of the best studied gliotransmitters. As mentioned above, astrocytes are responsible for most glutamate uptake at synapses, as well as extracellular glutamate regulation (Danbolt, 2001). mHTT aggregation in astrocytes leads to decreased glutamate uptake and modified glutamate signaling (Maragakis & Rothstein, 2004). Glutamate-mediated excitotoxicity is thought to cause selective neuronal injury. Increased glutamate release through Ca^{2+}-dependent exocytosis was found in astrocytes from BACHD mice. Increased levels of the astrocyte-specific, mitochondrial enzyme pyruvate carboxylase, resulting in increased de novo glutamate synthesis, and unchanged levels of astrocyte-specific enzyme GS, resulting in no change in the conversion rate of glutamate to glutamine, may help explain the increased availability of cytosolic glutamate in BACHD astrocytes (Lee et al., 2013). Dysregulation of glutamate plays a key role in the breakdown of cortico-striatal communication. This process is regulated by ascorbate, which is found in high levels in the striatum. Up-regulation of glutamate transport by drug administration or viral-vector delivery was shown to improve ascorbate homeostasis and neurobehavioral processing in HD mice (Rebec, 2018).

GABA is another highly studied gliotransmitter in HD. Electrophysiological recordings on cortical slices have shown that both GABAA-subunit containing (postsynaptic) and GABAB-subunit containing (presynaptic) currents are lower in R6/2 and zQ175 HD mice compared to control mice, resulting in a lower GABA-mediated tonic inhibition of striatal neurons. Pharmacological manipulation of the GABA transporter-3 (GAT-3), which is preferentially expressed by astrocytes, indicated that HD astrocytes have an impaired capacity to release GABA through GAT-3 (Wójtowicz et al., 2013).

Astrocytes secrete various trophic factors to neurons, such as growth factors (e.g., ciliary neurotrophic factor [CNTF], brain-derived neurotrophic factor [BDNF], and nerve growth factor [NGF]), neurosteroids and adhesion molecules involved in neurite outgrowth. Inadequate synthesis and release of such factors may contribute to neuronal toxicity in HD (Ben Haim et al., 2015a). For example, mHTT accumulation has been shown to modify exosome and BDNF release from astrocytes. mHTT expression in primary cultures of cortical astrocytes impaired BDNF production in those astrocytes, resulting in low levels of mature BDNF in the medium, thereby limiting neurite development of primary cortical neurons. Restoration of BDNF release from astrocytes subsequently displayed neuroprotective effects (Hong et al., 2016; Wang et al., 2012).

Astrocytes also play an important role in the defense against ROS by expressing detoxifying enzymes and transporters (Allaman et al., 2011). They produce high levels of antioxidants for neurons, including ascorbic acid (AA, also known as vitamin C), glutathione and its precursors. The release of AA by astrocytes has been shown to be altered in HD. Extracellular AA levels were reported to be lower in R6/2 mice than in age-matched wild-type mice during behavioral activation (Rebec et al., 2002).

The soluble form of mHTT is considered to be the most toxic form; it has been proposed that its aggregation may be a protective cellular mechanism preventing mHTT from interfering with intracellular process (Ross & Poirier, 2004). mHTT is mainly degraded by autophagy and the UPS, but these two major intracellular pathways are altered in HD. mHTT aggregation is mainly found in neurons, suggesting that astrocytes are more efficient at handling toxic protein aggregates via autophagy and ubiquitination, compared to neurons (Jansen et al., 2014). It was found that blocking astrocyte reactivity by over-expressing SOCS3 significantly promoted the formation of mHTT aggregation in the mouse striatum (Ben Haim et al., 2015a, b). Pearce et al. also reported that phagocytic glia regulate steady-state numbers of HTT aggregates in neurons through a clearance mechanism. Following this, some of the engulfed mHTT aggregates displayed prion-like conversion of soluble, wild-type HTT in the glial cytoplasm (Pearce et al., 2015). These results suggest that reactive astrocytes may regulate the synthesis and aggregation levels of mHTT in neurons, but the exact mechanism needs further exploration.

6 GFAP as a Potential Biomarker in Neurodegenerative Diseases

Astrocyte reactivity is an early feature of HD progression. Positive GFAP immunostaining has been detected in the striatum of pre-symptomatic HD carriers and increases with disease progression (Faideau et al., 2010). In one HD clinical study, plasma GFAP levels were significantly increased in Chinese HD mutation carriers, compared to healthy controls (You et al., 2021). Further, plasma GFAP levels were linearly correlated with levels of plasma neurofilament light (NfL), a marker for degenerating neurons. Importantly, plasma GFAP concentrations were also significantly correlated with disease burden, Total Motor Score (TMS) and Total Functional Capacity (TFC) (You et al., 2021). As an emerging biomarker, CSF and blood GFAP levels have been identified to change significantly with disease progression in several other neurodegenerative diseases (discussed in detail below and Table 1); however, its role as a biomarker is controversial.

GFAP has most commonly been evaluated as a biomarker in Alzheimer's disease (AD). Pereira et al. found that the concentration of plasma GFAP was significantly, positively correlated with amyloid-β-positive pathology as determined by high amyloid-β positron emission tomography (PET) imaging signals in AD. Plasma GFAP was also associated with both longitudinal amyloid-β-PET signals and cognitive decline, which indicated that plasma GFAP might be an early and easily accessible marker associated with brain amyloid-β pathology (Pereira et al., 2021). In an observational, cross-sectional, multi-center study, plasma GFAP levels were significantly higher in individuals with preclinical AD in comparison with cognitively

Table 1 GFAP as a biomarker in neurodegenerative diseases

Neurodegenerative Diseases	Cohort	Source	Clinical measures	Relation
HD	15 preHD, 42 HD, 26 HC (You et al., 2021)	Plasma	UHDRS-TMS UHDRS-TFC	Positive
AD	71 CU Aβ+, 78 CI Aβ+, 63 CI Aβ-, 75 non-AD, 217 CU Aβ- (Pereira et al., 2021)	Plasma, CSF	Aβ-PET burden	Positive
			MMSE	Negative
ALS	114 ALS, 38 HC (Verde et al., 2023)	Serum	ALSFRS-R	No significance
			EMA	No significance
			ECAS memory score	Negative
			AHI, ODI	Positive
PD	1065 PD (Xie et al., 2021)	CSF	NA	No significance
MSA	322 MSA (Xie et al., 2021)	CSF	NA	No significance
SCA3	143 SCA3, 172 HC (Garcia-Moreno et al., 2022)	Plasma	NA	No significance

HD Huntington Disease, *preHD* premanifest Huntington disease, *HC* healthy controls, *UHDRS-TMS* Unified Huntington's Disease Rating Scale-Total Motor Score, *UHDRS-TFC* Unified Huntington's Disease Rating Scale-Total Functional Capacity, *AD* Alzheimer disease, *CU Aβ+* amyloid-β-positive cognitively unimpaired, *CI Aβ+* amyloid-β-positive cognitively impaired, *CI Aβ-* amyloid-β-negative cognitively impaired, *non-AD* non-Alzheimer's disease, *CU Aβ-* amyloid-β-negative cognitively unimpaired, *CSF* cerebrospinal fluid, *Aβ-PET* amyloid--PET, *MMSE* Mini-Mental State Examination, *ALS* amyotrophic lateral sclerosis, *ALSFRS-R* ALS Functional Rating Scale, *EMA* eye movement abnormalities, *ECAS* Edinburgh Cognitive and Behavioral ALS Screen, *AHI* apnea-hypopnea index, *ODI* oxygen desaturation index, *PD* Parkinson's Disease, *MSA* multiple system atrophy, *SCA3* spinocerebral ataxia type 3, *NA* none available

unimpaired Aβ-negative individuals. Plasma GFAP levels were also higher among individuals in symptomatic stages of the AD continuum. Combining plasma GFAP with other plasma biomarkers (either p-tau181 or NfL) was associated with improved accuracy to discriminate Aβ status (Benedet et al., 2021). In these two studies, plasma GFAP performed better than CSF GFAP. One possible explanation is that plasma GFAP is very stable to freeze-thaw cycles, whereas CSF GFAP is far more sensitive to deterioration over time (Abdelhak et al., 2019). Another possible explanation is that astrocytes are part of the glymphatic system, which is highly organized and clears insoluble proteins and metabolites of the brain by draining them into the venous system (Iliff et al., 2012). Therefore, plasma GFAP may better reflect the pathophysiological change in the body and is more suitable to be a potential biomarker.

Regarding other neurodegenerative diseases, both multiple system atrophy (MSA) (putamen > substantia nigra > caudate > frontal cortex) and progressive supranuclear palsy (PSP) (substantia nigra > caudate > putamen, frontal cortex) showed widespread but regionally specific patterns of increased immunoreactivity of GFAP (Tong et al., 2015). In ALS, serum GFAP was shown to be correlated with cognitive scores associated with ALS-non-specific functions, polysomnographic indices of oxygen desaturation, apnea-hypopnea and survival. However, serum GFAP was not associated with most of the motor phenotypic features of ALS, including site of onset, functional status, disease progression rate, disease stage, and indices of upper (UMN) and lower motor neuron (LMN) impairment. The author hypothesized that activation of astrocytes also may occur in brain areas not belonging to the motor system, where they contribute to neurodegeneration of CNS regions involved in cognition and respiratory function, which are outside the main facets of ALS pathology (e.g., motor neuron degeneration) (Verde et al., 2023). In a meta-analysis of Parkinson's disease (PD) and MSA, no difference in CSF GFAP levels were found (Xie et al., 2021). In spinocerebellar ataxia type 3 (SCA3), GFAP did not show higher levels in mutation carriers compared to healthy controls either (Table 1) (Garcia-Moreno et al., 2022).

7 Conclusions

In conclusion, astrocytes play an important role in the neuropathology of neurodegenerative diseases, including HD. Some studies have supported the potential for astrocyte biomarkers, such as GFAP, to be used as biomarkers in HD, while others reported no detectable association with disease pathology. Although no clear evidence of astrocyte reactivity exists in most HD models, astrocytes still show functional alterations in HD even in the absence of astrocyte reactivity (hypertrophy and high GFAPGlial fibrillary acidic protein (GFAP) expression) (Ben Haim et al., 2015a; Serrano-Pozo et al., 2013). Given the robust effects of astrocyte markers in other neurodegenerative diseases, more research should be conducted to delineate the role of astrocytes in HD pathogenesis and their potential to serve as biomarkers in this disease.

Acknowledgments This work was supported by the National Natural Science Foundation of China (Grant Numbers 81873751 and 82071255), the Guangdong Provincial Key Laboratory of Diagnosis and Treatment of Major Neurological Diseases (2020B1212060017), Guangdong Provincial Clinical Research Center for Neurological Diseases (2020B1111170002), the Southern China International Cooperation Base for Early Intervention and Functional Rehabilitation of Neurological Diseases (2015B050501003 and 2020A0505020004), Guangdong Provincial Engineering Center For Major Neurological Disease Treatment, Guangdong Provincial Translational Medicine Innovation Platform for Diagnosis and Treatment of Major Neurological Disease, Guangzhou Clinical Research and Translational Center for Major Neurological Diseases (201604020010).

References

Abdelhak, A., Hottenrott, T., Morenas-Rodríguez, E., Suárez-Calvet, M., Zettl, U. K., Haass, C., Meuth, S. G., Rauer, S., Otto, M., Tumani, H., & Huss, A. (2019). Glial activation markers in CSF and serum from patients with primary progressive multiple sclerosis: potential of serum GFAP as disease severity marker? *Front Neurology, 10*, 280.

Allaman, I., Bélanger, M., & Magistretti, P. J. (2011). Astrocyte-neuron metabolic relationships: For better and for worse. *Trends in Neurosciences, 34*(2), 76–87.

Ben Haim, L., Carrillo-de Sauvage, M. A., Ceyzériat, K., & Escartin, C. (2015a). Elusive roles for reactive astrocytes in neurodegenerative diseases. *Frontiers in Cellular Neuroscience, 9*, 278.

Ben Haim, L., Ceyzériat, K., Carrillo-de Sauvage, M. A., Aubry, F., Auregan, G., Guillermier, M., Ruiz, M., Petit, F., Houitte, D., Faivre, E., Vandesquille, M., Aron-Badin, R., Dhenain, M., Déglon, N., Hantraye, P., Brouillet, E., Bonvento, G., & Escartin, C. (2015b). The JAK/STAT3 pathway is a common inducer of astrocyte reactivity in Alzheimer' s and Huntington' s diseases. *The Journal of Neuroscience, 35*(6), 2817–2829.

Benedet, A. L., Milà-Alomà, M., Vrillon, A., Ashton, N. J., Pascoal, T. A., Lussier, F., Karikari, T. K., Hourregue, C., Cognat, E., Dumurgier, J., Stevenson, J., Rahmouni, N., Pallen, V., Poltronetti, N. M., Salvadó, G., Shekari, M., Operto, G., Gispert, J. D., Minguillon, C., Fauria, K., Kollmorgen, G., Suridjan, I., Zimmer, E. R., Zetterberg, H., Molinuevo, J. L., Paquet, C., Rosa-Neto, P., Blennow, K., Suárez-Calvet, M., & Translational Biomarkers in Aging and Dementia (TRIAD) study, Alzheimer's and Families (ALFA) study, and BioCogBank Paris Lariboisière cohort (Eds.). (2021). Differences between plasma and cerebrospinal fluid glial fibrillary acidic protein levels across the Alzheimer disease continuum. *JAMA Neurology, 78*(12), 1471–1483.

Boussicault, L., Hérard, A. S., Calingasan, N., Petit, F., Malgorn, C., Merienne, N., Jan, C., Gaillard, M. C., Lerchundi, R., Barros, L. F., Escartin, C., Delzescaux, T., Mariani, J., Hantraye, P., Beal, M. F., Brouillet, E., Véga, C., & Bonvento, G. (2014). Impaired brain energy metabolism in the BACHD mouse model of Huntington' s disease: Critical role of astrocyte-neuron interactions. *Journal of Cerebral Blood Flow and Metabolism, 34*(9), 1500–1510.

Bu, G. (2009). Apolipoprotein E and its receptors in Alzheimer' s disease: pathways, pathogenesis and therapy. *Nature Reviews. Neuroscience, 10*(5), 333–344.

Buffo, A., Rolando, C., & Ceruti, S. (2010). Astrocytes in the damaged brain: Molecular and cellular insights into their reactive response and healing potential. *Biochemical Pharmacology, 79*(2), 77–89.

Burda, J. E., & Sofroniew, M. V. (2014). Reactive gliosis and the multicellular response to CNS damage and disease. *Neuron, 81*(2), 229–248.

Carrero, I., Gonzalo, M. R., Martin, B., Sanz-Anquela, J. M., Arévalo-Serrano, J., & Gonzalo-Ruiz, A. (2012). Oligomers of β-amyloid protein (Aβ1-42) induce the activation of cyclooxygenase-2 in astrocytes via an interaction with interleukin-1β, tumour necrosis factor-α, and a nuclear factor κ-B mechanism in the rat brain. *Experimental Neurology, 236*(2), 215–227.

Crosio, C., Valle, C., Casciati, A., Iaccarino, C., & Carrì, M. T. (2011). Astroglial inhibition of NF-κB does not ameliorate disease onset and progression in a mouse model for amyotrophic lateral sclerosis (ALS). *PLoS One, 6*(3), e17187.

Danbolt, N. C. (2001). Glutamate uptake. *Progress in Neurobiology, 65*(1), 1–105.

Ehrnhoefer, D. E., Southwell, A. L., Sivasubramanian, M., Qiu, X., Villanueva, E. B., Xie, Y., Waltl, S., Anderson, L., Fazeli, A., Casal, L., Felczak, B., Tsang, M., & Hayden, M. R. (2018). HACE1 is essential for astrocyte mitochondrial function and influences Huntington disease phenotypes in vivo. *Human Molecular Genetics, 27*(2), 239–253.

Eidelberg, D., & Surmeier, D. J. (2011). Brain networks in Huntington disease. *The Journal of Clinical Investigation, 121*(2), 484–492.

Eng, L. F., Vanderhaeghen, J. J., Bignami, A., & Gerstl, B. (1971). An acidic protein isolated from fibrous astrocytes. *Brain Research, 28*(2), 351–354.

Eng, L. F., Ghirnikar, R. S., & Lee, Y. L. (2000). Glial fibrillary acidic protein: GFAP-thirty-one years (1969-2000). *Neurochemical Research, 25*(9–10), 1439–1451.

Faideau, M., Kim, J., Cormier, K., Gilmore, R., Welch, M., Auregan, G., Dufour, N., Guillermier, M., Brouillet, E., Hantraye, P., Déglon, N., Ferrante, R. J., & Bonvento, G. (2010). In vivo expression of polyglutamine-expanded huntingtin by mouse striatal astrocytes impairs glutamate transport: A correlation with Huntington' s disease subjects. *Human Molecular Genetics, 19*(15), 3053–3067.

Frakes, A. E., Ferraiuolo, L., Haidet-Phillips, A. M., Schmelzer, L., Braun, L., Miranda, C. J., Ladner, K. J., Bevan, A. K., Foust, K. D., Godbout, J. P., Popovich, P. G., Guttridge, D. C., & Kaspar, B. K. (2014). Microglia induce motor neuron death via the classical NF-κB pathway in amyotrophic lateral sclerosis. *Neuron, 81*(5), 1009–1023.

Garcia-Moreno, H., Prudencio, M., Thomas-Black, G., Solanky, N., Jansen-West, K. R., Hanna Al-Shaikh, R., Heslegrave, A., Zetterberg, H., Santana, M. M., Pereira de Almeida, L., Vasconcelos-Ferreira, A., Januário, C., Infante, J., Faber, J., Klockgether, T., Reetz, K., Raposo, M., Ferreira, A. F., Lima, M., Schöls, L., Synofzik, M., Hübener-Schmid, J., Puschmann, A., Gorcenco, S., Wszolek, Z. K., Petrucelli, L., & Giunti, P. (2022). Tau and neurofilament light-chain as fluid biomarkers in spinocerebellar ataxia type 3. *European Journal of Neurology, 29*(8), 2439–2452.

Hong, Y., Zhao, T., Li, X. J., & Li, S. (2016). Mutant huntingtin impairs BDNF release from astrocytes by disrupting conversion of Rab3a-GTP into Rab3a-GDP. *The Journal of Neuroscience, 36*(34), 8790–8801.

Hsiao, H. Y., Chen, Y. C., Chen, H. M., Tu, P. H., & Chern, Y. (2013). A critical role of astrocyte-mediated nuclear factor-κB-dependent inflammation in Huntington' s disease. *Human Molecular Genetics, 22*(9), 1826–1842.

Iliff, J. J., Wang, M., Liao, Y., Plogg, B. A., Peng, W., Gundersen, G. A., Benveniste, H., Vates, G. E., Deane, R., Goldman, S. A., Nagelhus, E. A., & Nedergaard, M. (2012). A paravascular pathway facilitates CSF flow through the brain parenchyma and the clearance of interstitial solutes, including amyloid β. *Science Translational Medicine, 4*(147), 147ra111.

Jansen, A. H., Reits, E. A., & Hol, E. M. (2014). The ubiquitin proteasome system in glia and its role in neurodegenerative diseases. *Frontiers in Molecular Neuroscience, 8*, 73.

Jeffrey, K. L., Camps, M., Rommel, C., & Mackay, C. R. (2007). Targeting dual-specificity phosphatases: Manipulating MAP kinase signalling and immune responses. *Nature Reviews. Drug Discovery, 6*(5), 391–403.

Kaminska, B., Gozdz, A., Zawadzka, M., Ellert-Miklaszewska, A., & Lipko, M. (2009). MAPK signal transduction underlying brain inflammation and gliosis as therapeutic target. *Anatomical Record (Hoboken), 292*(12), 1902–1913.

Khoshnan, A., Ko, J., Watkin, E. E., Paige, L. A., Reinhart, P. H., & Patterson, P. H. (2004). Activation of the IkappaB kinase complex and nuclear factor-kappaB contributes to mutant huntingtin neurotoxicity. *The Journal of Neuroscience, 24*(37), 7999–8008.

Lee, W., Reyes, R. C., Gottipati, M. K., Lewis, K., Lesort, M., Parpura, V., & Gray, M. (2013). Enhanced ca(2+)-dependent glutamate release from astrocytes of the BACHD Huntington' s disease mouse model. *Neurobiology of Disease, 58*, 192–199.

Leoni, V., & Caccia, C. (2014). Study of cholesterol metabolism in Huntington' s disease. *Biochemical and Biophysical Research Communications, 446*(3), 697–701.

Maher, P., Dargusch, R., Bodai, L., Gerard, P. E., Purcell, J. M., & Marsh, J. L. (2011). ERK activation by the polyphenols fisetin and resveratrol provides neuroprotection in multiple models of Huntington' s disease. *Human Molecular Genetics, 20*(2), 261–270.

Maragakis, N. J., & Rothstein, J. D. (2004). Glutamate transporters: Animal models to neurologic disease. *Neurobiology of Disease, 15*(3), 461–473.

Motori, E., Puyal, J., Toni, N., Ghanem, A., Angeloni, C., Malaguti, M., Cantelli-Forti, G., Berninger, B., Conzelmann, K. K., Götz, M., Winklhofer, K. F., Hrelia, S., & Bergami, M. (2013). Inflammation-induced alteration of astrocyte mitochondrial dynamics requires autophagy for mitochondrial network maintenance. *Cell Metabolism, 18*(6), 844–859.

Ortega, Z., & Lucas, J. J. (2014). Ubiquitin-proteasome system involvement in Huntington' s disease. *Frontiers in Molecular Neuroscience, 29*, 77.

Palpagama, T. H., Waldvogel, H. J., Faull, R. L. M., & Kwakowsky, A. (2019). The role of microglia and astrocytes in Huntington' s disease. *Frontiers in Molecular Neuroscience, 25*, 258.

Pearce, M. M. P., Spartz, E. J., Hong, W., Luo, L., & Kopito, R. R. (2015). Prion-like transmission of neuronal huntingtin aggregates to phagocytic glia in the drosophila brain. *Nature Communications, 6*, 6768.

Pellerin, L., & Magistretti, P. J. (1994). Glutamate uptake into astrocytes stimulates aerobic glycolysis: A mechanism coupling neuronal activity to glucose utilization. *Proceedings of the National Academy of Sciences of the United States of America, 91*(22), 10625–10629.

Pereira, J. B., Janelidze, S., Smith, R., Mattsson-Carlgren, N., Palmqvist, S., Teunissen, C. E., Zetterberg, H., Stomrud, E., Ashton, N. J., Blennow, K., & Hansson, O. (2021). Plasma GFAP is an early marker of amyloid-β but not tau pathology in Alzheimer' s disease. *Brain, 144*(11), 3505–3516.

Potokar, M., Morita, M., Wiche, G., & Jorgačevski, J. (2020). The diversity of intermediate filaments in astrocytes. *Cell, 9*(7), 1604.

Rebec, G. V. (2018). Corticostriatal network dysfunction in Huntington' s disease: Deficits in neural processing, glutamate transport, and ascorbate release. *CNS Neuroscience & Therapeutics, 24*(4), 281–291.

Rebec, G. V., Barton, S. J., & Ennis, M. D. (2002). Dysregulation of ascorbate release in the striatum of behaving mice expressing the Huntington' s disease gene. *The Journal of Neuroscience, 22*(2), RC202.

Ross, C. A., & Poirier, M. A. (2004). Protein aggregation and neurodegenerative disease. *Nature Medicine, 10*(Suppl), S10–S17.

Serrano-Pozo, A., Gómez-Isla, T., Growdon, J. H., Frosch, M. P., & Hyman, B. T. (2013). A phenotypic change but not proliferation underlies glial responses in Alzheimer disease. *The American Journal of Pathology, 182*(6), 2332–2344.

Shin, J. Y., Fang, Z. H., Yu, Z. X., Wang, C. E., Li, S. H., & Li, X. J. (2005). Expression of mutant huntingtin in glial cells contributes to neuronal excitotoxicity. *The Journal of Cell Biology, 171*(6), 1001. –12, 1012.

Sofroniew, M. V., & Vinters, H. V. (2010). Astrocytes: Biology and pathology. *Acta Neuropathologica, 119*(1), 7–35.

Stanek, L. M., Bu, J., & Shihabuddin, L. S. (2019). Astrocyte transduction is required for rescue of behavioral phenotypes in the YAC128 mouse model with AAV-RNAi mediated HTT lowering therapeutics. *Neurobiology of Disease, 129*, 29–37.

Szlachcic, W. J., Switonski, P. M., Krzyzosiak, W. J., Figlerowicz, M., & Figiel, M. (2015). Huntington disease iPSCs show early molecular changes in intracellular signaling, the expression of oxidative stress proteins and the p53 pathway. *Disease Models & Mechanisms, 8*(9), 1047–1057.

Tong, X., Ao, Y., Faas, G. C., Nwaobi, S. E., Xu, J., Haustein, M. D., Anderson, M. A., Mody, I., Olsen, M. L., Sofroniew, M. V., & Khakh, B. S. (2014). Astrocyte Kir4.1 ion channel deficits contribute to neuronal dysfunction in Huntington' s disease model mice. *Nature Neuroscience, 17*(5), 694–703.

Tong, J., Ang, L. C., Williams, B., Furukawa, Y., Fitzmaurice, P., Guttman, M., Boileau, I., Hornykiewicz, O., & Kish, S. J. (2015 Oct). Low levels of astroglial markers in Parkinson's disease: Relationship to α-synuclein accumulation. *Neurobiology of Disease, 82*, 243–253.

Valenza, M., Leoni, V., Karasinska, J. M., Petricca, L., Fan, J., Carroll, J., Pouladi, M. A., Fossale, E., Nguyen, H. P., Riess, O., MacDonald, M., Wellington, C., DiDonato, S., Hayden, M., & Cattaneo, E. (2010). Cholesterol defect is marked across multiple rodent models of Huntington' s disease and is manifest in astrocytes. *The Journal of Neuroscience, 30*(32), 10844–10850.

Verde, F., Milone, I., Maranzano, A., Colombo, E., Torre, S., Solca, F., Doretti, A., Gentile, F., Manini, A., Bonetti, R., Peverelli, S., Messina, S., Maderna, L., Morelli, C., Poletti, B., Ratti,

A., Silani, V., & Ticozzi, N. (2023). Serum levels of glial fibrillary acidic protein in patients with amyotrophic lateral sclerosis. *Annals of Clinical Translational Neurology, 10*(1), 118–129.

Wang, L., Lin, F., Wang, J., Wu, J., Han, R., Zhu, L., Difiglia, M., & Qin, Z. (2012). Expression of mutant N-terminal huntingtin fragment (htt552-100Q) in astrocytes suppresses the secretion of BDNF. *Brain Research, 1449*, 69–82.

Wójtowicz, A. M., Dvorzhak, A., Semtner, M., & Grantyn, R. (2013). Reduced tonic inhibition in striatal output neurons from Huntington mice due to loss of astrocytic GABA release through GAT-3. *Frontiers in Neural Circuits, 7*, 188.

Xie, D., Feng, L., Huang, H., Zhao, Q., Ning, P., Shen, Q., Lu, H., Xu, F., & Xu, Y. (2021). Cerebrospinal fluid biomarkers in multiple system atrophy relative to Parkinson' s disease: A meta-analysis. *Behavioural Neurology, 2021*, 5559383.

You, H., Wu, T., Du, G., Huang, Y., Zeng, Y., Lin, L., Chen, D., Wu, C., Li, X., Burgunder, J. M., & Pei, Z. (2021). Evaluation of blood glial fibrillary acidic protein as a potential marker in Huntington's disease. *Frontiers in Neurology, 19*, 779890.

Mitochondrial/Oxidative Stress Biomarkers in Huntington's Disease

Kateřina Vodičková Kepková and Petr Vodička

Abbreviations

3-NP	3-nitropropionic acid
4-HNE	4-hydroxy-2-nonenal
8-OHdG	8-hydroxy-2′-deoxyguanosine
8-oxodG	7,8-dihydro-8-oxo-2′-deoxyguanosine
8-oxoG	8-oxo-7,8-dihydroguanine
AD	Alzheimer's disease
ARE	Antioxidant response element
AT	Ataxia telangiectasia
CAG	Cytosine-adenine-guanine
CAP	CAG length/age product
CAT	Catalase
CNS	Central nervous system
CoQ	Coenzyme Q
CSF	Cerebrospinal fluid
Cu/Zn-SOD	Cu/Zn-superoxide dismutase
CUL3	Cullin3
DNA	Deoxyribonucleic acid
DNPH	2,4-dinitrophenylhydrazone
DSB	Double-strand breaks
ELISA	Enzyme-linked immuno sorbent assay
ETC	Electron transport chain
F2-isoPs	F2-isoprostanes
GAPDH	Glyceraldehyde-3-phosphate dehydrogenase
GC	Gas chromatography
GC-MS	Gas chromatography mass spectrometry
GFAP	Glial fibrillary acidic protein

K. Vodičková Kepková · P. Vodička (✉)
Laboratory of Applied Proteome Analyses, Institute of Animal Physiology and Genetics, Czech Academy of Sciences, Libechov, Czech Republic
e-mail: vodicka@iapg.cas.cz

E. A. Thomas, G. M. Parkin (eds.), *Biomarkers for Huntington's Disease*, Contemporary Clinical Neuroscience, https://doi.org/10.1007/978-3-031-32815-2_13

GPx	Glutathione peroxidase
GPx1	Glutathione peroxidase 1
GR	Glutathione reductase
GSH	Glutathione
GSSG	Glutathione disulfide (oxidized glutathione)
H_2O_2	Hydrogen peroxide
HD	Huntington's disease
HPLC	High-performance liquid chromatography
HPLC-MS	High performance liquid chromatography-mass spectrometry
HTT	Huntingtin
Keap1	Kelch-like ECH-associated protein 1
KI	Knock-in
L•	Lipid radical
LC-ECD	Liquid chromatography with electrochemical detection
LC-MS/MS MRM	Liquid chromatography with tandem mass spectrometry-multiple reaction monitoring
LC-MS/MS	Liquid chromatography with tandem mass spectrometry
LOO•	Lipid peroxyl radical
LOOH	Lipid hydroperoxide
MAPK	Mitogen-activated protein kinase
MDA	Malondialdehyde
mHTT	Mutant huntingtin
Mn-SOD	Manganese superoxide dismutase
MRI	Magnetic resonance imaging
MRM	Multiple reaction monitoring
MRS	Magnetic resonance spectroscopy
mtDNA	Mitochondrial DNA
mtND2	Mitochondria-encoded NADH dehydrogenase 2
NADPH	Nicotinamide adenine dinucleotide phosphate
nDNA	Nuclear DNA
Nfl	Neurofilament light chain
NHEJ	Non-homologous end joining
NOX	NADPH oxidase
Nrf2	Nuclear factor erythroid 2-related factor 2
•OH	Hydroxyl radical
O_2	Oxygen
O_2•−	Superoxide anion radical
OXPHOS	Oxidative Phosphorylation
PBMC	Peripheral blood mononuclear cells
PGC-1α	Peroxisome proliferator-activated receptor-gamma coactivator 1 alpha
PET	Positron emission tomography
PreHD	Premanifest HD
PUFA	Polyunsaturated fatty acids
R•	Radical

ROC	Receiver operating characteristic
ROS	Reactive oxygen species
rRNA	Ribosomal ribonucleic acid
RTL	Relative telomere length
RTQ-PCR	Real-time quantitative polymerase chain reaction
SOD1	Cu/Zn-superoxide dismutase
SOD2	Mn superoxide dismutase
T/S	T-telomere length/S-single copy gene signal
TBA	Thiobarbituric acid
TBA-MDA	Thiobarbituric acid- malondialdehyde
TBARS	Thiobarbituric acid reactive substance
TFC	Total functional capacity
tRNA	Transfer RNA
UA	Uric acid
UHDRS	Unified Huntington's Disease Rating Scale

1 Introduction

HD

Huntington's disease (HD) is an autosomal dominant hereditary neurodegeneration, caused by the mutation in the *HTT* gene on chromosome 4 (MacDonald et al., 1993). Amplification of CAG triplet in the exon 1 of *HTT* over the threshold of 36 repeats leads to the disease, with full penetrance at repeat length exceeding 39 (Bates et al., 2015). Clinically, HD usually manifests in the fourth decade of life through a mix of motor and cognitive symptoms, as well as psychiatric disturbances. The age of onset is dependent on polyQ stretch length, with shorter repeats leading to the late onset of HD, while repeats over 60 CAG result in a juvenile onset form. Currently, no curative treatment is available, and the survival time is usually 15–20 years from the disease onset (Gusella & MacDonald, 2006). Although HTT is widely expressed throughout the body, GABAergic medium spiny neurons of the striatum are particularly vulnerable, followed at later disease stages by cortical neurons (Guedes-Dias et al., 2016). On a molecular level, the mutation mainly acts as a toxic gain of function, with mutated huntingtin (mHTT) protein, inclusive of an extended stretch of glutamines (polyQ) at its N-terminus, often cleaved into shorter N-terminal fragments that contribute to the phenotype (Sathasivam et al., 2013). Previous research has documented a wide range of cellular dysregulation caused by the mutation, mainly in neurons, including transcriptional dysregulation, mHTT protein aggregation, abnormal synaptic transmission, defects in cellular trafficking, proteostasis, and energy metabolism (Guedes-Dias et al., 2016; Tabrizi et al., 2020).

Endogenous ROS Sources

Reactive oxygen species (ROS) are unstable oxygen containing molecules that readily react with other cellular components. While they play many important roles in cell signaling and the immune system (Sies & Jones, 2020), their overproduction results in damage to nucleic acids, proteins and lipids (Oladele et al., 2021). Mitochondria are one of the major ROS sources within cells (Mailloux, 2020), and enzymes in the NADPH oxidase (NOX) family are significant producers of ROS within the cytosol and extracellularly (Pecchillo Cimmino et al., 2023). The superoxide anion radical O_2•- is produced by the electron transport chain (ETC) during oxidative phosphorylation (OXPHOS) in mitochondria (Murphy, 2009), and by the membrane bound NOX enzymes in the cytosol, among others. Hydrogen peroxide H_2O_2 is produced by superoxide dismutases (SODs) from O_2•- (Wang et al., 2018), directly by ETC in mitochondria, and by other oxidases with various cellular localization, e.g., in endoplasmic reticulum or peroxisomes. Highly reactive hydroxyl radical (•OH), another important ROS, can be produced from H_2O_2 via the Fenton reaction in the presence of transitional metal ions (Fe^{2+}, Cu^{+}). Unlike O_2•- and H_2O_2, •OH is not removed via enzymatic reactions, which makes it very damaging. Elevated levels of ROS over a normal homeostatic state are called oxidative stress.

Antioxidant Defense Mechanisms

The antioxidant defense mechanism is essential in protecting the cell against ROS and oxidative damage and involves both enzymatic and non-enzymatic components (Fig. 1). While enzymatic antioxidants directly catalyze the conversion of ROS into less reactive compounds, non-enzymatic antioxidants are capable of scavenging or quenching free radicals and other ROS or removing compounds inducing ROS production e.g., sequestering metal ions participating in Fenton reactions (Mirończuk-Chodakowska et al., 2018). SODs are enzymes converting superoxide anion radical (O_2•−) to hydrogen peroxide as part of the defense against ROS damage. Cu/Zn-SOD (SOD1) acts in the cytoplasm, while Mn-SOD (SOD2) serves the same purpose in mitochondria (Wang et al., 2018). Both catalase (CAT) and glutathione peroxidase (GPx) enzymes convert H_2O_2 into H_2O and oxygen (Sies, 1997). Tripeptide glutathione (GSH) is the most abundant thiol in animal cells and in its reduced GSH form serves as one of the main non-enzymatic antioxidants. GSH levels and the ratio of reduced GSH to oxidized glutathione (glutathione disulfide; GSSG) are often used as a measure of oxidative stress. Uric acid (UA) is a final product of purine catabolism that acts as a scavenger of oxygen radicals and a potent natural antioxidant. As the levels of UA in humans are higher than in other mammals, UA is the most abundant antioxidant in human blood (Mirończuk-Chodakowska et al., 2018). The fact, that it can be easily measured in blood plasma, urine, and saliva, makes it an attractive candidate biomarker.

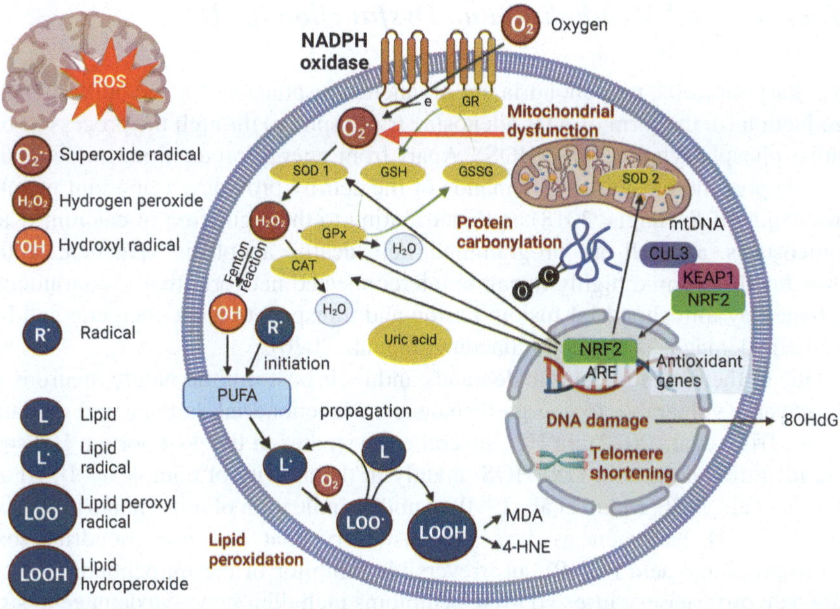

Fig. 1 Intracellular sources of ROS, products of oxidative stress damage, and antioxidant defense mechanisms. Reactive oxygen species (red circles) are excessively produced in mitochondria and cytosol under oxidative stress conditions and can lead to damage to lipids (lipid radicals/peroxides; in blue), proteins (protein carbonyls) and nucleic acids (direct DNA damage, telomere shortening). Antioxidants (yellow) act as part of an antioxidant defense mechanism and are partially regulated by the NRF2 pathway. Both products of ROS damage (dark red labels) and levels of antioxidants can serve as biomarkers of oxidative stress. (Created with BioRender.com)

Because damage by ROS represents an acute danger to many cellular components, response to oxidative stress is tightly regulated. The KEAP1-NRF2-ARE system controls the expression of a number of antioxidant defense system proteins. Nuclear factor erythroid 2-related (NRF2) is a transcription factor of the cap'n'collar (CNC) family (Moi et al., 1994). Under normal conditions, NRF2 is kept in the cytoplasm by interaction with Kelch-like ECH-associated protein-1 (KEAP1), which forms ubiquitin E3 ligase complex with CULLIN3 (CUL3), and polyubiquitinates NRF2, ensuring its rapid degradation by the proteasome (Furukawa & Xiong, 2005). KEAP1 contains several cysteine residues, which serve as a thiol-base sensor of ROS. Modification of KEAP1 cysteines in an oxidative environment reduces its E3 ubiquitin ligase activity, resulting in the accumulation of NRF2 and its translocation into the nucleus. In the nucleus, NRF2 forms heterodimers with other transcription factors, and binds to antioxidant response element (ARE) sequences, resulting in the transcription of phase II antioxidant and detoxifying genes, including CAT, GPx, and SOD1 (Yamamoto et al., 2018).

Mitochondrial Metabolism and Dysfunction in HD

In eukaryotic cells, mitochondria are organelles responsible for the bulk of energy production (in the form of ATP, adenosine triphosphate) through the process of oxidative phosphorylation (OXPHOS). Apart from energy production, mitochondria also support the biosynthetic demands of the cell by providing important metabolites (Spinelli & Haigis, 2018) and contributing to the regulation of calcium (Ca^{2+}) homeostasis as well as programmed cell death, apoptosis (Duchen, 2000). Mitochondria form a highly dynamic interconnected network, that is continuously reshaped by mitochondrial fusion, fission and transport to meet energetic and biosynthetic demands of the cell (Giacomello et al., 2020).

Due to their high metabolic demands and their post-mitotic nature, neurons are particularly vulnerable to energy disbalance, mitochondrial damage and oxidative stress (Trigo et al., 2022). In HD, several early studies of the postmortem HD striatum identified defects in OXPHOS, mainly in the activity of complexes II, III and IV (Gu et al., 1996; Mann et al., 1990). Another indication of mitochondria involvement in HD pathogenesis was the discovery that the mitochondrial toxin 3-nitropropionic acid (3-NP), an irreversible inhibitor of the mitochondrial succinate dehydrogenase, causes HD-like symptoms including selective damage to striatal medium spiny neurons (Brouillet et al., 2005). Using KI and transgenic mouse models, Cui et al. (2006) showed that mutant huntingtin (mHtt) represses the expression of peroxisome proliferator-activated receptor-gamma coactivator (PGC)-1α, an important regulator of the mitochondrial biogenesis, and that the lentiviral-mediated delivery of PGC-1α to the striatum of R6/2 mouse HD model prevented neuronal atrophy. Siddiqui et al. (2012) described higher levels of ROS generated by mitochondria in STHdhQ111 striatal cells, as well as lower spare respiratory capacity and higher mitochondrial DNA (mtDNA) damage when compared with STHdhQ7 cell line. Shirendeb et al. (2012) reported direct interaction of mHTT but not WT HTT with dynamin-1-like protein (DRP1), with DRP1 implicated in mitochondrial fission. All these data suggest that both mitochondrial energy metabolism itself and mitochondria dynamics might be affected by the presence of mHTT. Recently, Neueder et al. (2022) created a large scale omic dataset of changes in peripheral tissue (skin, skeletal muscle, and adipose tissue) of pre-manifest and early manifest HD patients compared to healthy controls. Among the changes identified, dysregulation of lipid and glucose metabolism in muscle tissue by transcriptomics and lipid metabolism in adipose tissue by proteomics indicates HD pathology related changes in energy metabolism include peripheral tissues.

Biomarkers

Biomarkers are quantifiable and reproducible indicators of a biological process, pathology, or response to treatment. Ideally, they should be readily accessible, reliable, relatively cheap to measure, and first of all, sensitive and disease/biological process specific. Biomarkers can be classified based on their mode of acquisition and intended use. Biomarkers can be non-invasive, such as **imaging** approaches, minimally invasive, such as **wet/biofluid biomarkers** and invasive, requiring tissue **biopsy** for histological or biochemical evaluation. Within the context of HD, imaging approaches including magnetic resonance imaging (MRI) morphometry of brain regions, magnetic resonance spectroscopy (MRS) of lactate levels or positron emission tomography (PET) are examples of non-invasive biomarkers (Wild & Tabrizi, 2008; Zeun et al., 2019). Levels of mHTT or neurofilament light chain (NfL) in plasma or cerebrospinal fluid (CSF) (Rodrigues et al., 2020) are examples of wet/biofluid biomarkers, while invasive biopsy based biomarkers are not commonly applied to HD, as the central nervous system (CNS) is the main affected region. Based on their use, biomarkers can be **diagnostic**, allowing identification of disease state or establishing time of onset, **prognostic**, allowing prediction of future disease development, or markers of therapeutic **efficacy**, allowing monitoring of response to treatments.

In this chapter, we review the application of oxidative stress biomarkers in human HD research. Mitochondrial dysfunction and oxidative stress are common not only in neurodegenerative diseases (Chen et al., 2012; Guo et al., 2013), but also in many other common pathological conditions, including cardiovascular disease (Senoner & Dichtl, 2019), metabolic syndrome (Raut & Khullar, 2023) and cancer (Sosa et al., 2013), which makes attributing them as biomarkers to a particular condition difficult. As the brain is the first and most important organ affected by HD pathology, extracting relevant information about the disease state from measures performed in the periphery (e.g., from blood plasma) may be complicated by these other possible comorbidities. Methods for direct measurement of ROS are available (Murphy et al., 2022), but due to the high reactivity of ROS and rapid changes in their levels in response to many physiological conditions, we have focused mainly on readouts that could potentially provide information about longer term oxidative stress load, such as the products of ROS damage to various biomolecules and components of the antioxidant defense system.

2 Biomarkers of Oxidative DNA Damage

8-Hydroxy-2′-Deoxyguanosine (8-OHdG)

Guanine is the most easily oxidized of the four DNA bases and the presence of its oxidative products, mainly 8-oxo-7,8-dihydroguanine (8-oxoG) can have functional consequences, as it induces G:C-T:A and G:C-C:G transversions in DNA structure (Kino et al., 2017). Although 8-hydroxy-2′-deoxyguanosine (8-OHdG) nomenclature is most used in the literature, the prevalent form in aqueous solution is the tautomeric 8-oxo-7,8-dihydro-2′-deoxyguanosine (8-oxodG), sometimes named also 8-oxodeoxyguanosine (Fig. 2) (Kasai & Kawai, 2016). Levels of 8-OHdG in

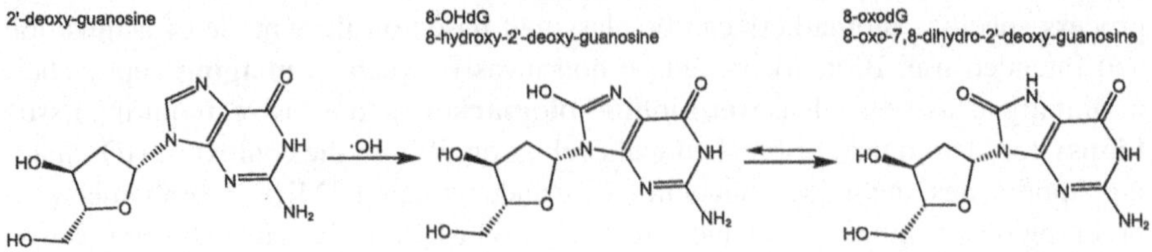

Fig. 2 Formation of 8-hydroxy-2′-deoxy-guanosine and its tautomer 8-oxo-7,8-dihydro-2′-deoxy-guanosine

biological fluids and tissues are commonly used as a biomarker of elevated oxidative stress, as it is one of the most abundant metabolites of oxidative damage to DNA. Several methodological approaches to analyze 8-OHdG levels have been developed, with the most commonly used being liquid chromatography with electrochemical detection (LC-ECD) (Bogdanov et al., 1999), enzyme-linked immuno sorbent assay (ELISA) (Gmitterová et al., 2009) and liquid chromatography with tandem mass spectrometry (LC-MS/MS) in multiple reaction monitoring (MRM) mode (Borowsky et al., 2013).

Elevation of 8-OHdG was reported in several neurodegenerative diseases, including Parkinson's disease (PD) (Gmitterová et al., 2009) and Alzheimer's disease (AD) (Bradley-Whitman et al., 2014), however, studies in HD patient populations have so far provided conflicting results (Table 1). Hersch et al. (2006) reported the elevation of baseline levels of 8-OHdG in serum as measured by LC-ECD in HD patients compared to matched healthy controls. The levels of this oxidative stress marker significantly decreased in the HD group after 16-week creatine supplementation relative to placebo. The study by Long et al. (2012) in a premanifest HD cohort divided into 3 groups based on CAG length/age product (CAP) as a disease burden proxy (Zhang et al., 2011) showed both levels of 8-OHdG at baseline and rate of change over time in longitudinal study increasing with CAP score. However, no significant difference between the same groups was detected in a cross-sectional study. LC-ECD was used as the detection method for the longitudinal study and LC-MS/MS for the cross-sectional study, which together with differences in sample handling could explain this discrepancy. Significantly increased level of 8-OHdG as measured by LC-ECD was observed in leukocytes of HD patients compared to healthy controls (Chen et al., 2007), but no difference was identified between subgroups of HD when divided based on Unified Huntington's Disease Rating Scale (UHDRS) scores (HD1 $N = 8$ with score ≤ 70, HD2 $N = 8$ with score 80–100), although this could be caused by overall small sample size. Similarly, increased levels of 8-OHdG were reported in plasma of both moderate severity (UHDRS >70) and mild severity (UHDRS <70) HD patients compared to controls, with no significant change observed related to disease severity (Túnez et al., 2011). In a study by Ciancarelli et al. (2014), 8-OHdG was found to be elevated in HD patients serum, but again showed no correlation to disease severity. Of note is that both (Túnez et al., 2011) and (Ciancarelli et al., 2014) used ELISA for 8-OHdG detection and reported values in the range of tens of ng/ml, compared to values in tens of pg/ml

Table 1 Overview of studies on DNA oxidative damage/8-OHdG in peripheral HD samples

Sample	Method	Stage of HD	Results	Reference
Serum	LC-ECD	HD (N = 32)	45.29 pg/ml baseline	Hersch et al. (2006)
		Control (N = 30)	13.5 pg/ml baseline	
Plasma	LC-ECD	CAP High (N = 31)	Linear mixed effects model shows increase levels at baseline in CAP high vs Control and significant increase over time in CAP High group.	Long et al. (2012)
		CAP Medium (N = 18)		
		CAP Low (N = 8)		
		Control (N = 20)		
	LC–MS/MS (MRM)	Separate Cross-sectional study (N = 16, 6, 38, 20 for CAP Low, Medium, High, Control)	No significant difference	
Leukocyte	LC-ECD	HD (N = 16)	42.82 per 10^6 dG	Chen et al. (2007)
		Control (N = 36)	11.96 per 10^6 dG	
Plasma	ELISA	HD1, moderate severity (N = 12)	71.5 ng/ml	Túnez et al. (2011)
		HD2, mild severity (N = 7)	65.70 ng/ml	
		HD1 and HD2 together (N = 19)	69.53 ng/ml	
		Control (N = 19)	45.51 ng/ml	
Serum	ELISA	HD (N = 18)	60.98 ng/ml	Ciancarelli et al. (2014)
		Control (N = 10)	3.35 ng/ml	
Plasma	LC-ECD	Manifest HD (N = 14)	18.70 pg/ml (median)	Biglan et al. (2012)
		Control (N = 6)	17.70 pg/ml (median)	
Plasma	LC-ECD LC–MS/MS (MRM)	Early symptomatic (HD1, HD2, N = 32 each)	~20 pg/ml by LC-MS assay, ~15 ng/ml by LC-ECD assay, no significant differences between groups at baseline and no differences between baseline and 24 m follow-up	Borowsky et al. (2013)
		Premanifest (PreHD-A, B, N = 32 each)		
		Control (N = 32)		

LC-ECD liquid chromatography with electrochemical detection. *LC-MS/MS* liquid chromatography with tandem mass spectrometry detection, *MRM* multiple reaction monitoring, targeted LC-MS/MS technique for measurement of specific analytes. CAP – CAG length/age product – a proxy for time to HD diagnosis developed by Zhang et al. (2011). The number of 8-OHdG molecules per 10^6 dG molecules is reported for Chen et al. (2007). Estimates from graphical representation are provided in the table where exact numbers are not listed in the study (labeled as ~)

reported in studies using LC-ECD or LC-MS/MS, suggesting caution when interpreting results obtained by different techniques. Biglan et al. (2012) reported no significant difference in baseline 8-OHdG levels between controls and HD patients, but significant decrease in response to coenzyme Q (CoQ) treatment in both groups. To date, the largest and most rigorous study of 8-OHdG in a HD population, which utilized a TRACK-HD cohort, was conducted by Borowsky et al. (2013). This study validated a new LC-MS/MS MRM assay for 8-OHdG, and established it as a more accurate than LC-ECD. No significant difference in 8-OHdG levels as measured by both LC-ECD and LC-MS/MS was identified among any of the TRACK-HD cohort groups or controls at baseline or 24 month follow-up, with the conclusion that 8-OHdG is not a suitable biomarker for HD state or progression (Borowsky et al., 2013).

Overall, while 8-OHdG is established biomarker of oxidative DNA damage, its relevance for monitoring HD state and progression in biofluids or cellular samples collected in the periphery remains unconvincing. Results by (Hersch et al., 2006) and (Biglan et al., 2012) point to its potential use in monitoring the efficacy of interventions intended to reduce oxidative stress.

Telomere Length

In eukaryotes, telomeric repeated DNA sequences are located at the end of linear chromosomes. Telomers are essential for replication, protection of chromosome ends, and genome stability (Blackburn & Szostak, 1984). Human telomeres consist of long stretches of hexanucleotide TTAGGG repeats (Moyzis et al., 1988) and six specific proteins that form the shelterin complex (Lange, 2005). The repetitive sequence acts as a buffer that prevents loss of genetic information, as telomeres shorten each DNA replication cycle and shelterin protects telomere integrity and prevents recognition of chromosome ends as damaged DNA (end-protection problem) (Sfeir & de Lange, 2012). Many studies have shown that telomere shortening in length over an organism's lifetime is linked to normal DNA replication; however, can be accelerated due to aging, oxidative stress (Zglinicki, 2002), DNA damage, and disorders linked to deficiencies of DNA repair, such as dementia, Alzheimer's disease (AD) (Coppede & Migliore, 2009), and ataxia telangiectasia (AT) (Metcalfe et al., 1996).

Elevated levels of ROS cause DNA damage, including ROS-induced double-strand breaks (DSB), and result in premature telomere shortening (Bar-Or et al., 2001; Henrikus et al., 2019). Protein Ku70/80 is important for telomere maintenance, including the prevention of telomere-telomere fusion events (Chai et al., 2002). Expression of mHTT induces DSB and DNA repair dysfunction, where one of the implicated mechanisms is an interaction between Ku70 and mHTT, leading to DSB accumulation through subsequent impairment of the DNA-dependent protein kinase function of Ku70 in non-homologous end joining (NHEJ) (Enokido et al., 2010).

Telomere shortening can be a marker of neuropathology (Kota et al., 2015) during cellular senescence (Bernadotte et al., 2016). Telomere length can be measured in blood leucocytes using the real-time quantitative polymerase chain reaction (RTQ-PCR) method. RTQ-PCR is used to measure relative telomere length (RTL). The detection method is based on the T/S ratio where T is the telomer length measured, and S is a single copy gene signal within the reaction (Cawthon, 2009). RTQ-PCR and T/S ratio were used in all subsequent research studies on samples of HD patients and their healthy controls covered in this section.

Kota et al. (2015) demonstrated telomer length attrition in leukocytes of symptomatic HD patients (median 0.21 T/S) compared to healthy controls (median 0.79 T/S) and similar telomer length attrition was observed in several other neurodegenerative disease, when compared to healthy controls. Another study included the pre-manifest stage of HD and compared the leucocyte telomere length to that of manifest HD patients and healthy controls. Significant differences in telomer length were observed between all groups. In the control sample group, the median T/S was 0.92, in pre-manifest HD it was 0.76, and at the manifest stage it was 0.56. Based on these results, including the relationship between age, telomere length, and disease onset, Scarabino et al. (2019) discussed the use of telomere length from blood leucocytes as a reliable biomarker of HD progression. The study by PerezGrovas-Saltijeral et al. (2020) also included all 3 groups, premanifest HD, symptomatic HD, and control, and confirmed a decrease in T/S ratio in the HD group (0.553). However, in contrast to Scarabino et al. (2019) the T/S ratio in the premanifest group (0.861) was not significantly different from the control (0.818). The difference in results in the premanifest HD stage may be due to the low age of this group (mean 33 years, compared to 47 in HD and 38 in control). In a follow-up study, Scarabino et al. (2022) analyzed the T/S ratio in leukocyte samples from premanifest HD subjects (2.5 years before clinical onset, T0) compared to samples from the same population collected one year after the clinical onset of HD (T1). At T0, a significant reduction of telomere length by 22% in comparison to controls was detected. At T1, a further decrease of 36% from the control group and 14% from T0 was observed. The median value of T/S in the control group was 0.99, 0.77 in HD gene carrier cohort at T0 and 0.66 after disease onset. The previously described relationship between telomere shortening and CAG repeat length and age of onset (Scarabino et al., 2019) was not confirmed in this study, but authors speculate that this could be due to the similar stage of disease progression in the premanifest HD group (2.5 years from clinical onset). Receiver Operating Characteristic curve (ROC) analysis confirmed the suitability of T/S ratio in leukocytes as a biomarker for discriminating premanifest HD from healthy controls and even from manifest HD. AUC for pre-HD (T0) vs controls was 0.997 (Sensitivity 1, Specificity 0.956), while AUC for pre-HD (T0) vs HD (T1) was 0.979 (Sensitivity 0.911, Specificity 0.978). A study in juvenile HD patients showed significantly reduced telomere length (median T/S 0.645) compared to their neurologically healthy control group (median T/S 0.958). The difference between these groups was not dependent on age, length of CAG repeat, and time of disease onset (PerezGrovas-Saltijeral et al., 2023).

In summary, significant telomere length shortening in peripheral blood cells was shown in all studies in manifest HD patient populations relative to healthy controls (Table 2). However, in the premanifest HD stage, where biomarkers of disease progression and predictors of expected age of onset would be most useful, the difference in telomere length wasn't replicated in the study by PerezGrovas-Saltijeral et al. (2020). Still, with several studies on relatively large groups, telomere length is one of the more promising molecular biomarkers of HD and further investigation of its relation to disease progression is warranted.

Table 2 Overview of studies on telomere length in peripheral HD patient samples

Sample	Methods	Stage of HD	Results (medians)	Reference
Leukocyte	RTQ-PCR. T/S ratio	HD (5 years from onset, N = 35)	0.21	Kota et al. (2015)
		Controls for HD (N = 40)	0.79	
		Dementia (N = 70)	0.48	
		Controls for dementia (N = 55)	0.91	
		AT (N = 9)	0.31	
		Controls for AT, N = 10	1.07	
Leukocyte	RTQ-PCR. T/S ratio	PreHD (N = 38)	0.76	Scarabino et al. (2019)
		HD (N = 62)	0.56	
		Control (N = 76)	0.92	
Leukocyte	RTQ-PCR. T/S ratio	PreHD (N = 29)	0.861	PerezGrovas-Saltijeral et al. (2020)
		HD (N = 71)	0.553	
		Control (N = 106)	0.818	
PBMCs	RTQ-PCR. T/S ratio	PreHD, T0 (N = 45)	0.77	Scarabino et al. (2022)
		HD, T1, (N = 45)	0.66	
		Control (N = 45)	0.99	
Leukocyte	RTQ-PCR. T/S ratio	Juvenile HD (N = 69)	0.645	PerezGrovas-Saltijeral et al. (2023)
		Control (N = 54)	0.958	

PBMC peripheral blood mononuclear cells, *RTQ-PCR* real-time quantitative PCR, *PreHD* pre-manifest *HD, HD* manifest HD, *AT* ataxia telangiectasia

mtDNA

As mitochondria are the major source of ROS production within cells, mitochondrial DNA (mtDNA) is exposed to a high risk of oxidative damage. Each mitochondrion contains 4–10 copies of 16.5 kb circular mtDNA, coding for 37 genes (2 rRNAs, 22 tRNAs and 13 proteins of OXPHOS respiratory chain) (Taanman, 1999). Due to the presence of several mtDNA copies within each mitochondrion, mutations of mtDNA, including point mutations, nucleic acid modifications (such as 8-OHdG) and large scale deletions can lead to heteroplasmy and subsequently to mitochondrial dysfunction (Stewart & Chinnery, 2015).

A study using the R6/2 mouse HD model showed increased mtDNA damage over time from 7–12 weeks of age. This damage was greater in the striatum compared to the cortex, and damage to mtDNA was greater than damage to nuclear DNA (nDNA)(Acevedo-Torres et al., 2009). Polidori et al. (1999) quantified oxidative damage to mtDNA through LC-ECD measurement of 8-OHdG levels in post-mortem tissue, and reported an increase in 8-OHdG levels in the parietal cortex of HD subjects compared to controls. Post-mortem human striata of HD patients also displayed increased levels of mtDNA damage and significant mtDNA depletion, and skin fibroblasts from HD patients had reduced respiratory capacity *in vitro*, suggesting that mitochondrial dysfunction can also to some extent be detected also in peripheral tissues (Siddiqui et al., 2012). Liu et al. (2008) studied mtDNA copy number changes in lymphocytes of polyQ disease patients (HD, spinal cerebellar atrophy 2/3, spinal bulbar muscular atrophy) and healthy controls. They found a significant decrease in all polyQ diseases studied, which was larger in symptomatic patients when compared to asymptomatic carriers (726 ± 135 copy/cell vs. 392 ± 114 copy/cell, $P = 0.041$). Leukocyte mtDNA copy number also negatively correlated with polyQ tract length. In contrast, Chen et al. (2007) reported a significant increase in the mtDNA copy number, and in 4799 bp deleted mtDNA copy numbers, in HD patient leukocytes as compared to healthy controls. A study by Askeland et al. (2018) showed no mtDNA damage in peripheral blood mononuclear cells (PBMCs) of HD patients compared to healthy controls, while nDNA damage was significantly increased in HD (both measured as inhibition of TaqαI restriction enzyme by DNA modifications). However, other mitochondrial functions were disturbed in HD PBMCs, including the reduced activity of citrate synthase.

Disatnik et al. (2016) showed that levels of mtDNA in plasma, as measured by RTQ-PCR of mitochondria-encoded NADH dehydrogenase 2 (mtND2), were significantly higher in young R6/2 mice (5w) compared to WT and rapidly decreased over time until the end stage of the disease (13w), at which point they were almost half the level of WT mice. The mtDNA levels in plasma also responded to treatment with mitochondrial fragmentation inhibitor P110 and returned to WT levels in treated R6/2 mice, suggesting the potential use of mtDNA plasma levels as a biomarker of response to treatment targeting mitochondria. As the R6/2 mouse model displays a very aggressive phenotype compared to the common HD progression timeline in humans, it is unclear whether the significant reduction of mtDNA levels in the late stage R6/2 mouse is relevant to human patients. In humans, plasma levels of mtDNA were elevated in both pre-manifest HD and HD subjects compared to healthy controls, similar to the situation in young R6/2 mice (Disatnik et al., 2016).

Conflicting results (reported increase, decrease and no change) of mtDNA copy number in peripheral blood cells (Table 3) suggest that the use of cellular mtDNA copy number as HD biomarker is currently not possible and will require further studies with higher N and better classification of involved patient populations. A better mechanistic understanding of peripheral changes related to HD progression and their influence on mtDNA damage is also needed.

3 Protein Carbonylation

Another common outcome of elevated ROS levels is the oxidation of proteins in an environment of increased oxidative stress. This results in the production of, protein carbonyls (>C=O), with amino acids histidine, arginine, lysine, and proline particularly sensitive to oxidation. In addition, other modifications specific to side chains of individual amino acids are also possible (Butterfield & Kanski, 2001). This can result in dysfunctional proteins and cellular damage. Methods for measuring the

Table 3 Overview of studies on mitochondrial DNA damage in peripheral HD patient samples

Sample	Methods	Stage of HD	Results (medians)	Reference
Leukocyte	RTQ-PCR	HD (N = 17)	232 ± 48 copies/cell	Liu et al. (2008)
		Control (N = 50)	804 ± 72 copies/cell	
Leukocyte	RTQ-PCR for total and 4977 bp deleted (Δ) mtDNA	HD (N = 16)	210 ± 1.2 (Δ 0.011 ± 0.003) copies/cell	Chen et al. (2007)
		Control (N = 19)	117 ± 1.17 (Δ 0.003 ± 0.001) copies/cell	
PBMC	inhibition of TaqαI restriction enzyme	HD (N = 21)	~0.6 FC HD vs WT	Askeland et al. (2018)
		Control (N = 16)		
Plasma	RTQ-PCR (mtND2)	HD (N = 6)	~20 FC to GAPDH	Disatnik et al. (2016)
		PreHD (N = 5)	~20 FC to GAPDH	
		Control (N = 5)	~5 FC to GAPDH	

PBMC peripheral blood mononuclear cells, *RTQ-PCR* real-time quantitative PCR, *PreHD* pre-manifest *HD, HD* manifest HD. Estimates from graphical representation are provided in the table where exact numbers are not listed in the manuscript (labeled as ~)

extent of protein carbonylation, assayed by the formation of the 2,4-dinitrophenylhydrazone (DNPH) covalent adduct, were developed as early as 1990 and can serve as a proxy for levels of oxidative stress in a given environment (Levine et al., 1990).

An early study by Alam et al. (2000) did not identify elevated levels of protein carbonylation or other oxidative damage (8-OHdG and other modifications of DNA, lipid peroxidation) in caudate, putamen or frontal cortex of HD patients as compared to controls. However, 9 out of 10 control samples were obtained from subjects with myocardial infarction as a cause of death, which could potentially lead to oxidative damage of brain tissue in this group as well. In two studies, Sorolla et al. used 2D-gel electrophoresis followed by either western blot with anti-DNP antibodies (2008) or MALDI-TOF peptide fingerprinting and LC-MS/MS (2010) to identify carbonylated proteins in striatal tissue of HD patients and healthy controls. Eighteen carbonylated proteins were identified in the first study, with glial fibrillary acidic protein (GFAP) being the major oxidized protein in HD. In the second study, an additional 13 oxidized proteins were identified, of which 7 were involved in energy metabolism (3 glycolytic, 4 mitochondrial); three of the mitochondrial enzymes displayed reduced activity in HD samples when investigated further. So far, the only study to describe protein carbonylation in peripheral tissue in human HD subjects was performed by Túnez et al. (2011) who reported increased levels of carbonylated proteins in both plasma samples (controls 0.36 ± 0.14 nM, HD 0.571 ± 0.24 nM) and erythrocytes (controls 0.46 ± 0.04 nM HD 0.66 ± 0.19 nM). Protein carbonylation requires further studies in larger cohorts before being considered as a reliable marker of oxidative damage in HD.

4 Lipid Peroxidation Biomarkers

Excessive lipid peroxidation is one of the hallmarks of oxidative stress in neurodegenerative disorders (Angelova et al., 2021). Oxidative stress damages membrane lipids, leading to changes in their attributes, such as altered membrane structure, decreased membrane fluidity (Borst et al., 2000), and enzyme inactivation. Lipids containing carbon-carbon double bonds, and preferentially long-chain fatty acids, (polyunsaturated fatty acids; PUFAs) are attacked by oxidants in the form of free radicals and this initiates lipid peroxidation. The oxidation of PUFA generates highly reactive aldehyde products, malondialdehyde (MDA), 4-hydroxy-2-nonenal (4-HNE) (Esterbauer et al., 1991; Gaschler & Stockwell, 2017), and isoprostanes (IsoPs) (Milne et al., 2011). MDA and 4-HNE are generated via lipid hydroperoxides and IsoPs via oxidation of arachidonic acid into isoprostane endoperoxide (Altomare et al., 2021).

The toxic effects of these secondary lipid peroxidation products play an important role in the pathogenesis of neurodegenerative diseases. Both MDA and 4-HNE react with various biomolecules, including phospholipids, amino acids, and proteins, to generate stable products and induce cross-linking (Slatter et al., 2004). MDA is covalently attached to proteins by Schiff-base type and conjugated imine-type adducts (Uchida et al., 1997), while 4-HNE and proteins react preferentially via Michael addition pathway (Bruenner et al., 1995). These reactions influence many biological activities in neurons. A study by Neely et al. (1999) shows 4-HNE cytotoxicity on cytoskeleton and neuronal motility. 4-HNE in higher concentration leads to disruption of neuronal microtubes, microtubule loss, and inhibition of neurite outgrowth. Cheng et al. (2011) described MDA influence on cultured cortical neurons, where external MDA treatment induced protein cross-linking, intracellular ROS accumulation, loss of mitochondrial membrane potential, and induced injury via MAPK-dependent pathway.

The products of lipid peroxidation can be detected in blood, urine, and other tissue and therefore have potential as oxidative stress biomarkers. F2-isoprostanes (F2-IsoPs) class of IsoPs are often used as biomarkers as they are relatively chemically stable and detectable in most body fluids and tissues. The most common methods for detection of lipid peroxidation products include gas or high performance liquid chromatography (GC, HPLC respectively) coupled to UV, fluorescence or mass spectrometry (MS) detection, immunological techniques (ELISA) and spectrophotometry of the reaction products of lipid peroxidation derived carbonyls with thiobarbituric acid (TBA) – TBARS assay for MDA/4-HNE (Altomare et al., 2021). Of these GC-MS and HPLC-MS/MS methods are the gold standard, as both immunoassays and TBARS chromogenic assay lack selectivity and can suffer from interference or unspecific reaction with other substances.

MDA and 4-HNE

No difference in lipid peroxidation was identified in the caudate, putamen, and frontal cortex of HD patients using ferrous ion oxidation–xylenol orange assay or HPLC analysis of TBA-MDA adducts by Alam et al. (2000). However, this study also failed to identify any evidence of oxidative stress in HD brains as aforementioned 9/10 control subjects had myocardial infarction which may confound these results. Lee et al. (2011) examined the levels of 4-HNE in the striatum and superior frontal cortex samples of human HD brains using immunochemistry, and levels of MDA + 4-HNE using the chromogenic assay. Both methods demonstrated an increase in lipid peroxidation in HD samples, with 4-HNE and MDA levels reaching 2.64 μM in HD brains and 2.13 μM in the control samples (Table 4). A similar increase in 4-HNE immunoreactivity was observed in R6/2 mouse brains, where it colocalized with mHTT inclusions. In addition, the authors also reported that treatment with the antioxidant and inhibitor of lipid peroxidation, nordihydroguaiaretic acid, resulted in a reduction in 4-HNE staining, improvement in mitochondrial function and increased survival of R6/2 mice (Lee et al., 2011). In a study by Peña-Sánchez et al. (2015), MDA concentration in blood serum was significantly increased in HD patients (0.7 μM) compared to the control group (0.5 μM), as measured by chromogenic assay. Similarly, Chen et al. (2007) showed that the level of plasma MDA was significantly higher in HD patients (1.77 μM) compared to the control group (1.33 μM), when measured by a HPLC-fluorescence method. HD patients were divided based on UHDRS score into two subgroups (HD1, $N = 8$ with score ≤ 70, HD2, $N = 8$ with score 80–100), and a significant difference in MDA levels was detected between these subgroups, with 2.1 μM MDA in HD1 group and 1.44 μM MDA in HD2 group. The HD1 group included patients with higher mean age (54.1 vs. 42.6 in HD2) and motor score (56.5 vs. 18.5 in HD2) but lower mean independence scale (48.8 vs. 91.2 in HD2) and total functional capacity scores (means 5.25 vs. 12.2 in HD2), as well as longer time from HD onset (8.9 vs. 2.63 years in HD2). The level of plasma MDA significantly correlated with the motor scale of UHDRS, but not with other variables, including age at onset and CAG repeat length (Chen et al., 2007). Klepac et al. (2007) reported elevated MDA/4-HNE levels as measured by chromogenic TBA assay in both premanifest HD and early symptomatic HD patients compared to matched controls and no significant difference between premanifest and HD group. Increased levels of MDA/4-HNE measured by commercial chromogenic assay were also reported by Stoy et al. (2005), but the data were provided as 6.98 ng/ml for the HD group and 0.88 μM in the control group, which makes their reliability questionable (particular assay used is calibrated for 0–20 μM MDA or MDA/4-HNE concentration). Data from the same patients (including the same unit discrepancy) are also included in the study by Christofides et al. (2006), which additionally shows similar MDA/4-HNE elevation in patients with a traumatic brain injury (reported in μM, suggesting these are indeed correct units).

Table 4 Overview of studies on lipid peroxidation in HD patient samples

Analyte	Sample	Methods	Stage of HD	Results	Reference
4-HNE	Striatum and the superior frontal cortex	IHC/densitometry	HD (N = 5)	Increase in caudate and putamen of HD vs Control (P < 0.01, DF = 8)	Lee et al. (2011)
			Control (N = 5)		
MDA/4-HNE		chromogenic/TBA	HD (N = 5)	2.64 ± 0.2 μM	
			Control (N = 5)	2.13 ± 0.12 μM	
MDA	Serum	chromogenic/TBA	HD (N = 14)	0.7 (0.4–1.65) μM	Peña-Sánchez et al. (2015)
			Control (N = 29)	0.5 (0.7–0.86)* μM	
MDA	Plasma	HPLC-fluorescence of MDA-TBA	HD, N = 16 (HD1 + HD2)	1.77 ± 0.16 μM	Chen et al. (2007)
			HD1 (N = 8)	2.1 ± 0.25 μM	
			HD2 (N = 8)	1.44 ± 0.11 μM	
			Control (N = 36)	1.33 ± 0.06 μM	
MDA/4-HNE	Plasma	chromogenic/TBA	PreHD (N = 11)	13.0 (9.2–14.2) mM	Klepac et al. (2007)
			Control (N = 22)	10.3 (8.41–12.1) mM	
			HD (N = 19)	13.3 (8.95–15.9) mM	
			Control (N = 47)	10.2 (7.3–13.7) mM	
MDA/4-HNE	Serum	chromogenic/TBA	HD (N = 11)	6.98 ± 1.16 ng/ml**	(Stoy et al. (2005), Christofides et al. (2006)
			Control (N = 15)	0.88 ± 0.09 μM	
F2-IsoP	CSF	GC-MS	early HD (N = 20), Control (N = 23)	35% increase in HD vs Control	Montine et al. (1999a, b)
F2-IsoP	Plasma	GC-MS	HD (N = 19, N = 20)	212 ± 29 pg/ml bound, 40 ± 5 pg/ml free	Montine et al. (2000)
			Control (N = 22)	165 ± 21 pg/ml bound, 41 ± 5 pg/ml free	
	Urine		HD (N = 11)	1.8 ± 0.2 pg/ml	
			Control (N = 22)	2.4 ± 0.5 pg/ml	

MDA malondialdehyde, *4-HNE* 4-hydroxy-2-nonenal, *TBA* thiobarbituric acid, *GC-MS* gas chromatography – mass spectrometry, *HPLC* high performance liquid chromatography, *PreHD* premanifest *HD, HD* manifest HD

*Values in Peña-Sánchez et al. (2015) should represent the median (10–90 percentile), values for the control group are probably incorrect.

**Values in Stoy et al. (2005), Christofides et al. (2006) are from the same samples. Values for the HD group are listed as ng/ml in both studies, but probably are in μM as all other groups reported.

F2-IsoPs

F2-IsoPs were quickly established as biomarkers of oxidative stress due to their detectability in blood plasma, urine, and most tissues. However, many subclasses of F2-IsoPs exist, and while all can be created by non-enzymatic peroxidation, some can also be created by enzymatic reactions (Praticó & FitzGerald, 1996). In neurodegenerative diseases, elevated levels of F2-IsoPs were demonstrated in CSF of probable AD patients (Montine et al., 1999b) and brains of early stage HD patients (N = 20, 35% increase relative to controls) (Montine et al., 1999a). However, no change in plasma and urinary F2-IsoPs levels was detected in a subsequent study between controls and probable AD or diagnosed HD patients (Montine et al., 2000). This suggests that elevated levels of lipid peroxidation are not system-wide, but restricted to the affected tissue, which limits the potential use of F2-IsoPs as easily accessible biomarkers of these neurodegenerations. No further studies reporting F2-IsoPs levels in human HD patients have been published since the early 2000s.

5 Antioxidants as Biomarkers

As the levels of antioxidants, as well as their oxidative state, respond to the oxidative environment, they can be useful as indirect biomarkers of oxidative stress. For CNS disorders, antioxidants can't be easily measured directly in the tissue of interest, but can be analyzed in CSF, blood serum or plasma, and blood cells. Additionally, UA can also be measured in urine, and saliva.

Enzymatic Antioxidants

In a proteomic study by Sorolla et al. (2008), both levels and activity of Mn-SOD were significantly increased in the striatum and cortex of HD patients compared to controls, while levels of Cu/Zn-SOD were not significantly changed. Elevated levels of Mn-SOD, but not Cu/Zn-SOD, CAT or GPx were also identified in the human HD fibroblasts compared to the controls (Jędrak et al., 2018). I studies by Chen et al. (2007) and (Túnez et al., 2011), Cu/Zn-SOD activity was found to be significantly reduced in erythrocytes of HD patients compared to healthy controls, while Peña-Sánchez et al. (2015) reported no significant difference in this parameter (Table 5). Klepac et al. (2007) analyzed Cu/Zn-SOD activity in blood plasma and also didn't identify significant changes. On the other hand, Ciancarelli et al. (2014) reported increased levels of serum Cu/Zn-SOD in the HD patients compared to the controls. These conflicting results both in blood fluids and erythrocytes indicate that further studies are needed before considering peripheral Cu/Zn-SOD levels or activity as a marker of oxidative stress in HD, while Mn-SOD might be a more interesting target.

Table 5 Overview of studies on levels of antioxidants in periphery in HD patient samples

Analyte	Sample	Methods	Stage of HD	Results	Reference
GSH	Plasma	Bioxytech GSH-420 and GSH-400 kits	HD1, moderate severity (N = 12)	24.03 ± 8.47 μM	Túnez et al. (2011)
			HD2, mild severity (N = 7)	24.68 ± 5.25 μM	
			HD1 and HD2 together (N = 19)	24.63 ± 7.44 μM	
			Control (N = 19)	30.19 ± 4.84 μM	
GSSG			HD1, moderate severity (N = 12)	13.73 ± 9.32 μM	
			HD2, mild severity (N = 7)	11.59 ± 5.06 μM	
			HD1 and HD2 together (N = 19)	13.06 ± 8.11 μM	
			Control (N = 19)	9.95 ± 5.9 μM	
GSH	Erythrocytes	Bioxytech GSH-420 and GSH-400 kits	HD1, moderate severity (N = 12)	126.2 ± 121.26 μM/g Hb	
			HD2, mild severity (N = 7)	114.72 ± 18.8 μM/g Hb	
			HD1 and HD2 together (N = 19)	122.8 ± 14.87 μM/g Hb	
			Control (N = 19)	141.5 ± 30.6 μM/g Hb	
GSSG			HD1, moderate severity (N = 12)	46.19 ± 16.27 μM/g Hb	
			HD2, mild severity (N = 7)	40.58 ± 15.74 μM/g Hb	
			HD1 and HD2 together (N = 19)	44.95 ± 15.29 μM/g Hb	
			Control (N = 19)	42.47 ± 29.3 μM/g Hb	
GPx		L. Flohé and W.A. Günzler (1984)	HD1, moderate severity (N = 12)	0.6 ± 0.01 (U/mL)/g Hb	
			HD2, mild severity (N = 7)	0.5 ± 0.017 (U/mL)/g Hb	
			HD1 and HD2 together (N = 19)	0.6 ± 0.14 (U/mL)/g Hb	
			Control (N = 19)	0.20 ± 0.1 (U/mL)/g Hb	
SOD		BioVision Colorimetric assay	HD1, moderate severity (N = 12)	1.29 ± 0.74 (U/mL)/g Hb	
			HD2, mild severity (N = 7)	1.86 ± 0.78 (U/mL)/g Hb	
			HD1 and HD2 together (N = 19)	1.64 ± 0.21 (U/mL)/g Hb	
			Control (N = 19)	2.24 ± 0.18 (U/mL)/g Hb	

(continued)

Table 5 (continued)

Analyte	Sample	Methods	Stage of HD	Results	Reference
GPx	Erythrocytes	D.E. Paglia and W.N. Valentine (1967)	HD (N = 16)	49.9 ± 4.17 U/g Hb	Chen et al. (2007)
			Control (N = 36)	70.14 ± 3.22 U/g Hb	
Cu/Zn-SOD		RANSOD kit	HD (N = 16)	1063.2 ± 91.9 U/g Hb	
			Control (N = 36)	1300.8 ± 40.0 U/g Hb	
Cu/Zn-SOD	Serum	ELISA Bender MedSystems GmbH	HD (N = 34)	51.61 ± 4.20 ng/ml	Ciancarelli et al. (2014)
			Control (N = 10)	30.06 ± 3.32 ng/ml	
GSH	Plasma	by Lang	PreHD (N = 11)	8.48 (7.01–14.1) µmol/mL	Klepac et al. (2007)
			Controls (N = 22)	13.8 (6.1–18.5) µmol/mL	
			HD (N = 19)	8.21 (3.54–13.6) µmol/mL	
			Controls (N = 47)	11.4 (6.1–19.2) µmol/mL	
CAT		by Johansson and Borg	PreHD (N = 11)	40.4 (19.2–54.8) µmol/mg	
			Controls (N = 22)	27.0 (22.0–55.0) µmol/mg	
			HD (N = 19)	35.3 (19.4–56.3) µmol/mg	
			Controls (N = 47)	29.9 (19.2–68.1) µmol/mg	
SOD		by Marklund	PreHD (N = 11)	0.89 (0.24–2.89) U/mg	
			Controls (N = 22)	1.29 (0.02–4.91) U/mg	
			HD (N = 19)	0.96 (0.09–2.68) U/mg	
			Controls (N = 47)	0.96 (0.02–3.95) U/mg	

GSH	Plasma	BIOXYTECH® GSH/GSSG-412 kit	HD (N = 14)	663.4 (45–1109.3) μM	Peña-Sánchez et al. (2015)
			Control (N = 29)	749.2 (481–953.5) μM	
GPx	Erythrocytes	D.E. Paglia and W.N. Valentine (1967)	HD (N = 14)	326.6 (170.5–514.5) mU/mgHb	
			Control (N = 29)	215.0 (71.9–364.4) mU/mgHb	
GR		by Carlberg and Mannervik	HD (N = 14)	1.5 (0.6–4.3) mU/mgHb	
			Control (N = 29)	4.2 (1.7–18.5) mU/mgHb	
Cu/Zn-SOD		by Marklund and Marklund	HD (N = 14)	2626.1 (1971.5–4016.9) U/gHb	
			Control (N = 29)	2816.8 (1425.3–4428.9) U/gHb	
CAT		method of Aebi	HD (N = 14)	1383.5 (1080.9–2190.5) U/gHb	
			Control (N = 29)	1365.0 (856.1–2190.3) U/gHb	
UA	Plasma	Colorimetric enzymatic reaction kit (1–3802 Salimetrics)	PreHD (N = 31, 18:13 F/M)	~4.5 mg/dl F, 6 mg/dl M	Corey-Bloom et al. (2020)
			HD (N = 38, 22:16 F/M)	~4.5 mg/dl F, 5.4 mg/dl M	
			Control (N = 38, 18:20 F/M)	~5.9 mg/dl F, 5.5 mg/dl M	
	Saliva		PreHD (N = 49, 26:23 F/M)	~3 mg/dl F, 5 mg/dl M	
			HD (N = 45, 29:16 F/M)	~2.6 mg/dl F, 2 mg/dl M	
			Control (N = 84, 41:43 F/M)	~3.8 mg/dl F, 4.2 mg/dl M	
	Postmortem brain homogenate supernatant		Control (N = 10, 3:7 F/M)	~35 ng/mg protein	
			HD (N = 10, 3:7 F/M)	~15 ng/mg protein	

GSH reduced glutathione, *GSSG* oxidized glutathione, *GPx* glutathione peroxidase, *GR* glutathione reductase, *Cu/Zn-SOD* copper zinc superoxide dismutase, *SOD* superoxide dismutase, *UA* uric acid, *PreHD* premanifest HD, *HD* manifest HD. Values in parentheses in (Klepac et al. 2007) represent range, while in Peña-Sánchez et al. (2015) they represent 10–90 percentile. Estimates from graphical representation are provided in the table where exact numbers are not listed in the manuscript (labeled as ~)

Regarding CAT, activity was found to be increased by approximately 50% in the human HD striatum and cortex (Sorolla et al., 2010). However, neither CAT plasma levels (Klepac et al., 2007) or plasma activity (Peña-Sánchez et al., 2015) were significantly changed in human HD blood samples. GPx1 levels were increased in both the cortex and striatum, while GPx6 was increased in the striatum only and total GPx activity was higher in both regions in HD brains compared to the controls (Sorolla et al., 2010). The GPx levels were found to be significantly lower in the HD erythrocytes by Chen et al. (2007), but the erythrocyte GPx activity was significantly higher in the HD than in the controls in two other studies (Peña-Sánchez et al., 2015; Túnez et al., 2011).

Non-enzymatic Antioxidants

Túnez et al. (2011) observed decreased GSH levels and GSH/GSSG ratio, and increased GSSG levels in the plasma of HD patients compared to controls, while only a decrease in GSH was detected in erythrocytes in the same study. In a study by Klepac et al. (2007), plasma GSH levels were reduced by 28% in manifest HD patients and by 27% in premanifest HD individuals compared to their matched controls. No significant difference was observed between the premanifest and the manifest HD group. Peña-Sánchez et al. (2015) reported no significant difference in the plasma GSH levels between the controls and HD, while glutathione reductase (GR) activity was reduced in HD erythrocytes.

A study by Auinger et al. (2010) analyzed baseline serum UA in 347 HD subjects and classified them into quintiles based on the UA level. Change in total functional capacity (TFC) score over 30 months from baseline was related to UA quintiles and a significant association between higher baseline UA level and slower HD progression was identified. Corey-Bloom et al. (2020) analyzed UA levels in blood plasma and saliva of a cohort of premanifest HD, manifest HD, and control subjects. While in the control group there was no difference in UA levels between genders, in both premanifest HD and HD, plasma and saliva UA levels were significantly higher in males. For this reason, authors compared UA levels across diagnostic groups for both genders separately. In females, the plasma UA levels were significantly lower in both the premanifest HD, and the HD group compared to controls. No significant change was detected in male subjects across the groups. UA levels were significantly reduced in the saliva of the premanifest HD and manifest HD females. In contrast in males, only the manifest HD group displayed a significant reduction in UA levels compared to both the premanifest and control groups. In a separate study, authors showed a significant reduction in UA levels in the prefrontal cortex of HD patients compared to the normal subjects. Given that significant differences in UA levels based on HD status were measured in saliva, a biofluid accessible via non-invasive collection, UA warrants further investigation as a biomarker of HD.

6 Conclusion and Future Directions

Many studies implicate mitochondrial dysfunction and oxidative stress as important players in HD pathogenesis. However, in HD and other neurodegenerative diseases, associated tissue damage predominantly occurs in the CNS, which is relatively isolated from the rest of the body. This complicates the search for reliable biomarkers, as these should be specific to a given pathological process, reflect disease stage/progression, and be relatively non-invasive to collect and measure. Currently, the most specific and least invasive approaches in HD are brain imaging methods and analysis of NfL in blood plasma.

In this review, we focused on biofluid oxidative stress biomarkers measured in peripheral fluids, which could be divided into two basic groups: 1. Products of ROS damage to biomolecules – nucleic acids, proteins, and lipids; 2. Components of the antioxidant defense system, which reflect the response to elevated oxidative stress. From the first group, 8-OHdG, reflecting both nDNA and mtDNA damage, provides conflicting results; the largest study to date by Borowsky et al. (2013) concluded it can't be used as a reliable HD biomarker. mtDNA copy number should reflect mitochondrial damage and thus could be interesting, but few studies have been conducted so far, and with a limited number of subjects; these studies have also provided conflicting results. The most promising of DNA damage markers is the measure of telomere shortening. In all studies reported, telomeres of manifest HD subjects, measured using blood cells, were significantly shorter compared to healthy subjects. The situation in premanifest HD is less clear, but data suggest that telomere length is not sensitive enough early in the disease process to be useful for predicting the time to symptom onset. It is unclear, whether telomere length responds to treatments targeting oxidative stress or mHTT lowering. Protein carbonylation is understudied as HD biomarker in peripheral fluids and no clear conclusion can be made from the available data. Markers of lipid peroxidation MDA and 4-HNE were increased in the plasma/serum of HD subjects compared to healthy controls in all studies, which is promising. The biggest challenge for the routine use of these lipid peroxidation products as biomarkers is methodological. GC and HPLC based methods are precise and specific, but require expensive instrumentation, complicated sample preparation and have relatively low throughput. On the other hand, chromogenic TBARS assays are cheap and fast, but results can be affected by interference from the detection of a wide range of other products, and these assays therefore lack both precision and specificity. Further methodological developments in this area are needed. From the second group of potential biomarkers, components of the antioxidative defense system, non-enzymatic antioxidants GSH and UA are promising. Both markers seem to decrease in HD compared to healthy controls, are relatively easy to measure and UA can be analyzed in saliva, making it completely non-invasive.

Oxidative stress is elevated under many physiological and pathological conditions and so its biomarkers alone can't be used as reliable readouts for HD. However, in combination with more specific biomarkers, they could still be useful for monitoring disease progression or response to various treatments. For future progress in this area, methodological improvements are needed to make the analysis of individual markers more specific and increase sample throughput. Development of composite scores reflecting overall oxidative stress load and further studies testing biomarker potential of such composite scores in larger premanifest and symptomatic HD cohorts and in HTT lowering or oxidative stress targeting therapies in animal models is needed.

Acknowledgments This work was supported by the Czech Science Foundation project 22-24983S.

References

Acevedo-Torres, K., Berríos, L., Rosario, N., Dufault, V., Skatchkov, S., Eaton, M. J., Torres-Ramos, C. A., & Ayala-Torres, S. (2009). Mitochondrial DNA damage is a hallmark of chemically induced and the R6/2 transgenic model of Huntington's disease. *DNA Repair, 8*, 126–136. https://doi.org/10.1016/j.dnarep.2008.09.004

Alam, Z. I., Halliwell, B., & Jenner, P. (2000). No evidence for increased oxidative damage to lipids, proteins, or DNA in Huntington's disease. *Journal of Neurochemistry, 75*, 840–846. https://doi.org/10.1046/j.1471-4159.2000.0750840.x

Altomare, A., Baron, G., Gianazza, E., Banfi, C., Carini, M., & Aldini, G. (2021). Lipid peroxidation derived reactive carbonyl species in free and conjugated forms as an index of lipid peroxidation: Limits and perspectives. *Redox Biology, 42*, 101899. https://doi.org/10.1016/j.redox.2021.101899

Angelova, P. R., Esteras, N., & Abramov, A. Y. (2021). Mitochondria and lipid peroxidation in the mechanism of neurodegeneration: Finding ways for prevention. *Medicinal Research Reviews, 41*, 770–784. https://doi.org/10.1002/med.21712

Askeland, G., Dosoudilova, Z., Rodinova, M., Klempir, J., Liskova, I., Kuśnierczyk, A., Bjørås, M., Nesse, G., Klungland, A., Hansikova, H., & Eide, L. (2018). Increased nuclear DNA damage precedes mitochondrial dysfunction in peripheral blood mononuclear cells from Huntington's disease patients. *Scientific Reports, 8*, 9817. https://doi.org/10.1038/s41598-018-27985-y

Auinger, P., Kieburtz, K., & McDermott, M. P. (2010). The relationship between uric acid levels and Huntington's disease progression. *Movem Disord, 25*, 224–228. https://doi.org/10.1002/mds.22907

Bar-Or, D., Thomas, G. W., Rael, L. T., Lau, E. P., & Winkler, J. V. (2001). Asp-Ala-His-Lys (DAHK) inhibits copper-induced oxidative DNA double strand breaks and telomere shortening. *Biochemical and Biophysical Research Communications, 282*, 356–360. https://doi.org/10.1006/bbrc.2001.4533

Bates, G. P., Dorsey, R., Gusella, J. F., Hayden, M. R., Kay, C., Leavitt, B. R., Nance, M., Ross, C. A., Scahill, R. I., Wetzel, R., Wild, E. J., & Tabrizi, S. J. (2015). Huntington disease. *Nature Reviews. Disease Primers, 1*, 15005. 10/gfsnsr.

Bernadotte, A., Mikhelson, V. M., & Spivak, I. M. (2016). Markers of cellular senescence. Telomere shortening as a marker of cellular senescence. *Aging (Albany NY), 8*, 3–11.

Biglan, K. M., Dorsey, E. R., Evans, R. V. V., Ross, C. A., Hersch, S., Shoulson, I., Matson, W., Kieburtz, K., & Investigators, for the H.S.G.P. (2012). Plasma 8-hydroxy-2′-deoxyguanosine levels in Huntington disease and healthy controls treated with coenzyme Q10. *Journal of Huntington's Disease, 1*, 65–69. https://doi.org/10.3233/JHD-2012-120007

Blackburn, E. H., & Szostak, J. W. (1984). The molecular structure of centromeres and telomeres. *Annual Review of Biochemistry, 53*, 163–194. https://doi.org/10.1146/annurev.bi.53.070184.001115

Bogdanov, M. B., Beal, M. F., McCabe, D. R., Griffin, R. M., & Matson, W. R. (1999). A carbon column-based liquid chromatography electrochemical approach to routine 8-hydroxy-2′-deoxyguanosine measurements in urine and other biologic matrices: A one-year evaluation of methods. *Free Radical Biology & Medicine, 27*, 647–666. https://doi.org/10.1016/S0891-5849(99)00113-6

Borowsky, B., Warner, J., Leavitt, B. R., Tabrizi, S. J., Roos, R. A. C., Durr, A., Becker, C., Sampaio, C., Tobin, A. J., & Schulman, H. (2013). 8OHdG is not a biomarker for Huntington disease state or progression. *Neurology, 80*, 1934–1941. https://doi.org/10.1212/WNL.0b013e318293e1a1

Borst, J. W., Visser, N. V., Kouptsova, O., & Visser, A. J. W. G. (2000). Oxidation of unsaturated phospholipids in membrane bilayer mixtures is accompanied by membrane fluidity changes. *Biochimica et Biophysica Acta (BBA), 1487*, 61–73. https://doi.org/10.1016/S1388-1981(00)00084-6

Bradley-Whitman, M. A., Timmons, M. D., Beckett, T. L., Murphy, M. P., Lynn, B. C., & Lovell, M. A. (2014). Nucleic acid oxidation: An early feature of Alzheimer's disease. *Journal of Neurochemistry, 128*, 294–304. https://doi.org/10.1111/jnc.12444

Brouillet, E., Jacquard, C., Bizat, N., & Blum, D. (2005). 3-Nitropropionic acid: A mitochondrial toxin to uncover physiopathological mechanisms underlying striatal degeneration in Huntington's disease. *Journal of Neurochemistry, 95*, 1521–1540. https://doi.org/10.1111/j.1471-4159.2005.03515.x

Bruenner, B. A., Jones, A. D., & German, J. B. (1995). Direct characterization of protein adducts of the lipid peroxidation product 4-Hydroxy-2-nonenal using electrospray mass spectrometry. *Chemical Research in Toxicology, 8*, 552–559. https://doi.org/10.1021/tx00046a009

Butterfield, D. A., & Kanski, J. (2001). Brain protein oxidation in age-related neurodegenerative disorders that are associated with aggregated proteins. *Mechanisms of Ageing and Development, 122*, 945–962. https://doi.org/10.1016/S0047-6374(01)00249-4

Cawthon, R. M. (2009). Telomere length measurement by a novel monochrome multiplex quantitative PCR method. *Nucleic Acids Research, 37*, e21. https://doi.org/10.1093/nar/gkn1027

Chai, W., Ford, L. P., Lenertz, L., Wright, W. E., & Shay, J. W. (2002). Human Ku70/80 associates physically with telomerase through interaction with hTERT. *The Journal of Biological Chemistry, 277*, 47242–47247. https://doi.org/10.1074/jbc.M208542200

Chen, C.-M., Wu, Y.-R., Cheng, M.-L., Liu, J.-L., Lee, Y.-M., Lee, P.-W., Soong, B.-W., & Chiu, D. T.-Y. (2007). Increased oxidative damage and mitochondrial abnormalities in the peripheral blood of Huntington's disease patients. *Biochemical and Biophysical Research Communications, 359*, 335–340. https://doi.org/10.1016/j.bbrc.2007.05.093

Chen, X., Guo, C., & Kong, J. (2012). Oxidative stress in neurodegenerative diseases. *Neural Regeneration Research, 7*, 376–385. https://doi.org/10.3969/j.issn.1673-5374.2012.05.009

Cheng, J., Wang, F., Yu, D.-F., Wu, P.-F., & Chen, J.-G. (2011). The cytotoxic mechanism of malondialdehyde and protective effect of carnosine via protein cross-linking/mitochondrial dysfunction/reactive oxygen species/MAPK pathway in neurons. *European Journal of Pharmacology, 650*, 184–194. https://doi.org/10.1016/j.ejphar.2010.09.033

Christofides, J., Bridel, M., Egerton, M., Mackay, G. M., Forrest, C. M., Stoy, N., Darlington, L. G., & Stone, T. W. (2006). Blood 5-hydroxytryptamine, 5-hydroxyindoleacetic acid and melatonin levels in patients with either Huntington's disease or chronic brain injury. *Journal of Neurochemistry, 97*, 1078–1088. https://doi.org/10.1111/j.1471-4159.2006.03807.x

Ciancarelli, I., De Amicis, D., Di Massimo, C., Di Scanno, C., Pistarini, C., D'Orazio, N., & Tozzi Ciancarelli, M. G. (2014). Peripheral biomarkers of oxidative stress and their limited potential in evaluation of clinical features of Huntington's patients. *Biomarkers, 19*, 452–456. https://doi.org/10.3109/1354750X.2014.935955

Coppede, F., & Migliore, L. (2009). DNA damage and repair in Alzheimers disease. *Current Alzheimer Research, 6*, 36–47. https://doi.org/10.2174/156720509787313970

Corey-Bloom, J., Haque, A., Aboufadel, S., Snell, C., Fischer, R. S., Granger, S. W., Granger, D. A., & Thomas, E. A. (2020). Uric acid as a potential peripheral biomarker for disease features in Huntington's patients. *Frontiers in Neuroscience, 14*. https://doi.org/10.3389/fnins.2020.00073

Cui, L., Jeong, H., Borovecki, F., Parkhurst, C. N., Tanese, N., & Krainc, D. (2006). Transcriptional repression of PGC-1α by Mutant Huntingtin leads to mitochondrial dysfunction and neurodegeneration. *Cell, 127*, 59–69. 10/chn36j. https://doi.org/10.1016/j.cell.2006.09.015
de Lange, T. (2005). Shelterin: The protein complex that shapes and safeguards human telomeres. *Genes & Development, 19*, 2100–2110. https://doi.org/10.1101/gad.1346005
Disatnik, M.-H., Joshi, A. U., Saw, N. L., Shamloo, M., Leavitt, B. R., Qi, X., & Mochly-Rosen, D. (2016). Potential biomarkers to follow the progression and treatment response of Huntington's disease. *The Journal of Experimental Medicine, 213*, 2655–2669. https://doi.org/10.1084/jem.20160776
Duchen, M. R. (2000). Mitochondria and calcium: From cell signalling to cell death. *The Journal of Physiology, 529*, 57–68. https://doi.org/10.1111/j.1469-7793.2000.00057.x
Enokido, Y., Tamura, T., Ito, H., Arumughan, A., Komuro, A., Shiwaku, H., Sone, M., Foulle, R., Sawada, H., Ishiguro, H., Ono, T., Murata, M., Kanazawa, I., Tomilin, N., Tagawa, K., Wanker, E. E., & Okazawa, H. (2010). Mutant huntingtin impairs Ku70-mediated DNA repair. *The Journal of Cell Biology, 189*, 425–443. https://doi.org/10.1083/jcb.200905138
Esterbauer, H., Schaur, R. J., & Zollner, H. (1991). Chemistry and biochemistry of 4-hydroxynonenal, malonaldehyde and related aldehydes. *Free Radical Biology & Medicine, 11*, 81–128. https://doi.org/10.1016/0891-5849(91)90192-6
Furukawa, M., & Xiong, Y. (2005). BTB protein Keap1 targets antioxidant transcription factor Nrf2 for ubiquitination by the Cullin 3-Roc1 Ligase. *Molecular and Cellular Biology, 25*, 162–171. https://doi.org/10.1128/MCB.25.1.162-171.2005
Gaschler, M. M., & Stockwell, B. R. (2017). Lipid peroxidation in cell death. *Biochemical and Biophysical Research Communications, 482*, 419–425. https://doi.org/10.1016/j.bbrc.2016.10.086
Giacomello, M., Pyakurel, A., Glytsou, C., & Scorrano, L. (2020). The cell biology of mitochondrial membrane dynamics. *Nature Reviews. Molecular Cell Biology, 21*, 204–224. https://doi.org/10.1038/s41580-020-0210-7
Gmitterová, K., Heinemann, U., Gawinecka, J., Varges, D., Ciesielczyk, B., Valkovic, P., Benetin, J., & Zerr, I. (2009). 8-OHdG in cerebrospinal fluid as a marker of oxidative stress in various neurodegenerative diseases. *NDD, 6*, 263–269. https://doi.org/10.1159/000237221
Gu, M., Gash, M. T., Mann, V. M., Javoy-Agid, F., Cooper, J. M., & Schapira, A. H. (1996). Mitochondrial defect in Huntington's disease caudate nucleus. *Annals of Neurology, 39*, 385–389. 10/dxjzjn.
Guedes-Dias, P., Pinho, B. R., Soares, T. R., de Proença, J., Duchen, M. R., & Oliveira, J. M. A. (2016). Mitochondrial dynamics and quality control in Huntington's disease. *Neurobiology of Disease, 90*, 51–57. https://doi.org/10.1016/j.nbd.2015.09.008
Guo, C., Sun, L., Chen, X., & Zhang, D. (2013). Oxidative stress, mitochondrial damage and neurodegenerative diseases. *Neural Regeneration Research, 8*, 2003–2014. https://doi.org/10.3969/j.issn.1673-5374.2013.21.009
Gusella, J. F., & MacDonald, M. E. (2006). Huntington's disease: Seeing the pathogenic process through a genetic lens. *Trends in Biochemical Sciences, 31*, 533–540.
Henrikus, S. S., Henry, C., McDonald, J. P., Hellmich, Y., Bruckbauer, S. T., Cherry, M. E., Wood, E. A., Woodgate, R., Cox, M. M., van Oijen, A. M., Ghodke, H., & Robinson, A. (2019). DNA double-strand breaks induced by reactive oxygen species promote DNA polymerase IV activity in Escherichia coli. https://doi.org/10.1101/533422.
Hersch, S. M., Gevorkian, S., Marder, K., Moskowitz, C., Feigin, A., Cox, M., Como, P., Zimmerman, C., Lin, M., Zhang, L., Ulug, A. M., Beal, M. F., Matson, W., Bogdanov, M., Ebbel, E., Zaleta, A., Kaneko, Y., Jenkins, B., Hevelone, N., Zhang, H., Yu, H., Schoenfeld, D., Ferrante, R., & Rosas, H. D. (2006). Creatine in Huntington disease is safe, tolerable, bioavailable in brain and reduces serum 8OH2′dG. *Neurology, 66*, 250–252. https://doi.org/10.1212/01.wnl.0000194318.74946.b6
Jędrak, P., Mozolewski, P., Węgrzyn, G., & Więckowski, M. R. (2018). Mitochondrial alterations accompanied by oxidative stress conditions in skin fibroblasts of Huntington's disease patients. *Metabolic Brain Disease, 33*, 2005–2017. https://doi.org/10.1007/s11011-018-0308-1

Kasai, H., & Kawai, K. (2016). 8-Hydroxyguanine, an oxidative DNA and RNA modification. In S. Jurga, V. A. Erdmann (Deceased), & J. Barciszewski (Eds.), *Modified nucleic acids in biology and medicine, RNA Technologies* (pp. 147–185). Springer International Publishing. https://doi.org/10.1007/978-3-319-34175-0_7

Kino, K., Hirao-Suzuki, M., Morikawa, M., Sakaga, A., & Miyazawa, H. (2017). Generation, repair and replication of guanine oxidation products. *Genes Environment, 39*, 1–8. https://doi.org/10.1186/s41021-017-0081-0

Klepac, N., Relja, M., Klepac, R., Hećimović, S., Babić, T., & Trkulja, V. (2007). Oxidative stress parameters in plasma of Huntington's disease patients, asymptomatic Huntington's disease gene carriers and healthy subjects: A cross-sectional study. *Journal of Neurology, 254*, 1676–1683. https://doi.org/10.1007/s00415-007-0611-y

Kota, L. N., Bharath, S., Purushottam, M., Moily, N. S., Sivakumar, P. T., Varghese, M., Pal, P. K., & Jain, S. (2015). Reduced telomere length in neurodegenerative disorders may suggest shared biology. *JNP, 27*, e92–e96. https://doi.org/10.1176/appi.neuropsych.13100240

Lee, J., Kosaras, B., Del Signore, S. J., Cormier, K., McKee, A., Ratan, R. R., Kowall, N. W., & Ryu, H. (2011). Modulation of lipid peroxidation and mitochondrial function by nordihydroguaiaretic acid (NDGA) improves neuropathology in Huntington's disease mice. *Acta Neuropathologica, 121*, 487–498. https://doi.org/10.1007/s00401-010-0788-5

Levine, R. L., Garland, D., Oliver, C. N., Amici, A., Climent, I., Lenz, A.-G., Ahn, B.-W., Shaltiel, S., & Stadtman, E. R. (1990). Determination of carbonyl content in oxidatively modified proteins. In *Methods in enzymology, oxygen radicals in biological systems part B: Oxygen radicals and antioxidants* (pp. 464–478). Academic. https://doi.org/10.1016/0076-6879(90)86141-H

Liu, C.-S., Cheng, W.-L., Kuo, S.-J., Li, J.-Y., Soong, B.-W., & Wei, Y.-H. (2008). Depletion of mitochondrial DNA in leukocytes of patients with poly-Q diseases. *Journal of the Neurological Sciences, 264*, 18–21. https://doi.org/10.1016/j.jns.2007.07.016

Long, J. D., Matson, W. R., Juhl, A. R., Leavitt, B. R., & Paulsen, J. S. (2012). 8OHdG as a marker for Huntington disease progression. *Neurobiology of Disease, 46*, 625–634. https://doi.org/10.1016/j.nbd.2012.02.012

MacDonald, M. E., Ambrose, C. M., Duyao, M. P., Myers, R. H., Lin, C., Srinidhi, L., Barnes, G., Taylor, S. A., James, M., Groot, N., MacFarlane, H., Jenkins, B., Anderson, M. A., Wexler, N. S., Gusella, J. F., Bates, G. P., Baxendale, S., Hummerich, H., Kirby, S., North, M., Youngman, S., Mott, R., Zehetner, G., Sedlacek, Z., Poustka, A., Frischauf, A.-M., Lehrach, H., Buckler, A. J., Church, D., Doucette-Stamm, L., O'Donovan, M. C., Riba-Ramirez, L., Shah, M., Stanton, V. P., Strobel, S. A., Draths, K. M., Wales, J. L., Dervan, P., Housman, D. E., Altherr, M., Shiang, R., Thompson, L., Fielder, T., Wasmuth, J. J., Tagle, D., Valdes, J., Elmer, L., Allard, M., Castilla, L., Swaroop, M., Blanchard, K., Collins, F. S., Snell, R., Holloway, T., Gillespie, K., Datson, N., Shaw, D., & Harper, P. S. (1993). A novel gene containing a trinucleotide repeat that is expanded and unstable on Huntington's disease chromosomes. *Cell, 72*, 971–983. https://doi.org/10.1016/0092-8674(93)90585-E

Mailloux, R. J. (2020). An update on mitochondrial reactive oxygen species production. *Antioxidants, 9*, 472. https://doi.org/10.3390/antiox9060472

Mann, V. M., Cooper, J. M., Javoy-Agid, F., Agid, Y., Jenner, P., & Schapira, A. H. V. (1990). Mitochondrial function and parental sex effect in Huntington's disease. *Lancet, 336*, 749. https://doi.org/10.1016/0140-6736(90)92242-A

Metcalfe, J. A., Parkhill, J., Campbell, L., Stacey, M., Biggs, P., Byrd, P. J., & Taylor, A. M. R. (1996). Accelerated telomere shortening in ataxia telangiectasia. *Nature Genetics, 13*, 350–353. https://doi.org/10.1038/ng0796-350

Milne, G. L., Yin, H., Hardy, K. D., Davies, S. S., & Roberts, L. J. (2011). Isoprostane generation and function. *Chemical Reviews, 111*, 5973–5996. https://doi.org/10.1021/cr200160h

Mirończuk-Chodakowska, I., Witkowska, A. M., & Zujko, M. E. (2018). Endogenous non-enzymatic antioxidants in the human body. *Advances in Medical Sciences, 63*, 68–78. https://doi.org/10.1016/j.advms.2017.05.005

Moi, P., Chan, K., Asunis, I., Cao, A., & Kan, Y. W. (1994). Isolation of NF-E2-related factor 2 (Nrf2), a NF-E2-like basic leucine zipper transcriptional activator that binds to the tandem NF-E2/AP1 repeat of the beta-globin locus control region. *Proceedings of the National Academy of Sciences, 91*, 9926–9930. https://doi.org/10.1073/pnas.91.21.9926

Montine, T. J., Beal, M. F., Robertson, D., Cudkowicz, M. E., Biaggioni, I., O'Donnell, H., Zackert, W. E., Roberts, L. J., & Morrow, J. D. (1999a). Cerebrospinal fluid F2-isoprostanes are elevated in Huntington's disease. *Neurology, 52*, 1104–1104. https://doi.org/10.1212/WNL.52.5.1104

Montine, T. J., Sidell, K. R., Crews, B. C., Markesbery, W. R., Marnett, L. J., Roberts, L. J., & Morrow, J. D. (1999b). Elevated CSF prostaglandin E2 levels in patients with probable AD. *Neurology, 53*, 1495–1495. https://doi.org/10.1212/WNL.53.7.1495

Montine, T. J., Shinobu, L., Montine, K. S., Roberts, L. J., II, Kowall, N. W., Beal, M. F., & Morrow, J. D. (2000). No difference in plasma or urinary F2-isoprostanes among patients with Huntington's disease or Alzheimer's disease and controls. *Annals of Neurology, 48*, 950–950. https://doi.org/10.1002/1531-8249(200012)48:6<950::AID-ANA23>3.0.CO;2-F

Moyzis, R. K., Buckingham, J. M., Cram, L. S., Dani, M., Deaven, L. L., Jones, M. D., Meyne, J., Ratliff, R. L., & Wu, J. R. (1988). A highly conserved repetitive DNA sequence, (TTAGGG) n, present at the telomeres of human chromosomes. *Proceedings of the National Academy of Sciences of the United States of America, 85*, 6622–6626.

Murphy, M. P. (2009). How mitochondria produce reactive oxygen species. *The Biochemical Journal, 417*, 1–13. https://doi.org/10.1042/BJ20081386

Murphy, M. P., Bayir, H., Belousov, V., Chang, C. J., Davies, K. J. A., Davies, M. J., Dick, T. P., Finkel, T., Forman, H. J., Janssen-Heininger, Y., Gems, D., Kagan, V. E., Kalyanaraman, B., Larsson, N.-G., Milne, G. L., Nyström, T., Poulsen, H. E., Radi, R., Van Remmen, H., Schumacker, P. T., Thornalley, P. J., Toyokuni, S., Winterbourn, C. C., Yin, H., & Halliwell, B. (2022). Guidelines for measuring reactive oxygen species and oxidative damage in cells and in vivo. *Nature Metabolism, 4*, 651–662. https://doi.org/10.1038/s42255-022-00591-z

Neely, M. D., Sidell, K. R., Graham, D. G., & Montine, T. J. (1999). The lipid peroxidation product 4-hydroxynonenal inhibits neurite outgrowth, disrupts neuronal microtubules, and modifies cellular tubulin. *Journal of Neurochemistry, 72*, 2323–2333. https://doi.org/10.1046/j.1471-4159.1999.0722323.x

Neueder, A., Kojer, K., Hering, T., Lavery, D. J., Chen, J., Birth, N., Hallitsch, J., Trautmann, S., Parker, J., Flower, M., Sethi, H., Haider, S., Lee, J.-M., Tabrizi, S. J., & Orth, M. (2022). Abnormal molecular signatures of inflammation, energy metabolism, and vesicle biology in human Huntington disease peripheral tissues. *Genome Biology, 23*, 189. https://doi.org/10.1186/s13059-022-02752-5

Oladele, J. O., Oladiji, A. T., Oladele, O. T., & Oyeleke, O. M. (2021). Reactive oxygen species in neurodegenerative diseases: Implications in pathogenesis and treatment strategies, reactive oxygen species. *IntechOpen*. https://doi.org/10.5772/intechopen.99976

Pecchillo Cimmino, T., Ammendola, R., Cattaneo, F., & Esposito, G. (2023). NOX dependent ROS generation and cell metabolism. *International Journal of Molecular Sciences, 24*, 2086. https://doi.org/10.3390/ijms24032086

Peña-Sánchez, M., Riverón-Forment, G., Zaldívar-Vaillant, T., Soto-Lavastida, A., Borrero-Sánchez, J., Lara-Fernández, G., Esteban-Hernández, E. M., Hernández-Díaz, Z., González-Quevedo, A., Fernández-Almirall, I., Pérez-López, C., Castillo-Casañas, Y., Martínez-Bonne, O., Cabrera-Rivero, A., Valdés-Ramos, L., Guerra-Badía, R., Fernández-Carriera, R., Menéndez-Sainz, M. C., & González-García, S. (2015). Association of status redox with demographic, clinical and imaging parameters in patients with Huntington's disease. *Clinical Biochemistry, 48*, 1258–1263. https://doi.org/10.1016/j.clinbiochem.2015.06.014

PerezGrovas-Saltijeral, A., Ochoa-Morales, A., Miranda-Duarte, A., Martínez-Ruano, L., Jara-Prado, A., Camacho-Molina, A., & Hidalgo-Bravo, A. (2020). Telomere length analysis on leukocytes derived from patients with Huntington disease. *Mechanisms of Ageing and Development, 185*, 111189. https://doi.org/10.1016/j.mad.2019.111189

PerezGrovas-Saltijeral, A., Ochoa-Morales, A., Jara-Prado, A., Velázquez-Cruz, R., Rivera-Paredez, B., Dávila-OrtizdeMontellano, D., Benítez-Alonso, E. O., Santamaría-Olmedo, M., Sevilla-Montoya, R., Marfil-Marín, E., Valdés-Flores, M., Martínez-Ruano, L., Camacho-Molina, A., & Hidalgo-Bravo, A. (2023). Unraveling the role of relative telomere length and CAG expansion on initial symptoms of juvenile Huntington disease. *European Journal of Neurology, 30*, 612–621. https://doi.org/10.1111/ene.15644

Polidori, M. C., Mecocci, P., Browne, S. E., Senin, U., & Beal, M. F. (1999). Oxidative damage to mitochondrial DNA in Huntington's disease parietal cortex. *Neuroscience Letters, 272*, 53–56. https://doi.org/10.1016/S0304-3940(99)00578-9

Praticó, D., & FitzGerald, G. A. (1996). Generation of 8-epiprostaglandin F by human monocytes: Discriminate production by reactive oxygen species and Prostaglandin Endoperoxide Synthase-2 (∗). *The Journal of Biological Chemistry, 271*, 8919–8924. https://doi.org/10.1074/jbc.271.15.8919

Raut, S. K., & Khullar, M. (2023). Oxidative stress in metabolic diseases: Current scenario and therapeutic relevance. *Molecular and Cellular Biochemistry, 478*, 185–196. https://doi.org/10.1007/s11010-022-04496-z

Rodrigues, F. B., Byrne, L. M., Tortelli, R., Johnson, E. B., Wijeratne, P. A., Arridge, M., De Vita, E., Ghazaleh, N., Houghton, R., Furby, H., Alexander, D. C., Tabrizi, S. J., Schobel, S., Scahill, R. I., Heslegrave, A., Zetterberg, H., & Wild, E. J. (2020). Mutant huntingtin and neurofilament light have distinct longitudinal dynamics in Huntington's disease. *Science Translational Medicine, 12*, eabc2888. https://doi.org/10.1126/scitranslmed.abc2888

Sathasivam, K., Neueder, A., Gipson, T. A., Landles, C., Benjamin, A. C., Bondulich, M. K., Smith, D. L., Faull, R. L. M., Roos, R. A. C., Howland, D., Detloff, P. J., Housman, D. E., & Bates, G. P. (2013). Aberrant splicing of HTT generates the pathogenic exon 1 protein in Huntington disease. *Proceedings of the National Academy of Sciences, 110*, 2366–2370. https://doi.org/10.1073/pnas.1221891110

Scarabino, D., Veneziano, L., Peconi, M., Frontali, M., Mantuano, E., & Corbo, R. M. (2019). Leukocyte telomere shortening in Huntington's disease. *Journal of the Neurological Sciences, 396*, 25–29. https://doi.org/10.1016/j.jns.2018.10.024

Scarabino, D., Veneziano, L., Mantuano, E., Arisi, I., Fiore, A., Frontali, M., & Corbo, R. M. (2022). Leukocyte telomere length as potential biomarker of HD progression: A follow-up study. *International Journal of Molecular Sciences, 23*, 13449. https://doi.org/10.3390/ijms232113449

Senoner, T., & Dichtl, W. (2019). Oxidative stress in cardiovascular diseases: Still a therapeutic target? *Nutrients, 11*, 2090. https://doi.org/10.3390/nu11092090

Sfeir, A., & de Lange, T. (2012). Removal of Shelterin reveals the telomere end-protection problem. *Science, 336*, 593–597. https://doi.org/10.1126/science.1218498

Shirendeb, U. P., Calkins, M. J., Manczak, M., Anekonda, V., Dufour, B., McBride, J. L., Mao, P., & Reddy, P. H. (2012). Mutant huntingtin's interaction with mitochondrial protein Drp1 impairs mitochondrial biogenesis and causes defective axonal transport and synaptic degeneration in Huntington's disease. *Human Molecular Genetics, 21*, 406–420. https://doi.org/10.1093/hmg/ddr475

Siddiqui, A., Rivera-Sánchez, S., del Castro, R. M., Acevedo-Torres, K., Rane, A., Torres-Ramos, C. A., Nicholls, D. G., Andersen, J. K., & Ayala-Torres, S. (2012). Mitochondrial DNA damage is associated with reduced mitochondrial bioenergetics in Huntington's disease. *Free Radical Biology & Medicine, 53*. https://doi.org/10.1016/j.freeradbiomed.2012.06.008

Sies, H. (1997). Oxidative stress: Oxidants and antioxidants. *Experimental Physiology, 82*, 291–295. https://doi.org/10.1113/expphysiol.1997.sp004024

Sies, H., & Jones, D. P. (2020). Reactive oxygen species (ROS) as pleiotropic physiological signalling agents. *Nature Reviews. Molecular Cell Biology, 21*, 363–383. https://doi.org/10.1038/s41580-020-0230-3

Slatter, D. A., Avery, N. C., & Bailey, A. J. (2004). Identification of a new cross-link and unique histidine adduct from bovine serum albumin incubated with malondialdehyde*. *The Journal of Biological Chemistry, 279*, 61–69. https://doi.org/10.1074/jbc.M310608200

Sorolla, M. A., Reverter-Branchat, G., Tamarit, J., Ferrer, I., Ros, J., & Cabiscol, E. (2008). Proteomic and oxidative stress analysis in human brain samples of Huntington disease. *Free Radical Biology & Medicine, 45*, 667–678. https://doi.org/10.1016/j.freeradbiomed.2008.05.014

Sorolla, M. A., Rodríguez-Colman, M. J., Tamarit, J., Ortega, Z., Lucas, J. J., Ferrer, I., Ros, J., & Cabiscol, E. (2010). Protein oxidation in Huntington disease affects energy production and vitamin B6 metabolism. *Free Radical Biology & Medicine, 49*, 612–621. https://doi.org/10.1016/j.freeradbiomed.2010.05.016

Sosa, V., Moliné, T., Somoza, R., Paciucci, R., Kondoh, H., & LLeonart, M. E. (2013). Oxidative stress and cancer: An overview. *Ageing Research Reviews, 12*, 376–390. https://doi.org/10.1016/j.arr.2012.10.004

Spinelli, J. B., & Haigis, M. C. (2018). The multifaceted contributions of mitochondria to cellular metabolism. *Nature Cell Biology, 20*, 745–754. https://doi.org/10.1038/s41556-018-0124-1

Stewart, J. B., & Chinnery, P. F. (2015). The dynamics of mitochondrial DNA heteroplasmy: Implications for human health and disease. *Nature Reviews. Genetics, 16*, 530–542. https://doi.org/10.1038/nrg3966

Stoy, N., Mackay, G. M., Forrest, C. M., Christofides, J., Egerton, M., Stone, T. W., & Darlington, L. G. (2005). Tryptophan metabolism and oxidative stress in patients with Huntington's disease. *Journal of Neurochemistry, 93*, 611–623. https://doi.org/10.1111/j.1471-4159.2005.03070.x

Taanman, J.-W. (1999). The mitochondrial genome: Structure, transcription, translation and replication. *Biochimica et Biophysica Acta (BBA), 1410*, 103–123. https://doi.org/10.1016/S0005-2728(98)00161-3

Tabrizi, S. J., Flower, M. D., Ross, C. A., & Wild, E. J. (2020). Huntington disease: New insights into molecular pathogenesis and therapeutic opportunities. *Nature Reviews. Neurology, 16*, 529–546. https://doi.org/10.1038/s41582-020-0389-4

Trigo, D., Avelar, C., Fernandes, M., Sá, J., & da Cruz e Silva, O. (2022). Mitochondria, energy, and metabolism in neuronal health and disease. *FEBS Letters, 596*, 1095–1110. https://doi.org/10.1002/1873-3468.14298

Túnez, I., Sánchez-López, F., Agüera, E., Fernández-Bolaños, R., Sánchez, F., & Tasset, I. (2011). Important role of oxidative stress biomarkers in Huntington's disease. *Journal of Medicinal Chemistry, 54*, 5602–5606. https://doi.org/10.1021/jm200605a

Uchida, K., Sakai, K., Itakura, K., Osawa, T., & Toyokuni, S. (1997). Protein modification by lipid peroxidation products: Formation of malondialdehyde-DerivedNϵ-(2-Propenal)lysine in proteins. *Archives of Biochemistry and Biophysics, 346*, 45–52. https://doi.org/10.1006/abbi.1997.0266

von Zglinicki, T. (2002). Oxidative stress shortens telomeres. *Trends in Biochemical Sciences, 27*, 339–344. https://doi.org/10.1016/S0968-0004(02)02110-2

Wang, Y., Branicky, R., Noë, A., & Hekimi, S. (2018). Superoxide dismutases: Dual roles in controlling ROS damage and regulating ROS signaling. *The Journal of Cell Biology, 217*, 1915–1928. https://doi.org/10.1083/jcb.201708007

Wild, E. J., & Tabrizi, S. J. (2008). Biomarkers for Huntington's disease. *Expert Opinion on Medical Diagnostics, 2*, 47–62. https://doi.org/10.1517/17530059.2.1.47

Yamamoto, M., Kensler, T. W., & Motohashi, H. (2018). The KEAP1-NRF2 system: A thiol-based sensor-effector apparatus for maintaining redox homeostasis. *Physiological Reviews, 98*, 1169–1203. https://doi.org/10.1152/physrev.00023.2017

Zeun, P., Scahill, R. I., Tabrizi, S. J., & Wild, E. J. (2019). Fluid and imaging biomarkers for Huntington's disease. *Molecular and Cellular Neurosciences, 97*, 67–80. https://doi.org/10.1016/j.mcn.2019.02.004

Zhang, Y., Long, J. D., Mills, J. A., Warner, J. H., Lu, W., Paulsen, J. S., & PREDICT-HD Investigators and Coordinators of the Huntington Study Group. (2011). Indexing disease progression at study entry with individuals at-risk for Huntington disease. *American Journal of Medical Genetics. Part B, Neuropsychiatric Genetics, 156B*, 751–763. https://doi.org/10.1002/ajmg.b.31232

TAR DNA-Binding Protein 43 as a Potential Biomarker for Huntington's Disease

Jon Rodríguez-Antigüedad, Jesús Pérez-Pérez, and Jaime Kulisevsky

Abbreviations

ALS	Amyotrophic Lateral Sclerosis
ATXN1	Ataxin 1
ATXN2	Ataxin 2
bvFTLD	behavioral variant-Frontotemporal Lobar Dementia
CFTR	Cystic Fibrosis Transmembrane Conductance Regulator
CSF	Cerebrospinal fluid
DN	Dystrophic Neurites
FTLD	Frontotemporal Lobar Dementia
FUS	Fused in Sarcoma protein
GFNI	Granulofilamentous neuronal inclusions
GOF	toxic Gain-of-function
HD	Huntington's Disease
HIV-1	Human Immunodeficiency Virus type 1
Htt	Huntingtin protein
IBMPFD-ALS	Inclusion Body Myopathy, Paget's Disease, and FTLD-ALS
LATE	Limbic-predominant age-related TDP-43 encephalopathy
LOF	Loss-of-function
mHtt	mutant Huntingtin protein
MRI	Magnetic Resonance Imaging
mRNA	messenger RNA
NCI	Neural Cytoplasmatic Inclusions
nfPPA	non-fluent Primary Progressive Aphasia
NII	Neural Intranuclear Inclusions.

J. Rodríguez-Antigüedad · J. Pérez-Pérez · J. Kulisevsky (✉)
Movement Disorders Unit, Neurology Department, Hospital Sant Pau, Barcelona, Spain

Institut d'Investigacions Biomèdiques-Sant Pau (IIB-Sant Pau), Barcelona, Spain

Universitat Autònoma de Barcelona (UAB), Medicine Department, Barcelona, Spain
e-mail: jkulisevsky@santpau.cat

E. A. Thomas, G. M. Parkin (eds.), *Biomarkers for Huntington's Disease*,
Contemporary Clinical Neuroscience, https://doi.org/10.1007/978-3-031-32815-2_14

OIWM	Oligodendorglial Inclusions in White Matter
PGRN	Progranulin
PolyQ	Polyglutamine
pTDP-43	phosphorylated TDP-43
SCA	Spinocerebellar Ataxia
SIMOA	Single MOlecule Array
svPPA	semantic-variant Primary Progressive Aphasia
TDP-43	TAR DNA-binding protein of 43 kDa

1 Introduction

TAR DNA-binding protein of 43 KDa (TDP-43) is a predominantly nuclear ribonucleoprotein (RNA-binding protein) encoded by the TARDBP gene, located in chromosome 1 (1p36.22) (Chen-Plotkin et al., 2010). The expression of TDP-43 is ubiquitous in cells and tissues throughout the human body, where it plays a crucial role in gene regulation (Human Protein Atlas, 2023: https://www.proteinatlas.org).

It was first described in 1995 as a protein which binds to the pyrimidine-rich DNA motifs (also known as TAR) of the human immunodeficiency virus type 1 (HIV-1), downregulating HIV-1 transcription (Ou et al., 1995). Subsequent studies have demonstrated that TDP-43 can bind both to DNA and, more commonly, to RNA, to regulate many aspects of gene expression, such as exon splicing, messenger RNA (mRNA) stabilization, translation, transportation or degradation (Buratti & Baralle, 2001; Chen-Plotkin et al., 2010; Gao et al., 2018).

TDP-43 was described by Neumann in 2006 as the main protein component of ubiquitinated inclusions in many Frontotemporal Lobar Dementia (FTLD) and Amyotrophic Lateral Sclerosis (ALS) patients (Neumann et al., 2006). Since then, this protein has drawn the attention of the field of neurodegenerative diseases including Huntington's Disease (HD).There is increasing evidence suggesting that TDP-43, together with other proteins that are found aggregated in HD, may play a role in HD pathophysiology and potentially modulate its phenotype.

2 TDP-43 Proteinopathy

Altered levels of TDP-43 has been a pathological finding in studies of several neurodegenerative diseases. In addition to FTLD-ALS spectrum, TDP-43 has been the primary or secondary histopathological finding in other neurodegenerative diseases including HD (Table 1) (Chen-Plotkin et al., 2010). Furthermore, TDP-43 pathology can also be detected in cognitively normal elderly individuals (Arnold et al., 2013b).

TDP-43 is the pathological hallmark in 95–97% of sporadic ALS cases (Tziortzouda et al., 2021). Clinically, ALS is characterized by an upper and lower

Table 1 TDP-43 in neurodegenerative diseases

TDP-43 as a PRIMARY histopathological feature	Frontotemporal Lobar Dementia-TDP (FTLD-TDP) Amyotrophic Lateral Sclerosis (ALS) Overlap syndromes (features of FTLD-TDP and ALS) Limbic-predominant age-related TDP-43 encephalopathy (LATE) Perry syndrome FTLD–inclusion body myopathy–Paget syndrome
TDP-43 as a SECONDARY histopathological feature	Alzheimer's disease Parkinson's Disease Atypical parkinsonisms (Progressive Supranuclear Palsy, Corticobasal Degeneration, Dementia with Lewy Bodies) Huntington's disease Spinocerebellar ataxia (SCA) 1, SCA2 and SCA3 Guam ALS and Guam ALS-PD

Please refer to (Chen-Plotkin et al., 2010; Hart et al., 2012; Toyoshima & Takahashi, 2014; Nelson et al., 2019). Adapted from Chen-Plotkin et al. (2010)

motor neuron degeneration that produce muscle weakness, atrophy, spasticity, and fasciculations, among others. Pathological studies show accumulation of TDP-43 in neuronal and glial cytoplasmic inclusions (Toyoshima & Takahashi, 2014). These inclusions are found widespread in the central nervous system, affecting both motor neuron system and non-motor nuclei (Toyoshima & Takahashi, 2014).

Approximately 50% of cases with the neuropathological diagnosis of FTLD are characterized by inclusions containing TDP-43 in the cytoplasm of neurons and glia, with a concomitant loss of nuclear TDP-43. These pathological changes are believed to underlie the neuronal susceptibility and degeneration observed in the frontal and temporal cortices in FTLD (Tziortzouda et al., 2021).

Classification systems have been proposed for these diseases based on the location and morphology of TDP-43 aggregates in the nervous system. Two subtypes have been defined for ALS: Type 1, which is observed in classical ALS, and Type 2, which is observed in cases of ALS-dementia (Nishihira et al., 2008). Five different histopathological subtypes of FTLD-TDP (A-E) have been proposed and certain subtypes have been linked to specific phenotypes or genotypes (Fig. 1) (Lee et al., 2017): Type A is characterized by neuronal cytoplasmatic inclusions and dystrophic neurites in superficial cortical layers, as well as white matter pathology; Type B is characterized by neuronal cytoplasmatic inclusions in superficial and deeper cortical layers, and white matter pathology; Type C is characterized by long dystrophic neurites throughout cortical layers; Type D is characterized by lentiform neuronal intranuclear inclusions in superficial and deep layers; Type E is characterized by neuronal inclusions of granulofilamentous appearance and punctate neuropil staining in all cortical layers, as well as white matter pathology.

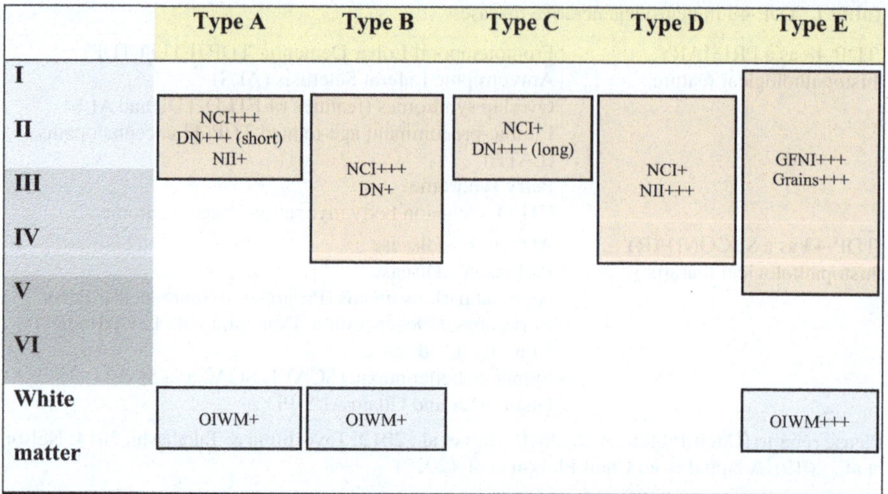

Location of pathology	Superficial	Superficial Deep	Superficial	Superficial Deep	Superficial Deep
Common phenotype	bvFTLD naPPA	bvFTLD MND	svPPA bvFTLD	IBMPFD- ALS	bvFTLD
Genetic associations	PGRN	C9orf72	-	VCP	Uncertain

Fig. 1 FTLD-TDP subtypes according to morphology and distribution of TDP-43 aggregates (Lee et al., 2017). I-VI Cortical layers I-VI, NCI Neural Cytoplasmatic Inclusions, DN Dystrophic Neurites, NII Neural intranuclear inclusions, OIWM Oligodendorglial inclusions in White matter, GFNI Granulofilamentous neuronal inclusions, Grains Dot-like neuropil aggregates, bvFTLD behavioral variant Frontotemporal Lobar Degeneration, nfPPA nonfluent Primary Progressive Aphasia, svPPA semantic variant PPA, IBMPFD-ALS Inclusion Body Myopathy, Paget's Disease, and Frontotemporal Dementia-ALS. (Adapted from Lee et al. (2017))

Pathophysiology of TDP-43

The pathophysiological relevance (pathogenic versus not pathogenic) of TDP-43 and the functional consequences (loss-of-function (LOF) versus toxic gain-of-function (GOF)) of its aggregation in neurodegenerative diseases are still a topic of debate, as are the mechanisms that precede the formation of TDP-43 inclusions (Lee et al., 2012). Most of the evidence of TDP-43 pathogenicity is derived from studies in ALS and FTLD models.

The main features of pathological TDP-43 found in the inclusions are: abnormal post translational modifications such as hyperphosphorylation, ubiquitination and aberrant cleavage; diminished solubility; and cytoplasmatic mislocalization (Sreedharan et al., 2008; Chen-Plotkin et al., 2010; Lee et al., 2012). Each feature suggests a different potential functional consequence. *Hyperphosphorylation* can lead to both a LOF or a toxic GOF, as already seen with tau in AD (Ballatore et al., 2007). *Ubiquitination*, which may represent a need of the cell to eliminate TDP-43 through proteasomal degradation, accumulation of *aberrantly cleaved* TDP-43 fragments, and *loss of solubility* of the protein, suggest a toxic GOF mechanism (Zhang et al., 2009). *Mislocalization* in the cytoplasm of a protein which participates in gene regulation suggests, at least to a certain degree, a LOF effect (Buratti & Baralle, 2009). Some studies, however, question the latter theory since they have shown that cytoplasmatic aberrant TDP-43 is sufficient to produce neurodegeneration, and nuclear TDP-43 depletion is not necessary for a toxic effect (Barmada et al., 2010; Arnold et al., 2013a).

Abnormal changes to the physiological properties of TDP-43 may also have functional consequences in neurodegenerative diseases. These include axonal transport, mitochondrial function, or RNA splicing, among others (Gao et al., 2018). For example, accumulated cytoplasmatic TDP-43 may interfere with *axonal transport* of mRNA, mitochondria, and other cargos, and may impair *mitochondrial bioenergetics*, since TDP-43 is also localized inside the mitochondria (Alami et al., 2014; Magrané et al., 2014). These two mechanisms are thought to be a consequence of a toxic GOF effect. On the other hand, the importance of TDP-43 in *RNA splicing* provide potential for a LOF mechanism. However, the absence of altered splicing in cystic fibrosis transmembrane conductance regulator (CFTR), a well-known target of TDP-43, challenges this theory (Buratti & Baralle, 2001).

The actual pathophysiological mechanisms and functional consequences of TDP-43 in neurodegenerative diseases are still unknown. However, the abovementioned evidence suggests a combination of both LOF and toxic GOF consequences.

3 TDP-43 in HD and Other PolyQ Disorders

Polyglutamine (PolyQ) Diseases are monogenic neurodegenerative disorders characterized by an abnormal expansion of the CAG trinucleotide repeat sequence in the corresponding gene, resulting in a protein with an abnormally long polyQ tract (Bunting et al., 2022). Although the precise roles of the polyQ tracts remain uncertain, it has been demonstrated that they display conformational diversity which may help the protein in its normal functions and interactions (Gerbich & Gladfelter, 2021). A pathological expansion results in increased DNA damage, intracellular aggregates, dysregulation of transcription, and proteasome dysregulation, among others, leading to a combination of motor and neuropsychiatric symptoms (Bunting et al., 2022). As of today, nine different PolyQ Diseases are known: HD, Spinocerebellar Ataxia (SCA) 1, 2, 3, 6, 7, and 17, Kennedy's Disease or Spinobulbar Muscular Atrophy, and Dentatorubral-pallidoluysian Atrophy.

Years after the discovery of the relevance of TDP-43 in ALS and FTLD, it was observed that some proteins related to polyQ diseases, such as huntingtin protein (Htt), Ataxin-1 (ATXN1) and ATXN2, were also associated with histopathological TDP-43 inclusions (Schwab et al., 2008; Toyoshima & Takahashi, 2014; Tazelaar et al., 2020). It has been observed that these proteins may not only show an association to TDP-43 but also interact with it contributing to its pathophysiology (Elden et al., 2010). This theory is supported by the fact that TDP-43 contains a glutamine/asparagine (Q/N) rich domain that binds to polyQ protein aggregates, which sequester TDP-43 in the cytoplasm impairing its functions (Fuentealba et al., 2010).

The Pathophysiology of HD

HD is a neurodegenerative disorder caused by the mutation in the HTT gene located in the short arm of chromosome 4 (Cr.4p 16.3) (Walker, 2007). The gene contains a CAG trinucleotide repeat sequence in the first exon, which encodes a polyQ domain. A trinucleotide expansion of 36–39 repeats will cause the disease with incomplete penetrance, and 40 or more repeats will cause the disease with onset guaranteed within the individual's lifetime. The expanded gene will lead to a mutant form of Htt (mHtt) that contains an abnormally long polyQ sequence (Bates et al., 2015). This aberrant protein has toxic properties and tends to aggregate in both nuclear and cytoplasmatic ubiquitin-positive inclusions, causing cellular dysfunction and death, particularly in the medium spiny neurons of the striatum (Ross & Tabrizi, 2011; Bates et al., 2015).

Clinically, HD has been described as a triad of motor, cognitive and psychiatric symptoms. The age of onset is related to the number of CAG repetitions; however, this only accounts for 50–70% of the variance (Wexler et al., 2004). Clinical presentation is also variable, and despite chorea presenting as the main and classical feature, not all the patients will develop chorea, and other motor phenotypes may be observed (Walker, 2007). It is well known that neuropsychiatric symptoms can appear as much as 15 years before HD motor diagnosis (Martinez-Horta et al., 2016). Neuropsychiatric symptoms present across a spectrum, with apathy as a more prominent and often earlier feature (Martinez-Horta et al., 2016). Irritability, obsessive compulsive behavior or psychosis are also found, but not every single patient will present with psychiatric disturbances. Cognitive decline can be detected even in the absence of motor symptoms, and its progression will lead to dementia, but this evolution is also very variable with rapid and slow progressors (Martinez-Horta et al., 2016, 2020). The cause of this variability of symptoms and of the age of presentation remains unclear, but genetic and environmental factors have been described (van Dellen & Hannan, 2004; Lee et al., 2022). Other concurrent "proteinopathies" can modulate the clinical phenotype of HD patients, such as tau, which has been described as a cognitive modifier in HD (Vuono et al., 2015).

The molecular mechanisms implicated in the pathophysiology of HD are not fully understood, but some have been proposed such as protein sequestration, synaptic dysfunction, mitochondrial dysfunction and axonal transport impairment, among others (Ross & Tabrizi, 2011; Bates et al., 2015). Htt has several roles in the cellular physiology such as cytoskeletal dynamics, endocytosis, vesicular trafficking, protein turnover or regulation of gene expression; in the presence of a pathological CAG expansion many of these functions are thought to be disrupted (Ross & Tabrizi, 2011; Saudou & Humbert, 2016). Additionally, it was observed that, like TDP-43 or Fused in sarcoma protein (FUS), Htt might have RNA binding properties and participate in the regulation of its own mRNA to avoid an excess of protein (Culver et al., 2016). Consequently, whether HD pathophysiology is caused by a toxic GOF of mHtt aggregates, LOF of Htt, or a combination of both mechanisms remains a topic of debate (Sturchio et al., 2023).

Beyond mHtt, increasing evidence suggest that neurodegeneration in HD, especially in the later stages of the disease, is the consequence of a mixture of aggregated proteins such as phosphorylated Tau, α-Synuclein, or TDP-43 (Tauffenberger et al., 2013; St-Amour et al., 2018). This co-pathology and the individual susceptibility may modulate the phenotype of HD patients.

Concurrent TDP-43 Pathology in HD

In 2008, Schwab et al. identified that TDP-43 colocalized with mHtt inclusions in HD brains, which has been replicated by other groups (Schwab et al., 2008; Coudert et al., 2019). Characteristically, these inclusions were predominantly cytoplasmic and contained phosphorylated TDP-43 (pTDP-43) (Schwab et al., 2008). These inclusions with both pTDP-43 and mHtt were particularly frequent in the lower layers of the cortex, where mHtt inclusions predominate (Schwab et al., 2008). This supports the notion that TDP-43 is, at least, a secondary hystopathological finding in HD, despite limited studies that did not find TDP-43 in HD *in vivo* models (Clippinger et al., 2013).

Potential clinical implications of TDP-43 in HD have been observed in some studies reporting individuals with mutations in the HTT gene and concurrent ALS or ALS-like phenotype. (Kanai et al., 2008; Sadeghian et al., 2011; Tada et al., 2012). Given that TDP-43 is a protein involved in the pathophysiology of ALS and also accumulates in HD, TDP-43 may serve as a potential link between these two diseases.

The histopathological study in two such patients by Tada et al. revealed neuronal cytoplasmatic inclusions of TDP-43 and mHtt inclusions in many regions of the nervous system including motor cortex and lower motor nuclei (Tada et al., 2012). The authors concluded that neuronal damage is contributed to by both a TPD-43-dependent and mHtt-dependent mechanism. However, contrary to the study conducted by Schwab et al., TDP-43 and mHtt inclusions were found in the same neurons but not in the same location within the neuron (Tada et al., 2012). The authors believe this could be due to the different methods in tissue preparation or a reflection of differences in the cases examined.

A recent HTT mutational screening study in FTLD/ALS and Dementia with Lewy Bodies cohorts identified the mutation in 7 pure FTLD/ALS cases with no apparent movement disorders, suggesting an additional phenotype for the disease (Dewan et al., 2021). The histopathological study revealed ALS-like features including loss of motor neurons of the anterior horns but scarce TDP-43 positive inclusions, as well as abundant mHtt co-pathology in the frontal lobes (Dewan et al., 2021). The authors suggest the consideration of genetic screening in sporadic FTD/ALS patients (Dewan et al., 2021).

Considered together, these studies suggest that mHtt in HD may contribute to the development of TDP-43 pathology, and the distribution of the latter may contribute to atypical FTLD/ALS-like HD phenotypes, thus reflecting its pathophysiological importance in the disease (Tada et al., 2012).

Pathophysiology of TDP-43 in HD

Even when a disease is caused by a single gene, it often reflects a disturbance in a complex intra/intercellular network due to the intricate relationship among the molecular components in the cells (Barabási et al., 2011). The fact that TDP-43 is not universally found in the brains of HD patients and, when present, is commonly observed aggregated along with mHtt inclusions, suggests that this protein is part of a complex molecular network in at least a subgroup of patients. Some studies have tried to disentangle the molecular role of TDP-43 in HD, but the literature is scarce, and several questions remain unsolved.

Does colocalization mean that both TDP-43 and Htt participate in common pathways, or is this just incidental protein colocalization after aggregation?

Human post-mortem studies of HD subjects have shown an abnormal accumulation of various neurodegenerative disease-related proteins such as TDP-43, tau or α-Synuclein (St-Amour et al., 2018). Moreover, TDP-43 protein levels have shown correlation with mHtt, tau and α-Synuclein, suggesting that this colocalization is not coincidental (St-Amour et al., 2018).

Do TDP-43 and Htt interact?

The presence of different neurodegenerative disease-related proteins, particularly in the later stages of the disease, supports the *"hypothesis of the rotten apple"* where Htt-pathology would act as a common trigger for a multiple proteinopathy (St-Amour et al., 2018). In a recent in vitro study using two different cellular models, the authors found that once mHtt reached a particular CAG length threshold, it acted as a pro-aggregating factor for TDP-43 (Coudert et al., 2019). It was hypothesized that CAG-repeat expansions can make RNA hairpin-like structures and sequester RNA binding proteins such as TDP-43.

A study using both worm and mammalian cell models showed that TPD-43 and FUS are conserved promoters of polyQ toxicity (Tauffenberger et al., 2013). While

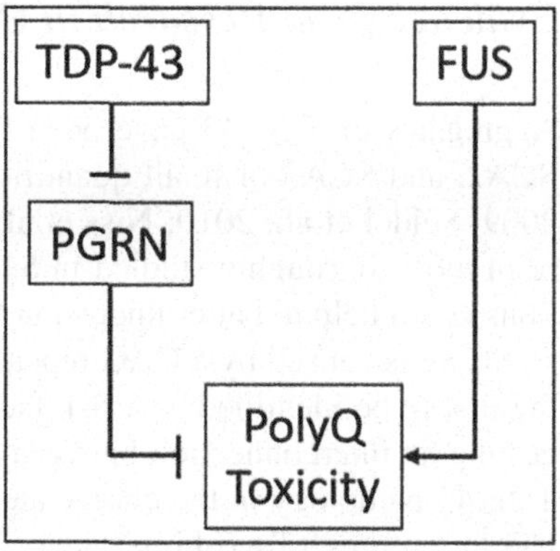

Fig. 2 Proposed relationship between TDP-43, PGRN, FUS and polyQ toxicity (Tauffenberger et al., 2013). TDP-43 TAR DNA-binding protein of 43 KDa, PGRN Progranulin, FUS Fused in sarcoma protein, PolyQ Polyglutamine. (Adapted from Tauffenberger et al. (2013))

FUS seems to have an independent mechanism through which it increases polyQ toxicity, TDP-43 acts by inhibiting the activity of the neurotrophic factor Progranulin (PGRN) (Fig. 2) (Tauffenberger et al., 2013). The directionality of the interaction between mHtt and TDP-43 is not clear, but it appears that they both participate in common pathways.

Does TDP-43 have a pathogenic role in HD or is it just incidental?

The age at onset and the heterogeneity of symptoms in HD are only partially explained by the CAG repeat expansion, suggesting that additional biological factors participate in both the onset and the progression (St-Amour et al., 2018). As already mentioned, pathologic TDP-43 can be detected in elderly individuals with intact cognition (Arnold et al., 2013b). However, the typical age of HD onset is 35–50 years and the normal lifespan after clinical onset is 15–20 years. Therefore, HD is a relatively young-adult disease, and the age alone would not explain the presence of significant TDP-43 aggregates in these patients.

Tauffenberger and colleagues showed in worm and mammalian cell models that experimental reductions in either TDP-43 or FUS resulted in diminished age-dependent neuronal dysfunction, neurodegeneration, and polyQ protein aggregation phenotypes (Tauffenberger et al., 2013). In vitro studies in yeast suggested that toxicity of aggregates occur only at extremely high levels of TDP-43 and these may be more toxic and more prone to aggregation than mHtt (Marte et al., 2022). Thus, the limited evidence available for HD suggests a toxic GOF mechanism of TDP-43. Two leading hypotheses have been proposed: the first one, a sequestration of proteins by TDP-43 impairs cell homeostasis; the second one, the interference of TDP-43 aggregates with chaperones or other protein degradation activities results in deficient protein quality control (Tauffenberger et al., 2013; Marte et al., 2022).

Additionally, the presence of phosphorylated, ubiquitinated, and cytoplasmic aggregates of TDP-43 in HD suggests that, similar to FTLD and ALS, not only toxic GOF but also LOF may be involved in the pathophysiology. However, these molecular mechanisms are still poorly understood, and further research will be needed.

Evidence from Research in Other PolyQ Diseases

Aggregates of TDP-43 have been found in other PolyQ Diseases such as SCA1, SCA2, and SCA3 in small quantities, and in cellular models of SCA7 (Tan et al., 2009; Seidel et al., 2010; Niss et al., 2022). Among these conditions, TDP-43 has been most thoroughly studied in SCA1 and SCA2. The results of these investigations could help to better understand the role of TDP-43 in HD.

SCA2 is caused by a CAG repeat expansion in the ATXN2 gene. This gene was the first to be identified as a risk factor of ALS and there is growing evidence suggesting an interconnection between ATXN2 and TDP-43. SCA2 patients can show TDP-43 pathology in the classic ataxic phenotypes, even in the absence of ALS or FTLD features (Toyoshima et al., 2011). In addition, it has been proposed that ATXN2 can lead to non-ataxic SCA2 phenotypes such as FTLD or ALS, also with TDP-43 pathology (Bäumer et al., 2014). It was suggested that these latter phenotypes are driven by the quantity and the distribution of TDP-43 pathology (Bäumer et al., 2014). That is, both ataxic and non-ataxic phenotypes of SCA2 show TDP-43 aggregates in pathology.

Several studies have tried to clarify this relationship between ATXN2 and TDP-43. Intermediate length polyQ tract expansions of ATXN2 (24–33 polyglutamines) have been associated with sporadic ALS, possibly contributing to the risk of developing the disease or modulating the age of disease onset (Elden et al., 2010; Daoud et al., 2011). ATXN2 forms a complex with TDP-43 in ALS patients' spinal cord neurons and act as a dose sensitive modulator of its toxicity in vitro and in vivo, although the exact mechanism is unknown (Elden et al., 2010). Actually, a series of sporadic ALS cases with and without ATXN2 intermediate length polyQ expansions showed that patients with the intermediate expansion presented with abundant typical filamentous TDP-43 inclusions, compared to round inclusions seen in patients without the ATXN2 intermediate expansion (Fig. 3) (Hart et al., 2012). This suggests two different pathological TDP-43 subtypes of ALS, likely with different pathogenesis, and provides support for hypotheses of interactions between TDP-43 and polyQ expansions, as well as the importance of TDP-43 in the phenotype of these diseases.

SCA1 is caused by a mutation in the ATXN1 gene which results in an expansion of the CAG repeat sequence located in exon 8 (Banfi et al., 1994). Similar to ATXN2, the expansion of CAG repeats in ATXN1 gene has been associated with increased risk of ALS, pathologically mislocalized TDP-43 in the cytoplasm, and more severe phenotypes of SCA2 in transgenic fly models (Tazelaar et al., 2020). This evidence suggests that there is also an interaction between the polyQ tract of ATXN2 and TDP-43, likely through the disturbance of nucleocytoplasmic transport.

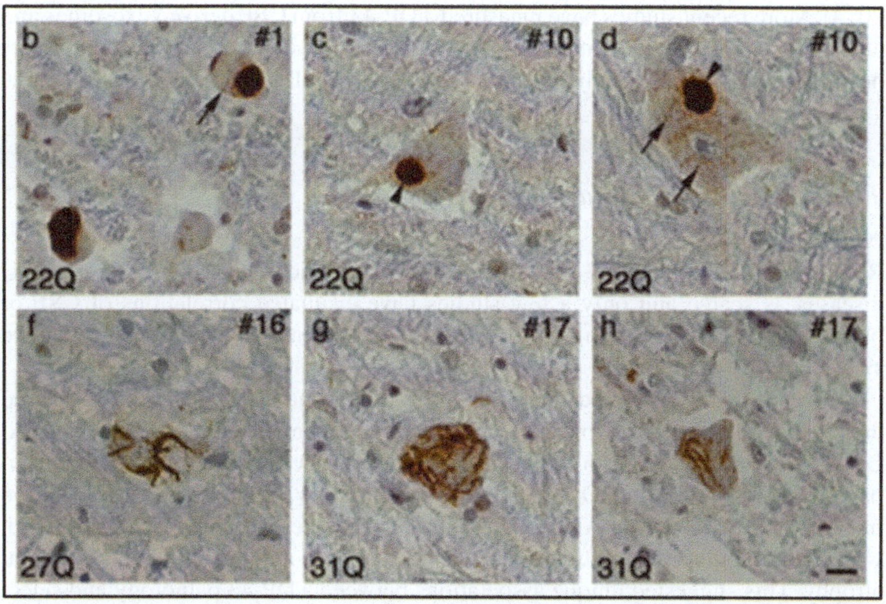

Fig. 3 TDP-43 pathology in ALS patients with and without ATXN2 intermediate polyQ expansions (24–33) (Hart et al., 2012). Cases with normal repeats of polyglutamine (top pictures) show round cytoplasmatic TDP-43 inclusions (arrowheads) and smaller round pre-inclusions (arrows). Cases with intermediate repeats of polyglutamine (bottom pictures) show cytoplasmatic filamentous and skein-like inclusions, and lack round inclusions. (Obtained and adapted with permission from Hart et al. (2012))

4 TDP-43 as a Potential Biomarker in HD

TDP-43 Related Biomarkers in Neurodegenerative Diseases

Most studies regarding TDP-43 as a biomarker have focused on the monitoring and diagnosis of FTLD and ALS. In 2008, plasma TDP-43 levels were found to be elevated in FTLD patients, suggesting that its detection in biofluids could indicate the presence and the progression of TDP-43 proteinopathies (Foulds et al., 2008). Antibody based immunological assays, such as ELISA, have been by far the most commonly used method for TDP-43 detection in biofluids (Steinacker et al., 2019). Different studies have detected elevated total or pTDP-43 levels both in blood and in cerebrospinal fluid (CSF) in patients with FTLD, ALS and a few other neurodegenerative diseases, but results have varied (Steinacker et al., 2019). Initially it was suggested that plasma TDP-43 correlated with FTLD pathology (Foulds et al., 2009); however, high variability in TDP-43 detection and direction of change across studies have prevented the advancement of TDP-43 as a biomarker for these diseases (Ren et al., 2021). CSF TDP-43, for example, is mostly blood-derived, and consequently does not adequately reflect brain TDP-43 pathology (Feneberg et al.,

2014). More recent studies have measured TDP-43 from CSF derived exosomes and anti-TDP-43 natural autoantibodies in serum, but these new potential biomarkers need further research (Feneberg et al., 2014; Conti et al., 2021).

TDP-43 Related Biomarkers in HD

Research regarding TDP-43 related biomarkers specifically in HD, or other polyQ diseases, is scarce and only a few studies have evaluated correlations of CSF or plasma TDP-43 levels with the evolution of the disease or with specific disease symptoms or phenotypes.

A recent study evaluated the association of plasma TDP-43 levels, measured using the ultrasensitive biomarker detection technique SIMOA (Quanterix, Lexington, MA, USA), with brain Magnetic Resonance Imaging (MRI) and neuropsychiatric symptoms in HD (Fig. 4a and b) (Sampedro et al., 2022). An association of plasma TDP-43 levels with frontotemporal damage and apathy was observed, which was not observed with plasma tau and Htt levels, and was unrelated to striatal atrophy. This suggested unique associations between TDP-43 pathology and select behavioral symptoms in HD. The study also compared TDP-43 levels with cognitive and motor performance, but no correlations were found.

However, these conclusions must be taken with caution considering the controversy discussed above about the actual accuracy of plasma TDP-43 levels in monitoring the TDP-43 pathology in the nervous system (Steinacker et al., 2019).

The authors of this chapter have assessed 51 HD gene expansion carriers and 29 controls ($n = 81$ total), with gene expansion carriers divided into premanifest ($n = 27$; far from onset = 10, near to onset = 17) and manifest ($n = 24$; early stage = 16 advanced stage = 8). No differences in plasma levels of TDP-43 were observed across disease stages ($p = 0.38$), and no correlation with disease burden score was found ($r = -0.007$ ($p = 0.9$)). These preliminary results reinforce the hypothesis that

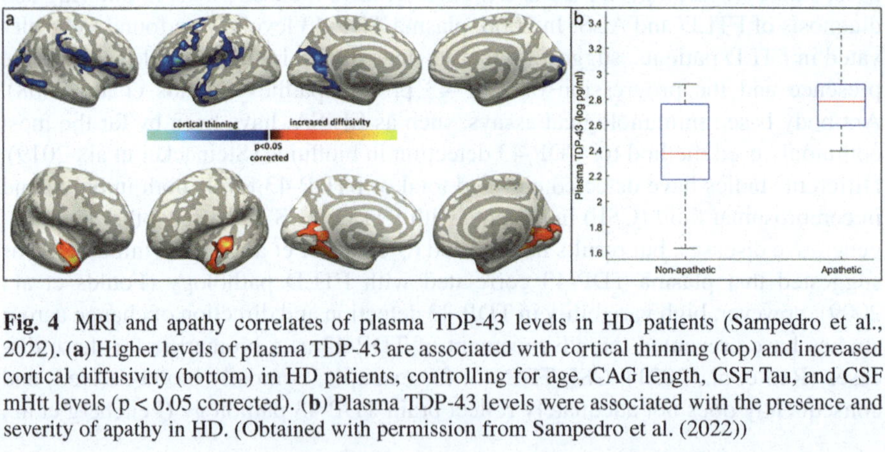

Fig. 4 MRI and apathy correlates of plasma TDP-43 levels in HD patients (Sampedro et al., 2022). (**a**) Higher levels of plasma TDP-43 are associated with cortical thinning (top) and increased cortical diffusivity (bottom) in HD patients, controlling for age, CAG length, CSF Tau, and CSF mHtt levels ($p < 0.05$ corrected). (**b**) Plasma TDP-43 levels were associated with the presence and severity of apathy in HD. (Obtained with permission from Sampedro et al. (2022))

TDP-43 may be associated with specific phenotypes, as seen with apathy in Sampedro et al. (2022), while in more heterogeneous HD cohorts, plasma TDP-43 level differences may be less apparent. However, additional longitudinal studies including clinico-pathologic correlations are needed to clarify the role of TDP-43 in HD phenotype as well as the utility of TDP-43 plasma levels in HD diagnosis, phenotyping, monitoring, and prognosis.

5 Conclusions

TDP-43 is frequently found as secondary pathology along with mHtt in the brain of individuals with HD, particularly in the advanced stages of the disease. The precise role of this protein in HD remains unclear, although studies in other neurodegenerative diseases suggest a combination of both toxic GOF and LOF consequences. Research in other PolyQ diseases show that TDP-43 may not only contribute to the pathophysiology of these diseases but also interact with the polyQ tracts of Htt and other proteins. Despite limited research on HD, it has been hypothesized that alterations in TDP-43 may be associated with specific disease phenotypes. A major challenge in this field is the difficulty of *in vivo* measuring of TDP-43 pathology, as the levels observed in biofluids may not reliably reflect the underlying nervous system pathology. Thus, TDP-43 cannot currently be used as a biomarker for HD. Further studies will be needed to clarify the role of TDP-43 in the pathogenesis and clinical phenotype of HD.

References

Alami, N. H., Smith, R. B., Carrasco, M. A., Williams, L. A., Winborn, C. S., Han, S. S. W., et al. (2014). Axonal transport of TDP-43 mRNA granules is impaired by ALS-causing mutations. *Neuron, 81*(3), 536–543.

Arnold, E. S., Ling, S. C., Huelga, S. C., Lagier-Tourenne, C., Polymenidou, M., Ditsworth, D., et al. (2013a). ALS-linked TDP-43 mutations produce aberrant RNA splicing and adult-onset motor neuron disease without aggregation or loss of nuclear TDP-43. *Proceedings of the National Academy of Sciences of the United States of America, 110*(8), E736–E745.

Arnold, S. J., Dugger, B. N., & Beach, T. G. (2013b). TDP-43 deposition in prospectively followed, cognitively normal elderly individuals: Correlation with argyrophilic grains but not other concomitant pathologies. *Acta Neuropathologica, 126*(1), 51–57.

Ballatore, C., Lee, V. M. Y., & Trojanowski, J. Q. (2007). Tau-mediated neurodegeneration in Alzheimer's disease and related disorders. *Nature Reviews. Neuroscience, 8*(9), 663–672.

Banfi, S., Servadio, A., Chung, M. Y., Kwiatkowski, T. J., McCall, A. E., Duvick, L. A., et al. (1994). Identification and characterization of the gene causing type 1 spinocerebellar ataxia. *Nature Genetics, 7*(4), 513–520.

Barabási, A. L., Gulbahce, N., & Loscalzo, J. (2011). Network medicine: A network-based approach to human disease. *Nature Reviews. Genetics, 12*(1), 56–68.

Barmada, S. J., Skibinski, G., Korb, E., Rao, E. J., Wu, J. Y., & Finkbeiner, S. (2010). Cytoplasmic mislocalization of TDP-43 is toxic to neurons and enhanced by a mutation associated with familial amyotrophic lateral sclerosis. *The Journal of Neuroscience, 30*(2), 639–649.

Bates, G. P., Dorsey, R., Gusella, J. F., Hayden, M. R., Kay, C., Leavitt, B. R., et al. (2015). Huntington disease. *Nature Reviews Disease Primers, 1*(1), 15005.

Bäumer, D., East, S. Z., Tseu, B., Zeman, A., Hilton, D., Talbot, K., et al. (2014). FTLD-ALS of TDP-43 type and SCA2 in a family with a full ataxin-2 polyglutamine expansion. *Acta Neuropathologica, 128*(4), 597–604.

Bunting, E. L., Hamilton, J., & Tabrizi, S. J. (2022). Polyglutamine diseases. *Current Opinion in Neurobiology, 72*, 39–47.

Buratti, E., & Baralle, F. E. (2001). Characterization and functional implications of the RNA binding properties of nuclear factor TDP-43, a novel splicing regulator of CFTR exon 9. *The Journal of Biological Chemistry, 276*(39), 36337–36343.

Buratti, E., & Baralle, F. E. (2009). The molecular links between TDP-43 dysfunction and neurodegeneration. *Advances in Genetics, 66*, 1–34.

Chen-Plotkin, A. S., Lee, V. M. Y., & Trojanowski, J. Q. (2010). TAR DNA-binding protein 43 in neurodegenerative disease. *Nature Reviews Neurology, 6*(4), 211–220.

Clippinger, A. K., D'Alton, S., Lin, W. L., Gendron, T. F., Howard, J., Borchelt, D. R., et al. (2013). Robust cytoplasmic accumulation of phosphorylated TDP-43 in transgenic models of tauopathy. *Acta Neuropathologica, 126*(1), 39–50.

Conti, E., Sala, G., Diamanti, S., Casati, M., Lunetta, C., Gerardi, F., et al. (2021). Serum naturally occurring anti-TDP-43 auto-antibodies are increased in amyotrophic lateral sclerosis. *Scientific Reports, 11*, 1978.

Coudert, L., Nonaka, T., Bernard, E., Hasegawa, M., Schaeffer, L., & Leblanc, P. (2019). Phosphorylated and aggregated TDP-43 with seeding properties are induced upon mutant Huntingtin (mHtt) polyglutamine expression in human cellular models. *Cellular and Molecular Life Sciences, 76*(13), 2615–2632.

Culver, B. P., DeClercq, J., Dolgalev, I., Yu, M. S., Ma, B., Heguy, A., et al. (2016). Huntington's Disease Protein Huntingtin Associates with its own mRNA. *Journal of Huntington's Disease, 5*(1), 39–51.

Daoud, H., Belzil, V., Martins, S., Sabbagh, M., Provencher, P., Lacomblez, L., et al. (2011). Association of long ATXN2 CAG repeat sizes with increased risk of amyotrophic lateral sclerosis. *Archives of Neurology, 68*(6), 739–742.

Dewan, R., Chia, R., Ding, J., Hickman, R. A., Stein, T. D., Abramzon, Y., et al. (2021). Pathogenic Huntingtin repeat expansions in patients with frontotemporal dementia and amyotrophic lateral sclerosis. *Neuron, 109*(3), 448–460.e4.

Elden, A. C., Kim, H. J., Hart, M. P., Chen-Plotkin, A. S., Johnson, B. S., Fang, X., et al. (2010). Ataxin-2 intermediate-length polyglutamine expansions are associated with increased risk for ALS. *Nature, 466*(7310), 1069–1075.

Feneberg, E., Steinacker, P., Lehnert, S., Schneider, A., Walther, P., Thal, D. R., et al. (2014). Limited role of free TDP-43 as a diagnostic tool in neurodegenerative diseases. *Amyotrophic Lateral Sclerosis and Frontotemporal Degeneration, 15*(5–6), 351–356.

Foulds, P., McAuley, E., Gibbons, L., Davidson, Y., Pickering-Brown, S. M., Neary, D., et al. (2008). TDP-43 protein in plasma may index TDP-43 brain pathology in Alzheimer's disease and frontotemporal lobar degeneration. *Acta Neuropathologica, 116*(2), 141–146.

Foulds, P. G., Davidson, Y., Mishra, M., Hobson, D. J., Humphreys, K. M., Taylor, M., et al. (2009). Plasma phosphorylated-TDP-43 protein levels correlate with brain pathology in frontotemporal lobar degeneration. *Acta Neuropathologica, 118*(5), 647–658.

Fuentealba, R. A., Udan, M., Bell, S., Wegorzewska, I., Shao, J., Diamond, M. I., et al. (2010). Interaction with polyglutamine aggregates reveals a Q/N-rich domain in TDP-43. *The Journal of Biological Chemistry, 285*(34), 26304–26314.

Gao, J., Wang, L., Huntley, M. L., Perry, G., & Wang, X. (2018). Pathomechanisms of TDP-43 in neurodegeneration. *Journal of Neurochemistry*. https://doi.org/10.1111/jnc.14327

Gerbich, T. M., & Gladfelter, A. S. (2021). Moving beyond disease to function: Physiological roles for polyglutamine-rich sequences in cell decisions. *Current Opinion in Cell Biology, 69*, 120–126.

Hart, M. P., Brettschneider, J., Lee, V. M. Y., Trojanowski, J. Q., & Gitler, A. D. (2012). Distinct TDP-43 pathology in ALS patients with ataxin 2 intermediate-length polyQ expansions. *Acta Neuropathologica, 124*(2), 221–230.

Kanai, K., Kuwabara, S., Sawai, S., Nakata, M., Misawa, S., Isose, S., et al. (2008). Genetically confirmed Huntington's disease masquerading as motor neuron disease. *Movement Disorders, 23*(5), 748–751.

Lee, E. B., Lee, V. M. Y., & Trojanowski, J. Q. (2012). Gains or losses: Molecular mechanisms of TDP43-mediated neurodegeneration. *Nature Reviews Neuroscience, 13*(1), 38–50.

Lee, E. B., Porta, S., Baer, G. M., Xu, Y., Suh, E., Kwong, L. K., et al. (2017). Expansion of the classification of FTLD-TDP: Distinct pathology associated with rapidly progressive frontotemporal degeneration. *Acta Neuropathologica, 134*(1), 65–78.

Lee, J. M., Huang, Y., Orth, M., Gillis, T., Siciliano, J., Hong, E., et al. (2022). Genetic modifiers of Huntington disease differentially influence motor and cognitive domains. *American Journal of Human Genetics, 109*(5), 885–899.

Magrané, J., Cortez, C., Gan, W. B., & Manfredi, G. (2014). Abnormal mitochondrial transport and morphology are common pathological denominators in SOD1 and TDP43 ALS mouse models. *Human Molecular Genetics, 23*(6), 1413–1424.

Marte, L., Boronat, S., Barrios, R., Barcons-Simon, A., Bolognesi, B., Cabrera, M., et al. (2022). Expression of Huntingtin and TDP-43 derivatives in fission yeast can cause both beneficial and toxic effects. *International Journal of Molecular Sciences, 23*(7), 3950.

Martinez-Horta, S., Perez-Perez, J., van Duijn, E., Fernandez-Bobadilla, R., Carceller, M., Pagonabarraga, J., et al. (2016). Neuropsychiatric symptoms are very common in premanifest and early stage Huntington's disease. *Parkinsonism & Related Disorders, 25*, 58–64.

Martinez-Horta, S., Sampedro, F., Horta-Barba, A., Perez-Perez, J., Pagonabarraga, J., Gomez-Anson, B., et al. (2020). Structural brain correlates of dementia in Huntington's disease. *Neuroimage Clinical, 28*, 102415.

Nelson, P. T., Dickson, D. W., Trojanowski, J. Q., Jack, C. R., Boyle, P. A., Arfanakis, K., et al. (2019). Limbic-predominant age-related TDP-43 encephalopathy (LATE): Consensus working group report. *Brain, 142*(6), 1503–1527.

Neumann, M., Sampathu, D. M., Kwong, L. K., Truax, A. C., Micsenyi, M. C., Chou, T. T., et al. (2006). Ubiquitinated TDP-43 in frontotemporal lobar degeneration and amyotrophic lateral sclerosis. *Science, 314*(5796), 130–133.

Nishihira, Y., Tan, C. F., Onodera, O., Toyoshima, Y., Yamada, M., Morita, T., et al. (2008). Sporadic amyotrophic lateral sclerosis: Two pathological patterns shown by analysis of distribution of TDP-43-immunoreactive neuronal and glial cytoplasmic inclusions. *Acta Neuropathologica, 116*(2), 169–182.

Niss, F., Piñero-Paez, L., Zaidi, W., Hallberg, E., & Ström, A. L. (2022). Key modulators of the stress granule response TIA1, TDP-43, and G3BP1 are altered by polyglutamine-expanded ATXN7. *Molecular Neurobiology, 59*(8), 5236–5251.

Ou, S. H., Wu, F., Harrich, D., García-Martínez, L. F., & Gaynor, R. B. (1995). Cloning and characterization of a novel cellular protein, TDP-43, that binds to human immunodeficiency virus type 1 TAR DNA sequence motifs. *Journal of Virology, 69*(6), 3584–3596.

Ren, Y., Li, S., Chen, S., Sun, X., Yang, F., Wang, H., et al. (2021). TDP-43 and phosphorylated TDP-43 levels in paired plasma and CSF samples in amyotrophic lateral sclerosis. *Frontiers in Neurology, 14*(12), 663637.

Ross, C. A., & Tabrizi, S. J. (2011). Huntington's disease: From molecular pathogenesis to clinical treatment. *Lancet Neurology, 10*(1), 83–98.

Sadeghian, H., O'Suilleabhain, P. E., Battiste, J., Elliott, J. L., & Trivedi, J. R. (2011). Huntington chorea presenting with motor neuron disease. *Archives of Neurology, 68*(5), 650–652.

Sampedro, F., Martínez-Horta, S., Pérez-Pérez, J., Pérez-González, R., Horta-Barba, A., Campolongo, A., et al. Plasma TDP-43 reflects cortical neurodegeneration and correlates with neuropsychiatric symptoms in Huntington's disease. *Clinical Neuroradiology* [Internet]. 2022 Mar 3 [cited 2022 Dec 10]; Available from: https://link.springer.com/10.1007/s00062-022-01150-5

Saudou, F., & Humbert, S. (2016). The biology of Huntingtin. *Neuron, 89*(5), 910–926.
Schwab, C., Arai, T., Hasegawa, M., Yu, S., & McGeer, P. L. (2008). Colocalization of transactivation-responsive DNA-binding protein 43 and Huntingtin in inclusions of Huntington disease. *Journal of Neuropathology and Experimental Neurology, 67*(12), 1159–1165.
Seidel, K., den Dunnen, W. F. A., Schultz, C., Paulson, H., Frank, S., de Vos, R. A., et al. (2010). Axonal inclusions in spinocerebellar ataxia type 3. *Acta Neuropathologica, 120*(4), 449–460.
Sreedharan, J., Blair, I. P., Tripathi, V. B., Hu, X., Vance, C., Rogelj, B., et al. (2008). TDP-43 mutations in familial and sporadic amyotrophic lateral sclerosis. *Science, 319*(5870), 1668–1672.
St-Amour, I., Turgeon, A., Goupil, C., Planel, E., & Hébert, S. S. (2018). Co-occurrence of mixed proteinopathies in late-stage Huntington's disease. *Acta Neuropathologica, 135*(2), 249–265.
Steinacker, P., Barschke, P., & Otto, M. (2019). Biomarkers for diseases with TDP-43 pathology. *Molecular and Cellular Neurosciences, 97*, 43–59.
Sturchio, A., Duker, A. P., Muñoz-Sanjuan, I., & Espay, A. J. Subtyping monogenic disorders: Huntington disease. In: *Handbook of clinical neurology* [Internet]. Elsevier; 2023 [cited 2023 Mar 12]. pp. 171–184. Available from: https://linkinghub.elsevier.com/retrieve/pii/B9780323855556000035
Tada, M., Coon, E. A., Osmand, A. P., Kirby, P. A., Martin, W., Wieler, M., et al. (2012). Coexistence of Huntington's disease and amyotrophic lateral sclerosis: A clinicopathologic study. *Acta Neuropathologica, 124*(5), 749–760.
Tan, C. F., Yamada, M., Toyoshima, Y., Yokoseki, A., Miki, Y., Hoshi, Y., et al. (2009). Selective occurrence of TDP-43-immunoreactive inclusions in the lower motor neurons in Machado-Joseph disease. *Acta Neuropathologica, 118*(4), 553–560.
Tauffenberger, A., Chitramuthu, B. P., Bateman, A., Bennett, H. P. J., & Parker, J. A. (2013). Reduction of polyglutamine toxicity by TDP-43, FUS and progranulin in Huntington's disease models. *Human Molecular Genetics, 22*(4), 782–794.
Tazelaar, G. H. P., Boeynaems, S., De Decker, M., van Vugt, J. J. F. A., Kool, L., Goedee, H. S., et al. (2020). ATXN1 repeat expansions confer risk for amyotrophic lateral sclerosis and contribute to TDP-43 mislocalization. *Brain Communications, 2*(2), fcaa064.
Toyoshima, Y., & Takahashi, H. (2014). TDP-43 pathology in polyglutamine diseases: With reference to amyotrphic lateral sclerosis: TDP-43 and polyQ. *Neuropathology, 34*(1), 77–82.
Toyoshima, Y., Tanaka, H., Shimohata, M., Kimura, K., Morita, T., Kakita, A., et al. (2011). Spinocerebellar ataxia type 2 (SCA2) is associated with TDP-43 pathology. *Acta Neuropathologica, 122*(3), 375–378.
Tziortzouda, P., Van Den Bosch, L., & Hirth, F. (2021). Triad of TDP43 control in neurodegeneration: Autoregulation, localization and aggregation. *Nature Reviews Neuroscience, 22*(4), 197–208.
van Dellen, A., & Hannan, A. J. (2004). Genetic and environmental factors in the pathogenesis of Huntington's disease. *Neurogenetics, 5*(1), 9–17.
Vuono, R., Winder-Rhodes, S., de Silva, R., Cisbani, G., Drouin-Ouellet, J., REGISTRY Investigators of the European Huntington's Disease Network, et al. (2015). The role of tau in the pathological process and clinical expression of Huntington's disease. *Brain, 138*(Pt 7), 1907–1918.
Walker, F. O. (2007). Huntington's disease. *Lancet, 369*(9557), 218–228.
Wexler, N. S., Lorimer, J., Porter, J., Gomez, F., Moskowitz, C., Shackell, E., et al. (2004). Venezuelan kindreds reveal that genetic and environmental factors modulate Huntington's disease age of onset. *Proceedings of the National Academy of Sciences of the United States of America, 101*(10), 3498–3503.
Zhang, Y. J., Xu, Y. F., Cook, C., Gendron, T. F., Roettges, P., Link, C. D., et al. (2009). Aberrant cleavage of TDP-43 enhances aggregation and cellular toxicity. *Proceedings of the National Academy of Sciences of the United States of America, 106*(18), 7607–7612.

DNA Methylation and Epigenetic Aging Biomarkers in Huntington's Disease

Elizabeth A. Thomas

Abbreviations

3-ma	3-methyl-adenosine
5-hmC	5-hydroxymethylcytosine
5-mC	5-methylcytosine
7-mG	N7-methyl-guanine
AID/APOBEC	Activation-induced deaminase apolipoprotein B editing complex
BER	Base excision repair
CAG	Cytosine-adenine-guanine
CGI	CpG islands
CLDN16	Claudin 16
COX4I2	Cytochrome c oxidase subunit 4I2
CpG	Cytosine-phosphate-guanine
DDC	Dopa decarboxylase
DMR	Differentially methylation regions
DNAJB8	DNAJ (Hsp40) chaperone family
DNMT	DNA methyltransferase
EWAS	Epigenome-wide association study
GRIK4	Glutamate ionotropic receptor kainate type subunit 4
HADS	Hospital Anxiety and Depression Scale
HD	Huntington's disease
HDAC4	Histone deacetylase 4
HES	Hairy and enhancer of split 4
hiPSCs	Induced pluripotent stem cells
HTT	Huntingtin

E. A. Thomas (✉)
Department of Neurobiology and Behavior, Institute for Interdisciplinary Salivary Bioscience Research, University of California, Irvine, CA, USA

Department of Neuroscience, The Scripps Research Institute, La Jolla, CA, USA
e-mail: eathoma1@uci.edu

E. A. Thomas, G. M. Parkin (eds.), *Biomarkers for Huntington's Disease*,
Contemporary Clinical Neuroscience, https://doi.org/10.1007/978-3-031-32815-2_15

LINE-1	Long interspersed nuclear elements-1
MBD	Methyl-CpG binding domain-based
MeDIP	Methyl-DNA immunoprecipitation
MSRE	Methylation-sensitive restriction enzymes
NXT2	Nuclear transport factor 2 like export factor
PEX14	Peroxisomal biogenesis factor 14
RRBS	Reduced representation bisulfite sequencing
TET	Ten-Eleven Translocation
WGBS	Whole-genome bisulfite sequencing

1 Introduction

Huntington's disease (HD) is caused by an abnormal cytosine-adenine-guanine (CAG) repeat expansion in the Huntington (*HTT*) gene (Gusella et al., 1983; Huntington Disease Collaborative Research Group, 1993). Despite HD being a single gene disorder, there is enormous variability in disease onset and severity, even in patients with the same CAG repeat number, which is itself a main contributor to disease onset (Andrew et al., 1993; Wexler et al., 2004; Andresen et al., 2007). Symptom severity and disease progression also vary among patients, necessitating the identification of biomarkers that could be developed to predict and/or track such features. The *HTT* gene encodes the protein huntingtin (HTT), a large protein of 3144 amino acids, which is involved in a wide range of cellular functions including gene transcription, vesicle transport and energy metabolism (Miller et al., 2010; Marques Sousa & Humbert, 2013). Pathogenesis in HD arises largely from the expression of mutant HTT (mHTT), which leads to the formation of toxic soluble protein oligomers as well as insoluble aggregates that contribute to the disruption of many intracellular pathways. Perhaps most notably, both HTT and mHTT are associated with a wide range of transcriptional abnormalities, ranging from interactions with specific transcription factors to interference with the core transcriptional machinery, as well as affecting epigenetic mechanisms, including altering posttranscriptional modifications of histones and DNA methylation (for review see (Thomas, 2016, 2019).

DNA methylation is the oldest known epigenetic mark, which was first reported in bacteria in 1925 (Johnson & Coghill, 1925); however, its biological importance would not be appreciated for another several decades. It is now known that methylated DNA exist across many species and is associated with a wide range of essential biological functions. For example, DNA methylation plays critical roles in the regulation of transcription, genomic imprinting, maintenance of X-chromosome inactivation, chromosomal maintenance, and genomic stability. Growing evidence suggests significant connections between DNA methylation patterns and various clinical outcomes in human disorders, underscoring the value of DNA methylation profiles to serve as novel clinical indicators. For example, DNA methylation is currently being used as a biomarker for cancer, where several clinical trials have investigated the use of DNA methylation-based tests for cancer diagnosis and monitoring of treatment response (Shirley, 2020; Loomans-Kropp et al., 2022).

Increasing evidence has also demonstrated that DNA methylation changes that occur throughout the lifespan contribute to the pathophysiology of aging and age-associated neurodegenerative diseases. DNA methylation estimators, also referred to as "epigenetic clocks", have been developed and have been used to accurately estimate the age of cell types and tissues (Horvath & Raj, 2018; Duan et al., 2022). Epigenetic clocks have been used in a wide range of studies to investigate the relationship between epigenetic aging and a variety of age-related diseases, such as cancer, cardiovascular disease, and neurodegenerative diseases, including HD (Noroozi et al., 2021).

This chapter will review the processes of DNA methylation, methods used to measure DNA methylation and differences in DNA methylation associated with HD, including studies that have investigated "epigenetic aging" in the context of HD.

2 Overview of DNA Methylation

Historical Perspective

The first discovery of DNA methylation in living cells was by Johnson and Coghill in 1925 (Johnson & Coghill, 1925). Twent-three years would pass before a second report on a methylated cytosine (called "epi-cytosine") would be discovered in calf thymus DNA (Hotchkiss, 1948); this finding was rapidly followed by a report demonstrating the presences of methylated cytosines in mammalian, insect and plant DNA (Wyatt, 1950). Paired with the realization that nucleic acids are the carriers of genetic information and the discovery of the DNA double helix structure by Watson and Crick (1953), investigations into DNA methylation rapidly expanded. Although the term 'epigenetics' was coined in 1942 (Waddington, 1942), the significance of DNA methylation was not fully appreciated at this time. Indeed, many years later, DNA methylation would become most intensely studied, epigenetic modifications in mammals. Characterizations of DNA methylation in the 1980s would implicate it as the main determining factor for X-chromosome inactivation, parental imprinting of genes, and silencing of transposable elements (Blasi & Toniolo, 1983; Reik et al., 1987). Around the same time, studies began to suggest that DNA methylation was a critical player in the proper regulation of gene expression and gene silencing (Razin & Riggs, 1980). Although the DNA methylation of gene promoter regions is the most well-known means of altering gene expression, other site of DNA methylation across the genome have aslo been identified as playing important roles in regulating gene expression (Stadler et al., 2011).

Methylation of DNA

DNA methylation involves the addition of a methyl group at the 5 position on the pyrimidine ring of a cytosine residue, resulting in the formation of 5-methylcytosine (5-mC) (Razin & Riggs, 1980) (Fig. 1). In humans, the formation of 5-mC most commonly occurs when cytosine is paired with guanine, at so-called cytosine-phosphate-guanine (CpG) sites, although non-CpG methylation also occurs (Guo et al., 2014). Approximately 4% of cytosines exist in CpG sites, and 60–80% of CpG cytosines are methylated, depending on the cell type or pathologic state. In particular, CpG sites in promoter regions of genes, called CpG islands, undergo dynamic methylation, which controls gene expression.

While cytosine methylation is the most well-characterized and most common DNA modification in the genome, other residues can also be methylated. For example, other modified bases include N7-methyl-guanine (7-mG) and 3-methyladenosine (3-mA). Additionally, another DNA modification, 5-hydroxymethylcytosine (5hmC), which is an oxidation product of 5mC, is also a potential epigenetic modifier that might have important roles in neurodegenerative diseases (Sherwani & Khan, 2015).

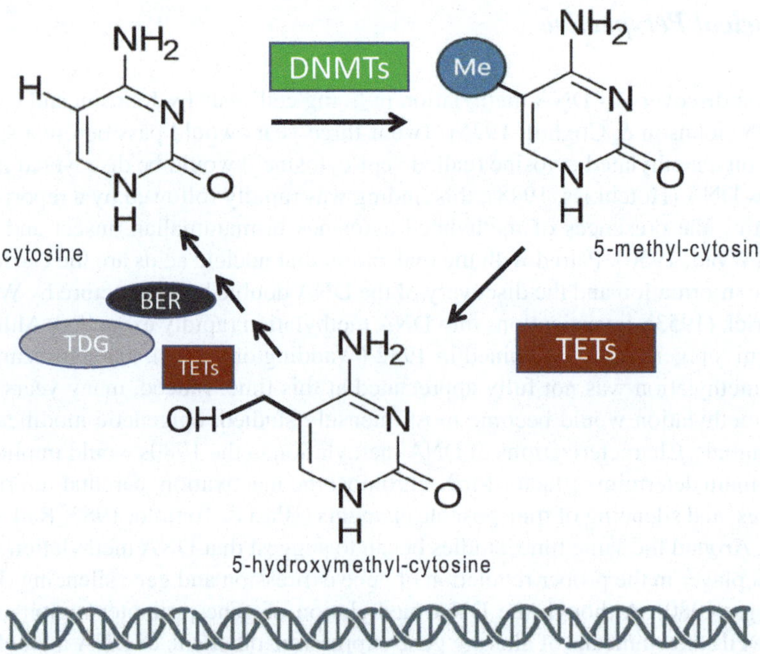

Fig. 1 Schematic picture of the formation and removal of 5-methylcytosine (5mC) marks. Cytosines are methylated by DNA methyltransferases (DNMTs). Once methylated, 5mC can be converted to 5-hydroxymethylcytosine (5hmC) by Ten-Eleven Translocation (TET) dioxygenase enzymes, and then further modified by a multistep process involving TET enzymes and base excision repair (BER) enzymes. (Figure created in part with BioRender.com)

DNA Methyltransferase Enzymes

It is now known that many important factors are responsible for the establishment, maintenance and removal of DNA methylation marks. DNA methyltransferase (DNMT) enzymes catalyze the transfer of a methyl group from *S*-adenosylmethionine to the fifth position on the cytosine ring (Wu & Santi, 1985) (Fig. 1), and this family of enzymes has five members, DNMT1, DNMT2, DNMT3a, DNMT3b, and DNMT3L. These enzymes are expressed throughout many cell types and tissues and play important roles neuronal function, survival and plasticity (Klose & Bird, 2006). The primary DNMT enzyme responsible for maintenance DNA methylation in mammals is DNMT1. Maintenance DNA methylation refers to the process of preserving the existing DNA methylation patterns during DNA replication, ensuring the accurate transmission of epigenetic marks to daughter cells. DNMT1 recognizes hemi-methylated DNA, which occurs after DNA replication when one strand of the DNA double helix is methylated, and the other strand is newly synthesized and unmethylated. DNMT1 then adds a methyl group to the newly synthesized strand, thus maintaining the methylation pattern.

The DNMT enzymes, DNMT3A and DNMT3B are involved in de novo DNA methylation. Both of these enzymes play crucial roles in establishing DNA methylation patterns during early development and in somatic cells. DNMT3A and DNMT3B enzymes catalyze the transfer of a methyl group onto a "naked" or an unmethylated DNA template (Okano et al., 1999). These enzymes also contribute to genomic imprinting, which is the epigenetic marking of specific genes based on parental origin. Two others family members, DNMT2 and DNMT3L, while showing sequence homology with the other family members, are thought to have distinct functions from cytosine methylation, although DNMT3L is thought to enhance the enzymatic activity of DNMT3A and DNMT3B, thereby can facilitate de novo DNA methylation (Goll et al., 2006).

DNA Demethylation

Traditionally, the school of thought was that DNA methylation was an irreversible process. The prevailing view was that once DNA was methylated, the methylation marks would persist and remain unchanged. One exception to this rule occurs in replicating cells through a process called passive demethylation, where the activity of DNMT1 is blocked during DNA synthesis; this mechanism does not work in post-mitotic, differentiated cells. However, a growing body of scientific literature has provided evidence for active, replication-independent DNA demethylation (Carey et al., 2011; Kohli & Zhang, 2013; Schafer, 2013). One crucial player in active DNA demethylation is a family of enzymes called ten-eleven translocation (TET) enzymes, consisting of TET1, TET2 and TET3 family members. These enzymes are capable of oxidizing 5-mC to 5hmC, which serves as the initial step in the active demethylation pathway (Fig. 1). Subsequently, 5-hmC can undergo further oxidation and be actively converted back to cytosine through a series of iterative oxidation reactions, facilitated by enzymes and mechanisms that involve thymine DNA glycosylase (TDG) and the BER (base excision repair) glycosylase family (Fig. 1) (van der Wijst et al., 2015). Additionally, there are other genes that encode proteins involved in DNA demethylation. For example, Methyl-CpG binding domain protein 3 (Mbd3), Growth arrest and DNA-damage-inducible 45 (Gadd45), and RING finger protein 4 (Rnf4)(Brown et al., 2008; Hu et al., 2010; Schafer, 2013).

3 Methods to Detect DNA Methylation

Due to rapid methodological progress, a wide range of techniques are available for determining DNA methylation status. An excellent review on this topic has been published by Singer (2019). Additionally, quantification of DNA methylation has become increasingly cost effective, which has allowed for epigenome wide approaches, either via microarrays or sequenced-based technologies, to dominate the literature in recent years. Nonetheless, the choice of which method to use ultimately depends on the research question, available resources, technical expertise and the desired levels of resolution and coverage. Different approaches will target either specific methylation sites in the genome, genome-wide methylation, or measure DNA methylation on a global level. By far, the most extensively used methods for DNA methylation quantification rely on bisulfite conversion of the DNA, which is a technique to discriminate between unmethylation vs. methylated cytosines. However, additional experimental techniques are available for DNA methylation quantification and these will also be mentioned below, including affinity enrichment-based approaches, restriction enzyme-based methods and global DNA methylation approaches.

Bisulfite Conversion

Sodium bisulfite conversion of genomic DNA is the most widely-used approach for DNA methylation analysis and, although somewhat labor intensive, is now considered to be the gold standard for quantifying DNA methylation. Bisulfite conversion involves the deamination of unmodified cytosines into uracil, while leaving the modified bases 5-mC (and 5-hmC) intact. After bisulfite conversion, when the DNA fragments undergo PCR amplification, uracils are replaced by thymines; however, the cytosines corresponding to 5-mC or 5-hmC residues remain unchanged, allowing them to be identified as cytosines in the DNA sequence. By comparing this sequence to a reference unconverted sequence, it becomes possible to determine the methylation status of individual cytosines in the original sample.

Bisulfite conversion of the DNA can then be followed by downstream approaches that can identify DNA methylation at specific sites. The most common targeted DNA methylation method is Pyrosequencing, in which PCR amplification of the target region is performed using specific primers, followed by sequencing of the PCR-amplified products. Genome-wide interrogation of methylation status at single-nucleotide resolution can be performed via array hybridization of bisulfite-converted DNA, using site-specific, bead-ligated probes that can distinguish between methylated and unmethylated loci based on their differential sequence after bisulfite treatment. Illumina Infinium methylation arrays are the most commonly used arrays in research and current versions of these arrays are capable of measuring methylation at up to 850,000 CpG sites (Moran et al., 2016). This approach is popular for large-scale human studies and often referred to as epigenome wide association studies (EWAS). Alternatively, whole-genome bisulfite sequencing (WGBS) offers a more comprehensive approach to genome-wide DNA methylation profiling (Lister et al., 2009). WGBS generates extensive amounts of data, which is sometimes unnecessary, and this has led to the development of reduced representation bisulfite sequencing (RRBS), which provides a enrichment approach that focuses only on CpG-rich regions of the genome (Meissner et al., 2005).

Affinity Enrichment Approaches

Two commonly used approaches to enrich for methylated DNA regions of the genome include methyl-DNA immunoprecipitation (MeDIP), which is based on enrichment of methylated DNA using antibodies specific for 5-mC (Weber et al., 2005, 2007; Hsu et al., 2014), or the capture of methylated DNA using a methyl-CpG binding domain (MBD) protein (Jørgensen et al., 2006; Aberg et al., 2018). Both of these methods involve the use of antibodies specific for 5-mC or MBD to immunoprecipitate methylated DNA fragments. The enriched methylated fractions can then be used in various downstream applications, including qualitative and quantitative PCR, microarray analysis, or next-generation sequencing technologies. These methods are low-cost and straightforward but require experience in chromatin immunoprecipitation sequencing or related techniques. Disadvantages of this approach include relatively low resolution, susceptibility to copy number variation bias, and a bias toward observations of methylated DNA compared with bisulfite-based approaches (Laird, 2010).

Restriction Enzyme Based Methods

Restriction enzyme-based approaches take advantage of methylation-sensitive restriction enzymes (MSREs), which cut DNA at specific sites, depending on whether the cytosine is unmethylated, or methylated. These restriction enzymes are not able to cleave cytosine residues that are methylated, hence leave methylated DNA intact. For example, the restriction enzyme, Msp I, which is known to cleave at a DNA sequence of CCGG when the internal cytosine residue is methylated, does not cleave when both cytosines are methylated. The resulting DNA fragments can be analyzed by gel electrophoresis, PCR or high-throughput sequencing to determine the methylation status at the specific restriction site.

Global DNA Methylation

Global DNA methylation levels can be quantified with a simple immunoassay using 5-mC antibodies. Several commercially-available kits allow for such measurements, however these approaches are known to show high variability and are not as informative as other approaches, given the enormity of DNA methylation changes that occur at specific sites. Nonetheless, they represent and quick and easy approach that can identify global changes in DNA methylation. A related, but better approach specifically quantifies the methylation levels of long interspersed nuclear elements-1 (LINE-1) retrotransposons, of which ~17% of the human genome is composed. This approach is more widely viewed as a better surrogate for global DNA methylation compared to approaches that use a 5-mC antibody (Ogino et al., 2008; Ohka et al., 2011).

4 Altered DNA Methylation in HD

Growing literature supports an involvement of DNA methylation changes in HD. Given the known functions of DNA methylation to regulate gene expression, changes in DNA methylation clearly represent a relevant and potential means of altering gene expression, which could thereby lead to neuronal dysfunction in this disease. Several studies with the aim of elucidating the epigenetic mechanisms underlying HTT gene regulation and HD pathogenesis have reported changes in DNA methylation in human HD brain and peripheral fluids (Table 1). Basic and pre-clinical research studies have also demonstrated changes in DNA methylation in HD cell culture systems and rodent models (Ng et al., 2013; Wang et al., 2013), but these will not be covered herein, as we aim to focus only on changes in human tissues/fluids in this chapter.

DNA Methylation Changes in Post-mortem Brain

An early study by Villar-Menendez and colleagues measured targeted DNA methylation of 5-mC, as well as the hydroxylated version, 5-hmC, using MeDIP in the 5′ untranslated region of the adenosine A2A receptor gene (*ADORA2A*) in the

Table 1 DNA methylation changes in human subjects with Huntington's disease

Source:	Approach	Technique:	Methylation site	Reference:
Studies from post-mortem brain				
Putamen	Targeted	MeDIP	5-mC, 5-hmC	Villar-Menendez et al. (2013)
Cortex	Targeted	MSRE	5-mC	Bai et al. (2015)
Visual and motor cortices, cingulate gyrus	Global	High performance liquid chromatography	7-mG	Thomas et al. (2013)
Cortex, liver	Genome-wide	Illumina Human 450K Bead Chip	5-mC	De Souza et al. (2016)
Frontal, parietal and occipital cortices	Genome-wide	Illumina Human 450K Bead Chip	5-mC	Horvath et al. (2016)
Source:	**Approach**	**Technique:**	**Methylation site**	**Reference:**
Studies from blood and cells				
Whole blood	Targeted	Pyrosequencing	5-mC	Gutierrez et al. (2019)
Whole blood	Genome-wide	Illumina Human 450K Bead Chip	5-mC	Zadel et al. (2018)
Whole Blood	Genome-wide	Illumina Infinium custom array; RRBS	5-mC	Lu et al. (2020)
Human Fibroblasts, lymphoblastoid cells	Genome-wide	Illumina Infinium Custom Array; RRBS	5-mC	Lu et al. (2020)
Human Fibroblasts	Genome-wide	Illumina Human 450K Bead Chip	5-mC	Jia et al. (2015)
Human iPSC cells	Genome-wide	Agilent 244K arrays	5-mC	Baronchelli et al. (2017)

MSRE methylation-sensitive restriction enzyme; *MeDIP* Methyl-DNA immunoprecipitation; *RRBS* reduced representation bisulfite sequencing; *5-mC* 5-methylcytosine; *5-hmC* 5- hydroxymethylcytosine; *7-mG* 7-methyl-guanine

post-mortem putamen of HD patients (Villar-Menendez et al., 2013). Increased levels of 5-mC were found in HD patients compared to age-matched controls, while reductions in 5-hmC were reported (Villar-Menendez et al., 2013). These findings suggested that altered methylation patterns of the *ADORA2A* gene were linked to the pathological decrease of *ADORA2A* expression levels found in HD (Villar-Menendez et al., 2013).

Another early study measured global levels of 7-mG in three different brain regions from HD subjects (Thomas et al., 2013). While no major differences in 7-mG were seen in the primary visual cortex and cingulate gyrus regions of HD brains, the motor cortex showed an increased methylation pattern that corresponded with the pathologic grade of disease (Thomas et al., 2013).

Most of the other studies carried out on human brain tissue have focused on genome-wide DNA methylation. In one of the first genome-wide studies, Horvath et al. examined brain samples from 26 HD cases and 39 controls using the Illumina HumanMethylation450K BeadChip array (Horvath et al., 2016). This array allows for the assessment of methylation at over 450,000 sites, both within and outside of CpG islands (CGIs), covering 99% of RefSeq genes. Through a genome-wide analysis adjusted for age and sex, the researchers identified 1467 CpGs whose methylation status showed an association with HD in combined samples from the frontal, parietal, and occipital lobes. Using weighted correlation network analysis, they further identified 54 co-methylation modules that were consistent across all three brain regions. Among these modules, 11 exhibited a significant correlation with HD status, with 5 being hypermethylated and six being hypomethylated. The genes within these HD-associated co-methylation modules were found to be enriched in Gene Ontology biological process terms, including the detection of chemical stimuli, nucleic acid metabolic processes, and RNA metabolic processes (Horvath et al., 2016).

Another study quantified DNA methylation in a cohort of age-matched HD and control human post-mortem cortex and post-mortem liver tissues using the same Illumina HumanMethylation450K arrays (De Souza et al., 2016). In this study however, minimal evidence of altered DNA methylation was observed in the cortex after correction for cell heterogeneity, which is necessary when complex tissues, such as brain tissue, are used. However, this study did identify a novel differentially methylated residue in the CTCF-binding site of the *HTT* proximal promoter, suggesting that DNA methylation may, in part, contribute to tissue-specific HTT transcription through differential CTCF occupancy. Furthermore, these investigators found a significant correlation between methylation state and age of disease onset in the HD cohort (De Souza et al., 2016). Finally, comparison of the matched liver–cortical sample pairs revealed 38 differentially methylated sites in the HTT gene, suggesting that differential CpG methylation might contribute to tissue-specific expression levels of HTT.

A study that was initially focused on genome-wide mapping of histone methylation in human HD cortex led to the identification of important DNA methylation differences on the Hairy and enhancer of split 4 (*HES4)* gene promoter (Bai et al., 2015). Loss of histone methylation at CpG-rich sequences on the *HES4* promoter was associated with increased DNA methylation, as determined by MSRE analysis, and corresponding altered expression of *HES4* and its target genes. Further, increased methylation of the *HES4* promoter was correlated with striatal degeneration and age of disease onset in HD patients (Bai et al., 2015), suggesting robust potential as a relevant epigenetic biomarker in HD.

DNA Methylation Changes in Human Cells

DNA methylation patterns in cultured human fibroblasts from normal and HD patients have been measured, also using the Infinium HumanMethylation450 BeadChip (Jia et al., 2015). Comparing the HD versus control cells, 16,431 CpG positions were found to be increased in methylation and 8118 positions decreased in methylation, using a multiple hypothesis corrected p-value of 0.001 (Jia et al., 2015).

Subsequently, another study investigated DNA methylation changes in an induced pluripotent stem cell model of HD (HD-hiPSCs) during differentiation towards a striatal neuron phenotype using Agilent 244 K Human CpG microarrays. Pluripotent HD-hiPSCs cells showed higher levels of DNA methylation compared to control human embryonic stem cells, and increased hypermethylation of DNA was observed during striatal differentiation. Differentially hypermethylated regions affected genes involved in axon guidance, neural crest differentiation, or neuroactive ligand–receptor interaction and signaling pathways of neuronal differentiation (Baronchelli et al., 2017) (Table 1).

Altered DNA Methylation Patterns in Blood from HD Patients

A targeted methylation study carried out on blood from HD patients focused on methylation of the brain-derived neurotrophic factor (BDNF) gene, given that it plays an important role in survival, growth and differentiation of neurons (Lu, 2003; Greenberg et al., 2009) and that decreased levels of BDNF, especially in the striatum, are thought to play a crucial role in HD pathogenesis (Ferrer et al., 2000; Zuccato et al., 2001, 2003). This study investigated targeted DNA methylation at promoter IV of the *BDNF* gene using pyrosequencing on DNA derived from whole blood of HD patients and healthy controls (Gutierrez et al., 2019) (Table 1). Results showed that four CpG sites were increased in methylation in HD patients compared to controls, however no changed in plasma levels of the BDNF protein were observed. Interestingly, methylation levels at three of these CpG sites was inversely correlated to the Hospital Anxiety and Depression Scale (HADS) scores, but no other associations were found with any motor or cognitive behaviors in HD patients (Gutierrez et al., 2019). These findings suggested that DNA methylation at *BDNF* promoter IV might represent a biomarker of psychiatric symptoms in HD patients.

A genome-wide study investigated DNA methylation patterns in whole blood in a cohort of HD patients and healthy controls using the Infinium® Human Methylation BeadChip microarray (Zadel et al., 2018). Several hundred differentially methylated sites were detected, however CpGs in only three genes, claudin 16 (*CLDN16*), dopa decarboxylase (*DDC*), and nuclear transport factor 2 like export factor 2 (*NXT2*), remained statistically significance after correcting for multiple tests (FDR < 0.05) (Zadel et al., 2018) (Table 1).

The largest DNA methylation study carried out to date generated methylation datasets from seven sources of DNA from three species, including a mouse *HTT* gene knock-in model, a transgenic *HTT* sheep model, and 2164 blood samples from individuals participating in the Enroll-HD and Registry-HD observational studies (Lu et al., 2020) (Table 1). This study reported 33 genome-wide significant CpGs associated with HD status, one of which, cg22982173, was located in the *first exon of the HTT* gene itself. Importantly, this altered CpG in the *HTT* gene was validated across seven sources of DNA (blood, lymphoblastoid cells, fibroblasts, striatum, cerebellum, cortex, and liver) and across all three species (Lu et al., 2020). Further, hypermethylation of this CpG site was associated with slower disease progression in manifest HD cases, as indicated by motor scores and cognitive function assessments. Other EWAS results implicated genes know to modify mutant HTT aggregation, such as DNAJB8, a member of the DNAJ (Hsp40) chaperone family and histone deacetylase 4 (HDAC4), which binds to mutant HTT in a polyglutamine length-dependent manner. Additionally investigating changes in DNA methylation associated with motor progression in manifest HD cases, these authors reported significant associations at three loci: cg26919387 near the Peroxisomal biogenesis factor 14 *(PEX14) gene*, cg12823408 near the Glutamate ionotropic receptor kainate type subunit 4 (*GRIK4)* gene and cg21497164 near the Cytochrome c oxidase subunit 4I2 (*COX4I2)* gene (Lu et al., 2020). These findings were not confounded by medications used for treating movement disorders (i.e. tetrabenazine). The gene, *GRIK4*, was thought to be of particular interest due to its specific expression in the cortex and striatum. This study concluded that HD is accompanied by profound changes of DNA methylation levels and that several of these sites were linked to disease progression.

5 DNA Methylation Age and the "Epigenetic" Clock

The aging process is associated with many alterations at the cellular level, including protein misfolding, genomic instability, telomere attrition and epigenetic alterations. With respect to the latter, a large body of literature, dating back to the 1960s, demonstrates that aging has a profound effect on genome-wide DNA methylation levels. DNA methylation patterns are known to change with aging in predictable ways, with certain regions of the genome becoming more or less methylated over time. These specific age-related DNA methylation patterns form the basis of DNA methylation age estimators, which are typically built by regressing the chronological age onto a selected set of age-predicting CpGs, using a supervised machine learning method. These DNA methylation age estimators are more commonly referred to as "epigenetic clocks" and can be used to compare an individual's chronological age, defined as the number of years a person is alive, compared to their biological, or "epigenetic", age, which describes the overall health of the individual. The difference between an individual's chronological age and their epigenetic age is known as "epigenetic age acceleration," and has been shown to be associated with a range of age-related diseases, including cardiovascular disease, Alzheimer's disease, cancer and even HD (Duan et al., 2022; Noroozi et al., 2021). Epigenetic "clocks" based on DNA methylation have revolutionized the field's ability to estimate biological age and more rapidly test the efficacy of interventions that slow aging and prevent diseases.

Different Types of Epigenetic Clocks

Several epigenetic clocks have been developed for use as biological age predictors and most have been generated using Illumina DNA methylation arrays (Table 2). Bocklandt and colleagues described the first DNA methylation age estimator, which was based on 88 CpGs and used DNA extracted from saliva (Bocklandt et al., 2011). Later that year, another study identified 75 CpGs specifically linked to aging in human fibroblasts (Koch et al., 2011). Subsequently, several additional epigenetic clocks have been developed and optimized and are currently more widely used by researchers. These include the Horvath and Hannum clocks, PhenoAge, GrimAge and the pediatric buccal clocks and are described below and in Table 2.

The most widely used epigenetic clock was described in 2013 by Steve Horvath, who was also a coauthor on the initial publication using saliva samples (Horvath, 2013). In this paper, Horvath developed a multivariate model that predicts chronological age based on DNA methylation patterns at 353 CpG sites across the genome. The model was trained using data from over 8000 samples from a variety of tissues and cell types, and was found to be highly accurate at predicting chronological age,

Table 2 Different types of DNA methylation age estimators (i.e. epigenetic clocks)

Clock Name:	Tissue type:	# of CpGs:	Age range of subjects used	Predictor:	Reference:
Bocklandt DNAm age	Saliva	88	18–70 years	Estimates biological age	Bocklandt et al. (2011)
Koch DNAm age	Mulitple	75	16–72	Estimates biological age	Koch et al. (2011)
Horvath clock	51 different tissues	353	0–101 years	Measures aging rate	Horvath (2013)
Horvath Skin & Blood clock	Skin/Blood	391	0–94 years	Measures age acceleration ex-vivo	Horvath et al. (2018)
Hannum clock	Whole blood	71	19–101 years	Measures aging rate	Hannum et al. (2013)
PhenoAge	Whoel blood and Phenotypic data	513	21–100 years	Predicts phenotypes of aging; mortality risks	Levine et al. (2018)
GrimAge	Blood proteins and smoking as surrogates	1030	46–78 years	Predicts lifespan and healthspan	Lu et al. (2019)
PedBe clock	Buccal cells	94	0–20 years	Measures biological age in children	McEwen et al. (2019)
DunedinPACE	Whole blood	173	26–45 years	Predictys morbidity and mortality	Belsky et al. (2022)

with a median error of only 3.6 years (Horvath, 2013). This method produces a pan-tissue "DNA methylation age". On the other hand, the Hannum clock used only 71 CpG sites and was trained on whole blood samples (Hannum et al., 2013). Because this clock was developed using blood from adult subjects, it leads to biased in estimates when applied to non-blood tissues or samples from children. Recognizing this latter limitation, Michael Kobor's laboratory developed an epigenetic clock from buccal cell DNA that is specific for measuring biological age in children (McEwen et al., 2019). This epigenetic tool is based on DNA methylation at 94 CpG sites from buccal cells (epithelial cells of the cheek) from younger subjects, specifically aged 0–20 years old (McEwen et al., 2019).

Another clock developed by Horvath and colleagues is the Horvath Skin and Blood clock, which is a highly robust DNA methylation age estimator (based on 391 CpGs) for human fibroblasts, keratinocytes, buccal cells, endothelial cells, lymphoblastoid cells, skin, blood, and saliva samples (Horvath et al., 2018) (Table 2). Overall, this biomarker is expected to become useful for forensic applications (e.g. blood or buccal swabs) and for a quantitative *ex vivo* human cell aging assay. The GrimAge DNA methylation estimator employs seven DNA methylation-based estimators of plasma proteins including those of plasminogen activator inhibitor 1 and growth differentiation factor and includes smoking in the formula (Lu et al., 2019). A newer clock, called Phenoage, was developed with a two-step process, and strongly outperformed previous measures in predicting a variety of aging outcomes, including all-cause mortality, cancers, healthspan, physical functioning, and Alzheimer's disease (Levine et al., 2018). While this biomarker was developed using data from whole blood, it correlates strongly with age in other tissues and cell tested. Another next-generation DNA methylation clock has recently been developed to determine the pace of aging, and this is referred to as the DunedinPACE (for Pace of Aging Calculated from the Epigenome) (Belsky et al., 2022). Additional studies using this clock are likely to add value to the literature on gerontology.

Epigenetic Aging in HD

While mutant HTT is present ubiquitously throughout life, HD onset typically occurs in mid-life, suggesting that aging may play an active role in pathogenesis. Strikingly, many of the hallmarks of aging, including telomere attrition, epigenetic alterations, altered intercellular communication and glial activation in the brain, are also hallmarks of HD pathology (Machiela & Southwell, 2020). Hence, it is not surprising that investigations into epigenetic aging have been carried out in this disease. In one study, DNA methylation data was generated for 475 brain samples from nine brain regions obtained from 26 HD manifest subjects and 39 controls. Overall, brains from HD subjects exhibited a significant epigenetic age acceleration effect that was confined to specific regions, including the frontal lobe, parietal lobe, and cingulate gyrus (Horvath et al., 2016). Specifically, the data suggested a biological age increase of 3.2 years in HD patients compared to controls. Further, a negative correlation was observed between *HTT* gene CAG repeat length and the epigenetic age of HD brain samples. However, in another study, DNA methylation age predictions from HD prefrontal cortex were not significantly different from control samples (De Souza et al., 2016).

Epigenetic aging has also been investigated in blood samples from HD patients. In a highly robust manner, five different human epigenetic clocks (see Table 2) were utilized: the Horvath's pan-tissue epigenetic age, the Hannum's blood-based clock the Horvath Skin and Blood clock, the DNAmPhenoAge clock and DNAmGrimAge. Epigenetic aging differences were observed in blood samples from manifest HD patients obtained from the Enroll-HD and Registry-HD cases compared to controls (Lu et al., 2020). Further, and of particular relevance, epigenetic age acceleration was positively correlated with the progression of motor symptoms in these patients (Lu et al., 2020).

Epigenetic aging was also investigated in a similar CAG-repeat disease, Spinocerebellar ataxia type 3, also known as Machado Joseph disease, which is a major type of dominantly inherited ataxia caused by a CAG repeat expansion of exon10 in the Ataxin 3 gene (Kawaguchi et al., 1994). This study found that DNA methylation age acceleration was associated with age of onset, suggesting that every 5 year increase in DNA methylation-age acceleration corresponded to a 6.3 year earlier disease onset (Li et al., 2022).

6 Summary and Conclusions

Overall, these studies suggest that DNA methylation changes are extensive in HD, not only in brain tissue, but also in blood samples. Importantly, changes in DNA methylation at the *HTT* promoter may play an important role in HD pathogenesis. Further research is needed to elucidate the precise mechanisms underlying *HTT* gene regulation by DNA methylation, and to determine whether DNA methylation changes at the *HTT* promoter could serve as potential therapeutic targets for HD. Another important area of investigation is whether methylation changes on *HTT* exon 1 have any impact on the somatic instability of the mutant CAG repeat itself. Recent large-scale human genetic studies revealed variants in mismatch repair enzymes (e.g., *MLH1*, *MSH2, MSH3*), which are known to modify the germline and somatic instability of expanded *HTT* CAG repeats in HD mouse models and can act as modifiers of the age of motor symptom onset in HD (Pinto et al., 2013).

Studies have suggested that HD is associated with an accelerated epigenetic age of specific brain regions and more broadly with substantial changes in brain methylation levels. Correspondingly, findings from studies related to the DNA methylation age clock could be translated into useful tools for predicting an individual's risk of age-related disease onset. The ability to detect epigenetic aging differences in a peripheral fluid provides promise for an easily accessible biomarker and these might be especially important given the associations detected between epigenetic aging and disease progression.

One issue that is important to remember is that the field of epigenetic ageing is relatively new, and its rapid incorporation into many research studies and disease conditions presents a challenge in keeping up with their implications. Importantly, even though past studies have highlighted and affirmed the ability of epigenetic clocks to capture biologically meaningful changes associated with age, they do not provide any information regarding the underlying mechanisms (Raj & Horvath, 2020). This will be an important area of research moving forward.

Although numerous studies have attempted to explore the role and significance of DNA methylation machinery in the development of neurological disorders, the application of this knowledge in clinical practice is still in its early stages. This is primarily because accessing the target tissue poses challenges, and in most cases, biomarkers can only be measured after death through brain biopsies. Furthermore, the specific nature of DNA methylation, which is tied to particular tissues or cells, complicates analysis as it necessitates the use of disease-affected tissues. To overcome this hurdle, many propose using blood samples as a substitute for affected brain tissue. However, there is a potential risk that disease-related epigenetic changes may differ in surrogate tissues. Alternatively, the DNA methylation patterns in disease-affected tissues might vary, rendering them undetectable in surrogate samples.

Despite the challenges mentioned above, recent evidence has shown that various DNA methylation alterations in HD are associated with disease activity, disease progression, and clinical outcomes and may have prognostic or diagnostic value. The fact that DNA methylation alterations are reversible also makes them a valuable target for therapeutic intervention.

References

Aberg, K. A., Chan, R. F., Xie, L., Shabalin, A. A., & van den Oord, E. (2018). Methyl-CpG-Binding domain sequencing: MBD-seq. *Methods in Molecular Biology, 1708*, 171–189.

Andresen, J. M., Gayan, J., Djousse, L., Roberts, S., Brocklebank, D., Cherny, S. S., et al. (2007). The relationship between CAG repeat length and age of onset differs for Huntington's disease patients with juvenile onset or adult onset. *Annals of Human Genetics, 71*(Pt 3), 295–301.

Andrew, S. E., Goldberg, Y. P., Kremer, B., Telenius, H., Theilmann, J., Adam, S., et al. (1993). The relationship between trinucleotide (CAG) repeat length and clinical features of Huntington's disease. *Nature Genetics, 4*(4), 398–403.

Bai, G., Cheung, I., Shulha, H. P., Coelho, J. E., Li, P., Dong, X., et al. (2015). Epigenetic dysregulation of hairy and enhancer of split 4 (HES4) is associated with striatal degeneration in postmortem Huntington brains. *Human Molecular Genetics, 24*(5), 1441–1456.

Baronchelli, S., La Spada, A., Ntai, A., Barbieri, A., Conforti, P., Jotti, G. S., et al. (2017). Epigenetic and transcriptional modulation of WDR5, a chromatin remodeling protein, in Huntington's disease human induced pluripotent stem cell (hiPSC) model. *Molecular and Cellular Neurosciences, 82*, 46–57.

Belsky, D. W., Caspi, A., Corcoran, D. L., Sugden, K., Poulton, R., Arseneault, L., et al. (2022). DunedinPACE, a DNA methylation biomarker of the pace of aging. *eLife, 11*, e73420.

Blasi, F., & Toniolo, D. (1983). DNA methylation and X-chromosome inactivation. *Molecular Biology & Medicine, 1*(2), 271–274.

Bocklandt, S., Lin, W., Sehl, M. E., Sánchez, F. J., Sinsheimer, J. S., Horvath, S., & Vilain, E. (2011). Epigenetic predictor of age. *PLoS One, 6*(6), e14821.

Brown, S. E., Suderman, M. J., Hallett, M., & Szyf, M. (2008). DNA demethylation induced by the methyl-CpG-binding domain protein MBD3. *Gene, 420*(2), 99–106.

Carey, N., Marques, C. J., & Reik, W. (2011). DNA demethylases: A new epigenetic frontier in drug discovery. *Drug Discovery Today, 16*(15–16), 683–690.

De Souza, R. A., Islam, S. A., McEwen, L. M., Mathelier, A., Hill, A., Mah, S. M., et al. (2016). DNA methylation profiling in human Huntington's disease brain. *Human Molecular Genetics, 25*(10), 2013–2030.

Duan, R., Fu, Q., Sun, Y., & Li, Q. (2022). Epigenetic clock: A promising biomarker and practical tool in aging. *Ageing Research Reviews, 81*, 101743.

Ferrer, I., Goutan, E., Marin, C., Rey, M. J., & Ribalta, T. (2000). Brain-derived neurotrophic factor in Huntington disease. *Brain Research, 866*(1–2), 257–261.

Goll, M. G., Kirpekar, F., Maggert, K. A., Yoder, J. A., Hsieh, C. L., Zhang, X., et al. (2006). Methylation of tRNAAsp by the DNA methyltransferase homolog Dnmt2. *Science, 311*(5759), 395–398.

Greenberg, M. E., Xu, B., Lu, B., & Hempstead, B. L. (2009). New insights in the biology of BDNF synthesis and release: Implications in CNS function. *The Journal of Neuroscience, 29*(41), 12764–12767.

Guo, J. U., Su, Y., Shin, J. H., Shin, J., Li, H., Xie, B., et al. (2014). Distribution, recognition and regulation of non-CpG methylation in the adult mammalian brain. *Nature Neuroscience, 17*(2), 215–222.

Gusella, J. F., Wexler, N. S., Conneally, P. M., Naylor, S. L., Anderson, M. A., Tanzi, R. E., et al. (1983). A polymorphic DNA marker genetically linked to Huntington's disease. *Nature, 306*(5940), 234–238.

Gutierrez, A., Corey-Bloom, J., Thomas, E. A., & Desplats, P. (2019). Evaluation of biochemical and epigenetic measures of peripheral Brain-Derived Neurotrophic Factor (BDNF) as a biomarker in Huntington's disease patients. *Frontiers in Molecular Neuroscience, 12*, 335.

Hannum, G., Guinney, J., Zhao, L., Zhang, L., Hughes, G., Sadda, S., et al. (2013). Genome-wide methylation profiles reveal quantitative views of human aging rates. *Molecular Cell, 49*(2), 359–367.

Horvath, S. (2013). DNA methylation age of human tissues and cell types. *Genome Biology, 14*(10), R115.

Horvath, S., & Raj, K. (2018). DNA methylation-based biomarkers and the epigenetic clock theory of ageing. *Nature Reviews. Genetics, 19*(6), 371–384.

Horvath, S., Langfelder, P., Kwak, S., Aaronson, J., Rosinski, J., Vogt, T. F., et al. (2016). Huntington's disease accelerates epigenetic aging of human brain and disrupts DNA methylation levels. *Aging (Albany NY), 8*(7), 1485–1512.

Horvath, S., Oshima, J., Martin, G. M., Lu, A. T., Quach, A., Cohen, H., et al. (2018). Epigenetic clock for skin and blood cells applied to Hutchinson Gilford Progeria Syndrome and ex vivo studies. *Aging (Albany NY), 10*(7), 1758–1775.

Hotchkiss, R. D. (1948). The quantitative separation of purines, pyrimidines, and nucleosides by paper chromatography. *The Journal of Biological Chemistry, 175*(1), 315–332.

Hsu, H. K., Weng, Y. I., Hsu, P. Y., Huang, T. H., & Huang, Y. W. (2014). Detection of DNA methylation by MeDIP and MBDCap assays: An overview of techniques. *Methods in Molecular Biology, 1105*, 61–70.

Hu, X. V., Rodrigues, T. M., Tao, H., Baker, R. K., Miraglia, L., Orth, A. P., et al. (2010). Identification of RING finger protein 4 (RNF4) as a modulator of DNA demethylation through a functional genomics screen. *Proceedings of the National Academy of Sciences of the United States of America, 107*(34), 15087–15092.

Huntington Disease Collaborative Research Group, D. (1993). A novel gene containing a trinucleotide repeat that is expanded and unstable on Huntington's disease chromosomes. The Huntington's Disease Collaborative Research Group. *Cell, 72*(6), 971–983.

Jia, H., Morris, C. D., Williams, R. M., Loring, J. F., & Thomas, E. A. (2015). HDAC inhibition imparts beneficial transgenerational effects in Huntington's disease mice via altered DNA and histone methylation. *Proceedings of the National Academy of Sciences of the United States of America, 112*(1), E56–E64.

Johnson, T. B., & Coghill, R. D. (1925). Researches on pyrimidines. C111. The discovery of 5-methyl-cytosine in tuberculinic acid, the nucleic acid of the tubercle bacillus. *Journal of the American Chemical Society, 47*, 2838–2844.

Jørgensen, H. F., Adie, K., Chaubert, P., & Bird, A. P. (2006). Engineering a high-affinity methyl-CpG-binding protein. *Nucleic Acids Research, 34*(13), e96.

Kawaguchi, Y., Okamoto, T., Taniwaki, M., Aizawa, M., Inoue, M., Katayama, S., et al. (1994). CAG expansions in a novel gene for Machado-Joseph disease at chromosome 14q32.1. *Nature Genetics, 8*(3), 221–228.

Klose, R. J., & Bird, A. P. (2006). Genomic DNA methylation: The mark and its mediators. *Trends in Biochemical Sciences, 31*(2), 89–97.

Koch, C. M., Suschek, C. V., Lin, Q., Bork, S., Goergens, M., Joussen, S., et al. (2011). Specific age-associated DNA methylation changes in human dermal fibroblasts. *PLoS One, 6*(2), e16679.

Kohli, R. M., & Zhang, Y. (2013). TET enzymes, TDG and the dynamics of DNA demethylation. *Nature, 502*(7472), 472–479.

Laird, P. W. (2010). Principles and challenges of genomewide DNA methylation analysis. *Nature Reviews Genetics, 11*(3), 191–203.

Levine, M. E., Lu, A. T., Quach, A., Chen, B. H., Assimes, T. L., Bandinelli, S., et al. (2018). An epigenetic biomarker of aging for lifespan and healthspan. *Aging (Albany NY), 10*(4), 573–591.

Li, J., Shu, A., Sun, Y., Yang, W., Tang, X., Pu, H., et al. (2022). DNA methylation age acceleration is associated with age of onset in Chinese spinocerebellar ataxia type 3 patients. *Neurobiology of Aging, 113*, 1–6.

Lister, R., Pelizzola, M., Dowen, R. H., Hawkins, R. D., Hon, G., Tonti-Filippini, J., et al. (2009). Human DNA methylomes at base resolution show widespread epigenomic differences. *Nature, 462*(7271), 315–322.

Loomans-Kropp, H. A., Song, Y., Gala, M., Parikh, A. R., Van Seventer, E. E., Alvarez, R., et al. (2022). Methylated Septin9 (mSEPT9): A promising blood-based biomarker for the detection and screening of early-onset colorectal cancer. *Cancer Research Communications, 2*(2), 90–98.

Lu, B. (2003). BDNF and activity-dependent synaptic modulation. *Learning & Memory, 10*(2), 86–98.

Lu, A. T., Quach, A., Wilson, J. G., Reiner, A. P., Aviv, A., Raj, K., et al. (2019). DNA methylation GrimAge strongly predicts lifespan and healthspan. *Aging (Albany NY), 11*(2), 303–327.

Lu, A. T., Narayan, P., Grant, M. J., Langfelder, P., Wang, N., Kwak, S., et al. (2020). DNA methylation study of Huntington's disease and motor progression in patients and in animal models. *Nature Communications, 11*(1), 4529.

Machiela, E., & Southwell, A. L. (2020). Biological aging and the cellular pathogenesis of Huntington's disease. *Journal of Huntington's Disease, 9*(2), 115–128.

Marques Sousa, C., & Humbert, S. (2013). Huntingtin: Here, there, everywhere! *Journal of Huntington's Disease, 2*(4), 395–403.

McEwen, L. M., O'Donnell, K. J., McGill, M. G., Edgar, R. D., Jones, M. J., MacIsaac, J. L., et al. (2019). The PedBE clock accurately estimates DNA methylation age in pediatric buccal cells. *Proceedings of the National Academy of Sciences of the United States of America*.

Meissner, A., Gnirke, A., Bell, G. W., Ramsahoye, B., Lander, E. S., & Jaenisch, R. (2005). Reduced representation bisulfite sequencing for comparative high-resolution DNA methylation analysis. *Nucleic Acids Research, 33*(18), 5868–5877.

Miller, J., Arrasate, M., Shaby, B. A., Mitra, S., Masliah, E., & Finkbeiner, S. (2010). Quantitative relationships between huntingtin levels, polyglutamine length, inclusion body formation, and neuronal death provide novel insight into Huntington's disease molecular pathogenesis. *The Journal of Neuroscience, 30*, 10541–10550.

Moran, S., Arribas, C., & Esteller, M. (2016). Validation of a DNA methylation microarray for 850,000 CpG sites of the human genome enriched in enhancer sequences. *Epigenomics, 8*(3), 389–399.

Ng, C. W., Yildirim, F., Yap, Y. S., Dalin, S., Matthews, B. J., Velez, P. J., et al. (2013). Extensive changes in DNA methylation are associated with expression of mutant huntingtin. *Proceedings of the National Academy of Sciences of the United States of America, 110*(6), 2354–2359.

Noroozi, R., Ghafouri-Fard, S., Pisarek, A., Rudnicka, J., Spólnicka, M., Branicki, W., et al. (2021). DNA methylation-based age clocks: From age prediction to age reversion. *Ageing Research Reviews, 68*, 101314.

Ogino, S., Nosho, K., Kirkner, G. J., Kawasaki, T., Chan, A. T., Schernhammer, E. S., et al. (2008). A cohort study of tumoral LINE-1 hypomethylation and prognosis in colon cancer. *Journal of the National Cancer Institute, 100*(23), 1734–1738.

Ohka, F., Natsume, A., Motomura, K., Kishida, Y., Kondo, Y., Abe, T., et al. (2011). The global DNA methylation surrogate LINE-1 methylation is correlated with MGMT promoter methylation and is a better prognostic factor for glioma. *PLoS One, 6*(8), e23332.

Okano, M., Bell, D. W., Haber, D. A., & Li, E. (1999). DNA methyltransferases Dnmt3a and Dnmt3b are essential for de novo methylation and mammalian development. *Cell, 99*(3), 247–257.

Pinto, R. M., Dragileva, E., Kirby, A., Lloret, A., Lopez, E., St Claire, J., et al. (2013). Mismatch repair genes Mlh1 and Mlh3 modify CAG instability in Huntington's disease mice: Genome-wide and candidate approaches. *PLoS Genetics, 9*(10), e1003930.

Raj, K., & Horvath, S. (2020). Current perspectives on the cellular and molecular features of epigenetic ageing. *Experimental Biology and Medicine (Maywood, N.J.), 245*(17), 1532–1542.

Razin, A., & Riggs, A. D. (1980). DNA methylation and gene function. *Science, 210*(4470), 604–610.

Reik, W., Collick, A., Norris, M. L., Barton, S. C., & Surani, M. A. (1987). Genomic imprinting determines methylation of parental alleles in transgenic mice. *Nature, 328*(6127), 248–251.

Schafer, A. (2013). Gadd45 proteins: Key players of repair-mediated DNA demethylation. *Advances in Experimental Medicine and Biology, 793*, 35–50.

Sherwani, S. I., & Khan, H. A. (2015). Role of 5-hydroxymethylcytosine in neurodegeneration. *Gene, 570*(1), 17–24.

Shirley, M. (2020). Epi proColon(®) for colorectal cancer screening: A profile of its use in the USA. *Molecular Diagnosis & Therapy, 24*(4), 497–503.

Singer, B. D. (2019). A practical guide to the measurement and analysis of DNA methylation. *American Journal of Respiratory Cell and Molecular Biology, 61*(4), 417–428.

Stadler, M. B., Murr, R., Burger, L., Ivanek, R., Lienert, F., Scholer, A., et al. (2011). DNA-binding factors shape the mouse methylome at distal regulatory regions. *Nature, 480*(7378), 490–495.

Thomas, E. A. (2016). DNA methylation in Huntington's disease: Implications for transgenerational effects. *Neuroscience Letters, 625*, 34–39.

Thomas, E. A. (2019). Epigenetic mechanisms in Huntington's disease. In O. Binda (Ed.), *Chromatin Signaling and Neurological Disorders* (Vol. 7, pp. 73–95). Elsevier.

Thomas, B., Matson, S., Chopra, V., Sun, L., Sharma, S., Hersch, S., et al. (2013). A novel method for detecting 7-methyl guanine reveals aberrant methylation levels in Huntington disease. *Analytical Biochemistry, 436*(2), 112–120.

van der Wijst, M. G., Venkiteswaran, M., Chen, H., Xu, G. L., Plösch, T., & Rots, M. G. (2015). Local chromatin microenvironment determines DNMT activity: From DNA methyltransferase to DNA demethylase or DNA dehydroxymethylase. *Epigenetics, 10*(8), 671–676.

Villar-Menendez, I., Blanch, M., Tyebji, S., Pereira-Veiga, T., Albasanz, J. L., Martin, M., et al. (2013). Increased 5-methylcytosine and decreased 5-hydroxymethylcytosine levels are associated with reduced striatal A2AR levels in Huntington's disease. *Neuromolecular Medicine, 15*(2), 295–309.

Waddington, C. H. (1942). The epigenotype. *Endeavor, 1*, 18–20.

Wang, F., Yang, Y., Lin, X., Wang, J. Q., Wu, Y. S., Xie, W., et al. (2013). Genome-wide loss of 5-hmC is a novel epigenetic feature of Huntington's disease. *Human Molecular Genetics, 22*(18), 3641–3653.

Watson, J. D., & Crick, F. H. (1953). Molecular structure of nucleic acids; A structure for deoxyribose nucleic acid. *Nature, 171*(4356), 737–738.

Weber, M., Davies, J. J., Wittig, D., Oakeley, E. J., Haase, M., Lam, W. L., & Schübeler, D. (2005). Chromosome-wide and promoter-specific analyses identify sites of differential DNA methylation in normal and transformed human cells. *Nature Genetics, 37*(8), 853–862.

Weber, M., Hellmann, I., Stadler, M. B., Ramos, L., Pääbo, S., Rebhan, M., & Schübeler, D. (2007). Distribution, silencing potential and evolutionary impact of promoter DNA methylation in the human genome. *Nature Genetics, 39*(4), 457–466.

Wexler, N. S., Lorimer, J., Porter, J., Gomez, F., Moskowitz, C., Shackell, E., et al. (2004). Venezuelan kindreds reveal that genetic and environmental factors modulate Huntington's disease age of onset. *Proceedings of the National Academy of Sciences of the United States of America, 101*(10), 3498–3503.

Wu, J. C., & Santi, D. V. (1985). On the mechanism and inhibition of DNA cytosine methyltransferases. *Progress in Clinical and Biological Research, 198*, 119–129.

Wyatt, G. R. (1950). Occurrence of 5-methylcytosine in nucleic acids. *Nature, 166*(4214), 237–238.

Zadel, M., Maver, A., Kovanda, A., & Peterlin, B. (2018). DNA methylation profiles in whole blood of Huntington's disease patients. *Frontiers in Neurology, 9*, 655.

Zuccato, C., Ciammola, A., Rigamonti, D., Leavitt, B. R., Goffredo, D., Conti, L., et al. (2001). Loss of huntingtin-mediated BDNF gene transcription in c disease. *Science, 293*(5529), 493–498.

Zuccato, C., Tartari, M., Crotti, A., Goffredo, D., Valenza, M., Conti, L., et al. (2003). Huntingtin interacts with REST/NRSF to modulate the transcription of NRSE-controlled neuronal genes. *Nature Genetics, 35*(1), 76–83.

MicroRNAs as Potential Biomarkers in Huntington's Disease

Huajing You and Zhong Pei

Abbreviations

AD	Alzheimer's disease
Ago2	Argonaute 2
ANXA1	Anexin-1A
BDNF	Brain Derived Neurotrophic Factor
CNS	Central Nervous System
CoREST	Repressor Element-1 Silencing Transcription Factor Corepressor
CREB1	CREB1, cAMP response element binding protein 1
EE	Early endosomes
Ex-miRNA	Exosomal-miRNA
HD	Huntington's Disease
HTT	Huntingtin
MeCP2	Methyl CpG binding protein 2
miRNA	MicroRNAs
mRNA	Messenger RNA
MVB	Multivesicular bodies
NfL	Neurofilament light
p250GAP	GTP-activated protein P250
PD	Parkinson's disease
PGC1	Proliferator-activated receptor gamma coactivator 1
Pol II	RNA polymerase II
Pre-miRNA	Precursor miRNA
Pri-miRNA	Primary transcript miRNA
PTB1	Polypyrimidine tract binding protein-1

H. You · Z. Pei (✉)
Department of Neurology, The First Affiliated Hospital, Sun Yat-sen University, Guangdong Provincial Key Laboratory of Diagnosis and Treatment of Major Neurological Diseases, National Key Clinical Department and Key Discipline of Neurology, Guangzhou, China
e-mail: peizhong@mail.sysu.edu.cn

E. A. Thomas, G. M. Parkin (eds.), *Biomarkers for Huntington's Disease*,
Contemporary Clinical Neuroscience, https://doi.org/10.1007/978-3-031-32815-2_16

REST	Repressor Element-1 Silencing Transcription Factor
RISC	RNA-induced silencing complex
SOX9	SRY-related HMG box transcription factor 9
TBP	TATA-binding protein
UTR	Untranslated region

1 Introduction

Huntington's disease (HD) is an autosomal dominant neurodegenerative disorder characterized by progressively deteriorative motor, psychiatric, and cognitive dysfunction. It is caused by abnormal expansion of a CAG repeat segment in the huntingtin protein (HTT) gene and striatal atrophy is the most prominent structural change in imaging (Ross et al., 2014). The age of onset is negatively correlated with the length of CAG expansion and most patients are first manifested with chorea, dystonia and bradykinesia (Ross et al., 2014). The clinical severity of HD mutation carriers is usually evaluated by clinicians through use of the internationally Unified Huntington's Disease Rating Scale (UHDRS). In recent years, some blood-based biomarkers such as plasma neurofilament light protein (NfL) have emerged to be reliable biomarkers in HD (Byrne et al., 2017). However, NfL is non-specific and has little reported relevance with HD pathogenesis.

MiRNAs are non-coding single-stranded small molecules approximately 21 nucleotides in length (Dong & Cong, 2021). MiRNAs regulate the expression of approximately 90% of genes in the body and are involved in cell proliferation, development, and senescence (Ying et al., 2018). By binding to the 3' untranslated region (UTR) of target mRNAs, the translation or degradation of target mRNAs is inhibited (Tafrihi & Hasheminasab, 2019). Dysregulated expression of miRNAs is associated with pathological processes of HD. Here, we will briefly introduce the biosynthesis pathway, secretion, and distribution of miRNAs. We will also describe the role of miRNAs in HD pathogenesis, and miRNAs associated with HD clinical severity, which have the potential to be specific biomarkers for HD diagnosis.

2 The Biosynthesis Pathway of MicroRNAs

MiRNA genes are encoded either in intergenic regions, transcribed by their own promoter, or in exonic regions, overlapping with coding regions and transcribed by the host promoter (Rodriguez et al., 2004). Over 40% of human miRNAs are found in clusters that are co-transcribed as polycistronic transcriptional units (Griffiths-Jones et al., 2008). In the nucleus, miRNA genes are transcribed by RNA polymerase II (pol II) into initial miRNAs (pri-miRNAs), which are several kilobases long. The pri-miRNAs are then processed by the microprocessor complex into precursor miRNAs (pre-miRNAs), which are approximately 70 nucleotides with a stem-ring structure. The microprocessor complex is comprised of the RNase III enzyme called Drosha and its binding partner DiGeorge syndrome critical region gene 8 (DGCR8), a double-stranded RNA-binding protein (Denli et al., 2004). Exportin-5 and Ran GTPase transport the pre-miRNAs to the cytoplasm. A second RNase III enzyme named Dicer then cleaves the pre-miRNAs into double-stranded miRNAs of approximately 21 nucleotides. The double strands are broken down into two single-stranded molecules, one of which binds to the Argonaute proteins and forms the RNA-induced silencing complex (RISC) to participate in mRNA regulation (Yates et al., 2013). The complementary strand is believed to be non-functional and is rapidly degraded. RISC binds to the 3'-UTR of target genes and inhibits their expression at the post-transcriptional level (Miranda et al., 2006).

In recent years, many miRNAs were found to be generated via alternative pathways. For example, three Drosha independent pathways called the mirtron pathways have been identified in mammals; the generation of miRNAs via these pathways are relatively uncommon but particularly important in the primate nervous system (Berezikov et al., 2007). The most common pathway replaces the microprocessor complex step outlined above with a splicing event to produce short hairpin introns known as mirtrons (Ruby et al., 2007). A single miRNA (miR-451) is known to be produced via direct pre-miRNA loading onto Argonaute 2 (Ago2) protein through another mirtron pathway, and miRNA-like small RNA sequences are generated from transfer RNAs, with RNaseZ cleavage of pre-miRNAs in place of Dicer, through the third mirtron pathway (Griffiths-Jones et al., 2008).

3 The Secretion and Distribution of MicroRNAs

MiRNAs play important roles in regulating neurodevelopment, cell proliferation, synaptic plasticity and learning and memory in the nervous system (Snyman et al., 2017). MiRNA expression is regulated by epigenetic mechanisms, including histone modifications and DNA methylation, or by enzymes that stabilize mature miRNAs (Chuang & Jones, 2007). MiRNAs are secreted in small vesicles, as well as external to vesicles (e.g., in peripheral blood, saliva, and urine) and bind to proteins or other molecules. For example, plasma miRNAs can bind to high-density lipoprotein to form stable structures in peripheral circulation, which is an excellent characteristic for an HD biomarker (Ritchie, 2017). Exosomal-miRNAs (ex-miRNAs) are relatively more stable than miRNAs in serum or plasma (Hu et al., 2012), and miRNAs actively packed into exosomes are enriched in those exosomes compared to cell free serum and plasma (Cheng et al., 2014). Exosomes are generated by early endosomes (EEs) formed from the inner plasma membrane. EEs interact with the Golgi complex to produce multivesicular bodies (MVBs), which are, in turn, degraded and released as exosomes into the extracellular space (Kourembanas, 2015). Exosomes can deliver specific miRNAs to target cells with specific surface ligands (Valadi et al., 2007). After being transferred into the recipient cells, miRNAs can modify the gene expression of those cells. The micro-environment of the nervous system can also influence the nature and composition of ex-miRNAs, suggesting that ex-miRNAs are valuable as early diagnostic markers in HD (Catanesi et al., 2020).

There are more than 1000 types of miRNAs in the human brain and their distribution has regional, tissue and spatiotemporal specificity. For example, miR-10b-5p, miR-145-5p, miR-378c and miR-217 are mainly expressed in the prefrontal cortex, miR-205 in the frontal cortex and miR-21-3p, miR-224 and miR-373-3p in the substantia nigra. Different types of miRNAs also have different functions. For example, miR-134 participates in the formation of dendritic spines, while miR-124 and miR-132 participate in axon growth (Dong & Cong, 2019). Fifty four miRNAs have been reported to be differentially expressed in the postmortem brains of HD patients, compared to controls, including 30 miRNAs with high expression and 24 miRNAs with low expression (Sinha et al., 2012). Studies on the expression of miRNAs in different brain regions of HD mouse models have revealed that the greatest number of miRNAs with differential expression was found in the striatum (number: 159), followed by the cerebellum (number: 102), hippocampus (number: 51) and cerebral cortex (number: 45). Based on the presence of Repressor Element-1 Silencing Transcription Factor (REST) binding sites in the genome, Johnson et al. identified a set of brain or neuron-specific REST-target miRNAs (miR-9-1, 9-3, 29a, 29b-1, 124a-1, 124a-2, 124a-3, 132, 135b, 139, 203, 204, 212, 330, and 346) that were differentially expressed in HD mice. However, only down-regulation of miR-132 was confirmed in parietal cortical tissue of HD patients (Johnson et al., 2008). By analyzing the expression profile of miRNAs in the frontal cortex and striatum of HD patients, Marti et al. found that 51 miRNAs such as miR-100, miR-151-3p, miR-16, miR-219-2-3p, miR-27b, miR-451, and miR-92a were increased in the frontal cortex and striatum of HD brains, while 36 miRNAs such as miR-128, miR-139-3p, miR-222, miR-382, miR-433, and miR-483-3p were decreased in the frontal cortex and striatum of HD brains (Martí et al., 2010). The expression of the miR-34 family members was investigated in the brain, liver, and skeletal muscle of the R6/2 HD mouse model. Expression of miR-34a-5p was significantly higher than miR-34-b/c isoforms in all three tissues (Reynolds et al., 2018). In addition, Kocerha et al. found that 11 miRNAs were significantly altered (P value <0.05) in the frontal cortex of HD monkeys (Kocerha et al., 2014).

4 The Role of MicroRNAs in HD Pathogenesis

The role of miRNAs in HD pathogenesis has been increasingly appreciated (Langfelder et al., 2018). MiRNA-mRNA networks are the most well-studied and and have been shown to regulate different phenotypes in cell and animal models of HD, including in STHdh(Q111)/Hdh(Q111) cells, HD-iPSC derived neurons, R6/2 mice, transgenic HD monkeys and human HD brains (Fig. 1) (Wu & Kuo, 2020). MicroRNAs and mRNAs appear to regulate levels of neurotoxicity in HD through different mechanisms, including transcriptional dysregulation, chromatin regulation, cell cycle regulation, inflammation and apoptosis (Kawahara, 2014).

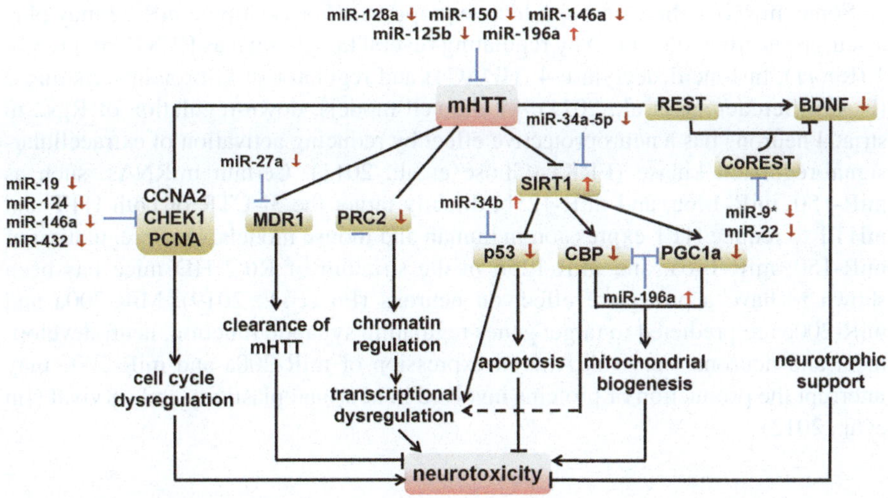

Fig. 1 Networks of microRNAs and mRNAs regulate the levels of neurotoxicity in HD via modulation of mutant HTT. mRNA are shown in yellow textboxes

MiR-212, miR-132, miR-128 and miR-218 have been reported to be associated with HTT CAG repeat expansion, while miR-125b and miR-150 target p53 to regulate expression of the p65 subunit of NF-κB (RelA/NF-κB) and miR-146a (Kocerha et al., 2014). MiR-146a, miR-125b and miR-150 could target the HTT gene. MiR-125b and miR-150 could also target the transcription regulator TP53 while miR-146a was reciprocally regulated by TP53 in HD (Sinha et al., 2012). mHTT was shown to reduce expression of miR-432, miR-146a, miR-19a, and miR-146a in the STsdh(Q111)/Hdh(Q111) cell model of HD. These miRNAs directly target cell cycle genes, such as proliferating cell nuclear antigen (PCNA), checkpoint kinase 1 (CHEK1) and CyclinA2 (CCNA2). Exogenous expression of miR-432 and miR-146a in STsdh(Q111)/Hdh(Q111) cells rescued cell cycle abnormalities and abnormally high levels of apoptosis (Das et al., 2015). MiR-34a positively interacted with cell cycle progression, cellular senescence, and apoptosis (Catanesi et al., 2020).

REST is a zinc finger protein that binds to the corresponding neuron-restricted silencing elements (NRSEs) and deacetylates histones to inhibit the expression of some neurogenic genes in non-neuronal cells (Soldati et al., 2013). Under normal conditions in neuronal cells, REST is retained in the cytoplasm by interacting with HTT. mHTT protein can inhibit the interaction between REST and HTT protein, thereby promoting REST aggregation in the nucleus of HD patients and inhibiting the related neuronal gene transcription (Wu & Xie, 2006). REST is also regulated by a variety of miRNAs including miR-9, miR-29a, miR-29b, miR-124a, miR-132 and miR-135b, miR-139, miR-203, miR-204, miR-212, miR-330, and miR-346 (Reed et al., 2018; Jovicic et al., 2013).

Some miRNAs show neuroprotective properties. For example, miR-22 may play a neuroprotective role in HD by regulating several targets such as REST corepressor 1 (Rcor1), histone deacetylase 4 (HDAC4) and regulator of G-protein signaling 2 (Rgs2) (Seredenina et al., 2011). In HD cell models, down-regulation of Rgs2 in striatal neurons has a neuroprotective effect by reducing activation of extracellular-signal-regulated kinase (ERK) (Ghose et al., 2011). Certain miRNAs, such as miR-150, miR-146a, and miR-125b, directly target the 3'-UTR of both HTT and mHTT to reduce HTT expression in human and mouse models. Downregulation of miR-150, miR-146a, and miR-125b in the striatum of R6/2 HD mice has been shown to have a protective effect on neurons (Fu et al., 2019). MiR-200a and miR-200c are predicted to target genes regulating synaptic function, neurodevelopment and neuronal survival. Altered expression of miR-200a and miR-200c may interrupt the production of proteins involved in neuronal plasticity and survival (Jin et al., 2012).

5 Specific MicroRNAs as Potential Biomarker in HD

The expression levels of miR-9 and miR-9*, which target REST and CoREST, are decreased in the cerebral cortex of HD patients. REST contains a miR-9 recognition element, which aligns with the original transcriptional sequence of miR-9 (Johnson & Buckley, 2009). Similarly, miR-9*, which is predicted to increase the expression of the corepressor of REST has been shown to be down-regulated in peripheral leukocytes of HD patients (Chang et al., 2017). Decreased levels of miR-9/9* in HD would therefore promote the transcription and accumulation of REST. This phenomenon is further magnified because miR-9/9* transcription also depends on REST (Johnson et al., 2009).

MiR-10b-5p is located in the HOXD cluster and targets HOXD4. MiR-10b-5p was significantly up-regulated in the prefrontal cortex and striatum of HD brain and this change may begin to occur pre-symptomatically (Hoss et al., 2015b). MiR-10b-5p has a binding target within the 3'-UTR of brain derived neurotrophic factor (BDNF) transcript and up-regulation of miR-10b-5p has been reported to mediate BDNF down-regulation in the prefrontal cortex of HD brains (Varendi et al., 2014). CREB1 is another target of miR-10b-5p. MiR-10b-5p binds to the 3'-UTR of CREB1 and down-regulates CREB1 expression in HD (Lewis et al., 2005). Moreover, miR-10b-5p presumably targets HTT by binding to two 3'-UTR sites (both 7mer-1A seeds, positions 2742-2748 and 3301-3307) and could decrease HTT expression via this mechanism (Ritchie, 2017).

MiR-124 is a neuronal-specific miRNA and is thought to be a crucial regulator of neuronal differentiation in HD (Liu et al., 2015). It is down-regulated in HD mouse models and REST was suggested to be a primary mediator of this effect (Das et al., 2013). Repression of SRY-related HMG box transcription factor 9 (SOX9) mediated by miR-124 is essential for neuron formation from neural stem cells

(Cheng et al., 2009). MiR-124 also has been shown to promote the differentiation process by facilitating neuron-specific splicing of mRNAs through repression of polypyrimidine tract binding protein-1 (PTB1) in developing neurons (Makeyev et al., 2007). Evidence suggested that miR-124 over-expression up-regulated BDNF and peroxisome proliferator-activated receptor gamma coactivator 1 (PGC1) expression and down-regulated SOX9 expression in the striatum of R6/2 mouse brain, indicating that miR-124 has the potential to affect HD progression (Meza-Sosa et al., 2012).

MiR-128a was down-regulated in a HD monkey model by the time of birth (Kocerha et al., 2014). Consistently, MiR-128a was also decreased in the brains of pre-symptomatic and symptomatic HD patients (Kocerha et al., 2014). Interestingly, miR-128a targets a panel of transcription factors, including Huntingtin Interaction Protein 1 (HIP1) and SP1, to suppress their 3'-UTR activity and promote HTT transcription; however, mir-128a is also able to target the 3'-UTR of HTT directly (Kocerha et al., 2014). Thus, down-regulation of miR-128a contributes to HTT related phenotypes by both direct and indirect mechanisms.

MiR-132 is a key component of the activity-dependent gene regulatory response in neurons (Chen et al., 2018). Methyl CpG binding protein 2 (MeCP2) increases BDNF levels in the brain, which in turn increases transcription of the miR-132 cluster. Inversely, miR-132 dysregulation also affects BDNF balance in the HD brain and failure to regulate MeCP2 levels might contribute to HD progression by impairing neurogenesis (Klein et al., 2007). REST could repress the expression of miR-132 by not only directly repressing its transcription, but also repressing BDNF in an repressive feed-forward loop (Bicker et al., 2014). In addition, the level of miR-132 is associated with Ago2-dependent HTT clearance and supplementation with miR-132 has been shown to improve the motor function and lifespan in R6/2 mice (Fukuoka et al., 2018).

Expression of miR-146a was directly or indirectly regulated by p53 (Ghose et al., 2011). Its expression was decreased in STHdh (Q111)/Hdh (Q111) cells, which was due to the low activity of p65 subunit of NF-κB (Ghose et al., 2011). MiR-146a may directly regulate TATA-binding protein (TBP) expression such that decreased miR-146a levels lead to high TBP expression levels and further affect cellular responses in HD cell models, contributing to HD pathogenesis (Sinha et al., 2010). MiR-146a also targets the human and mouse Htt gene and its overexpression reduced the expression of endogenous HTT (Sinha et al., 2011). Heat shock factor 1 (HSF1) can regulate miR-146a and suppress mHTT aggregates in cells (Das & Bhattacharyya, 2015).

Over-expression of miR-196a was detected in both animal and human HD brains, and has subsequently been shown to ameliorate cytotoxicity, apoptosis, neural cell growth, mitochondrial morphology and function in HD cell models (Cheng et al., 2013; Kunkanjanawan et al., 2016). For instance, miR-196a targets Anexin-1A (ANXA1) to promote cell survival and reduce apoptosis (Kunkanjanawan et al., 2016). Expression of BDNF was significantly evelated in HD-196a neural cells to promote neural cell growth (Kunkanjanawan et al., 2016). MiR-196a upregulation modulates mitochondrial function by its effects on CREB binding protein (CBP) and PGC-1α (Kunkanjanawan et al., 2016). Upregulated miR-196a could suppress mHTT expression at the protein and mRNA levels through alterations in the ubiquitin-proteasome system (UPS), gliosis, cAMP response element-binding protein, as well as through several neuronal regulatory pathways in the brain, inhibiting the neuropathological progression. In addition, miR-196a activity may be beneficial via inhibition of the protein synthesis and partial enhancement of protein degradation (Tan et al., 2015). Cheng et al. generated a double transgenic mouse model, which overexpressed miR-196a, as well as mHTT fused to green fluorescent protein (GFP). The levels of miR-196a and mHTT were up- and down-regulated in the brain of these mice, respectively (Cheng et al., 2013). Over-expression of MiR-196a also targets the 3'-UTR of RAN binding protein 10 (RANBP10), suppresses RANBP10 and increases the ability of β-tubulin polymerization, which enhances neuronal morphology and intracellular transport in a HD mouse model (Her et al., 2017).

MiR-214 targets the 3'-UTR of the HTT gene and its expression is increased in HD cell models (Kozlowska et al., 2013). MiR-214 expression could target mitofusin2 (MFN2), altering mitochondrial morphology and deregulating the cell cycle (Table 1) (Bucha et al., 2015).

Table 1 MiRNAs most commonly associated with HD

MiRNAs	Role in HD pathophysiology	Targets	Evidence
MiR-9/9*	Suppresses mHTT expression	REST and CoREST	↓ in the brain
MiR-10b-5p	Reduces mHTT expression	mHTT, BDNF and CREB1	↑in the prefrontal cortex
MiR-124	Participates in the axon growth and neuronal differentiation in HD	SOX9, PTB1 and PGC1	↓ in the brain
MiR-132	Regulates the axonal growth	p250GAP, MeCP2 and REST	↓ in the cortices
MiR-146a	A major regulator of the NF-κB pathway; acts as an inflammatory response repressor in the CNS	mHTT and TBP	↑ in the brain
MiR-196a	Suppresses mHTT expression; cytoskeleton modification and RANBP10 regulation	mHTT, ANXA1 and BDNF	↑ in the prefrontal cortex

HD Huntington disease, *REST* Repressor Element-1 Silencing Transcription Factor, *CoREST* Repressor Element-1 Silencing Transcription Factor Corepressor, *BDNF* Brain Derived Neurotrophic Factor, *CREB1* cAMP response element binding protein 1, *SOX9* SRY-related HMG box transcription factor 9, *PTB1* polypyrimidine tract binding protein-1, *PGC1* proliferator-activated receptor gamma coactivator 1, *p250GAP* GTP-activated protein P250, *MeCP2* Methyl CpG binding protein 2, *TBP* TATA-binding protein, *ANXA1* Anexin-1A

6 The Association of MiRNAs with HD Clinical Severity

MiRNAs are considered to be ideal biomarker candidates for HD. Firstly, it is easier to discover new miRNAs through amplification and detection by real-time quantitative PCR (qPCR), which provide higher throughput than mass spectrometry, the main tool for protein and metabolite identification. Secondly, the detection method is simple, easy and minimally invasive, which is ideal for early screening (Li & Kowdley, 2012; Majdi et al., 2016). By detecting the sequence, structure, types and number of miRNAs in peripheral blood, it might be possible to predict different HD stages. The potential for miRNAs to serve as biomarkers for diagnosis and prognosis is gradually being applied to clinical practice.

Several miRNAs have been detected from exosomes, blood and cerebrospinal fluid (CSF) samples of HD patients and proved to be significantly associated with CAG expansion, age of onset or clinical severity. For example, MiR-10b-5p level showed significant linear increases from healthy controls, to asymptomatic HD gene carriers to manifest HD patients, indicating that it would be a promising clinical diagnostic marker to predict the clinical severity of HD (Hoss et al., 2015a; Packer et al., 2008). MiR-10b-5p expression was also associated with CAG length-adjusted age of onset and the extent of striatal degeneration (Hoss et al., 2015b). Packer et al. have investigated whether miRNAs correlate with disease progression in HD brain samples. They analyzed the expression profile of miRNAs that were predicted to be regulated by REST in Brodmann's area 4 (BA4) brain regions, which were isolated from healthy controls and HD patients at clinical stages 1–4. The expression of five miRNAs (miR-9, miR-9*, miR-29b, miR-124a, and miR-132) were significantly changed with increasing clinical stage of HD (Packer et al., 2008). In another study, the expression levels of miR-29a and miR-330 were increased but miR-124a did not differ between HD and control subjects (Hu et al., 2012). The authors showed that elevated expression of miR-34b appeared prior to symptom onset. Additionally, the miRNA reverse transcription and PCR detection was affordable, only requiring the plasma volume as small as 10 μL (Gaughwin et al., 2011). In another study, CSF samples were collected from 60 PREDICT-HD study participants (15 symptomatic, 30 pre-symptomatic, and 15 controls) to evaluate miRNA levels and a total of 2081 miRNAs were detected. MiR-135b-3p, miR-140-5p, miR-520f-3p, miR-3928-5p, miR-4317 and miR-8082 levels were the highest in symptomatic HD patients, followed by pre-manifest HD patients, with the lowest levels detected in healthy controls. These results indicated that an increase in miRNA levels was positively correlated with the risk of HD onset (Reed et al., 2018). MiR-122-5p levels were found to be negatively associated with higher Unified Huntington's Disease Rating Scale-total motor scores (UHDRS-TMS) (Table 2) (Díez-Planelles et al., 2016). Moreover, miR-100-5p was positively correlated with Unified Huntington's Disease Rating Scale-total capacity score (UHDRS-TFC), whereas miR-330-5p and miR-641 was negatively correlated with UHDRS-TFC (Díez-Planelles et al., 2016). MiR-196a-5p was shown to be significantly related to CAG repeat length and age of onset in HD patients, which indicates its potential to be a biomarker for HD (Ghose et al., 2011).

Table 2 MiRNAs as biomarkers in HD

MiRNAs	Source	Cohort	Clinical measures	Target	Relation
miR-9*	Peripheral leukocytes (Chang et al., 2017)	8 PreHD, 36 HD, 28 HC	UHDRS	HTT, CoREST	Negative
miR-520f-3p	CSF (Reed et al., 2018)	30 PreHD, 15 HD, 15 HC	UHDRS	NA	Positive
miR-135b-3p					
miR-4317					
miR-3928-5p					
miR-8082					
miR-140-5p					
miR-34b	Plasma (Gaughwin et al., 2011)	27 HD, 12 HC	UHDRS, TFC	HTT	Positive
miR-10b-5p	Plasma (Díez-Planelles et al., 2016)	4 PreHD, 26 HD, 8 HC	NA	HTT, BDNF	Positive
miR-486-5p				NA	
miR-122-5p	Plasma (Díez-Planelles et al., 2016)	15 HD, 7 HC	UHDRS	HTT, HIP1, SP1, etc	Negative
miR-100-5p			TFC	AACS, ADAM10, BCL2, etc	Positive
miR-330-3p				NA	Negative
miR-641				NA	

UHDRS Unified Huntington's Disease Rating Scale, *TFC* Total Functional Capacity, *HTT* Huntingtin, *BDNF* Brain-Derived Neurotrophic Factor, *NA* Not assessed

7 Limitations in MiRNAs Biomarker Studies and Possible Solutions

MiRNAs have been identified as a potential biomarker in many central nervous system (CNS) diseases. However, study limitations that apply to other CNS diseases can also be applied to the analysis of miRNAs in HD. First of all, there is limited overlap in miRNAs identified in different studies using the same biological fluids. This may be due to several reasons, including a lack of standardization in the choice of control populations, source of miRNAs (e.g., serum/plasma/urine/saliva, exosomal/non-exosomal), isolation protocol, analytical approach (e.g., using microarray versus RT-qPCR) and/or sample size (Nik Mohamed Kamal & Shahidan, 2020). Secondly, studies are rarely able to conduct detailed and conclusive analyses focused on a particular disease type in its entirety because of lack of financing, facilities and sample sources, etc. Thirdly, most of the studies collected circulating miRNAs in blood. However, human blood contains a variety of cell types, hence making it difficult to identify the cell origin of particular miRNAs (Nik Mohamed Kamal & Shahidan, 2020).

MiRNAs in Other Neurodegenerative Disorders-Lessons Learned

Studies of miRNAs in other CNS diseases can help us better understand the function of miRNAs in association with HD. Combinations of miRNAs, utilization of miRNAs from various sources within the same study, and studying miRNAs related to the pathogenesis of disease have been recommended to increase the accuracy and potential of miRNAs to be used as biomarkers for other diseases. For example, in Parkinson's disease (PD), miR-19b-3p was shown to be down-regulated in serum and exosomes, while being up-regulated in CSF (Dong et al., 2016). Patil et al. demonstrated the combination of serum miRNAs such as miR-335-5p/miR-3613-3p/miR-6865-3p could increase the discriminatory accuracy of miRNAs as biomarkers (Patil et al., 2019). In Alzheimer disease (AD), some miRNAs target AD-related genes, such as amyloid precursor protein (APP) and Presenilin −1 and −2 (PSEN1&2) (Wu et al., 2016). In addition, miR-29a-3p and miR-29b-3p target β-site APP-cleaving enzyme (BACE1), which contribute to the formation of Aβ plaques. Accordingly, decreased expression of these miRNAs were associated with an abnormally high expression of BACE1 in AD patients (Hébert et al., 2008). A panel of at 10 miRNAs (miRNA-26b, miRNA-30e, miRNA-34a, miRNA-34c, miRNA-107, miRNA-125b, miRNA-146a, miRNA-200c, miRNA-210, and hsa-miRNA-485), related to different processes in the pathogenesis of AD, were hypothesized to be de-regulated early in AD and up to nearly 20 years before onset of clinical symptoms (Swarbrick et al., 2019). Yang et al. found that both serum exosomal miR-135a and miR-384 were up-regulated while miR-193b was down-regulated in AD patients compared with that of normal controls. Importantly, the combination of miR-135a, −193b, and −384 was proved to better predict early AD diagnosis than any single miRNA alone (Yang et al., 2018).

Advantages of Exosomal miRNAs-Another Possible Solution

MiRNAs are delivered mainly by exosomes. Exosomal miRNAs in the body fluids of subjects can exist stably due to the fact that they are free from degradation by ribonuclease (RNase) and can stably be stored at 4 °C in vitro for 48 h (Weber et al., 2010). The internal exosomal environment reflect the status of parental cells and an individual's health status (e.g., high temperature, very low or high PH levels), which enables the analysis of low-abundance molecules of interest in a less complex environment (Nik Mohamed Kamal & Shahidan, 2020). Actually, most of the literature preferred exosomal miRNAs rather than non-exosomal miRNAs to be biomarkers owing to their quantity, quality and stability (Nik Mohamed Kamal & Shahidan, 2020). Of course, the potential of other biomarkers for disease diagnosis must also be acknowledged. The analysis of exosomal miRNAs could therefore complement other biomarkers discussed in this book to make disease diagnosis more efficient and timely.

Acknowledgments This work was supported by the National Natural Science Foundation of China (Grant Numbers 81873751 and 82071255), the Guangdong Provincial Key Laboratory of Diagnosis and Treatment of Major Neurological Diseases (2020B1212060017), Guangdong Provincial Clinical Research Center for Neurological Diseases (2020B1111170002), the Southern China International Cooperation Base for Early Intervention and Functional Rehabilitation of Neurological Diseases (2015B050501003 and 2020A0505020004), Guangdong Provincial Engineering Center For Major Neurological Disease Treatment, Guangdong Provincial Translational Medicine Innovation Platform for Diagnosis and Treatment of Major Neurological Disease, Guangzhou Clinical Research and Translational Center for Major Neurological Diseases (201604020010).

References

Berezikov, E., Chung, W. J., Willis, J., Cuppen, E., & Lai, E. C. (2007). Mammalian mirtron genes. *Molecular Cell, 28*(2), 328–336.

Bicker, S., Lackinger, M., Weiß, K., & Schratt, G. (2014). MicroRNA-132, −134, and −138: A microRNA troika rules in neuronal dendrites. *Cellular and Molecular Life Sciences, 71*(20), 3987–4005.

Bucha, S., Mukhopadhyay, D., & Bhattacharyya, N. P. (2015). Regulation of mitochondrial morphology and cell cycle by microRNA-214 targeting Mitofusin2. *Biochemical and Biophysical Research Communications, 465*(4), 797–802.

Byrne, L. M., Rodrigues, F. B., Blennow, K., Durr, A., Leavitt, B. R., Roos, R. A. C., Scahill, R. I., Tabrizi, S. J., Zetterberg, H., Langbehn, D., & Wild, E. J. (2017). Neurofilament light protein in blood as a potential biomarker of neurodegeneration in Huntington's disease: A retrospective cohort analysis. *Lancet Neurology, 16*(8), 601–609.

Catanesi, M., d'Angelo, M., Tupone, M. G., Benedetti, E., Giordano, A., Castelli, V., & Cimini, A. (2020). MicroRNAs dysregulation and mitochondrial dysfunction in neurodegenerative diseases. *International Journal of Molecular Sciences, 21*(17), 5986.

Chang, K. H., Wu, Y. R., & Chen, C. M. (2017). Down-regulation of miR-9* in the peripheral leukocytes of Huntington's disease patients. *Orphanet Journal of Rare Diseases, 12*(1), 185.

Chen, D., Hu, S., Wu, Z., Liu, J., & Li, S. (2018). The role of MiR-132 in regulating neural stem cell proliferation, differentiation and neuronal maturation. *Cellular Physiology and Biochemistry, 47*(6), 2319–2330.

Cheng, L. C., Pastrana, E., Tavazoie, M., & Doetsch, F. (2009). miR-124 regulates adult neurogenesis in the subventricular zone stem cell niche. *Nature Neuroscience, 12*(4), 399–408.

Cheng, P. H., Li, C. L., Chang, Y. F., Tsai, S. J., Lai, Y. Y., Chan, A. W., Chen, C. M., & Yang, S. H. (2013). miR-196a ameliorates phenotypes of Huntington disease in cell, transgenic mouse, and induced pluripotent stem cell models. *American Journal of Human Genetics, 93*(2), 306–312.

Cheng, L., Sharples, R. A., Scicluna, B. J., & Hill, A. F. (2014). Exosomes provide a protective and enriched source of miRNA for biomarker profiling compared to intracellular and cell-free blood. *Journal of Extracellular Vesicles, 26*, 3.

Chuang, J. C., & Jones, P. A. (2007). Epigenetics and microRNAs. *Pediatric Research, 61*(5 Pt 2), 24R–29R.

Das, S., & Bhattacharyya, N. P. (2015). Heat shock factor 1-regulated miRNAs can target Huntingtin and suppress aggregates of mutant huntingtin. *Microrna, 4*(3), 185–193.

Das, E., Jana, N. R., & Bhattacharyya, N. P. (2013). MicroRNA-124 targets CCNA2 and regulates cell cycle in STHdh(Q111)/Hdh(Q111) cells. *Biochemical and Biophysical Research Communications, 437*(2), 217–224.

Das, E., Jana, N. R., & Bhattacharyya, N. P. (2015). Delayed cell cycle progression in STHdh(Q111)/Hdh(Q111) cells, a cell model for Huntington's disease mediated by microRNA-19a, microRNA-146a and microRNA-432. *Microrna, 4*(2), 86–100.

Denli, A. M., Tops, B. B., Plasterk, R. H., Ketting, R. F., & Hannon, G. J. (2004). Processing of primary microRNAs by the Microprocessor complex. *Nature, 432*(7014), 231–235.

Díez-Planelles, C., Sánchez-Lozano, P., Crespo, M. C., Gil-Zamorano, J., Ribacoba, R., González, N., Suárez, E., Martínez-Descals, A., Martínez-Camblor, P., Álvarez, V., Martín-Hernández, R., Huerta-Ruíz, I., González-García, I., Cosgaya, J. M., Visioli, F., Dávalos, A., Iglesias-Gutiérrez, E., & Tomás-Zapico, C. (2016). Circulating microRNAs in Huntington's disease: Emerging mediators in metabolic impairment. *Pharmacological Research, 108*, 102–110.

Dong, X., & Cong, S. (2019). The emerging role of microRNAs in polyglutamine diseases. *Frontiers in Molecular Neuroscience, 12*, 156.

Dong, X., & Cong, S. (2021). MicroRNAs in Huntington's disease: Diagnostic biomarkers or therapeutic agents? *Frontiers in Cellular Neuroscience, 15*, 705348.

Dong, H., Wang, C., Lu, S., Yu, C., Huang, L., Feng, W., Xu, H., Chen, X., Zen, K., Yan, Q., Liu, W., Zhang, C., & Zhang, C. Y. (2016). A panel of four decreased serum microRNAs as a novel biomarker for early Parkinson's disease. *Biomarkers, 21*(2), 129–137.

Fu, J., Peng, L., Tao, T., Chen, Y., Li, Z., & Li, J. (2019). Regulatory roles of the miR-200 family in neurodegenerative diseases. *Biomedicine & Pharmacotherapy, 119*, 109409.

Fukuoka, M., Takahashi, M., Fujita, H., Chiyo, T., Popiel, H. A., Watanabe, S., Furuya, H., Murata, M., Wada, K., Okada, T., Nagai, Y., & Hohjoh, H. (2018). Supplemental treatment for Huntington's disease with miR-132 that is deficient in Huntington's disease brain. *Molecular Therapy--Nucleic Acids, 11*, 79–90.

Gaughwin, P. M., Ciesla, M., Lahiri, N., Tabrizi, S. J., Brundin, P., & Björkqvist, M. (2011). Hsa-miR-34b is a plasma-stable microRNA that is elevated in pre-manifest Huntington's disease. *Human Molecular Genetics, 20*(11), 2225–2237.

Ghose, J., Sinha, M., Das, E., Jana, N. R., & Bhattacharyya, N. P. (2011). Regulation of miR-146a by RelA/NFkB and p53 in STHdh(Q111)/Hdh(Q111) cells, a cell model of Huntington's disease. *PLoS One, 6*(8), e23837.

Griffiths-Jones, S., Saini, H. K., van Dongen, S., & Enright, A. J. (2008). miRBase: Tools for microRNA genomics. *Nucleic Acids Research, 36*(Database issue), D154–D158.

Hébert, S. S., Horré, K., Nicolaï, L., Papadopoulou, A. S., Mandemakers, W., Silahtaroglu, A. N., Kauppinen, S., Delacourte, A., & De Strooper, B. (2008). Loss of microRNA cluster miR-29a/b-1 in sporadic Alzheimer's disease correlates with increased BACE1/beta-secretase expression. *Proceedings of the National Academy of Sciences of the United States of America, 105*(17), 6415–6420.

Her, L. S., Mao, S. H., Chang, C. Y., Cheng, P. H., Chang, Y. F., Yang, H. I., Chen, C. M., & Yang, S. H. (2017). miR-196a enhances neuronal morphology through suppressing RANBP10 to provide neuroprotection in Huntington's disease. *Theranostics, 7*(9), 2452–2462.

Hoss, A. G., Lagomarsino, V. N., Frank, S., Hadzi, T. C., Myers, R. H., & Latourelle, J. C. (2015a). Study of plasma-derived miRNAs mimic differences in Huntington's disease brain. *Movement Disorders, 30*(14), 1961–1964.

Hoss, A. G., Labadorf, A., Latourelle, J. C., Kartha, V. K., Hadzi, T. C., Gusella, J. F., MacDonald, M. E., Chen, J. F., Akbarian, S., Weng, Z., Vonsattel, J. P., & Myers, R. H. (2015b). miR-10b-5p expression in Huntington's disease brain relates to age of onset and the extent of striatal involvement. *BMC Medical Genetics, 8*, 10.

Hu, G., Drescher, K. M., & Chen, X. M. (2012). Exosomal miRNAs: Biological properties and therapeutic potential. *Frontiers in Genetics, 20*(3), 56.

Jin, J., Cheng, Y., Zhang, Y., Wood, W., Peng, Q., Hutchison, E., Mattson, M. P., Becker, K. G., & Duan, W. (2012). Interrogation of brain miRNA and mRNA expression profiles reveals a molecular regulatory network that is perturbed by mutant huntingtin. *Journal of Neurochemistry, 123*(4), 477–490.

Johnson, R., & Buckley, N. J. (2009). Gene dysregulation in Huntington's disease: REST, microRNAs and beyond. *Neuromolecular Medicine, 11*(3), 183–199.

Johnson, R., Zuccato, C., Belyaev, N. D., Guest, D. J., Cattaneo, E., & Buckley, N. J. (2008). A microRNA-based gene dysregulation pathway in Huntington's disease. *Neurobiology of Disease, 29*(3), 438–445.

Johnson, R., Teh, C. H., Jia, H., Vanisri, R. R., Pandey, T., Lu, Z. H., Buckley, N. J., Stanton, L. W., & Lipovich, L. (2009). Regulation of neural macroRNAs by the transcriptional repressor REST. *RNA, 15*(1), 85–96.

Jovicic, A., Zaldivar Jolissaint, J. F., Moser, R., Silva Santos Mde, F., & Luthi-Carter, R. (2013). MicroRNA-22 (miR-22) overexpression is neuroprotective via general anti-apoptotic effects and may also target specific Huntington's disease-related mechanisms. *PLoS One, 8*(1), e54222.

Kawahara, Y. (2014). Human diseases caused by germline and somatic abnormalities in microRNA and microRNA-related genes. *Congenital Anomalies, 54*(1), 12–21.

Klein, M. E., Lioy, D. T., Ma, L., Impey, S., Mandel, G., & Goodman, R. H. (2007). Homeostatic regulation of MeCP2 expression by a CREB-induced microRNA. *Nature Neuroscience, 10*(12), 1513–1514.

Kocerha, J., Xu, Y., Prucha, M. S., Zhao, D., & Chan, A. W. (2014). microRNA-128a dysregulation in transgenic Huntington's disease monkeys. *Molecular Brain, 7*, 46.

Kourembanas, S. (2015). Exosomes: Vehicles of intercellular signaling, biomarkers, and vectors of cell therapy. *Annual Review of Physiology, 77*, 13–27.

Kozlowska, E., Krzyzosiak, W. J., & Koscianska, E. (2013). Regulation of huntingtin gene expression by miRNA-137, −214, −148a, and their respective isomiRs. *International Journal of Molecular Sciences, 14*(8), 16999–17016.

Kunkanjanawan, T., Carter, R. L., Prucha, M. S., Yang, J., Parnpai, R., & Chan, A. W. (2016). miR-196a ameliorates cytotoxicity and cellular phenotype in transgenic Huntington's disease monkey neural cells. *PLoS One, 11*(9), e0162788.

Langfelder, P., Gao, F., Wang, N., Howland, D., Kwak, S., Vogt, T. F., Aaronson, J. S., Rosinski, J., Coppola, G., Horvath, S., & Yang, X. W. (2018). MicroRNA signatures of endogenous Huntingtin CAG repeat expansion in mice. *PLoS One, 13*(1), e0190550.

Lewis, B. P., Burge, C. B., & Bartel, D. P. (2005). Conserved seed pairing, often flanked by adenosines, indicates that thousands of human genes are microRNA targets. *Cell, 120*(1), 15–20.

Li, Y., & Kowdley, K. V. (2012). MicroRNAs in common human diseases. *Genomics, Proteomics & Bioinformatics, 10*(5), 246–253.

Liu, T., Im, W., Mook-Jung, I., & Kim, M. (2015). MicroRNA-124 slows down the progression of Huntington's disease by promoting neurogenesis in the striatum. *Neural Regeneration Research, 10*(5), 786–791.

Majdi, A., Mahmoudi, J., Sadigh-Eteghad, S., Farhoudi, M., & Shotorbani, S. S. (2016). The interplay of microRNAs and post-ischemic glutamate excitotoxicity: An emergent research field in stroke medicine. *Neurological Sciences, 37*(11), 1765–1771.

Makeyev, E. V., Zhang, J., Carrasco, M. A., & Maniatis, T. (2007). The MicroRNA miR-124 promotes neuronal differentiation by triggering brain-specific alternative pre-mRNA splicing. *Molecular Cell, 27*(3), 435–448.

Martí, E., Pantano, L., Bañez-Coronel, M., Llorens, F., Miñones-Moyano, E., Porta, S., Sumoy, L., Ferrer, I., & Estivill, X. (2010). A myriad of miRNA variants in control and Huntington's disease brain regions detected by massively parallel sequencing. *Nucleic Acids Research, 38*(20), 7219–7235.

Meza-Sosa, K. F., Valle-García, D., Pedraza-Alva, G., & Pérez-Martínez, L. (2012). Role of microRNAs in central nervous system development and pathology. *Journal of Neuroscience Research, 90*(1), 1–12.

Miranda, K. C., Huynh, T., Tay, Y., Ang, Y. S., Tam, W. L., Thomson, A. M., Lim, B., & Rigoutsos, I. (2006). A pattern-based method for the identification of MicroRNA binding sites and their corresponding heteroduplexes. *Cell, 126*(6), 1203–1217.

Nik Mohamed Kamal, N. N. S. B., & Shahidan, W. N. S. (2020). Non-exosomal and exosomal circulatory microRNAs: Which are more valid as biomarkers? *Frontiers in Pharmacology, 10*, 1500.

Packer, A. N., Xing, Y., Harper, S. Q., Jones, L., & Davidson, B. L. (2008). The bifunctional microRNA miR-9/miR-9* regulates REST and CoREST and is downregulated in Huntington's disease. *The Journal of Neuroscience, 28*(53), 14341–14346.

Patil, K. S., Basak, I., Dalen, I., Hoedt, E., Lange, J., Lunde, K. A., Liu, Y., Tysnes, O. B., Forsgren, L., Aarsland, D., Neubert, T. A., Larsen, J. P., Alves, G., & Møller, S. G. (2019). Combinatory microRNA serum signatures as classifiers of Parkinson's disease. *Parkinsonism & Related Disorders, 64*, 202–210.

Reed, E. R., Latourelle, J. C., Bockholt, J. H., Bregu, J., Smock, J., Paulsen, J. S., & Myers, R. H. (2018). PREDICT-HD CSF ancillary study investigators. MicroRNAs in CSF as prodromal biomarkers for Huntington disease in the PREDICT-HD study. *Neurology, 90*(4), e264–e272.

Reynolds, R. H., Petersen, M. H., Willert, C. W., Heinrich, M., Nymann, N., Dall, M., Treebak, J. T., Björkqvist, M., Silahtaroglu, A., Hasholt, L., & Nørremølle, A. (2018). Perturbations in the p53/miR-34a/SIRT1 pathway in the R6/2 Huntington's disease model. *Molecular and Cellular Neurosciences, 88*, 118–129.

Ritchie, W. (2017). microRNA Target Prediction. *Methods in Molecular Biology, 1513*, 193–200.

Rodriguez, A., Griffiths-Jones, S., Ashurst, J. L., & Bradley, A. (2004). Identification of mammalian microRNA host genes and transcription units. *Genome Research, 14*(10A), 1902–1910.

Ross, C. A., Aylward, E. H., Wild, E. J., Langbehn, D. R., Long, J. D., Warner, J. H., Scahill, R. I., Leavitt, B. R., Stout, J. C., Paulsen, J. S., Reilmann, R., Unschuld, P. G., Wexler, A., Margolis, R. L., & Tabrizi, S. J. (2014). Huntington disease: Natural history, biomarkers and prospects for therapeutics. *Nature Reviews. Neurology, 10*(4), 204–216.

Ruby, J. G., Jan, C. H., & Bartel, D. P. (2007). Intronic microRNA precursors that bypass Drosha processing. *Nature, 448*(7149), 83–86.

Seredenina, T., Gokce, O., & Luthi-Carter, R. (2011). Decreased striatal RGS2 expression is neuroprotective in Huntington's disease (HD) and exemplifies a compensatory aspect of HD-induced gene regulation. *PLoS One, 6*(7), e22231.

Sinha, M., Ghose, J., Das, E., & Bhattarcharyya, N. P. (2010). Altered microRNAs in STHdh(Q111)/Hdh(Q111) cells: miR-146a targets TBP. *Biochemical and Biophysical Research Communications, 396*(3), 742–747.

Sinha, M., Ghose, J., & Bhattarcharyya, N. P. (2011). Micro RNA -214,-150,-146a and-125b target Huntingtin gene. *RNA Biology, 8*(6), 1005–1021.

Sinha, M., Mukhopadhyay, S., & Bhattacharyya, N. P. (2012). Mechanism(s) of alteration of micro RNA expressions in Huntington's disease and their possible contributions to the observed cellular and molecular dysfunctions in the disease. *NeuroMolecular Medicine, 14*(4), 221–243.

Snyman, M. C., Solofoharivelo, M. C., Souza-Richards, R., Stephan, D., Murray, S., & Burger, J. T. (2017). The use of high-throughput small RNA sequencing reveals differentially expressed microRNAs in response to aster yellows phytoplasma-infection in Vitis vinifera cv. 'Chardonnay'. *PLoS One, 12*(8), e0182629.

Soldati, C., Bithell, A., Johnston, C., Wong, K. Y., Stanton, L. W., & Buckley, N. J. (2013). Dysregulation of REST-regulated coding and non-coding RNAs in a cellular model of Huntington's disease. *Journal of Neurochemistry, 124*(3), 418–430.

Swarbrick, S., Wragg, N., Ghosh, S., & Stolzing, A. (2019). Systematic review of miRNA as biomarkers in Alzheimer's disease. *Molecular Neurobiology, 56*(9), 6156–6167.

Tafrihi, M., & Hasheminasab, E. (2019). MiRNAs: Biology, biogenesis, their web-based tools, and databases. *Microrna., 8*(1), 4–27.

Tan, L., Yu, J. T., & Tan, L. (2015). Causes and consequences of microRNA dysregulation in neurodegenerative diseases. *Molecular Neurobiology, 51*(3), 1249–1262.

Valadi, H., Ekström, K., Bossios, A., Sjöstrand, M., Lee, J. J., & Lötvall, J. O. (2007). Exosome-mediated transfer of mRNAs and microRNAs is a novel mechanism of genetic exchange between cells. *Nature Cell Biology, 9*(6), 654–659.

Varendi, K., Kumar, A., Härma, M. A., & Andressoo, J. O. (2014). miR-1, miR-10b, miR-155, and miR-191 are novel regulators of BDNF. *Cellular and Molecular Life Sciences, 71*(22), 4443–4456.

Weber, J. A., Baxter, D. H., Zhang, S., Huang, D. Y., Huang, K. H., Lee, M. J., Galas, D. J., & Wang, K. (2010). The microRNA spectrum in 12 body fluids. *Clinical Chemistry, 56*(11), 1733–1741.

Wu, Y. Y., & Kuo, H. C. (2020). Functional roles and networks of non-coding RNAs in the pathogenesis of neurodegenerative diseases. *Journal of Biomedical Science, 27*(1), 49.

Wu, J., & Xie, X. (2006). Comparative sequence analysis reveals an intricate network among REST, CREB and miRNA in mediating neuronal gene expression. *Genome Biology, 7*(9), R85.

Wu, H. Z., Ong, K. L., Seeher, K., Armstrong, N. J., Thalamuthu, A., Brodaty, H., Sachdev, P., & Mather, K. (2016). Circulating microRNAs as biomarkers of Alzheimer's disease: A systematic review. *Journal of Alzheimer's Disease, 49*(3), 755–766.

Yang, T. T., Liu, C. G., Gao, S. C., Zhang, Y., & Wang, P. C. (2018). The serum exosome derived microRNA-135a, −193b, and −384 were potential Alzheimer's disease biomarkers. *Biomedical and Environmental Sciences, 31*(2), 87–96.

Yates, L. A., Norbury, C. J., & Gilbert, R. J. (2013). The long and short of microRNA. *Cell, 153*(3), 516–519.

Ying, S. Y., Chang, D. C., & Lin, S. L. (2018). The MicroRNA. *Methods in Molecular Biology, 1733*, 1–25.

Part VI
Biomarker Considerations for Clinical Trial Design

Considerations and Advances in Huntington's Disease Clinical Trial Design

Georgia M. Parkin and Jody Corey-Bloom

Abbreviations

AAV	Adeno-associated virus
ASO	Antisense Oligonucleotide
BDNF	Brain-Derived Neurotrophic Factor
CAP score	CAG-Age Product score
CNS	Central Nervous System
CSF	Cerebrospinal fluid
DCL	Diagnostic Confidence Level
HTT	Huntingtin
HD	Huntington's disease
HD-ISS	Huntington's Disease Integrated Staging System
miRNA	microRNA
MMSE	Mini-Mental State Examination
MoCA	Montreal Cognitive Assessment
NfL	Neurofilament light
OLE	Open-Label Extension
PIN score	Prognostic Index Normed score
RNAi	RNA interference
shRNA	short hairpin RNA
siRNA	short interfering RNA
SDMT	Symbol Digit Modalities Test
SNP	Single Nucleotide Polymorphism
TFC	Total Functional Capacity
TMS	Total Motor Score
UHDRS	Unified Huntington's Disease Rating Scale

G. M. Parkin (✉) · J. Corey-Bloom
Department of Neurosciences, University of California, San Diego, CA, USA
e-mail: gparkin@health.ucsd.edu

E. A. Thomas, G. M. Parkin (eds.), *Biomarkers for Huntington's Disease*,
Contemporary Clinical Neuroscience, https://doi.org/10.1007/978-3-031-32815-2_17

1 Standard Clinical Trial Design

Standard clinical trial design necessitates a selection of necessary key features, including use of unbiased randomization into treatment or placebo control groups, often with stratification or blocking to allow for a balance of clinical and/or demographic characteristics across cohorts, as well as single-, double-, triple- or quadruple- blind masking. This masking – the process of keeping the allocation to treatment or control groups hidden – applies to the participant for single blind, the participant and the treatment administrator/investigator for double blind, the participant, treatment administrator/investigator, and outcome assessor for triple blind, and all of the above plus caregiver for quadruple blind studies. Much of clinical trial design, such as the selection of an initial dosing schedule, and selection of key clinical goals or "primary endpoints", are governed by pre-clinical laboratory testing in cells and mice, as well as advice from clinicians, scientists, statisticians, and HD families. If a clinical trial does not meet its primary endpoints, it is considered a failed trial; consequently, the selection of appropriate primary endpoints is paramount. However, as evidenced by several studies covered in this chapter, trials can still be beneficial to the Huntington's Disease community even if such endpoints are not met.

The design, length, and target recruitment number for a clinical trial is also determined in part by its phase, which itself is governed by what is currently known about the treatment in question (Fig. 1). In brief summary, after the completion of preclinical or animal studies, investigational treatments transition to the first stage of human trials – Phase 1. Phase 1 trials investigate how a new drug interacts with the human body, through assessments of safety and the potential for side effects when administered to healthy volunteers. Phase 1 trials also monitor drug pharmacokinetics – how a drug is metabolized – and determine an appropriate dose range. Next, Phase 2 trials investigate the drug or treatment in a larger population of affected individuals for whom the treatment is targeted, to measure efficacy, as well

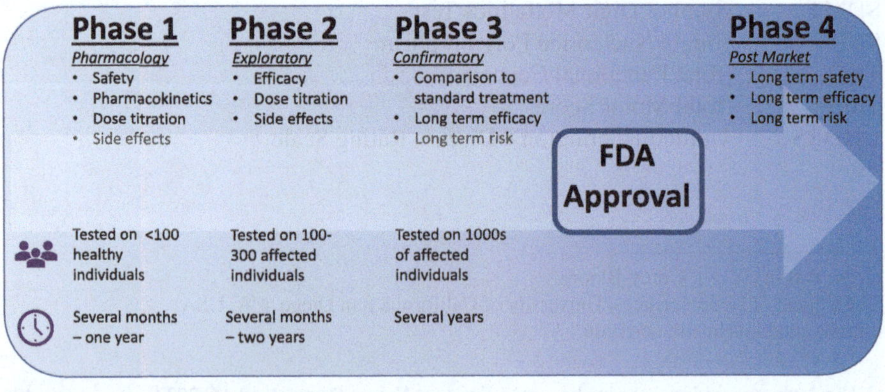

Fig. 1 Phases of clinical trial design

as confirm dose titration and potential side effects. Phase 3 trials compare the efficacy of the new drug or treatment to standard treatment, or what is already available to the affected population. Following successful completion of a Phase 3 trial, the new drug is submitted to the U.S Food and Drug Administration (FDA) for approval, after which it may be marketed for consumer use. Phase 4 clinical trials occur after FDA approval, when the drug is already available and used by the public, with the aim of monitoring long-term safety, efficacy and risk of a drug in real life scenarios.

2 Lessons Learned from Previous HD Clinical Trial Design

The design of HD clinical trials has changed drastically in recent years. Earlier trials were much smaller, with a shorter duration, and without consistent outcome measures. These drawbacks have been summarized in (Sampaio et al., 2014) and are outlined briefly below.

Shorter Trial Length

Early clinical trials investigated the potential to therapeutically treat HD by targeting aberrant post-translational modifications; however, these trials were short in duration, which may have affected the perceived usefulness of these treatments. For example, the PHEND-HD Phase 2 trial (NCT00212316) assessed 4 weeks of treatment with phenylbutyrate in 60 patients with HD, followed by a 12-week open-label phase. The primary outcome measure was the proportion of subjects able to complete treatment, whereas secondary outcome measures included safety, tolerability, performance on Unified Huntington's Disease Rating Scale (UHDRS) measures, and unspecified biomarker analyses (Hersch, 2008). This trial, which reported that phenylbutyrate was safe and well-tolerated, has not been succeeded by another (Gray, 2011).

Another example of a short duration study is the Phase 2 trial (NCT01521585) by Siena Biotech, which enrolled 144 patients treated for 12 weeks with either the silent information regulator T1 (sirT1) inhibitor selisistat (SEN0014196) or placebo. SirT1 is one of a limited number of deacetylases capable of deacetylating mutant HTT, resulting in increased levels; inhibition of SirT1 was shown to downregulate mHTT protein levels in transgenic mice and *Drosophila* (Süssmuth et al., 2015). In the human trial, no changes in any UHDRS measures were observed. This Phase 2 trial followed the completion of a Phase 1b trial conducted in 63 participants, who were treated for 2 weeks. These studies raised the question of trial length for the HD research field, as an investigational drug would need to have a large magnitude treatment effect to produce a significant UHDRS measure change in such short time. Notably, regarding the incorporation of biomarkers into clinical trial design, Siena's Phase 1 trial also assessed total Huntingtin (HTT) protein, interleukin (IL)-1, IL-6, IL-8 and tumor necrosis factor alpha (TNFa), and their Phase 2 trial assessed mutant HTT levels in peripheral blood mononuclear cells. These were two of the first clinical trials to quantify these biomarkers, and to register an observed change in mutant HTT levels within the context of a clinical trial (Sampaio et al., 2014; Reilmann et al., 2014b).

Smaller Sample Sizes

Previous clinical trials of HD had much smaller sample sizes, with many recruiting under 100 participants regardless of treatment and primary outcome measures (Sampaio et al., 2014). For example, the first clinical trial of cysteamine involved 5 participants in 1986, preceding studies into its mechanism of action in HD by almost two decades (2002). This was followed by a Phase 1 unblinded trial involving 9 participants and 2–5 weeks of treatment in 2005, the provision of orphan drug designation in 2008, and a Phase 2 double-blind, placebo-controlled trial (CYST-HD) involving 96 participants over 18 months of treatment in 2010 (Sampaio et al., 2014). Results from this latter Phase 2 trial showed a 32% slowing in Total Motor Score (TMS) progression in the treated cohort that was not statistically significant; however, a post-hoc analysis of TMS in a sub-group of 66 participants (32 treated, 34 placebo) who were not on concomitant tetrabenazine elevated this to a statistically significant 58% (Sampaio et al., 2014). These positive findings in a smaller non-prespecified sub-group may be considered spurious and highlight the limitation of conducting unplanned post-hoc analyses in sub-groups of already small cohorts. Regarding the analysis of biomarkers, cysteamine was previously shown to increase levels of brain derived neurotrophic factor (BDNF) (Besouw et al., 2013) and subsequently, blood levels of BDNF were also quantified in this Phase 2/3 trial as a secondary outcome measure.

Variable Outcomes

Clinical trials specify primary outcome measures, which determine whether the trial is considered a success or failure, and secondary outcome measures, which are more 'exploratory' investigations allowing researchers to fully understand the effect of a treatment. However, in a number of cases outlined in this chapter, trials that failed to meet their primary outcome measure reported significant, or near-significant, exploratory results. These exploratory findings may change the focus of future clinical trials using the same treatment. For example, the use of pridopidine for treatment of HD was investigated through the MermaiHD Phase 3 trial (NCT00665223; primary outcome: modified UHDRS TMS score), HART-HD Phase 2/3 trial (NCT00724048; primary outcome: UHDRS modified TMS); PRIDE-HD Phase 2 trial (NCT02006472; primary outcome: UHDRS TMS) and PROOF-HD Phase 3 trial (NCT04556656; primary outcome: UHDRS Total Functional Capacity (TFC)) (see section "Teva Pharmaceutical Industries, and Prilenia Therapeutics" for more information on pridopidine trials).

While use of the UHDRS has become standard, some trials selected other measures based on the expected effect of treatment. For example, Medivation's Phase 2 HD clinical trial of orally administered Dimebon/ Latrepirdine (NCT00497159) enrolled 91 patients for 90 days of treatment or placebo and used the UHDRS, Alzheimer's Disease Assessment Scale-cognition scale, and Mini-Mental State Examination (MMSE) as outcome measures. The selection of these outcome measures was determined by preliminary pre-clinical and clinical findings, including a successful Phase 2 Alzheimer's Disease trial. In the HD study, the treatment group showed improvement on the MMSE at 90 days (Bezprozvanny, 2010; Kieburtz et al., 2010). This study was followed by a Phase 3 clinical trial of Dimebon ('HORIZON'; NCT00920946), which recruited 403 participants for 26 weeks of treatment, and did not replicate the Phase 2 trial's significant MMSE- related findings (HORIZON Investigators of the Huntington Study Group and European Huntington's Disease Network, 2013).

3 Huntingtin-Lowering Strategies and Other Treatments Currently Used in Clinical Trials

Recent and current clinical trials for the treatment of HD utilize a range of biological strategies to reduce full-length and/or mutant HTT levels or modulate pathways downstream of the mutation. However, the targeting of both full-length or wildtype and mutant HTT is a concession, rather than a goal. As the *HTT* mutation is caused by an expanded CAG tract, current HTT-lowering therapies, with the exception of a select few (see sections "Wave Life Sciences and Vico Therapeutics"), are unable to distinguish between normal, and expanded, *HTT* CAG lengths. Consequently, clinical trials must consider the impact of reducing both full-length HTT, which is necessary for normal neuronal function (Cattaneo et al., 2005; Rook & Southwell, 2022), in addition to mutant HTT protein. The impact that non-allele specific treatment has on the expression of wildtype HTT is particularly important, given that HTT is ubiquitously expressed throughout the brain and body, whereas areas affected by HD are often region- and disease-stage specific. It is therefore necessary to monitor clinical and biological markers indicative of treatment and adverse outcome during a clinical trial, as well as once a treatment is approved by the FDA for consumer sale.

The role and measurement of HTT protein has been covered elsewhere in this book; the following section will expand on current HTT-lowering strategies, as well as other treatments that target processes downstream of HTT dysfunction, and the observed effects that these treatments have had on select biomarker levels. Several FDA-approved therapies and strategies to reduce symptom burden in HD patients, such as the use of dopaminergic antagonists and anti-glutamatergics to reduce chorea, and other antipsychotic medication and antidepressants to reduce psychiatric and behavioral symptoms, are already available to consumers. The focus of the current chapter is on clinical trial design and advances in HD, and therefore these approved treatments will not be covered in detail in this chapter (Ferguson et al., 2022).

Antisense Oligonucleotides

Antisense oligonucleotides (ASOs), and RNA interference (RNAi) compounds (covered in the next section), act by binding directly to target RNA (Fig. 2). ASOs are single-stranded, chemically modified DNA sequences that use Watson-Crick base pairing to bind to pre-messenger and messenger RNA, either to modify splicing or block translation, and label it for degradation by RNase H. ASO therapies are administered via intrathecal injection into the cerebrospinal fluid (CSF), have a long half-life due to chemical modifications, and are easily taken up by cells of the central nervous system (CNS). While most ASOs are non-selective, and target both wildtype and mutant HTT, the identification of single nucleotide polymorphisms (SNPs) in linkage disequilibrium with HD-causing CAG repeat lengths provides a means to develop a predominantly allele-specific, or allele-preferencing, ASO therapeutic (Rook & Southwell, 2022).

Hoffman-La Roche

No doubt the most well-known HD clinical trial to use an ASO is the Phase 3 tominersen/IONIS-HTTRx trial, GENERATION-HD1 (NCT03761849)) by Hoffman-La Roche and Ionis Pharmaceuticals. Preliminary data offered the HD community much hope for the tominersen trial; however, the trial was halted in early 2022 following recommendation from an unblinded Independent Data Monitoring Committee. Preliminary analyses of trial data showed that participants randomized to receive treatment were performing worse than placebo on the primary outcome measures, the composite UHDRS and UHDRS TFC. This trial had enrolled almost 800 participants, who received either placebo, or 120 mg of tominersen every 8 or 16 weeks. At the termination of the trial, 70% of participants had reached week 17 out of a planned 25 (Kingwell, 2021). This disappointing news followed relative success in the Phase 1/2a trial for tominersen (NCT02519036): of the 46 participants enrolled, all completed the trial, and CSF mutant HTT levels were lowered by 20–42% on average, suggesting successful target engagement (Rook & Southwell, 2022). However, this Phase 1/2a trial also showed that, in participants who received 90–120 mg of tominersen, CSF levels of neurofilament light (NfL) and tau increased 1–2 months after the first injection and did not return to baseline until 7–27 months later. NfL and tau are widely considered as markers of neurodegeneration, with NfL exclusive to neurons, and tau found in both neurons and glia. In this trial, ventricular volume additionally increased at days 113 and 197 in participants across all dose titration ranges (Rook & Southwell, 2022). The cause for these side effects is not known.

The Phase 1/2a tominersen trial was subsequently followed by an Open-Label Extension (OLE) Study (GEN-EXTEND; NCT03842969), designed to investigate the safety and tolerability of 120 mg tominersen at dosing intervals of 4 or 8 weeks. Again, CSF mHTT and NfL were monitored throughout the trial. Participants dosed every 8 weeks showed a reduction in CSF mHTT (44%) within the target range of 30–50%, whereas those dosed every 4 weeks showed a mean reduction of 70%. Participants dosed every 4 weeks also showed an elevation in CSF NfL at 5 months, higher than those dosed every 8 weeks; however, both arms showed a subsequent decrease in CSF NfL with continued treatment. Information from the Phase 1/2a trial, as well as OLE Study, allowed for the development of a pharmacokinetic/ pharmacodynamic (PK/PD) model and revision of the dosing schedule, which was implemented in the Phase 3 trial (Rook & Southwell, 2022).

Despite the ultimately negative outcome of the Phase 3 trial, unplanned analyses of the data collected suggested that some participants – those younger, in earlier stages of HD progression, and receiving a less frequent dose – did not show a worsening of symptoms over the course of the trial. These findings have led to the development of a new Phase 2 trial for tominersen by Hoffman-La Roche (GENERATION-HD2; NCT05686551), with participation limited to this cohort of younger, earlier disease stage participants. These cumulative trial findings highlight the importance of monitoring biofluid and neuroimaging biomarker changes within the context of a clinical trial, as well as the potential to gain knowledge from the conduct of clinical trials, even in the absence of promising primary outcome data. The implications of selection criteria, and particularly of targeting participants who are younger or earlier in their disease progression, will be discussed in more detail later in this chapter (Sect. 5).

Wave Life Sciences

Two additional, allele-specific ASO-based clinical trials, PRECISION-HD1 (NCT03225833) and PRECISION-HD2 (NCT03225846), both Phase 1b/2a and developed by Wave Life Sciences, while originally showing promise, were similarly halted. These allele-specific ASOs, called WVE-120101 and WVE-120102, targeted two SNPs in the *HTT* gene, rs362307 and rs362331, which were estimated to be present in 36–70% of HD patients based on patient genotype. While both drugs were overall well tolerated, these trials were halted as there was no observable target effect – the ASOs did not lower CSF HTT protein as intended. Specifically, WVE-120101 resulted in a trend reduction CSF mHTT that did not reach statistical significance, and WVE-120102 resulted in a mean reduction in CSF mHTT of 12.4%. Neither treatment resulted in observed changes in CSF wild-type HTT or NfL levels (Rook & Southwell, 2022). These biomarker changes were observed in cohorts administered up to 16 mg of drug, and could not be improved upon with initiation of a 32 mg treatment arm (Drew et al., 2021). In addition, while PRECISION-HD1 and PRECISION-HD2 trial participants were assigned to dosage levels of either 2, 4, 8, 16 or 32 mg of treatment, reductions in levels of CSF mHTT were consistent across treatment groups, suggesting the lack of a dose-dependent response and therefore failure to engage the treatment target (Drew et al., 2021).

Wave Life Sciences are now enrolling participants in a new Phase 1b/2a clinical trial, SELECT-HD (NCT05032196), to investigate a new version of their ASO treatment, WVE-003, which targets a unique SNP estimated to occur in 40% of the HD patient population (Ferguson et al., 2022). In September 2022, Wave Life Sciences provided an update from the SELECT-HD trial, reporting that treatment with 30 mg or 60 mg of WVE-003 led to a lowering of CSF mHTT protein by 20–30%, without affecting levels of the wild-type HTT protein. Interestingly, some participants in this trial also showed a trend increase in CSF NfL levels, mirroring observations reported in the early stages of Hoffman-La Roche's tominersen trials. It is worth reiterating that these NfL changes went away over time in the tominersen trial, and while the same may occur with time in SELECT-HD, the cause for increased NfL following treatment is not known.

Vico Therapeutics

On April 3rd, 2023, Vico Therapeutics announced the dosing of their first patient with the mutant HTT allele-preferential ASO VO659 (NCT05822908), as part of a Phase 1/2a basket trial assessing safety, tolerability, and dosage. Vico Therapeutics is the only company in this chapter to employ a basket trial design; the treatment of multiple diseases – in this case, HD, spinocerebellar ataxia type 1 (SCA1) and type 3 (SCA3) – in the same study. Pre-clinically, VO659 produced dose-dependent reductions in mHTT, and improvements in motor function in HD mouse models, as well as allele-preferential reductions in mHTT in a HD patient cell model.

RNA Interference Compounds

RNA interference (RNAi) compounds, such as short interfering RNA (siRNA), short hairpin RNA (shRNA) and micro-RNA (miRNA), target messenger RNA sequences (Fig. 2b). Unlike ASOs, RNAi compounds do not readily diffuse into the CNS, and require stable transfection methods, such as a viral vector, coupled with an intracerebral stereotactic injection. While the use of a viral vector may allow for long-term treatment, this is a double-edged sword, as adverse effects observed during a clinical trial would not as easily be halted or reversed.

Voyager Therapeutics

The VY-HTT01 investigational therapy, developed by Voyager Therapeutics, Sanofi-Genzyme and the CHDI Foundation, was to be one such RNAi (miRNA) compound delivered into neurons via an adeno-associated viral (AAV) vector to target *HTT* mRNA. Following promising preclinical findings in mice and non-human primates, Voyager Therapeutics had originally planned and registered a Phase 1/2 clinical trial, called VYTAL (NCT04885114), in 2021. Registration of this trial was ultimately withdrawn, governed by a decision to focus their efforts on 'recent breakthroughs' regarding the company's proprietary AAV capsids with the potential for intravenous administration.

While no miRNA compounds are currently used in clinical trials, UniQure Biopharma B.V.'s gene therapy treatment, AMT-130, involves the administration of AAV-enclosed single-strand DNA which encodes *HTT* mRNA-targeting miRNA (Stimming et al., 2022). AMT-130 is covered in more detail in section "Gene Therapy".

Alternate RNA Splicing

When RNA is first transcribed from DNA, it contains both coding sequences called exons, as well as non-coding sequences interspersed within (introns) and on either side (5′- and 3′- untranslated regions) of those coding sequences. This pre-mRNA sequence is then processed to remove or splice out the introns, resulting in an mRNA sequence that is ready to be translated into a functional protein. Therapeutic treatment which utilizes splice modulators aim to modify the sections of the pre-mRNA sequence that are removed, thereby making the resulting RNA sequence unusable and unable to be translated into protein (Fig. 2a).

Novartis

In what began as exciting news in 2022, Novartis initiated a Phase 2b, randomized, double-blind, placebo-controlled trial, the VIBRANT-HD trial (NCT05111249), involving oral administration of the splicing modulator drug, Branaplam. Branaplam was originally developed to treat another disease called spinal muscular dystrophy (SMD), with the aim of increasing levels of the survival motor neuron (SMN) protein through correction of a pre-mRNA splicing defect (Cheung et al., 2018). However, during preclinical and clinical studies for SMD (NCT02268552), Novartis also found that Branaplam reduced levels of the HTT protein. Prior to the start of VIBRANT-HD, and as evidenced by the other trials covered in this chapter, experimental HTT-lowering treatment for HD had been administered via either intracerebral or intrathecal injection. Having shown promise in animal models and an early safety trial, Branaplam and the VIBRANT-HD study offered the first oral

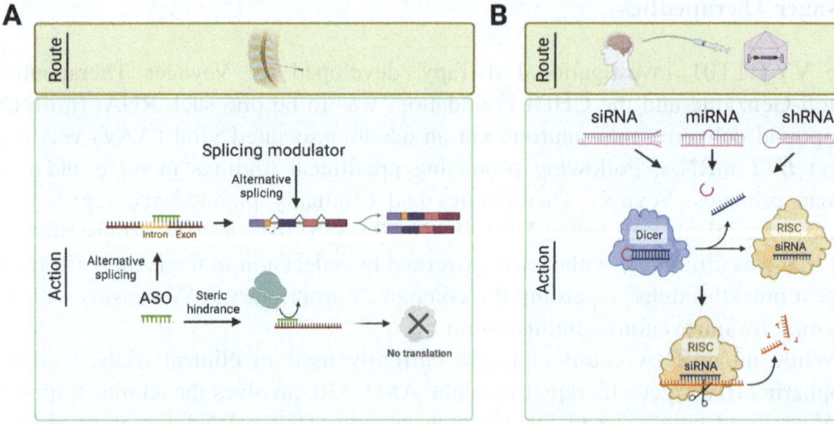

Fig. 2 Routes of administration and mechanism of action for antisense oligonucleotides (ASOs), splicing modulators, and RNA interference (RNAi) compounds. (**a**) ASOs and splicing modulators modify pre-messenger RNA splicing; ASOs may also bind to messenger RNA and block translation via steric hindrance. (**b**) RNAi compounds are delivered to the target site via intracerebral injection of a AAV viral vector, which carries the therapeutic compound. A RNA III enzyme, Dicer, then cleaves the RNAi double-stranded compounds into single strands; one of these strands forms the RNA-induced silencing complex (RISC), whereas the other strand is believed to be degraded. RISC, with the RNAi compound bound, is then able to degrade target mRNA. (Created with BioRender.com)

HTT-lowering treatment for HD, which would be safer, less expensive, and less invasive for patients. Unfortunately, the trial was halted in August 2022, as several of the first 20 participants provided with a low dose of Branaplam showed signs of peripheral neuropathy. Some participants also showed increases in serum NfL, which was regarded as a sign of neuronal stress. Given that Branaplam was developed for SMD, and serendipitously found to lower HTT protein levels, it is possible that Branaplam has other unforeseen ‘off-target’ effects that result in peripheral neuropathy, and the halting of the VIBRANT-HD clinical trial should be seen as a delay but not a failure for the development of oral HTT-lowering treatments.

PTC Therapeutics

In fact, another clinical trial for HTT-lowering oral treatment is also underway. PTC Therapeutics hope to lower HTT protein through the drug PTC-518, which similarly acts by changing how the *HTT* RNA is processed. Through a Phase 1 trial, PTC Therapeutics have so far shown that the amount of HTT protein reduced in the brain following treatment is directly proportional to the amount of PTC-518 given to patients and, importantly, that the effects of the drug are reversible. PTC Therapeutics is currently recruiting for a Phase 2 randomized, placebo-controlled trial, the PIVOT-HD trial (NCT05358717), which will involve two different doses of the drug – 5 mg and 10 mg. Currently, enrolment in the USA has been paused as the FDA have requested additional data to allow it to proceed, however recruitment is active and ongoing in Europe.

Zinc Finger Transcriptional Repressors

Zinc-finger proteins (ZFPs) bind directly to DNA to modulate transcription; for treatment of HD, this means repressing production of the *HTT* RNA via direct binding to the expanded CAG tract in exon 1 of the *HTT* gene. For example, Isalan and colleagues have incorporated the *KOX1* domain into the investigational drug ZF-KOX1, which recruits cellular components to cause long-term gene suppression through the formation of heterochromatin at the DNA region of interest (Garriga-Canut et al., 2012; Agustín-Pavón et al., 2016). Like RNAi compounds, synthetically designed ZFPs require an AAV vector and intracerebral injection for delivery to the CNS. Currently, there are no ZFP-based clinical trials, however it is worth highlighting two preclinical models as a means of indicating their future potential. The first is the aforementioned ZF-KOX1 (Agustín-Pavón et al., 2016; Garriga-Canut et al., 2012), which was shown to reduce *mHTT* RNA, and protein, in the striata of R6/2 transgenic mice. The second is TAK-686, developed by Takeda and Sangamo Therapeutics. TAK-686 has been shown to selectively reduce mHTT in HD patient fibroblasts, neural stem cells, neurons, knock-in HdhQ50 mouse models of HD – that is, mice that have a *HTT* allele with 48 CAG repeats inserted into their genome to mimic HD – as well as R6/2 transgenic mice with 120 CAG repeats (Zeitler et al., 2019).

Notably, ZF-KOX1 was developed using the human *KOX1* gene, and the introduction of a synthetic protein or one that is not native to the host (in this case, mouse) can produce an immune response that leads to neuronal death. Isalan and colleagues have since modified the original ZF-KOX1 to develop a host-matched analogue (called mZF-KRAB), using a mouse gene (Agustín-Pavón et al., 2016). While mZF-KRAB has been developed specifically for a mouse host, it raises an important consideration: even the most promising preclinical treatment may be ineffective or deleterious in human clinical trials if the (human) host immune system does not accept the drug – this is the aim of Phase 1 clinical trials.

Gene Therapy

Gene therapy – a one-time treatment which repairs or alters an individual's affected genome, shows promise for hereditary diseases where the gene mutation is known, such as HD. At present, gene therapy predominantly uses an AAV vector, which has an excellent safety profile and does not integrate directly into the recipient's genome, thereby reducing the potential for DNA damage and unforeseen side effects

(Mendell et al., 2021). AAV is a virus that is capable of encoding 9 distinct proteins; however, it requires co-infection with a second virus, the *Adenovirus,* to replicate. These characteristics make recombinant AAVs, which are modified to remove the viral DNA and instead contain DNA sequences of interest, advantageous for gene therapy, as they may be used to deliver and express therapeutic genes in a (human) host body, without undesirable replication.

UniQure Biopharma B.V

UniQure utilizes an AAV viral vector paired with single-stranded DNA that is then transcribed into *HTT* mRNA-targeted miRNA once administered. This drug is delivered via MRI-guided neurosurgical injection into the brain's striatum (Stimming et al., 2022), with the aim of targeting and reducing levels of both normal and mutant HTT protein. Currently, a small Phase 1/2 trial to test the safety of the drug, known as AMT-130 (HD-GENETRX-1; NCT04120493), is underway. Patients receive either a low dose or a high dose of the drug; to date, those receiving the low dose have not shown any significant drug-related safety issues. Similar to what had been seen with other therapies, levels of CSF NfL spiked one month after receiving the drug, but returned to baseline within 12 months (Stimming et al., 2022).

Stem Cell Therapy and Cell Replacement Therapy

Cellavita HD

Stem cell therapy utilizes the introduction of stem cells, which have the potential to differentiate into different cell types, with the aim of replacing neuronal loss or treating neuronal damage characteristic of HD (Maucksch et al., 2013). Currently, the most advanced stem cell therapy for HD, Cellavita HD, has been developed by the Brazilian companies Cellavita and Azidus, through the use of human immature dental pulp stem cells (hIDPSC) isolated from deciduous teeth. A non-randomized, open label Phase 1 trial (SAVE-DH, NCT02728115), involving three intravenous injections of Cellavita HD, comprises a reported 6 participants recruited with a 5 year follow up period (completion expected in 2023) (Ferguson et al., 2022). An interim analysis at 2 years indicated that the treatment was well tolerated, with 5 patients showing improved motor function (Macedo et al., 2021). Consequently, this study has been followed by a randomized, triple blind Phase 2 trial to test dose-response relationships for Cellavita HD (ADORE-DH; NCT03252535), with the intention of recruiting 35 participants treated with 9 low or high doses of Cellavita HD or placebo. A Phase 2/3 open-label extension study (ADORE-EXT; NCT04219241) has also been registered but is not actively recruiting (Ferguson et al., 2022). As Cellavita HD is administered intravenously, this treatment does not specifically target the *HTT* mutation.

Cardiff University

Another more targeted example of stem cell treatment is the TRIal designs for DElivery of Novel Therapies for neurodegeneration (TRIDENT) study (ISRCTN52651778), a Phase 1 single-center, open-label study of cell replacement therapy operated by the Center for Trials Research at Cardiff University (Drew et al., 2021). The TRIDENT study plans to recruit up to 30 participants with manifest HD, of which one participant deemed most eligible for surgery, and four participants eligible for surgery and randomly selected, will receive 11–22 million foetal stem cells surgically inserted into the striatum, followed by 12 months of immunosuppression therapy. The TRIDENT study, which should conclude by the end of 2023, will assess safety at 4 weeks post-surgery, as well as feasibility and long-term safety.

4 Targeting Other Proteins/Antibody Therapy

Annexon Biosciences

A number of clinical trials target RNA, DNA or protein complexes other than *HTT.* For example, Annexon Biosciences aims to inhibit the function of complement C1q protein with the monoclonal antibody drug ANX005. Annexon Biosciences' open-label Phase 2 study (ANX005-HD-01; NCT04514367) is assessing the safety, tolerability, pharmacokinetics and pharmacodynamics of an intravenous injection of ANX005 in participants at, or at risk for, manifest HD. C1q is involved in the cortical immune response, as well as neuronal development, via regulation of the complement cascade. The complement cascade, which acts first and foremost as a first-line defence against infection, is a key pathway that appears to be affected in premanifest HD, paired with chronic inflammation (Nayak et al., 2011). A review of inflammatory biomarkers in HD has been covered in more detail elsewhere in this book. So far, it appears that a sub-population of participants – those with excessive baseline complement activity – may benefit from ANX005 treatment.

SAGE Therapeutics

SAGE Therapeutics is testing the effectiveness and safety of a NMDA receptor positive allosteric modulator drug, SAGE-718, on cognitive and day-to-day functioning in HD. This line of investigation is being conducted through several Phase 2 clinical trials that are run under the umbrella of the PERSPECTIVE program. The first trial, which began recruitment in February 2022, is the DIMENSION Study

(NCT05107128), a randomized, placebo-controlled study involving nine visits over 4 months. Participants receive 1.2 mg of the SAGE-718 drug for the first month, followed by 0.9 mg for the next 2 months. Next, the SURVEYOR Study (NCT05358821), which opened for recruitment in August 2022, is a similarly randomized, placebo-controlled trial that involves 6 visits over 2 months. The SURVEYOR Study is recruiting both participants with HD, who will receive the drug or placebo, as well as healthy control volunteers, who will not receive drug or placebo. This study is intended to assess real-life functioning through the use of additional measures, such as a driving simulator and virtual reality functional capacity assessment.

Teva Pharmaceutical Industries, and Prilenia Therapeutics

Teva Pharmaceutical Industries, in collaboration with the Huntington Study Group (HSG), was the first company to target the sigma-1 receptor for treatment of HD, using the oral, small molecule drug, pridopidine. Pridopidine has been heralded as a "dopamine stabilizer", however research suggests that pridopidine may also act by rescuing mitochondria function and endoplasmic reticulum associated calcium levels (Naia et al., 2021; Ryskamp et al., 2017). Teva Pharmaceutical Industries began their investigation with a Phase 2 trial (PRIDE-HD; NCT02006472), which included 45 mg of pridopidine twice daily for the treatment of motor symptoms. While the primary endpoint – change in motor symptoms as measured by the UHDRS TMS – was not reached, the researchers did find that NfL levels, and UHDRS TFC scores, stabilized. This led to the start-up of a Phase 3 clinical trial by Prilenia Therapeutics and HSG, PROOF-HD (NCT04556656), to investigate the effect of pridopidine on functional, behavioral and motor symptoms of HD, using the same dosing schedule. All double-blind PROOF-HD study activities have been completed, with enrolled participants subsequently given the option to transition into an ongoing OLE study. Although preliminary findings from PROOF-HD suggest that the primary (UHDRS TFC) and secondary (composite UHDRS) endpoints were not met, participants who were not on neuroleptics or chorea medication showed clinically meaningful improvements in motor function, as measured by the Quantitative-Motor (Q-Motor) battery.

BrainVectis

BrainVectis, a subsidiary of Asklepios BioPharmaceutical, Inc. (AskBio) have recently received approval to begin a Phase I/II open-label, dose-finding study to investigate the efficacy, safety and tolerability of AB-1001 (NCT05541627), delivered via one-time intracerebral bilateral injection into the striatum. AB-1001 is a novel, AAV gene therapy vector that will result in the production of CYP46A1 protein, a crucial enzyme involved in the cortical elimination of cholesterol that appears to be reduced in HD. CYP46A1 also appears to have a downstream positive impact on mHTT aggregate clearance (Kacher et al., 2019).

Vaccinex

SIGNAL-HD (NCT02481674) was a randomized, placebo-controlled Phase 2 trial of VX15/2503 (pepinemab) conducted by Vaccinex in cohorts of late prodromal and early manifest HD patients. Twenty mg/kg pepinemab, which is a semaphorin D (SEMA4D) neutralizing antibody, or placebo, was administered monthly for 1 year via intravenous drip. SEMA4D is instrumental in regulating the transition of glial cells from a homeostatic, or resting state, to a reactive state, through binding to plexin receptors expressed on the surface of those glia cells. This SEMA4D-driven process is upregulated in stressed or damaged neurons. Unfortunately, while pepinemab was deemed safe and tolerable, HD participants treated with pepinemab did not perform any better on cognitive tests, or functionally as determined by a clinician, compared to the placebo group. On the other hand, pepinemab did appear to provide cognitive benefit for participants with cognitive impairment at baseline (a Montreal Cognitive Assessment (MoCA) score < 26) and reduce disease progression for a subgroup of participants with reduced functional capacity at baseline (a UHDRS TFC score of 11). Pepinemab also slowed or reversed decline in metabolic activity across 26 brain regions measured, as quantified by FDG-PET, reduced caudate atrophy as measured by MRI, and showed a trend reduction in ventricular expansion (Evans et al., 2022). These findings highlight the importance of quantifying biomarkers as outcome measures in clinical trials, as well as the significance of determining eligibility criteria for trial participation. This will be covered in more detail in Sect. 5.

Azevan Pharmaceuticals

Recently, the Phase 2 dose-finding trial, Safety, Tolerability, and Activity of SRX246 in Irritable HD patients (STAIR; NCT02507284), met primary and secondary endpoints and, through exploratory analyses, demonstrated clinical benefit in reducing aggressive or explosive behavior in HD. The vasopressin 1a receptor antagonist drug, SRX246, was consequently granted an Orphan Designation, providing Azevan Pharmaceuticals with exclusive marketing rights and tax incentives for 7 years in order to encourage further development (Brownstein et al., 2020; Maibach et al., 2022).

A number of additional clinical trial treatments, including natural remedies as well as compounds that target pathogenic cascades downstream of mHTT, are also under investigation and have not been included here (Jurcau & Jurcau, 2022).

Unfortunately, the emergence of the COVID-19 pandemic impacted clinical trial recruitment, resulting in early termination of some trials, such as EIP Pharma's proof-of-concept Phase 2 clinical trial for the small molecule, p38 alpha kinase Inhibitor, Neflamapimod (NCT03980938).

Overall, the nature of treatment used in a clinical trial will determine its route of administration – intrathecal, intracerebral, intravenous or oral, the resulting distribution throughout the body, whether desired or not, and region-specific efficacy. In addition, the regimen of administration – continuous, intermittent or one-time – must also be considered (Mestre & Sampaio, 2017). The ability to track biological changes following treatment, including the monitoring of pharmacokinetics and pharmacodynamics, determining target-engagement, and assessing the potential for off-target and/or down-stream effects, is ultimately dependent on the selection and adequate detection of appropriate biomarker levels.

5 Selecting Participants for Clinical Trials

Why We Need Clinical Trial Selection Criteria

To date, clinical trials have predominantly recruited manifest HD participants – those with a score of 4 on the UHDRS Diagnostic Confidence Level (DCL), indicating that the diagnosing clinician has 99% "diagnostic confidence" that the presence of motor signs indicates the onset of manifest HD. The decision to restrict clinical trial participation to individuals with a clinical diagnosis has primarily been governed by the anticipated risk/benefit profile of a cohort with observable disease progression, with the potential to display a clinically meaningful treatment effect, or absence of treatment effect. However, this limitation has also impeded the potential for significant findings or advancements in the treatment of HD, as patients with clinically diagnosed HD may already be too far in their disease progression for disease-modifying treatment to be effective (Rubinsztein & Orr, 2016; Langbehn & Hersch, 2020). Therefore, recently, concerted efforts have been made to include individuals who have the *HTT* genetic mutation but have not progressed to a clinical diagnosis of HD, otherwise known as prodromal HD participants. Unfortunately, HD disease progression is not a steady decline, and individuals may remain within the pre-symptomatic HD stage for many years before the emergence of prodromal symptoms. It is therefore necessary to find clinical trial selection criteria that would enrich trials with individuals who are near manifest HD onset, and therefore have the potential to display a clinically meaningful treatment effect, while excluding those who are far from manifest HD onset. To complicate matters further, clinical trials are reliant on the inclusion of enough participants in each arm (treatment and placebo), such that the trial has enough power to see a discernible difference between groups if it exists (avoid a type II error), and for researchers to be able to state with confidence that any observed findings have not occurred by chance (avoid a type I error). It is therefore also necessary to ensure that selection criteria are not so stringent that clinical trial recruitment cannot be completed within the estimated, or funded, study period.

Distinguishing Stages of Disease Progression: Implications for Clinical Trials

The UHDRS DCL Score

Until recently, participation in a clinical trial was restricted to individuals with a DCL score of 4. UHDRS scoring, including the determination of a DCL score, is based on an individual's symptom presentation at a single visit. Prior to a manifest HD diagnosis, a person may receive an alternate DCL score of 0 (no motor abnormalities), 1 (nonspecific motor abnormalities), 2 (motor abnormalities that may be signs of HD [50–89% confidence of a diagnosis]), or 3 (motor abnormalities that are likely signs of HD [90–98% confidence])(Reilmann et al., 2014a). A diagnosis could therefore depend on the minute circumstances surrounding a particular clinic appointment. In 2014, Reilmann and colleagues proposed taking into account previous examinations – considering the natural history of an individual's condition – to determine disease stage, which may assist with the incorporation of prodromal HD individuals into clinical trials. Further, the researchers proposed including participants with a DCL score of 3 or 4 – reducing the confidence level cut-off from 99% confidence to 90% confidence. By considering the natural history of the disease, they argue that *evidence of change*, rather than a one-size-fits all motor score cut-off, would be more appropriate (Reilmann et al., 2014a).

On the other hand, in 2013, investigators from the Neurobiological Predictors of Huntington's Disease (PREDICT-HD) study proposed utilizing a DCL score based on the *entirety* of the UHDRS, such that a person could be diagnosed with manifest HD, at least for research purposes, based on behavioral, instead of motor, symptom presentation (Biglan et al., 2013). The authors concluded by suggesting a staging system for prodromal HD that incorporates clinical and neuroimaging markers. Since the publication of the PREDICT-HD article in 2013, and significantly only last year, has such a fully-fleshed staging system for HD been proposed (Tabrizi et al., 2022). This will be covered in more detail later in this chapter. It is also important to note that some current clinical trials, such as the GENERATION-HD2 Phase 3 trial by Hoffman-La Roche, have already expanded inclusion criteria to include DCL 2 and 3, provided that the individual also meets other symptom criteria as well, such as a TMS >6 (NCT05686551).

The Disease Burden Score

The Disease Burden Score (DBS) was proposed by Penney and colleagues in 1997, following a comparative analysis of subject CAG repeat length, post-mortem striatal neuropathology, and age of death (Penney Jr et al., 1997). Penney found a linear correlation between CAG repeat length, starting at 35.5 repeats, and striatal atrophy divided by age of death. This comparison was simplified to *Disease Burden Scale* = (*Age* × [*CAG* – 35.5]).

The CAG-Age Product (CAP) Score

At the time of Penney and colleagues' findings in 1997, the relationship between CAG repeat length, an individual's age, and symptom presentation, was an area of active investigation (Penney Jr et al., 1997). In fact, the DBS was actively used in clinical research until 2011, when it began to be superseded by the highly similar CAG-Age Product (CAP) score (Zhang et al., 2011). The CAP score, defined as (*Age* × [*CAG* – 33.66]), is a measure of the cumulative exposure of CAG repeat length and age, or disease burden, and is commonly used to estimate disease progression for the purpose of clinical trial inclusion (Zhang et al., 2011). Recently, and as aforementioned in Sect. 3, the termination of the GENERATION-HD1 trial by Hoffman-La Roche came with a silver lining; a subgroup of participants who were younger (<48 years of age), with a CAP score under 500, and who were provided the drug tominersen less frequently – every 16 weeks compared to every 8 weeks – appeared to perform better than placebo on the cognitive, motor and functional measures incorporated into the primary outcome measure (Boak & Mccolgan, 2022). These post-hoc findings prompted Hoffman-La Roche to instigate GENERATION-HD2 (NCT05686551), which is restricted to individuals age 25–60 with CAP scores 400 to 500 (Boak & Mccolgan, 2022). Unfortunately, use of the CAP score is also limited by its simplicity; it is a measure of disease burden based solely on an individual's age and CAG repeat length, with no consideration of their current symptom presentation. Acknowledging this limitation, researchers developed the multivariate risk score, which also incorporated the UHDRS TMS, Symbol Digit Modalities Test (SDMT), DCL (replaced with '1' in the below equation if the patient fell within the corresponding DCL group, and '0' if not), and the three Stroop test measures – color (SC), word (SW), and Interference (SI) score (Paulsen et al., 2019; Long et al., 2015):

$$\begin{aligned} MRS = & -0.282 \times \mathrm{AGE} + 0.140 \times CAG + 0.565 \times TMS - 0.021 \times \mathrm{SDMT} \\ & + (0.347 \times DCL1) + (0.542 \times DCL2) + (1.086 \times DCL3) - 0.004 \times SC \\ & + 0.002 \times SW - 0.023 \times SI - 0.004 \times TMS2 - 0.010 \times TMS \times CAG \\ & + 0.009 \times \mathrm{AGE} \times CAG \end{aligned}$$

However, this formula was soon also superseded by the Prognostic Index Normed (PIN) score (Paulsen et al., 2019).

The Prognostic Index Normed (PIN) Score

The PIN score incorporates the SDMT score and UHDRS TMS, as well as participant CAG repeat length and age (Long et al., 2017). The PIN score is reported in standard deviation (SD) units around a 50% probability of 10-year survival, where survival equates to no diagnosis of manifest HD, and such that a PIN of 0.0 equals 10 years until this diagnosis, and −1.0 is one SD below that (Long et al., 2017). These PIN scores could then be used to stratify participants and determine eligibility criteria for clinical trials that included prodromal HD participants. For example, Langbehn and colleagues have shown through longitudinal modelling that prodromal HD cohort effect sizes for clinical trials could be amplified by limiting recruitment to participants with PIN score thresholds of >0.0 or >0.4 (Langbehn & Hersch, 2020). PIN score-related inclusion criteria have also been incorporated into some recent clinical trials (for example, a PIN score between 0.18 to 4.93, in addition to a UHDRS TFC score of 13 and UHDRS Independence Scale score of 100 for the PTC Therapeutics PTC-518 trial [NCT05358717]).

The HD-Integrated Staging System (HD-ISS)

The Huntington's Disease Integrated Staging System (HD-ISS) is a newly proposed measure which categorizes individuals with the HD mutation into one of four stages (Stage 0–3), based on age- and education- adjusted neuroimaging, cognitive, and functional markers (Fig. 3). Specifically, those who test positive for the HD mutation but are still asymptomatic are placed in Stage 0, those who are largely

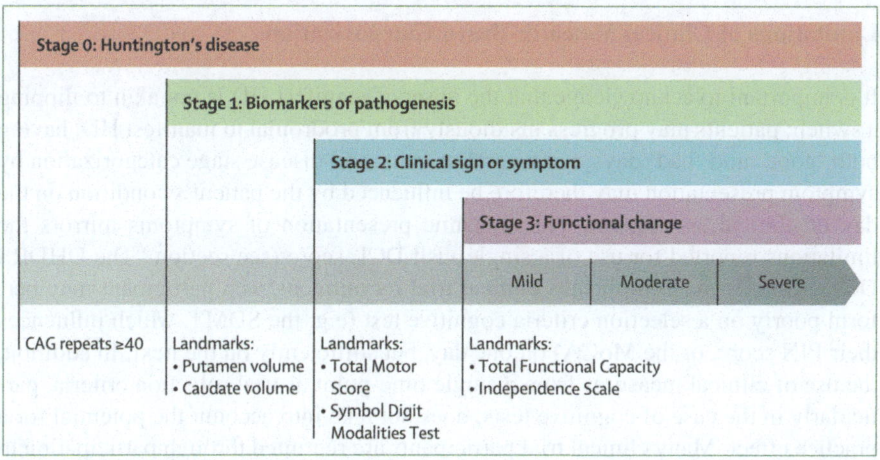

Fig. 3 Cumulative staging framework and landmarks of the Huntington's disease Integrated Staging System. Representation of the temporal sequence of stage progression and the associated landmark assessments that define stage entry (time not to scale). (Reprinted from Tabrizi et al. (2022), Copyright (2022), with permission from Elsevier)

asymptomatic but present with caudate and putamen volumetric changes as determined by quantitative MRI are placed in Stage 1, those with threshold cognitive and/or motor changes as determined by the SDMT and UHDRS TMS, respectively, are placed in Stage 2, and those with threshold functional changes as determined by the UHDRS TFC and Independence scores are placed in Stage 3 (Tabrizi et al., 2022). As full application of the HD-ISS necessitates the use of quantitative neuroimaging data, which is not feasible for many research sites, Long, Tabrizi and colleagues have also developed an imputation method, using a machine learning algorithm and data from multiple, large observational studies including TRACK-HD and PREDICT-HD (Long et al., 2023). This imputation method has been incorporated into the latest global, observational Enroll-HD (NCT01574053) periodic dataset (PDS6) (Long et al., 2023). While use of the imputation method is advised and preferred, the authors of the HD-ISS have also presented PIN and CAP score thresholds that provide the highest sensitivity and specificity for separating the HD-ISS stages. It is also important to note that the use of quantitative neuroimaging would only be required for categorization into Stage 1, and no clinical trial are currently designed to target such early disease progression; correspondingly, for the purpose of clinical trial recruitment, individuals may be categorized into ≤Stage 1, Stage 2 or Stage 3. Promisingly, for clinical studies that would require categorization of Stage 0 and 1 participants, research has also shown correlations between biomarkers, such as plasma NfL, and either cognitive scores and/or MRI-measured brain volume, suggesting that future inclusion of biofluid markers into the HD-ISS may improve its functionality. Indeed, investigations into the usefulness of plasma NfL in defining earlier HD-ISS disease stages are already published, and others are underway (Parkin et al., 2023).

Limitations of Clinical Measure-Based Categorization

It is important to acknowledge that the onset of manifest HD is not akin to flipping a switch; patients may progress insidiously from prodromal to manifest HD, having both 'good' and 'bad' days, for several months, and disease stage categorization by symptom presentation may therefore be influenced by the patient's condition on the day of clinical assessment. This dynamic presentation of symptoms mirrors the limitations reported for use of a single-visit DCL score (see section "The UHDRS DCL Score"), and complicates clinical trial recruitment, as a participant may perform poorly on a selection criteria cognitive test (e.g. the SDMT, which influences their PIN score, or the MoCA) on one day, but sufficiently on the next. In addition, the use of clinical measures from a single time-point as trial selection criteria, particularly in the case of cognitive tests, does not take into account the potential for a practice effect. Many clinical trial participants are recruited through participation in annual, observational studies, such as Enroll-HD (NCT01574053), which includes the SDMT and MMSE tests. Research sites may also administer the MoCA often for the purpose of screening observational study participants for clinical trial inclusion, once again resulting in a practice effect as individuals become familiar with the measures over time. Overall, the use of disease burden predictors alone (CAG repeat length x age), risks the categorization of individuals irrespective of symptom presentation, whereas the incorporation of cognitive measures, risks categorization based on a single moment in time.

Predicting Years to Manifest Onset for Clinical Trial Inclusion

To circumvent variability associated with day-to-day performance on clinical measures, it is important to marry disease stage categorization with an objective prediction of years to manifest HD onset. Indeed, a number of prediction formulae for years to manifest disease onset, which incorporate both patient age, and CAG repeat length, have been developed (Langbehn et al., 2010). While such formulae offer the potential to enrich clinical trials with participants predicted to be within 5–10 years of manifest onset, the aforementioned limitations of relying on CAG repeat length and age still remain. Recently, Long and colleagues also adapted the PIN score, which utilizes CAG repeat length, age, SDMT and UHDRS TMS, into a score for predicted years to manifest HD onset (Long et al., 2017). The PIN score-derived calculation for predicted years to onset once again balances disease burden predictors with the presentation of cognitive and motor symptoms, albeit at risk of introducing variability inherent in all of its components. That said, in terms of clinical trial inclusion, predicting age of manifest HD onset to the exact year is not necessary; the aim would be to *enrich* clinical trial selection criteria for prodromal HD participants, such that the majority of premanifest HD participants selected are near manifest symptom onset.

6 The Incorporation of Biomarkers

Prognostic Biomarkers

As aforementioned, new clinical trials for HD are being designed to recruit participants who are earlier in their disease progression. As such, it will be important to determine an objective indication of premanifest disease progression, such that included trial participants are those with the potential to show treatment-modified symptom progression, without disease burden too severe for a treatment effect.

Selective changes in regional brain volume and atrophy have been reported to occur up to 15 years before a diagnosis of manifest HD. On the other hand, white matter volume loss appeared to occur close to manifest symptom onset (Squitieri et al., 2009). The authors also found that the incorporation of CSF volume, which increases early on in the disease progression, improved a regression model of CAG repeat length and observed age to manifest HD onset (Squitieri et al., 2009). Studies have also reported reduced putamen and caudate volume, and subcortical activation, in participants with an estimated <12 years until manifest HD onset (Paulsen et al., 2004; Aylward, 2007). In another study, global metabolic pattern expression – quantified as increasing metabolic activity in the cerebellum, orbitofrontal cortex and pons, and decreasing metabolic activity in the cingulate cortex, prefrontal cortex, caudate, putamen and occipital lobe – correlated linearly with the number of years until manifest symptom onset (Tang et al., 2013). Given these findings, it is no surprise that quantitative neuroimaging changes – specifically caudate and putamen threshold values- are landmarks for membership into Stage 1 of the HD-ISS (Tabrizi et al., 2022). Overall, these studies suggest that quantitative neuroimaging measures are highly correlated with disease progression and may be used to predict manifest HD onset. One draw-back, however, is that such medical imaging is expensive and time-consuming, and this may limit accessibility and use as selection criteria for clinical trials, or to track longitudinal change.

On the other hand, biofluid markers are accessible, relatively inexpensive to measure, and – if analytically validated – could be implemented into any pathology laboratory. In particular, plasma NfL has garnered significant interest as a biomarker of disease progression in HD. Recent studies have shown significant correlations between both CSF and plasma NfL levels and predicted years to manifest HD onset (Parkin et al., 2021, 2022; Scahill et al., 2020; Byrne et al., 2017), as well as between plasma NfL and caudate and putamen volume (Byrne et al., 2017, 2018). While not as well established, a similar association with predicted years to manifest HD onset has been reported for salivary BDNF (Gutierrez et al., 2020). Serum BDNF levels have been correlated with fractional grey matter, with both showing non-linear decreases many years before manifest onset (Squitieri et al., 2009). These findings suggest the potential for multiple biomarkers, via a multivariate model, to assist in the determination of a prognostic outlook model for patients. Unfortunately, to date, all fluid biomarker findings have been determined using research-grade biological assays, obtained from a range of manufacturers. While discrepancies in absolute quantification values between assays from different manufacturers prevents implementation of these findings into clinical practice (Kuhle et al., 2016), with additional research, and following cross-site and cross-manufacturer standardization, these biomarkers may still be used in eligibility criteria for clinical trials.

Biomarkers as Outcome Measures

As evidenced by trials discussed in Sect. 3, changes in levels of CSF HTT and CSF NfL from baseline are already incorporated into current clinical trials as secondary or 'other' outcome measures. Measurements of CSF HTT, particularly for HTT-lowering therapeutic trials, are included as an indication of target effect - whether the drug is lowering HTT as intended. Measurements of CSF NfL are included due to its well-established associations with axonal degeneration and with the presentation of HD (Byrne et al., 2017). Consequently, a number of trials have reported an increase in CSF NfL from baseline and within a few months of treatment (GENERATION-HD1/ NCT03761849; GEN-EXTEND/ NCT03842969; HD-GENETRX-1/ NCT04120493; SELECT-HD/ NCT05032196; VIBRANT-HD/ NCT05111249 (serum NfL)), with overall return to baseline levels within approximately one year. In addition, levels of CSF tau increased within the same time-period in GENERATION-HD1. While it is believed that these increases are due to the administration of treatment, as all trials listed above (with the exception of VIBRANT-HD, which reported indications of peripheral neuropathy), utilized an intrathecal or intrastriatal route, the true implications of these changes are still unknown.

7 Conclusions and Future Direction

HD clinical trial outcomes have for the most part been disappointing. For as far back as 2014, concerns have been raised that the step-wise process from Phase 1 to Phase 3 trials, including the abandonment of unsuccessful programs, has been unsuccessful (Sampaio et al., 2014). This may in part be due to the selection of clinical trial primary outcomes (for example, a number of 'failed' pridopidine trials such as MermaiHD, HART, and most recently PROOF-HD, have reported significant secondary endpoints while failing to meet their primary outcome)(Sampaio et al., 2014). Incorporation of additional biomarker measures, which reflect pathophysiological change, may provide further objective support for trial success. Indeed, the inclusion of biomarker measures has become an essential requirement for clinical trial design in HD research. Quantification of CSF mHTT and NfL are now expected as secondary, or other, outcome measures; however, exploratory investigations of other fluid markers, and quantitative neuroimaging changes, may assist in characterizing the safety profile and efficacy of investigational drugs.

While the inclusion of biomarkers as either secondary or exploratory outcome measures is now well-established, research regarding their incorporation into prognostic models of disease progression requires more work. To date, much of the work has focused on CSF and plasma levels of NfL; results have been promising however, rigorous longitudinal data is still required to fully understand the association between NfL and HD disease progression. Furthermore, it is likely that quantitative neuroimaging and fluid biomarkers would be best utilized in a multivariate model, or in prognostic models that are specific to particular HD disease progression stages. For example, the incorporation of grey matter, CSF volume, and serum BDNF values into a prognostic model for individuals estimated to be far from disease onset, and white matter loss into a prognostic model for individuals closer to disease onset. Similarly, CSF NfL appears to be more useful in distinguishing and predicting disease progression for participants an estimated ≥18 years from clinical onset, whereas plasma NfL may have use for those closer to onset (Scahill et al., 2020; Parkin et al., 2021). Indeed, both CSF and plasma NfL - and likely other markers – may display non-linear trajectories across the HD disease progression spectrum

(Rodrigues et al., 2020), and thus one prognostic model is unlikely to best fit all stages. Overall, the utilization of measures to distinguish HD disease stages, such as the HD-ISS, in hand with prognostic models which incorporate biomarkers – which may be stage-specific – is likely to improve not only eligibility criteria and enrichment for clinical trials, but also the quality and accuracy of clinician-provided prognostic outlook and clinical care.

References

Agustín-Pavón, C., Mielcarek, M., Garriga-Canut, M., & Isalan, M. (2016). Deimmunization for gene therapy: Host matching of synthetic zinc finger constructs enables long-term mutant Huntingtin repression in mice. *Molecular Neurodegeneration, 11*, 1–16.

Aylward, E. H. (2007). Change in MRI striatal volumes as a biomarker in preclinical Huntington's disease. *Brain Research Bulletin, 72*, 152–158.

Besouw, M., Masereeuw, R., Van Den Heuvel, L., & Levtchenko, E. (2013). Cysteamine: An old drug with new potential. *Drug Discovery Today, 18*, 785–792.

Bezprozvanny, I. (2010). The rise and fall of Dimebon. *Drug News & Perspectives, 23*, 518.

Biglan, K. M., Zhang, Y., Long, J. D., Geschwind, M., Kang, G. A., Killoran, A., et al. (2013). Refining the diagnosis of Huntington disease: The PREDICT-HD study. *Frontiers in Aging Neuroscience, 5*, 12.

Boak, L. & Mccolgan, P. Understanding the treatment and post-treatment effects of tominersen in the Phase III GENERATION HD1 study. CHDI Foundation Annual Therapeutics Conference 28th February-3rd March, 2022

Brownstein, M. J., Simon, N. G., Long, J. D., Yankey, J., Maibach, H. T., Cudkowicz, M., et al. (2020). Safety and tolerability of srx246, a vasopressin 1a antagonist, in irritable Huntington's disease patients—a randomized phase 2 clinical trial. *Journal of Clinical Medicine, 9*, 3682.

Byrne, L. M., Rodrigues, F. B., Blennow, K., Durr, A., Leavitt, B. R., Roos, R. A., et al. (2017). Neurofilament light protein in blood as a potential biomarker of neurodegeneration in Huntington's disease: A retrospective cohort analysis. *The Lancet Neurology., 16*, 601–609.

Byrne, L. M., Rodrigues, F. B., Johnson, E. B., Wijeratne, P. A., De Vita, E., Alexander, D. C., et al. (2018). Evaluation of mutant huntingtin and neurofilament proteins as potential markers in Huntington's disease. *Science Translational Medicine, 10*, eaat7108.

Cattaneo, E., Zuccato, C., & Tartari, M. (2005). Normal huntingtin function: An alternative approach to Huntington's disease. *Nature Reviews Neuroscience, 6*, 919–930.

Cheung, A. K., Hurley, B., Kerrigan, R., Shu, L., Chin, D. N., Shen, Y., et al. (2018). *Discovery of small molecule splicing modulators of survival motor neuron-2 (SMN2) for the treatment of spinal muscular atrophy (SMA)*. ACS Publications.

Drew, C. J., Sharouf, F., Randell, E., Brookes-Howell, L., Smallman, K., Sewell, B., et al. (2021). Protocol for an open label: Phase I trial within a cohort of foetal cell transplants in people with Huntington's disease. *Brain Communications, 3*, fcaa230.

Evans, E., Fisher, T., Mishra, V., Boise, M., Foster, A., Smith, E., et al. (2022). *Clinical evidence that treatment with pepinemab, a novel regulator of neuroinflammation, provides cognitive benefit to patients with Huntington's and potentially other neurodegenerative diseases (P3–11.007)*. AAN Enterprises.

Ferguson, M. W., Kennedy, C. J., Palpagama, T. H., Waldvogel, H. J., Faull, R. L., & Kwakowsky, A. (2022). Current and possible future therapeutic options for Huntington's disease. *Journal of Central Nervous System Disease, 14*, 11795735221092517.

Garriga-Canut, M., Agustín-Pavón, C., Herrmann, F., Sánchez, A., Dierssen, M., Fillat, C., et al. (2012). Synthetic zinc finger repressors reduce mutant huntingtin expression in the brain of R6/2 mice. *Proceedings of the National Academy of Sciences, 109*, E3136–E3145.

Gray, S. G. (2011). Targeting Huntington's disease through histone deacetylases. *Clinical Epigenetics, 2*, 257–277.

Gutierrez, A., Corey-Bloom, J., Thomas, E. A., & Desplats, P. (2020). Evaluation of biochemical and epigenetic measures of peripheral brain-derived neurotrophic factor (BDNF) as a biomarker in Huntington's disease patients. *Frontiers in Molecular Neuroscience, 12*, 335.

Hersch, S. (2008). PHEND-HD: A safety, tolerability, and biomarker study of phenylbutyrate in symptomatic HD. *Neurotherapeutics, 2*, 363.

Horizon Investigators of the Huntington Study Group & European Huntington's Disease Network. (2013). A randomized, double-blind, placebo-controlled study of latrepirdine in patients with mild to moderate Huntington disease. *JAMA Neurology, 70*, 25–33.

Jurcau, A., & Jurcau, M. C. (2022). Therapeutic strategies in Huntington's disease: From genetic defect to gene therapy. *Biomedicine, 10*, 1895.

Kacher, R., Lamazière, A., Heck, N., Kappes, V., Mounier, C., Despres, G., et al. (2019). CYP46A1 gene therapy deciphers the role of brain cholesterol metabolism in Huntington's disease. *Brain, 142*, 2432–2450.

Kieburtz, K., Mcdermott, M. P., Voss, T. S., Corey-Bloom, J., Deuel, L. M., Dorsey, E. R., et al. (2010). A randomized, placebo-controlled trial of latrepirdine in Huntington disease. *Archives of Neurology, 67*, 154–160.

Kingwell, K. (2021). Double setback for ASO trials in Huntington disease. Nature reviews. *Drug Discovery*.

Kuhle, J., Barro, C., Andreasson, U., Derfuss, T., Lindberg, R., Sandelius, Å., et al. (2016). Comparison of three analytical platforms for quantification of the neurofilament light chain in blood samples: ELISA, electrochemiluminescence immunoassay and Simoa. *Clinical Chemistry and Laboratory Medicine (CCLM), 54*, 1655–1661.

Langbehn, D. R., & Hersch, S. (2020). Clinical outcomes and selection criteria for prodromal Huntington's disease trials. *Movement Disorders*.

Langbehn, D. R., Hayden, M. R., Paulsen, J. S., & Predict-Hd Investigators. (2010). CAG-repeat length and the age of onset in Huntington disease (HD): A review and validation study of statistical approaches. *American Journal of Medical Genetics Part B: Neuropsychiatric Genetics, 153*, 397–408.

Long, J. D., Paulsen, J. S., & Investigators, P. H. & Group, C. O. T. H. S. (2015). Multivariate prediction of motor diagnosis in Huntington's disease: 12 years of PREDICT-HD. *Movement Disorders, 30*, 1664–1672.

Long, J. D., Langbehn, D. R., Tabrizi, S. J., Landwehrmeyer, B. G., Paulsen, J. S., Warner, J., et al. (2017). Validation of a prognostic index for Huntington's disease. *Movement Disorders, 32*, 256–263.

Long, J. D., Gantman, E. C., Mills, J. A., Vaidya, J. G., Mansbach, A., Tabrizi, S. J., et al. (2023). Applying the Huntington's disease integrated staging system (HD-ISS) to observational studies. *Journal of Huntington's Disease, 12*, 57–69.

Macedo, J., Pagani, E., Wenceslau, C., Ferrara, L., & Kerkis, I. (2021). A phase I clinical trial on intravenous administration of immature human dental pulp stem cells (Nestacell HDTM) to Huntington's disease patients. *Cytotherapy, 23*, 1.

Maibach, H. T., Brownstein, M. J., Hersch, S. M., Anderson, K. E., Itzkowitz, D. E., Damiano, E. M., et al. (2022). The Vasopressin 1a Receptor Antagonist SRX246 Reduces Aggressive Behavior in Huntington's Disease. *Journal of Personalized Medicine, 12*, 1561.

Maucksch, C., Vazey, E. M., Gordon, R. J., & Connor, B. (2013). Stem cell-based therapy for Huntington's disease. *Journal of Cellular Biochemistry, 114*, 754–763.

Mendell, J. R., Al-Zaidy, S. A., Rodino-Klapac, L. R., Goodspeed, K., Gray, S. J., Kay, C. N., et al. (2021). Current clinical applications of in vivo gene therapy with AAVs. *Molecular Therapy, 29*, 464–488.

Mestre, T. A., & Sampaio, C. (2017). Huntington disease: Linking pathogenesis to the development of experimental therapeutics. *Current Neurology and Neuroscience Reports, 17*, 1–8.
Naia, L., Ly, P., Mota, S. I., Lopes, C., Maranga, C., Coelho, P., et al. (2021). The Sigma-1 receptor mediates pridopidine rescue of mitochondrial function in Huntington Disease models. *Neurotherapeutics, 18*, 1017–1038.
Nayak, A., Ansar, R., Verma, S. K., Bonifati, D. M., & Kishore, U. (2011). Huntington's disease: An immune perspective. *Neurology Research International, 2011.*
Parkin, G. M., Corey-Bloom, J., Snell, C., Castleton, J., & Thomas, E. A. (2021). Plasma neurofilament light in Huntington's disease: A marker for disease onset, but not symptom progression. *Parkinsonism & Related Disorders, 87*, 32–38.
Parkin, G. M., Corey-Bloom, J., Long, J. D., Snell, C., Smith, H., & Thomas, E. A. (2022). Associations between prognostic index scores and plasma neurofilament light in Huntington's disease. *Parkinsonism & Related Disorders.*
Parkin, G. M., Thomas, E. A., & Corey-Bloom, J. (2023). Plasma NfL as a prognostic biomarker for enriching HD-ISS stage 1 categorisation: A cross-sectional study. *eBioMedicine*, 93.
Paulsen, J. S., Zimbelman, J. L., Hinton, S. C., Langbehn, D. R., Leveroni, C. L., Benjamin, M. L., et al. (2004). fMRI biomarker of early neuronal dysfunction in presymptomatic Huntington's disease. *American Journal of Neuroradiology, 25*, 1715–1721.
Paulsen, J. S., Lourens, S., Kieburtz, K., & Zhang, Y. (2019). Sample enrichment for clinical trials to show delay of onset in huntington disease. *Movement Disorders, 34*, 274–280.
Penney, J. B., Jr., Vonsattel, J. P., Macdonald, M. E., Gusella, J. F., & Myers, R. H. (1997). CAG repeat number governs the development rate of pathology in Huntington's disease. *Annals of Neurology: Official Journal of the American Neurological Association and the Child Neurology Society, 41*, 689–692.
Reilmann, R., Leavitt, B. R., & Ross, C. A. (2014a). Diagnostic criteria for Huntington's disease based on natural history. *Movement Disorders, 29*, 1335–1341.
Reilmann, R., Squitieri, F., Priller, J., Saft, C., Mariotti, C., Suessmuth, S., et al. (2014b). *Safety and tolerability of selisistat for the treatment of Huntington's disease: Results from a randomized.*, double-blind, placebo-controlled phase II trial (S47. 004). AAN Enterprises.
Rodrigues, F. B., Byrne, L. M., Tortelli, R., Johnson, E. B., Wijeratne, P. A., Arridge, M., et al. (2020). Mutant huntingtin and neurofilament light have distinct longitudinal dynamics in Huntington's disease. *Science Translational Medicine, 12.*
Rook, M. E., & Southwell, A. L. (2022). Antisense oligonucleotide therapy: From design to the Huntington disease clinic. *BioDrugs, 1–15.*
Rubinsztein, D. C., & Orr, H. T. (2016). Diminishing return for mechanistic therapeutics with neurodegenerative disease duration? There may be a point in the course of a neurodegenerative condition where therapeutics targeting disease-causing mechanisms are futile. *BioEssays, 38*, 977–980.
Ryskamp, D., Wu, J., Geva, M., Kusko, R., Grossman, I., Hayden, M., et al. (2017). The sigma-1 receptor mediates the beneficial effects of pridopidine in a mouse model of Huntington disease. *Neurobiology of Disease, 97*, 46–59.
Sampaio, C., Borowsky, B., & Reilmann, R. (2014). Clinical trials in Huntington's disease: Interventions in early clinical development and newer methodological approaches. *Movement Disorders, 29*, 1419–1428.
Scahill, R. I., Zeun, P., Osborne-Crowley, K., Johnson, E. B., Gregory, S., Parker, C., et al. (2020). Biological and clinical characteristics of gene carriers far from predicted onset in the Huntington's disease Young Adult Study (HD-YAS): A cross-sectional analysis. *The Lancet Neurology., 19*, 502–512.
Squitieri, F., Cannella, M., Simonelli, M., Sassone, J., Martino, T., Venditti, E., et al. (2009). Distinct brain volume changes correlating with clinical stage, disease progression rate, mutation size, and age at onset prediction as early biomarkers of brain atrophy in Huntington's disease. *CNS Neuroscience & Therapeutics, 15*, 1–11.

Stimming, E. F., Sung, V., Testa, C., Kostyk, S., Ross, C. A., Samii, A., et al. (2022). Interim results from cohort 1 of the double-blind, dose-escalation phase I/II clinical trial of AMT-130 (HD-GENETRX-1) for early-stage Huntington's disease (HD). *Neurodegeneration, 1*, 2.

Süssmuth, S. D., Haider, S., Landwehrmeyer, G. B., Farmer, R., Frost, C., Tripepi, G., et al. (2015). An exploratory double-blind, randomized clinical trial with selisistat, a SirT1 inhibitor, in patients with H untington's disease. *British Journal of Clinical Pharmacology, 79*, 465–476.

Tabrizi, S. J., Schobel, S., Gantman, E. C., Mansbach, A., Borowsky, B., Konstantinova, P., et al. (2022). A biological classification of Huntington's disease: The Integrated Staging System. *The Lancet Neurology, 21*, 632–644.

Tang, C. C., Feigin, A., Ma, Y., Habeck, C., Paulsen, J. S., Leenders, K. L., et al. (2013). Metabolic network as a progression biomarker of premanifest Huntington's disease. *The Journal of Clinical Investigation, 123*, 4076–4088.

Zeitler, B., Froelich, S., Marlen, K., Shivak, D. A., Yu, Q., Li, D., et al. (2019). Allele-selective transcriptional repression of mutant HTT for the treatment of Huntington's disease. *Nature Medicine, 25*, 1131–1142.

Zhang, Y., Long, J. D., Mills, J. A., Warner, J. H., Lu, W., Paulsen, J. S., et al. (2011). Indexing disease progression at study entry with individuals at-risk for Huntington disease. *American Journal of Medical Genetics. Part B, Neuropsychiatric Genetics, 156*, 751–763.

Digital Measures in Huntington's Disease

Jamie L. Adams, Emma M. Waddell, Natalia Chunga, and Lori Quinn

Abbreviations

CAN	Central autonomic network
DAT	Digitalized Arithmetic Task
DHT	Digital health technology
ECG	Electrocardiogram
HC	Healthy control
HD	Huntington's disease
HD-CAB	HD-Cognitive Assessment Battery
HRV	Heart rate variability
IMU	Inertial measurement units
n.s.	Non-significant results
NITS	Neural Impairment Test Suite
PD	Parkinson's disease
PKG	Parkinson's Kinetigraph
Pre-HD	Premanifest Huntington's Disease
Q-Motor	Quantitative-Motor battery
REM	Rapid eye movement
TID	Time in bed
TST	Total sleep time
UHDRS	Unified Huntington's Disease Rating Scale
UST	Ultra short term
WASO	Wake after sleep onset

J. L. Adams (✉)
Department of Neurology, University of Rochester, Rochester, NY, USA

Center for Health + Technology, University of Rochester, Rochester, NY, USA
e-mail: jamie_adams@urmc.rochester.edu

E. M. Waddell
Center for Health + Technology, University of Rochester, Rochester, NY, USA

N. Chunga
Department of Neurology, University of Rochester, Rochester, NY, USA

L. Quinn
Teachers College, Columbia University, New York, NY, USA

E. A. Thomas, G. M. Parkin (eds.), *Biomarkers for Huntington's Disease*,
Contemporary Clinical Neuroscience, https://doi.org/10.1007/978-3-031-32815-2_18

1 Introduction

With the emergence of genetic and disease-modifying interventions, there is an urgent need for early and accurate assessment of disease-specific impairments in Huntington's disease (HD). Large longitudinal studies have revealed that HD can take at least 15 years to develop from the onset of neurodegeneration, and motor and cognitive symptoms are evident years before formal diagnosis (Paulsen et al., 2008; Tabrizi et al., 2012)·. While rating scales have improved, they still provide subjective, episodic, and largely insensitive assessments contributing to large, lengthy, expensive trials that are prone to failure (Dorsey et al., 2017). Moreover, scales like the Unified Huntington's Disease Rating Scale (UHDRS) have high inter-observer variability and have faced challenges in detecting disease progression in neuroprotective trials (Reilmann et al., 2015). In addition, these scales may not accurately assess the patient experience. Better measures could lead to more efficient, patient-centric, and timely evaluation of therapies.

Digital measures are a component of the broader term digital health technologies (DHT). DHTs use computing platforms, connectivity, software, and sensors for healthcare and related uses (see Fig. 1). Digital measures are specifically designed to quantifiably measure physiological and behavioral data through digital devices such as portables (e.g. smartphone apps) or wearables (e.g. wrist worn accelerometers). Many researchers now use the term biomarkers, which are indicators of biological or pathological processes, when referring to different digital measures. Digital measures can be diagnostic, detecting the presence of disease, pharmacodynamic, showing a biological response, or monitoring, assessing disease status (Vasudevan et al., 2022).

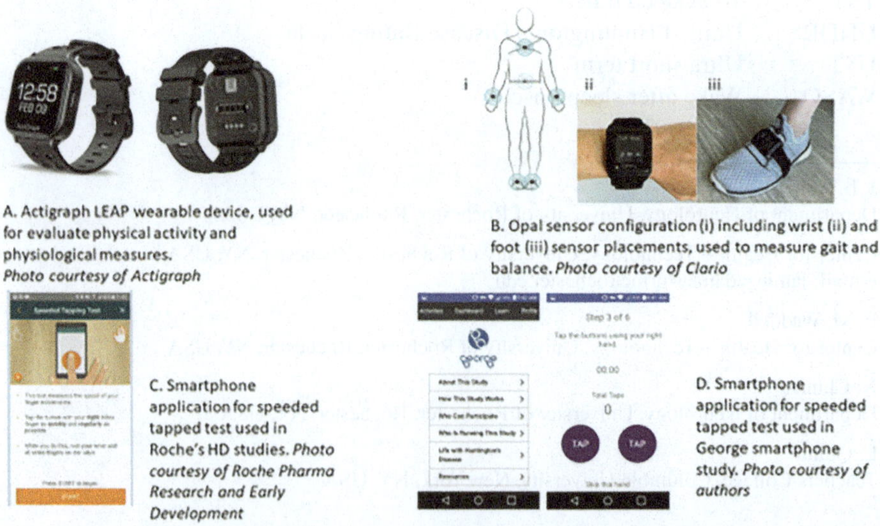

Fig. 1 Examples of digital health technologies used in HD studies

Digital measures have excellent potential to provide quantitative assessment of motor, cognitive, and speech tasks in individuals with HD (Tortelli et al., 2021). These measures can be collected with a range of devices and monitoring strategies, both in clinic and at home and using either passive or active monitoring. For instance, active monitoring can be accomplished using a smartphone application with various tasks (e.g. finger tapping, speech tasks) and passive monitoring is possible using wearable devices (e.g. step counts or sleep). Wearable sensors including inertial measurement units (IMUs) and body-worn accelerometers provide a comprehensive kinematic analysis of complex motor behavior and offer a non-invasive and efficient method for objective measurement of motor behavior. The high sampling rate allows for multiple data points from which to extract subtle movement abnormalities that often cannot be detected with the naked eye.

Clinical trials using digital measures are already underway in HD with promising results (Lipsmeier et al., 2022). Various studies have provided proof of principle for use of wearable sensors and app-based assessments to provide quantitative measures of motor and cognitive impairments in individuals with manifest HD (Lipsmeier et al., 2022; Adams et al., 2017; Dinesh et al., 2020; Muratori et al., 2021; Beckmann et al., 2018; Gaßner et al., 2020; Keren et al., 2021). Research has found that motor impairments in tasks such as finger tapping (speeded tapping inter-tap interval) can be a reliable indicator of HD impairment in premanifest and manifest stages and are linked to function and central nervous system structure in all stages of the disease (Reilmann et al., 2014; Tabrizi et al., 2013; Biglan et al., 2009; Bechtel et al., 2010). Chorea and postural control are important markers of disease progression, and both currently lack reliable and objective measurement tools. Preliminary results from these and other studies indicate that several measures show promise as quantitative measures of motor impairment in HD including truncal chorea index during sitting, postural sway in standing, and measures of gait and gait variability (Dinesh et al., 2020; Gaßner et al., 2020; Purcell et al., 2019; Gordon et al., 2019). In addition, measures of sedentary behavior, heart rate variability and sleep have been able to differentiate individuals with HD and healthy controls and can be assessed using actigraphy (Dinesh et al., 2020; Poile et al., 2018). Digital measures also show promise in detecting early changes in speech patterns, cognitive impairments and physiological functions such as sleep and heart rate variability. In prodromal HD, changes in voice patterns and cognitive impairments have shown promise as sensitive markers prior to the onset of a motor diagnosis (Vogel et al., 2012; Rusz et al., 2014; Chan et al., 2019; Paulsen et al., 2017). These quantitative markers may be meaningful endpoints that could be used in clinical assessments as well as outcome measures in clinical trials for individuals with premanifest and early manifest HD (see Fig. 2).

Throughout the nearly 50 years of research into digital assessments of motor function in HD, concurrent advancements in technology have occurred (see Fig. 3). Simple handheld devices and single accelerometers have given way to complex multimodal systems. For example, the Roche HD Monitoring Platform combines data from a smartphone and a smartwatch (Lipsmeier et al., 2022). In addition to prompting individuals to complete daily, structured tasks, the Roche HD Monitoring

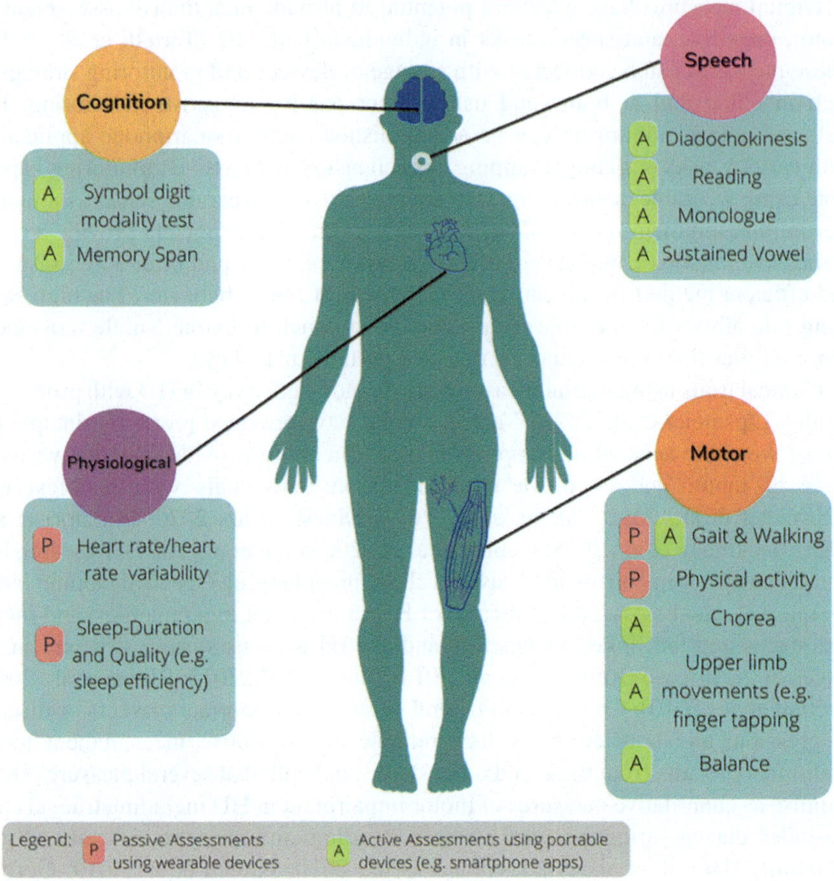

Fig. 2 Digital measures in HD by domain

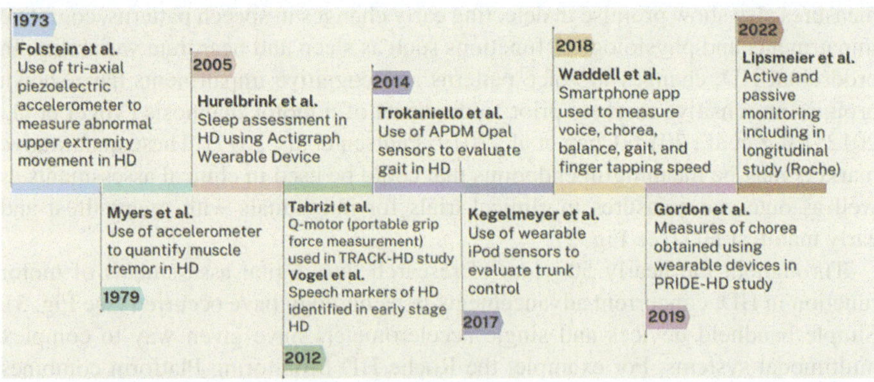

Fig. 3 Timeline of select digital measure studies in HD

Platform passively collects data about daily living from device-embedded sensors. The data is then transmitted via WiFi and analyzed remotely.

In this chapter, we will review digital measures in HD based on symptom domain: motor, speech, cognitive and physiological measures. We will review digital measures obtained from active and passive devices as well as those used for any purpose (diagnostic, pharmacodynamic or monitoring) and summarize the evidence to date in support of their use.

2 Motor

Huntington's disease is characterized by progressive disturbances of motor function resulting from degeneration of the basal ganglia (Bates et al., 2015). Individuals with HD can experience impairments in voluntary movement as well as involuntary movements, namely chorea. Defined as sudden, non-rhythmic, purposeless movement, chorea is the hallmark motor symptom of HD. The evaluation of chorea, as well as hand coordination, rigidity, bradykinesia, dystonia, gait, and balance, is a major component of the gold standard UHDRS (Huntington Study Group, 1996). A motor-specific subscore reflects the severity of such symptoms. However, as with many categorical scales administered in the clinic, the UHDRS Total Motor Score is limited by subjectivity among raters and does not capture daily symptom variability.

Digital measures may be used to derive more objective, frequent measures of motor function. Data from portable and wearable devices, some commercially available, and smartphone applications have proven effective in characterizing various motor symptoms including upper limb movements, chorea, balance, gait, and physical activity (see Table 1). Digital motor measures generally demonstrate good reproducibility and correlation to traditional rating scales like the UHDRS (Gordon et al., 2019). Furthermore, they are acceptable to the HD community. A survey of over 100 people across four Huntington's Disease Association meetings revealed an overall positive attitude toward wearable activity trackers (Morgan-Jones et al., 2022). Respondents felt that trackers are advantageous, understandable, enjoyable to use, and compatible with their lifestyle (Morgan-Jones et al., 2022). A 6-month study of participant compliance with a smartphone- and smartwatch-based protocol demonstrated high adherence to medication and symptom reporting, data streaming, and the performance of motor tasks at home (Cohen et al., 2018). A longer 12-month study involving physical activity tracking with a Fitbit revealed that the vast majority (82.6%) of participants with HD wore the device for at least 10 hours on greater than 70% of days (Quinn et al., 2022).

The following discussion of digital measures is structured by motor domain. However, it is important to point out that measures often evaluate combinations of motor signs and tasks, as is the case with the Quantitative-Motor battery (Q-Motor) (Reilmann & Schubert, 2017). The Q-Motor battery, which is derived from force transducers and three-dimensional position sensors, addresses finger tapping,

Table 1 Motor digital measures

Motor sign	Measures	Studies
Finger tapping	Tap rate, inter-tap interval, percentage decrement in tap rate	Folstein et al., Barker et al., Waddell et al., Lipsmeier et al., Michell et al., Collins et al., Antoniades et al.
Diadochokinesis	Velocity	Muller et al.
Typing	Cadence	Lang et al.
Chorea	Upper limb and truncal scores	Buruma et al., Lange et al., Waddell et al., Reilmann et al., Kotschet et al., Dinesh et al., Lipsmeier et al.
Balance	Total body sway, jerk, sway variability, truncal movements	Beckmann et al., Purcell et al., Mustafa et al., Lipsmeier et al., Grimbergen et al., Kegelmeyer et al.
Gait	Step and stride time and variability, smoothness, regularity	Beckmann et al., Kloos et al., Gabner et al., Keren et al., Lipsmeier et al., Grimbergen et al., Kegelmeyer et al., Dinesh et al.
Physical activity	Sedentary time, moderate-to-vigorous physical activity, step count	Cohen et al., Lipsmeier et al., van Vugt et al., Dinesh et al., Andrzeiewski et al.

alternating pronation and supination of the hands, grasping and lifting, choreiform movements, foot tapping, and tongue protrusion.

Upper Limb Movement Measures

One of the most straightforward digital measures of motor function in HD is finger tapping. Manual dexterity is reduced in HD, and the degree of impairment can be quantified in a timed task. A device as simple as a box with two buttons spaced 6 cm apart can be connected to computer software to quantitatively describe performance. Using such a set-up, one research group showed that individuals with HD tapped at a slower rate and had greater variability of inter-tap interval compared to controls (Michell et al., 2008). A few of the same researchers later expanded on this research to show that tapping declines linearly with disease progression at an average rate of 5.1 taps per year and more rapidly with a greater number of CAG repeats (Collins et al., 2014).

Finger tapping can be measured outside of the clinic, too. Smartphone applications like GEORGE and the Roche HD monitoring app collect daily finger tapping counts on-the-go in as little as 60 seconds. In a pilot study of the GEORGE application, tap rate was significantly slower among participants with HD compared to controls and correlated strongly to UHDRS finger tapping score (Waddell et al., 2021). Similarly, the finger tapping test within the Roche HD monitoring app demonstrated strong correlation to an in-clinic speeded tapping test (Lipsmeier et al., 2022).

In addition to finger tapping, fine motor skills like writing and typing are impaired in HD and amenable to capture by technology. One study used participants' personal keyboards to demonstrate significantly higher typing cadence inconsistency between HD and control cohorts. The authors also noted a trend toward higher inconsistency among HD gene expansion carriers, suggesting that digital measures may help identify early disease manifestations (Lang et al., 2021). Individuals with HD also demonstrate discoordination of diadochokinesis, or rapid, alternating hand movements, reflected in lower peak velocities and longer intervals for performance of diadochokinetic tasks compared to controls (Müller et al., 2013). A key limitation of the task-based evaluation of all upper limb movements is practice effect, change that results from repetition rather than ability (Holm et al., 2022).

Chorea Measures

Derived from the Latin "choreus" for "dancing," chorea is both unpredictable and difficult to control. Compared to finger tapping, chorea is more challenging to quantify. Some studies have sought to track the number of involuntary movements in a set time frame. As early as 1982, a group of Scandinavian researchers used a Doppler radar to quantify choreatic movements (Buruma et al., 1982). Their method correlated with clinical assessments and was suggested to be preferable on account of its objectivity.

Other studies have used changes in position and orientation during static activities as a proxy for chorea (Reilmann et al., 2011). Novel technology-based algorithms rely on such data to score chorea. Both the GEORGE application and the Roche HD Monitoring Platform include a 30-second task that instructs participants to hold the smartphone still in an outstretched hand. GEORGE demonstrated a significant difference in chorea between individuals with and without HD (Waddell et al., 2021). Gordon et al. recorded 890 at-home assessments of chorea while participants sat and stood at rest wearing a smartphone at the waist and a smartwatch. Their application also asked people to report on their symptoms and found that the digital chorea level correlated highly to self-rated chorea severity (Gordon et al., 2019).

Finally, wearable sensors have been used to measure chorea over a longer duration. Dinesh et al. developed a chorea index from sensors worn for 48 continuous hours at a time and showed that chorea had substantial intra-day variability but minimal change over the 12-month study period, possibly due to small sample size (Dinesh et al., 2020). More recently, a wearable sensor system called the Parkinson's Kinetigraph (PKG) and its associated dyskinesia score were applied to evaluate chorea over a minimum of 4 days. Measures derived from the PKG demonstrated good correlation to bradykinesia and chorea scores within the UHDRS (Kotschet et al., 2022).

Of note, chorea has a wide variety of presentations and can involve different body parts; thus, sensor location is an important variable to consider. The algorithm introduced by Dinesh et al. relies on truncal accelerometer data from a sensor worn on the chest (Dinesh et al., 2020). Research involving smartwatches usually requires participants to wear the sensors on their chorea-dominant upper extremity though this placement can introduce noise from idiosyncratic behaviors like tooth brushing.

Balance Measures

In addition to chorea, wearable technology can characterize balance dysfunction. Postural control, the ability to maintain body alignment in space, is impaired even in early HD. Such control is key to static (e.g., sitting, standing), dynamic, and reactive movements. Wearable and portable devices can measure features of postural stability such as total sway, sway variability, and jerk, or the number of postural corrections. Static measures like postural sway, or the amount and frequency of body movement within the base of support, has been shown to be greater in people with HD compared to control participants in multiple studies. Purcell et al. used a lumbar sensor to show that individuals with HD have worse postural control under environmentally challenging circumstances (e.g., multiple tasks, limited vision) (Purcell et al., 2019). Recently, in 2022, a Nintendo Wii balance board demonstrated significantly different total body sway between three cohorts – manifest HD, prodromal HD, and controls – suggesting a potential role in early disease detection (Mustafa et al., 2022).

Postural instability is associated with an increased fall risk. In a study of 45 individuals with HD, angular velocity sensors worn on the lower back identified significantly greater trunk sway in the medio-lateral direction among participants who had fallen at least twice in the previous year ("fallers") compared to non-fallers (Grimbergen et al., 2008). Not only can digital tools recognize those at greater risk of injury, but they may be an option for prevention and rehabilitation. Wearable iPod sensors equipped with biofeedback programming can monitor trunk stability and cue individuals in real-time to make postural corrections (Kegelmeyer et al., 2017).

Gait Measures

There are many technologies to assess gait in HD. Commonly used tools include the GAITRite walkway, shoe insoles, and inertial sensors attached to footwear, but recent studies have shown that wrist-worn wearable sensors, smartwatches, and smartphones can also assess gait. The magnetic and inertial measurement unit is a device that is often used to assess gait due to its ability to measure accelerations, angular velocities, and local magnetic fields (Tortelli et al., 2021). Similarly, gait can be described by a variety of qualitative and quantitative endpoints (Keren et al., 2021). Examples of quantitative measures that have been studied in HD include step count, gait velocity, stride and stance time and variability, and walking bouts per day. Qualitative measures like cadence and regularity reflect smoothness of walking. Generally, individuals with HD walk less and more slowly and have greater gait variability compared to a control population (Dinesh et al., 2020). There is conflicting data as to whether individuals with HD spend less time walking overall compared to non-HD peers (Keren et al., 2021).

Physical Activity Measures

Passive monitoring affords the opportunity to observe patterns in overall motor activity. Wearable devices can distinguish between sedentary behavior and physical activity, stratified further by vigorousness. Such data may be collected over a long period of time, illustrated by an average 10.5 hours of daily smartwatch use in a 6-month study period (Cohen et al., 2018).

Like walking frequency, activity level is variably reported in the literature. Wrist-worn monitors identified significantly decreased daytime motor activity among individuals with HD, especially those on neuroleptic medications, compared to controls (van Vugt et al., 1996). One study of wearable sensors revealed that individuals with HD spend more time lying down during the day (Dinesh et al., 2020). However, another study of wearable sensors demonstrated a statistically equal level of physical activity between individuals with HD and controls, and a third reported no difference in walking time per day between the same groups (Keren et al., 2021; Andrzejewski et al., 2016). Further research, potentially with multimodal platforms, will hopefully better elucidate the effects of HD on general daytime habits.

3 Speech and Language

Spoken language is a complex process that involves multiple systems, including coordination of motor, cognitive, and behavioral functions, and the striatum is believed to be part of the cortico-subcortical pathway involved in this process. It is thus expected that individuals with HD develop progressive deficits in speech and language, and prodromal changes might be valuable to track the progression of disease. As such, researchers are using digital measures to more objectively and frequently monitor speech and language in HD (see Table 2).

Different methods and tools have been proposed to study speech and language characteristics. For instance, voice samples can be obtained from spontaneous narrative discourse (e.g., storytelling), reading passages, automatic speech (e.g., saying the months of the year), and standardized tasks (e.g., syllable repetition, sustained phonation) (Rusz et al., 2021).

The acquisition of voice samples has become easier and more accessible, and the use of smart devices and computer-based analysis are promising for the study of speech and language. In the laboratory or office settings, voice samples can be

Table 2 Speech and language digital measures

Speech tasks (example)	Speech digital measures	Lab-based studies in HD*	Remote methods proposed for speech study in HD
Syllable repetition Single syllables (/pa/) Multiple syllables (/pataka/)	*Articulation* 1. Diadochokinesis 2. Stability of repetition 3. Vowel articulation	*Premanifest HD* Kaploun et al., 2011: Increased variability and reduced repetition rate Skodda et al., 2016: Increased variability and repetition rate *HD* Skodda et al., 2014: Increased variability	Fahed et al. (2022) Task: syllable repetition (/pa/, /ta/, /ka/, /pataka/, and /pati/) Device: smartphone or tablet microphone in a lab setting Mobile devices provided speech samples that were comparable to gold standard headset microphone
Sustained phonation Sustained vowel (/a/) Sustained fricative (/s/)	*Phonation* 1. Voice signal frequency 2. Perturbations: shimmer or jitter 3. Noise measures	*Premanifest HD* Kaploun et al., 2011: Increased shimmer, jitter, and noise Rusz et al., 2014: Increased noise and aperiodicity *HD* Rusz et al., 2013: Increased shimmer, jitter, noise, and aperiodicity	Waddell et al. (2021) Task: sustained vowel (/a/) Device: GEORGE app; smartphone microphone Voice samples from sustained vowel phonation were collected but not analyzed yet Fahed et al. (2022) Task: sustained vowel (/a/) Device and results as above
	Respiration 1. Maximum phonation	*Premanifest HD* Kaploun et al., 2011: Reduced phonation duration Rusz et al., 2014: n.s. Skodda et al., 2016: n.s. *HD* Rusz et al., 2013: Reduced phonation duration and faster first voice break Skodda et al., 2014: Reduced phonation duration	
	Resonance 1. Nasality properties	*HD* Novotný et al., 2016: Increased hypernasality (vs PD and HC)	
Narrative discourse Reading a selected paragraph Automated speech (days of the week)	*Prosody* 1. Speech time and rate 2. Silence time and rate	*Premanifest HD* Vogel et al., 2012: n.s. Saft et al., 2014: Increased velocity reading and reduced speech pauses Skodda et al., 2016: Increased speech rate	Lauraitis et al. (2020) Task: reading pre-defined poem Device: NITS app; smartphone or tablet microphone NITS differentiated healthy from diseased subjects (7 with neurological conditions, including 3 with HD)

(continued)

Table 2 (continued)

Speech tasks (example)	Speech digital measures	Lab-based studies in HD*	Remote methods proposed for speech study in HD
Monologue		*HD* Vogel et al., 2012: Increased silence time and reduced speech rate in HD (vs premanifest HD and HC) Rusz et al., 2014: Reduced speech pauses and speech rate Skodda et al., 2014: Increased speech pauses	Fahed et al. (2022) Task: reading the Rainbow passage Device and results as above

*All the studies used a digital microphone to record the speech samples
HD Huntington's disease, *HC* healthy control, *PD* Parkinson's disease, *n.s.* non-significant results, *NITS* Neural Impairment Test Suite

obtained by using audio recording devices and microphones, and guidelines for speech recording and acoustic analyses have been proposed by the International Parkinson and Movement Disorders Society (Rusz et al., 2021). To date, most of the studies in HD evaluating speech digital measures have been conducted in a controlled laboratory setting using high-quality microphones (see Table 2).

In addition, voice samples can be obtained remotely. Fahed et al. showed that quantitative speech measures derived from mobile devices (smartphone or tablet) in a lab setting had good agreement with data obtained from gold-standard headset microphones in healthy controls and HD subjects, supporting the potential use of mobile devices for remote monitoring of speech in HD (Fahed et al., 2022). A couple of studies have shown the potential utility of smartphone-based applications for remote voice recording in HD using the device microphone. The Neural Impairment Test Suite (NITS) mobile app includes a voice recorder task in which pre-defined poems are read, and a study conducted in 15 subjects (7 with neurological disease, including 3 with HD) showed that this mobile app had a 94% accuracy in differentiating healthy from diseased individuals (Lauraitis et al., 2020). The GEORGE mobile app includes a sustained phonation task as one of the four activities to monitor HD symptoms; in the pilot study, voice recordings were successfully obtained, but data has not been analyzed yet (Waddell et al., 2021). Similarly, the Roche PD mobile app includes monotonicity and jitter measures, which have been shown to correlate with speech impairment clinical scores in Parkinson's disease (PD) (Lipsmeier et al., 2022). Telephone and web-based voice recordings of sustained phonation have also been proposed as possible screening tools for PD under acoustically non-controlled conditions (Arora & Tsanas, 2021; Rahman et al., 2021). These methods could potentially be applied in HD as well.

Voice analyses were initially conducted by obtaining recordings that were manually transcribed and rated by judges (e.g. speech language pathologists) using standardized scores (Hartelius et al., 2003). Over time, algorithms for automatic voice recognition and processing have been developed. Speech characteristics can be analyzed with available audio acoustics softwares, such as WaveLab (https://www.steinberg.net/wavelab/) and PRAAT (https://www.fon.hum.uva.nl/praat/), which have been used in previous studies in HD (Vogel et al., 2012; Rusz et al., 2014; Skodda et al., 2016). In PD, the DARTH group software (https://www.darth-group.com/software) has been used for speech analysis, which might be suitable for HD studies as well (Arora & Tsanas, 2021). Computer-based software for the analysis of language characteristics is also available (Pezold et al., 2020); however, the applicability and validity of these methods in HD or other neurodegenerative diseases is still to be studied. Alternatively, ad-hoc algorithms or software can be developed for specific studies or device applications.

Speech Measures

Speech changes are noted early in HD, and they appear before other motor, cognitive and psychiatric symptoms. In fact, subtle alterations in the quantity and quality of speech have been reported in individuals with premanifest gene-positive HD, making these promising markers of disease onset and progression (Chan et al., 2019). In addition, obtaining speech samples is relatively easy and simple, adding to the benefits of speech function as a marker.

Different acoustic characteristics have been proposed as potential markers in premanifest and manifest HD (Chan et al., 2019), including the articulation, phonation, respiration, resonance, and prosody of speech (see Table 2). However, it still remains unclear which are the most suitable to differentiate premanifest HD from healthy individuals, and a combination of measures might be needed for an appropriate evaluation.

Speech articulation can be measured by the evaluation of diadochokinesis, stability of repetition, and vowel articulation, and articulatory deficits appear to start in the premanifest HD stage (Skodda et al., 2016; Kaploun et al., 2011). Phonation can be assessed by measuring voice signal frequencies, perturbations (shimmer and jitter) and noise, and previous studies have shown phonatory dysfunction in premanifest and manifest HD (Rusz et al., 2014; Kaploun et al., 2011; Rusz et al., 2013). Respiration and resonance properties can be evaluated by measuring the maximum phonation time and nasality characteristics, respectively. Studies have shown respiratory abnormalities in manifest HD (Rusz et al., 2013; Skodda et al., 2014), but there is not enough evidence to support changes in the premanifest stage (Rusz et al., 2014; Skodda et al., 2016; Kaploun et al., 2011). Speech resonance has only been studied in manifest HD by Novotny et al., who showed increased nasality levels and variability (Novotný et al., 2016). Finally, prosody can be studied by assessing the speech and silence time and rate, which are known to be affected in HD (Rusz et al., 2014; Skodda et al., 2014), but only a few studies have been completed in premanifest HD with significant but conflicting results (Vogel et al., 2012; Skodda et al., 2016; Saft et al., 2014).

Language Measures

Language is a complex system that not only serves to communicate with each other in daily life, but also to form and convey our thinking. Aside from the studies looking at the acoustic characteristics of speech, there is still scarce evidence regarding the utility of language as a biomarker in HD. However, its role in everyday social interactions, thinking processes, and behaviors highlights the potential value language has in identifying early signs of the disease – in particular, cognitive and psychiatric changes, as multiple systems are required for language production (Hinzen et al., 2018). Moreover, the way language influences our thinking is not fully understood yet; however, a possible interaction might partially explain the difficulties individuals with HD appear to have in understanding the mental state of others (theory of mind) and other behavioral changes (Brüne et al., 2011; de Villiers, 2007).

The study of spontaneous language stands out as a unique way to evaluate the different subsystems of this function given its closer association with other cognitive functions, when compared to standardized tasks or automatic language. In individuals with HD, reduction of spontaneous communication with use of fewer words, simpler sentences and short syntactic forms, as well as more grammatical errors and paraphasias have been described (Chenery et al., 2002).

Although the quantity of language is reduced in individuals with HD, Hinzen et al. reported that individuals with premanifest HD had quantitative domain results comparable to healthy individuals, when evaluating spontaneous narrative language. On the other hand, they noted that the connectivity and reference domains in premanifest HD differed significantly from both healthy controls and manifest HD, underscoring its potential role for identification of disease onset. Fluency and connectivity also showed a gradual pattern of impairment, with more severe deficits in the HD group, but differences were not statistically significant (Hinzen et al., 2018). Digital health technologies could be useful to both collect and analyze language remotely and continuously.

4 Cognition

Cognitive impairment is one of the three main clinical characteristics of HD. The growing use of digital measures brings with it the opportunity for remote monitoring of cognition in daily living environments. Thereby, well-known cognitive assessments have been incorporated into mobile devices and applications for the evaluation of cognitive symptoms in HD (see Table 3). Nevertheless, there is still scarce information about the utility of these tools. Furthermore, in 2018, a review of the available pen and paper cognitive scales for HD showed that none of them were sufficiently validated to be robustly recommended in this population, raising the need for further study of cognitive tools in HD (Mestre et al., 2018). This effort is

Table 3 Cognitive digital measures

Cognitive domain	Studies using digital devices in HD
Perception and visuospatial	McLaren et al. (2021) Task: Object identity and location learning Device: HD-Mobile app (smartphone) Lipsmeier et al. (2022) Task: Symbol digit modalities test Device: Roche HD app (smartphone)
Motor control and construction	Schubert et al. (2016) Tasks: Force matching, Morse code mimicry, spiral drawing Device: Q-Cog (force transducer and 6-axis position sensor) Lauraitis et al. (2020) Tasks: Sequential touch, clock drawing, 3D figure construction, spiral drawing, trail making Device: NITS app (smartphone or tablet) Muratori et al. (2021) Task: Dual tasking (standing balance, timed up & go, and walking without and with cognitive demand) Device: Body worn inertial sensors McLaren et al. (2021) Task: Speeded tapping Device: HD-Mobile app (smartphone) Lipsmeier et al. (2022) Task: Symbol digit modalities test Device: Roche HD app (smartphone)
Attention and concentration	Lipsmeier et al. (2022) Tasks: Symbol digit modalities test, Stroop word reading Device: Roche HD app (smartphone)
Memory	Ghilardi et al. (2008) Task: Sequence learning (moving cursor on a tablet) Device: Digital tablet Lauraitis et al. (2020) Tasks: Memorize a phrase, orientation (current date) Device: NITS app (smartphone or tablet) McLaren et al. (2021) Task: Object identity and location learning Device: HD-Mobile app (smartphone)
Executive functioning	Lauraitis et al. (2020) Task: Trail making, calculations, problem solving (forming a new shape) Device: NITS app (smartphone or tablet) McLaren et al. (2021) Task: Two-choice discrimination Device: HD-Mobile app (smartphone) Lunven et al. (2021) Task: Multiplication and substraction problems Device: Computer-based
Language and verbal skills	Lauraitis et al. (2020) Task: Picture naming, verbal fluency (writing 12 items of a given category) Device: NITS app (smartphone or tablet)

HD Huntington's disease, *NITS* Neural Impairment Test Suite

underway and newer scales including the HD-Cognitive Assessment Battery (HD-CAB) have been developed with potential for digital administration (Stout et al., 2017).

Cognitive Measures

Dual-tasking is impaired in individuals with HD due to cognitive-motor interference, which has been described with complex cognitive, motor-cognitive, and motor-motor dual-task. Using body worn inertial sensors, Muratori et al. showed that individuals with HD had worsening postural sway during dual tasking and demonstrated motor priority with greater dual-task effect on cognitive performance (Muratori et al., 2021).

Schubert et al. have also proposed 'Q-Cog' as a digital tool for cognitive evaluation in HD. This tool assesses cognitive function using tasks with high cognitive load, including force matching, trail making, spiral drawing and Morse code mimicry, while simultaneously using high-accuracy motor sensors to evaluate motor dysfunction (Schubert et al., 2016). The pilot phase showed that this test is applicable in HD, but additional results on its utility for symptom monitoring across the disease stages and over time are still pending.

Other cognitive measures in HD include the Stroop word reading (SWR) test and the symbol digit modalities test (SDMT), which have digital versions included in the Roche HD monitoring app with high test-retest reliability and strong correlation with in-clinic measurements. The cognitive scores from this app showed that the HD group differed significantly from the premanifest HD and healthy control groups, but there was no significant difference between the two latter (Lipsmeier et al., 2022). Similarly, 'HD-Mobile' is a smartphone application that evaluates several cognitive domains, including decision-making capacity by testing two-choice discrimination, as well as visual learning and recognition skills with the use of an object identity and location learning task. HD-Mobile also assesses psychomotor and motor control, based on the performance of speeded tapping. With the use of the HD-Mobile app, significant differences were noted between healthy controls, individuals with premanifest and early manifest HD (McLaren et al., 2021). Also, Ghilardi et al. used a tablet-based test to evaluate sequence learning in the presymptomatic stage, and showed significant impairment of implicit and explicit learning in premanifest HD compared to healthy control participants (Ghilardi et al., 2008).

The Neural Impairment Test Suite (NITS) is another mobile-based app that aims to help monitor individuals with neurodegenerative disorders and includes fourteen tasks that assess cognition, including clock drawing, verbal fluency, naming, memory, modified trails and problem-solving tasks. The NITS mobile app has shown an accuracy of more than 90% in differentiating healthy from diseased individuals with various neurological disorders (Lauraitis et al., 2020); however, only three individuals with HD were included in this study, so further studies with a larger number of

HD participants are needed to determine the utility of this app for this specific population.

The Digitalized Arithmetic Task (DAT) is an in-clinic computer-based assessment that has been proposed for the evaluation of cognition in early-stage HD. DAT discriminated between healthy controls and HD individuals and detected cognitive decline at 1-year follow-up; in addition, declining performance was associated with reduced striatal volume on neuroimaging, suggesting a correlation with the underlying pathological changes (Lunven et al., 2021). This tool could potentially be applied remotely, although further studies are needed to ensure its validity and reliability for remote use.

Finally, digital devices allow the use of computerized adaptive testing, which is a promising test delivery method for other commonly used cognitive assessments in HD, such as the Neuro-QOL quality of life measure, with potentially better reliability measures (Lai et al., 2018).

5 Sleep and Physiological Measures

Sleep Measures

Sleep problems are prevalent in individuals with HD, with up to 90% of individuals in one study reporting problems (Goodman & Barker, 2010). Sleep problems can be evident from the earliest disease stages, including in premanifest stages (Goodman et al., 2011). Sleep problems are wide ranging and include disorders related to REM sleep stage.

Several studies have demonstrated problems with sleep using polysomnography, the gold standard for sleep assessments (Zhang et al., 2019). Studies in individuals with HD have reported decreased sleep efficiency, percentage of slow wave sleep and rapid eye movement (REM) sleep; increased percentage of N1 sleep, wake time after sleep onset, motor activity during sleep, and rapid eye movement sleep latency; frequent nocturnal awakenings; and excessive daytime sleepiness (Goodman et al., 2011; Zhang et al., 2019; Arnulf et al., 2008; Wiegand et al., 1991). However, polysomnography has significant limitations, including the complexity of data collection and analysis, time requirements and expense (Herzog–Krzywoszanska & Krzywoszanski, 2019).

Advances in wearable devices have enabled assessment of sleep disturbances outside the laboratory setting in the general population as well as individuals with HD (Smith et al., 2018). Several studies have used wrist-worn actigraphy to measure sleep variables, including time in bed (TIB), total sleep time, sleep period, sleep latency, wake after sleep onset (WASO) and sleep efficiency (see Table 4). Actigraphy evaluates wakefulness and sleep based on physical activity or movement. Studies have shown that individuals with HD have more movements and increased activity during sleep compared to controls (Hurelbrink et al., 2005;

Table 4 Sleep digital measures

Sleep measure	Description
Time in Bed (TIB)	TIB is simply the interval from the time a person went to bed (bedtime) to the time the person arose (rise time)
Total Sleep Time	The amount of time that a person spends actually sleeping during a planned sleep episode. TST is the sum of all REM and NREM sleep in a sleep episode
Sleep Period	The sleep period also referred to as the true sleep period, is defined as the interval (within TIB) from sleep onset to sleep offset
Mean activity during time in bed	Average activity score or frequency of wrist movement (ZCM) per minute while a person is in bed
Sleep Latency	The length of time it takes a person to fall asleep (after turning the lights off). On average, a healthy person takes between 10 and 20 minutes to fall asleep
Wake after sleep onset	WASO refers to the number of minutes a person was awake between sleep onset and sleep offset (O–O interval)
Sleep efficiency	Sleep efficiency is the total amount of time a person was asleep on a given night divided by the total amount

Morton et al., 2005; Bartlett et al., 2019). Thus, sleep disturbances measured by actigraphy have the potential to be exaggerated or less reliable. Continued research validating actigraphy in HD will ultimately lead to improved algorithms for sleep detection.

While it is beyond the scope of this review to discuss data preprocessing in depth, it is an important component of understanding the meaningfulness of sleep data obtained from wearable devices. Wearables generate large amounts of data, which varies on the frequency of data collection. These data must be processed to identify non-wear periods and identify time in bed. For example, sleep estimation algorithms use the activity data from the wrist to determine whether an individual is awake or asleep, usually on a minute-by-minute basis (Fekedulegn et al., 2020).

In addition to actigraphy, a recent study in people with HD used a biosensor, which incorporates ECGs, accelerometers and a thermistor to measure temperature (Antikainen et al., 2022). Such sensors can provide a multi-modal assessment of sleep as well as physiological measures such as heart rate, and have potential for future research in HD.

Physiological Measures

Individuals with HD are known to have autonomic dysfunction as part of the disease manifestation (Aziz & Roos, 2011; Schultz et al., 2022). People have reported symptoms consistent with autonomic dysfunction including dizziness when standing up, excessive perspiration and tachycardia (Aziz & Roos, 2011). The underlying mechanism might be damage caused by progressive apoptosis of nerve cells in the

Table 5 Physiologic digital measures

Physiological measures	Description
Heart rate	Number of heart beats per minute
Heart Rate Variability	The fluctuation in the time intervals between adjacent heartbeat. An optimal level of HRV is associated with health and self-regulatory capacity, and adaptability or resilience. Can be described during 24 h, short-term (ST, ~5 min) or brief, and ultra-short-term (UST, <5 min) HRV using time-domain, frequency-domain, and non-linear measurements.
R to R interval	Distance between 2 consecutive R waves measures as time in milliseconds
Respiratory rate	Number of breaths per minute

central autonomic network (CAN) that performs high-level autonomic control, including the insular cortex and hypothalamus (Abildtrup & Shattock, 2013).

Measures of autonomic function that can be evaluated using wearable devices include standard physiological measures such as heart rate, heart rate variability, and respiratory rate (see Table 5) (Debnath et al., 2021; McCraty & Shaffer, 2015). These measures can be evaluated through chest worn devices (e.g. Polar® heart rate monitors) or more recently wrist worn devices with acceptable accuracy in healthy populations (Nuuttila et al., 2021; Graham et al., 2019; Hoog Antink et al., 2021). However, no studies to date have evaluated reliability or validity of heart rate or other measures using wearable devices in HD. These measures can be taken at rest or during exercise, or during activities such as squat test, cold pressor test, diving reflex test, deep breathing, and Valsalva maneuver (Debnath et al., 2021).

Studies have shown that individuals with HD have higher mean heart rate than controls during wake and sleep (Marotta et al., 2021). Several studies have also demonstrated impairments in heart rate variability, including during rest and sleep, in individuals with premanifest HD (Schultz et al., 2022; Kobal et al., 2022; Sharma et al., 1999). Reduced parasympathetic heart rate variability values have also been associated with risk of falls (Terroba-Chambi et al., 2021). While still in its infancy, use of wearables to measure physiological function has the potential to provide important insights into autonomic dysfunction in people with HD.

6 Conclusions

Traditional rating scales in movement disorders are subjective, episodic, and typically limited to in-person visits. Digital tools can provide objective, sensitive, real-world measures and an opportunity to frequently and remotely assess individuals with HD. Indeed, the disease lends itself well to digital measures, as it results in impaired mobility, activity, sleep, and abnormal movements, all of which can be measured by wearable or portable devices. These devices can capture rare and fluctuating events and uncover new details about disease. These tools have the potential to help with accurate diagnosis, disease characterization, and evaluation of disease progression and response to treatments. Moreover, studies incorporating digital measures have proven to be feasible, even in large cohorts, are well accepted by patients and research participants, and generally have high adherence rates (Silva de Lima et al., 2017; Botros et al., 2019).

Despite the advantages and opportunities for digital measures, there remain challenges in the use of DHTs in HD. There are many devices available, with variable sampling rates and algorithms, which makes comparisons difficult and limits generalizability of the findings. Furthermore, simple and reliable algorithms that can monitor more complex movements like chorea are not widely accessible. The amount of data collected by wearable sensors is immense and can be difficult to meaningfully condense or interpret, particularly data gathered in unsupervised settings, which can also be hard to validate (Fasano & Mancini, 2020; Warmerdam et al., 2020). Comparisons to "gold standard" traditional assessments are often sought, but given the subjective nature of these scales, the utility and meaning of these comparisons is unclear. While most DHTs have proven to be well-accepted and feasible to use, compliance in individuals with HD can require additional support. Finally, many studies to date have low sample size, which limits generalizability of the results. More research is needed with larger sample sizes, remote monitoring periods, and comparison to patient reported outcomes. In addition, more longitudinal studies are required, particularly in individuals with early disease where change may be slow and more easily detected by wearable devices (Adams et al., 2021).

For use as endpoints in clinical trials, digital measures should be sensitive to change. Longitudinal digital measures studies are limited in HD and more research is needed to help determine each measure's sensitivity to change. Digital measures should also be meaningful and efforts to include patient voice in digital measures studies in HD is paramount. The ability to assess patient-perceived relevance of new digital health technologies and capabilities of these measures to monitor personally meaningful symptoms is critical to future regulatory approval processes and the development of new therapeutics to treat HD (Mammen et al., 2022a, b).

Digital measures hold immense promise as objective, sensitive measures of HD. Gait, balance, activity, fine motor coordination, chorea, speech, cognition, and sleep are some of the domains that have been studied using digital technologies in HD. These measures may be obtained by active and/or passive monitoring, in clinic or remotely. Larger, longer studies are needed as are qualitative efforts to determine the meaningfulness of digital measures to individuals with HD. Ultimately, digital measures may help speed therapeutic development for this devastating disease.

Acknowledgements The authors would like to thank Dipti Wani, PhD, Maya Campbell and Tishya Ravikumar for their assistance with this chapter.

References

Abildtrup, M., & Shattock, M. (2013). Cardiac dysautonomia in Huntington's disease. *Journal of Huntington's Disease, 2*(3), 251–261. https://doi.org/10.3233/JHD-130054

Adams, J. L., Dinesh, K., Xiong, M., et al. (2017). Multiple wearable sensors in Parkinson and Huntington disease individuals: A pilot study in clinic and at home. *Digital Biomarkers, 1*(1), 52–63.

Adams, J. L., Lizarraga, K. J., Waddell, E. M., et al. (2021). Digital technology in movement disorders: Updates, applications, and challenges. *Current Neurology and Neuroscience Reports, 21*(4), 16. https://doi.org/10.1007/s11910-021-01101-6

Andrzejewski, K. L., Dowling, A. V., Stamler, D., et al. (2016). Wearable sensors in Huntington disease: A pilot study. *Journal of Huntington's Disease, 5*(2), 199–206.

Antikainen, E., Njoum, H., Kudelka, J., et al. (2022). Assessing fatigue and sleep in chronic diseases using physiological signals from wearables: A pilot study. *Frontiers in Physiology, 13*, 968185. https://doi.org/10.3389/fphys.2022.968185

Arnulf, I., Nielsen, J., Lohmann, E., et al. (2008). Rapid eye movement sleep disturbances in Huntington disease. *Archives of Neurology, 65*(4), 482–488. https://doi.org/10.1001/archneur.65.4.482

Arora, S., & Tsanas, A. (2021). Assessing Parkinson's disease at scale using telephone-recorded speech: Insights from the Parkinson's voice initiative. *Diagnostics, 11*(10), 1892. https://doi.org/10.3390/diagnostics11101892

Aziz, N. A., & Roos, R. A. C. (2011). Autonomic symptoms in Huntington's disease – current understanding and perspectives for the future. *European Neurological Review, 5*(2), 46–48. https://doi.org/10.17925/ENR.2010.05.02.46

Bartlett, D. M., Reyes, A., Zaenker, P., et al. (2019). Investigating the relationships between hypothalamic volume and measures of circadian rhythm and habitual sleep in premanifest Huntington's disease. *Neurobiology of Sleep and Circadian Rhythms, 6*, 1–8. https://doi.org/10.1016/j.nbscr.2018.07.001

Bates, G. P., Dorsey, R., Gusella, J. F., et al. (2015). Huntington disease. *Nature Reviews Disease Primers, 1*, 15005. https://doi.org/10.1038/nrdp.2015.5

Bechtel, N., Scahill, R. I., Rosas, H. D., et al. (2010). Tapping linked to function and structure in premanifest and symptomatic Huntington disease. *Neurology, 75*(24), 2150–2160. https://doi.org/10.1212/WNL.0b013e3182020123

Beckmann, H., Bohlen, S., Saft, C., et al. (2018). Objective assessment of gait and posture in premanifest and manifest Huntington disease - a multi-center study. *Gait & Posture, 62*, 451–457. https://doi.org/10.1016/j.gaitpost.2018.03.039

Biglan, K. M., Ross, C. A., Langbehn, D. R., et al. (2009). Motor abnormalities in premanifest persons with Huntington's disease: The PREDICT-HD study. *Movement Disorders, 24*(12), 1763–1772. https://doi.org/10.1002/mds.22601

Botros, A., Schütz, N., Camenzind, M., et al. (2019). Long-term home-monitoring sensor technology in patients with Parkinson's disease-acceptance and adherence. *Sensors, 19*(23), 5169. https://doi.org/10.3390/s19235169

Brüne, M., Blank, K., Witthaus, H., & Saft, C. (2011). "Theory of mind" is impaired in Huntington's disease. *Movement Disorders, 26*(4), 671–678. https://doi.org/10.1002/mds.23494

Buruma, O. J., Kemp, B., Roos, R. A., Franzen, J. M., & van der Velde, E. A. (1982). Quantification of choreatic movements by Doppler radar. *Acta Neurologica Scandinavica, 66*(3), 363–368. https://doi.org/10.1111/j.1600-0404.1982.tb06857.x

Chan, J. C. S., Stout, J. C., & Vogel, A. P. (2019). Speech in prodromal and symptomatic Huntington's disease as a model of measuring onset and progression in dominantly inherited neurodegenerative diseases. *Neuroscience and Biobehavioral Reviews, 107*, 450–460. https://doi.org/10.1016/j.neubiorev.2019.08.009

Chenery, H. J., Copland, D. A., & Murdoch, B. E. (2002). Complex language functions and subcortical mechanisms: Evidence from Huntington's disease and patients with non-thalamic subcortical lesions. *International Journal of Language & Communication Disorders, 37*(4), 459–474. https://doi.org/10.1080/1368282021000007730

Cohen, S., Waks, Z., Elm, J. J., et al. (2018). Characterizing patient compliance over six months in remote digital trials of Parkinson's and Huntington disease. *BMC Medical Informatics and Decision Making, 18*(1), 1–10.

Collins, L. M., Lazic, S. E., & Barker, R. A. (2014). A retrospective analysis of hand tapping as a longitudinal marker of disease progression in Huntington's disease. *BMC Neurology, 14*, 35. https://doi.org/10.1186/1471-2377-14-35

de Villiers, J. (2007). The interface of language and theory of mind. *Lingua, 117*(11), 1858–1878. https://doi.org/10.1016/j.lingua.2006.11.006

Debnath, S., Levy, T. J., Bellehsen, M., et al. (2021). A method to quantify autonomic nervous system function in healthy, able-bodied individuals. *Bioelectronic Medicine, 7*(1), 13. https://doi.org/10.1186/s42234-021-00075-7

Dinesh, K., Snyder, C. W., Xiong, M., et al. (2020). A longitudinal wearable sensor study in Huntington's disease. *Journal of Huntington's Disease, 9*(1), 69–81. https://doi.org/10.3233/JHD-190375

Dorsey, E. R., Papapetropoulos, S., Xiong, M., & Kieburtz, K. (2017). The first frontier: Digital biomarkers for neurodegenerative disorders. *Digital Biomarkers, 1*(1), 6–13. https://doi.org/10.1159/000477383

Fahed, V. S., Doheny, E. P., Busse, M., Hoblyn, J., & Lowery, M. M. (2022, November 12). Comparison of acoustic voice features derived from mobile devices and studio microphone recordings. *Journal of Voice*. https://doi.org/10.1016/j.jvoice.2022.10.006

Fasano, A., & Mancini, M. (2020). Wearable-based mobility monitoring: The long road ahead. *Lancet Neurology, 19*(5), 378–379. https://doi.org/10.1016/S1474-4422(20)30033-8

Fekedulegn, D., Andrew, M. E., Shi, M., Violanti, J. M., Knox, S., & Innes, K. E. (2020). Actigraphy-based assessment of sleep parameters. *Annals of Work Exposures and Health, 64*(4), 350–367. https://doi.org/10.1093/annweh/wxaa007

Gaßner, H., Jensen, D., Marxreiter, F., et al. (2020). Gait variability as digital biomarker of disease severity in Huntington's disease. *Journal of Neurology, 267*(6), 1594–1601. https://doi.org/10.1007/s00415-020-09725-3

Ghilardi, M. F., Silvestri, G., Feigin, A., et al. (2008). Implicit and explicit aspects of sequence learning in pre-symptomatic Huntington's disease. *Parkinsonism & Related Disorders, 14*(6), 457–464. https://doi.org/10.1016/j.parkreldis.2007.11.009

Goodman, A. O. G., & Barker, R. A. (2010). How vital is sleep in Huntington's disease? *Journal of Neurology, 257*(6), 882–897. https://doi.org/10.1007/s00415-010-5517-4

Goodman, A. O., Rogers, L., Pilsworth, S., et al. (2011). Asymptomatic sleep abnormalities are a common early feature in patients with Huntington's disease. *Current Neurology and Neuroscience Reports, 11*(2), 211–217.

Gordon, M. F., Grachev, I. D., Mazeh, I., et al. (2019). Quantification of motor function in Huntington disease patients using wearable sensor devices. *Digital Biomarkers, 3*(3), 103–115. https://doi.org/10.1159/000502136

Graham, S. A., Jeste, D. V., Lee, E. E., et al. (2019). Associations between heart rate variability measured with a wrist-worn sensor and older adults' physical function: Observational study. *JMIR mHealth and uHealth, 7*(10), e13757. https://doi.org/10.2196/13757

Grimbergen, Y. A. M., Knol, M. J., Bloem, B. R., Kremer, B. P. H., Roos, R. A. C., & Munneke, M. (2008). Falls and gait disturbances in Huntington's disease. *Movement Disorders, 23*(7), 970–976. https://doi.org/10.1002/mds.22003

Hartelius, L., Carlstedt, A., Ytterberg, M., Lillvik, M., & Laakso, K. (2003). Speech disorders in mild and moderate Huntington disease: Results of dysarthria assessments of 19 individuals. *Journal of Medical Speech-Language Pathology, 11*(1), 1–15.

Herzog–Krzywoszanska, R., & Krzywoszanski, L. (2019). Sleep disorders in Huntington's disease. *Frontiers in Psychiatry, 10*, 221. https://doi.org/10.3389/fpsyt.2019.00221

Hinzen, W., Rosselló, J., Morey, C., et al. (2018). A systematic linguistic profile of spontaneous narrative speech in pre-symptomatic and early stage Huntington's disease. *Cortex, 100*, 71–83. https://doi.org/10.1016/j.cortex.2017.07.022

Holm, S. P., Wolfer, A. M., Pointeau, G. H. S., Lipsmeier, F., & Lindemann, M. (2022). Practice effects in performance outcome measures in patients living with neurologic disorders - a systematic review. *Heliyon, 8*(8), e10259. https://doi.org/10.1016/j.heliyon.2022.e10259

Hoog Antink, C., Mai, Y., Peltokangas, M., Leonhardt, S., Oksala, N., & Vehkaoja, A. (2021). Accuracy of heart rate variability estimated with reflective wrist-PPG in elderly vascular patients. *Scientific Reports, 11*(1), 8123. https://doi.org/10.1038/s41598-021-87489-0

Huntington Study Group. (1996). Unified Huntington's Disease Rating Scale: Reliability and consistency. *Movement disorders, 11*(2), 136–142. https://doi.org/10.1002/mds.870110204

Hurelbrink, C. B., Lewis, S. J., & Barker, R. A. (2005). The use of the Actiwatch–Neurologica® system to objectively assess the involuntary movements and sleep–wake activity in patients with mild–moderate Huntington's disease. *Journal of Neurology, 252*(6), 642–647. https://doi.org/10.1007/s00415-005-0709-z

Kaploun, L. R., Saxman, J. H., Wasserman, P., & Marder, K. (2011). Acoustic analysis of voice and speech characteristics in presymptomatic gene carriers of Huntington's disease: Biomarkers for preclinical sign onset? *Journal of Medical Speech-Language Pathology, 19*(2), 49–65.

Kegelmeyer, D. A., Kostyk, S. K., Fritz, N. E., et al. (2017). Quantitative biomechanical assessment of trunk control in Huntington's disease reveals more impairment in static than dynamic tasks. *Journal of the Neurological Sciences, 376*, 29–34. https://doi.org/10.1016/j.jns.2017.02.054

Keren, K., Busse, M., Fritz, N. E., et al. (2021). Quantification of daily-living gait quantity and quality using a wrist-worn accelerometer in Huntington's disease. *Frontiers in Neurology, 12*, 719442. https://doi.org/10.3389/fneur.2021.719442

Kobal, J., Cankar, K., & Melik, Z. (2022). Heart rate variability in Huntington disease. A long term longitudinal study. *Neurodegenerative Diseases, 22*(1), 1–6. https://doi.org/10.1159/000525415

Kotschet, K., Osborn, S., & Horne, M. (2022). Measurement of bradykinesia and chorea in Huntington's Disease using ambulatory monitoring. *Clinical Parkinsonism & Related Disorders, 8*, 100179. https://doi.org/10.1016/j.prdoa.2022.100179

Lai, J. S., Goodnight, S., et al. (2018). Evaluating cognition in individuals with Huntington disease: Neuro-QoL cognitive functioning measures. *Quality of Life Research, 27*(3), 811–822. https://doi.org/10.1007/s11136-017-1755-6

Lang, C., Gries, C., Lindenberg, K. S., et al. (2021). Monitoring the motor phenotype in Huntington's disease by analysis of keyboard typing during real life computer use. *Journal of Huntington's Disease, 10*(2), 259–268. https://doi.org/10.3233/JHD-200451

Lauraitis, A., Maskeliūnas, R., Damaševičius, R., & Krilavičius, T. (2020). A mobile application for smart computer-aided self-administered resting of cognition, speech, and motor impairment. *Sensors, 20*(11), E3236. https://doi.org/10.3390/s20113236

Lipsmeier, F., Simillion, C., Bamdadian, A., et al. (2022). A remote digital monitoring platform to assess cognitive and motor symptoms in Huntington disease: Cross-sectional validation study. *Journal of Medical Internet Research, 24*(6), e32997. https://doi.org/10.2196/32997

Lunven, M., Hamet Bagnou, J., Youssov, K., et al. (2021). Cognitive decline in Huntington's disease in the Digitalized Arithmetic Task (DAT). *PLoS One, 16*(8), e0253064. https://doi.org/10.1371/journal.pone.0253064

Mammen, J., et al. (2022a, January 1). Mapping meaningful symptoms and impacts of early Parkinson's disease to digital outcome measures. *College of Nursing Faculty Publications*. https://digitalcommons.uri.edu/nursing_facpubs/343

Mammen, J, et al. (2022b, October 1). Relevance of digital measures to people with early Parkinson's for monitoring meaningful symptoms in the WATCH-PD study. *College of Nursing Faculty Publications*. https://digitalcommons.uri.edu/nursing_facpubs/345

Marotta, J., Piano, C., Brunetti, V., et al. (2021). Heart Rate Variability during wake and sleep in Huntington's disease patients: An observational, cross-sectional, cohort study. *Neurodegenerative Diseases, 21*(3–4), 79–86. https://doi.org/10.1159/000520754

McCraty, R., & Shaffer, F. (2015). Heart rate variability: New perspectives on physiological mechanisms, assessment of self-regulatory capacity, and health risk. *Global Advances in Health and Medicine, 4*(1), 46–61. https://doi.org/10.7453/gahmj.2014.073

McLaren, B., Andrews, S. C., Glikmann-Johnston, Y., et al. (2021). Feasibility and initial validation of 'HD-Mobile', a smartphone application for remote self-administration of performance-based cognitive measures in Huntington's disease. *Journal of Neurology, 268*(2), 590–601. https://doi.org/10.1007/s00415-020-10169-y

Mestre, T. A., Bachoud-Lévi, A. C., Marinus, J., et al. (2018). Rating scales for cognition in Huntington's disease: Critique and recommendations. *Movement Disorders, 33*(2), 187–195. https://doi.org/10.1002/mds.27227

Michell, A. W., Goodman, A. O. G., Silva, A. H. D., Lazic, S. E., Morton, A. J., & Barker, R. A. (2008). Hand tapping: A simple, reproducible, objective marker of motor dysfunction in Huntington's disease. *Journal of Neurology, 255*(8), 1145–1152. https://doi.org/10.1007/s00415-008-0859-x

Morgan-Jones, P., Jones, A., Busse, M., et al. (2022). Monitoring and managing lifestyle behaviors using wearable activity trackers: Mixed methods study of views from the Huntington disease community. *JMIR Formative Research, 6*(6), e36870. https://doi.org/10.2196/36870

Morton, A. J., Wood, N. I., Hastings, M. H., Hurelbrink, C., Barker, R. A., & Maywood, E. S. (2005). Disintegration of the sleep-wake cycle and circadian timing in Huntington's disease. *The Journal of Neuroscience, 25*(1), 157–163. https://doi.org/10.1523/jneurosci.3842-04.2005

Müller, T., Saft, C., Andrich, J., & Harati, A. (2013). Diadochokinetic movements differ between patients with Huntington's disease and controls. *NeuroRehabilitation, 33*(4), 649–655. https://doi.org/10.3233/NRE-130992

Muratori, L. M., Quinn, L., Li, X., Youdan, G., Busse, M., & Fritz, N. E. (2021). Measures of postural control and mobility during dual-tasking as candidate markers of instability in Huntington's disease. *Human Movement Science, 80*, 102881. https://doi.org/10.1016/j.humov.2021.102881

Mustafa, A. I., Corey-Bloom, J., Howell, S., Gilbert, P. E., & Goble, D. J. (2022). Differences in body sway can be identified in Huntington's disease using a practical balance assessment device. *Parkinsonism & Related Disorders, 103*, 115–118. https://doi.org/10.1016/j.parkreldis.2022.02.024

Novotný, M., Rusz, J., Čmejla, R., Růžičková, H., Klempíř, J., & Růžička, E. (2016). Hypernasality associated with basal ganglia dysfunction: Evidence from Parkinson's disease and Huntington's disease. *PeerJ, 4*, e2530. https://doi.org/10.7717/peerj.2530

Nuuttila, O. P., Korhonen, E., Laukkanen, J., & Kyröläinen, H. (2021). Validity of the wrist-worn polar vantage V2 to measure heart rate and heart rate variability at rest. *Sensors, 22*(1), 137. https://doi.org/10.3390/s22010137

Paulsen, J. S., Langbehn, D. R., Stout, J. C., et al. (2008). Detection of Huntington's disease decades before diagnosis: The Predict-HD study. *Journal of Neurology, Neurosurgery, and Psychiatry, 79*(8), 874–880. https://doi.org/10.1136/jnnp.2007.128728

Paulsen, J. S., Miller, A. C., Hayes, T., & Shaw, E. (2017). Cognitive and behavioral changes in Huntington disease before diagnosis. *Handbook of Clinical Neurology, 144*, 69–91. https://doi.org/10.1016/B978-0-12-801893-4.00006-7

Pezold, M. J., Imgrund, C. M., & Storkel, H. L. (2020). Using computer programs for language sample analysis. *Language, Speech, and Hearing Services in Schools, 51*(1), 103–114. https://doi.org/10.1044/2019_LSHSS-18-0148

Poile, V., Youdan, G., Sheeran, L., Quinn, L., & Busse, M. (2018). and the iWEARHD study management group. F64 Wearable technologies for assessment of physical activity in HD: Preliminary analysis of movement variability and wear time. *Journal of Neurology, Neurosurgery, and Psychiatry, 89*(Suppl 1), A62–A62. https://doi.org/10.1136/jnnp-2018-EHDN.165

Purcell, N. L., Goldman, J. G., Ouyang, B., Bernard, B., & O'Keefe, J. A. (2019). The effects of dual-task cognitive interference and environmental challenges on balance in Huntington's disease. *Movement Disorders Clinical Practice, 6*(3), 202–212. https://doi.org/10.1002/mdc3.12720

Quinn, L., Playle, R., Drew, C. J. G., et al. (2022). Physical activity and exercise outcomes in Huntington's disease (PACE-HD): Results of a 12-month trial-within-cohort feasibility study of a physical activity intervention in people with Huntington's disease. *Parkinsonism & Related Disorders, 101*, 75–89. https://doi.org/10.1016/j.parkreldis.2022.06.013

Rahman, W., Lee, S., Islam, M. S., et al. (2021). Detecting Parkinson disease using a web-based speech task: Observational study. *Journal of Medical Internet Research, 23*(10), e26305. https://doi.org/10.2196/26305

Reilmann, R., & Schubert, R. (2017). Motor outcome measures in Huntington disease clinical trials. *Handbook of Clinical Neurology, 144*, 209–225. https://doi.org/10.1016/B978-0-12-801893-4.00018-3

Reilmann, R., Bohlen, S., Kirsten, F., Ringelstein, E. B., & Lange, H. W. (2011). Assessment of involuntary choreatic movements in Huntington's disease--toward objective and quantitative measures. *Movement Disorders, 26*(12), 2267–2273. https://doi.org/10.1002/mds.23816

Reilmann, R., Leavitt, B. R., & Ross, C. A. (2014). Diagnostic criteria for Huntington's disease based on natural history. *Movement Disorders, 29*(11), 1335–1341. https://doi.org/10.1002/mds.26011

Reilmann, R., Rouzade-Dominguez, M. L., Saft, C., et al. (2015). A randomized, placebo-controlled trial of AFQ056 for the treatment of chorea in Huntington's disease. *Movement Disorders, 30*(3), 427–431. https://doi.org/10.1002/mds.26174

Rusz, J., Klempíř, J., Baborová, E., et al. (2013). Objective acoustic quantification of phonatory dysfunction in Huntington's disease. *PLoS One, 8*(6), e65881. https://doi.org/10.1371/journal.pone.0065881

Rusz, J., Saft, C., Schlegel, U., Hoffman, R., & Skodda, S. (2014). Phonatory dysfunction as a preclinical symptom of Huntington disease. *PLoS One, 9*(11), e113412. https://doi.org/10.1371/journal.pone.0113412

Rusz, J., Tykalova, T., Ramig, L. O., & Tripoliti, E. (2021). Guidelines for speech recording and acoustic analyses in dysarthrias of movement disorders. *Movement Disorders, 36*(4), 803–814. https://doi.org/10.1002/mds.28465

Saft, C., Schlegel, U., Hoffmann, R., & Skodda, S. (2014). I02 impaired motor speech performance in premotor stages of Huntington's disease. *Journal of Neurology, Neurosurgery, and Psychiatry, 85*(Suppl 1), A58–A58. https://doi.org/10.1136/jnnp-2014-309032.164

Schubert, R., Weber, N., & Reilmann, R. (2016). F1 Q-COG – a cognitive assessment battery for Huntington's disease. *Journal of Neurology, Neurosurgery, and Psychiatry, 87*(Suppl 1), A48–A49. https://doi.org/10.1136/jnnp-2016-314597.136

Schultz, J. L., Heinzerling, A. E., Brinker, A. N., et al. (2022). Autonomic changes in Huntington's disease correlate with altered central autonomic network connectivity. *Brain Communications, 4*(5), fcac253. https://doi.org/10.1093/braincomms/fcac253

Sharma, K. R., Romano, J. G., Ayyar, D. R., Rotta, F. T., Facca, A., & Sanchez-Ramos, J. (1999). Sympathetic skin response and heart rate variability in patients with Huntington disease. *Archives of Neurology, 56*(10), 1248–1252. https://doi.org/10.1001/archneur.56.10.1248

Silva de Lima, A. L., Hahn, T., Evers, L. J. W., et al. (2017). Feasibility of large-scale deployment of multiple wearable sensors in Parkinson's disease. *PLoS One, 12*(12), e0189161. https://doi.org/10.1371/journal.pone.0189161

Skodda, S., Schlegel, U., Hoffmann, R., & Saft, C. (2014). Impaired motor speech performance in Huntington's disease. *Journal of Neural Transmission, 121*(4), 399–407. https://doi.org/10.1007/s00702-013-1115-9

Skodda, S., Grönheit, W., Lukas, C., et al. (2016). Two different phenomena in basic motor speech performance in premanifest Huntington disease. *Neurology, 86*(14), 1329–1335. https://doi.org/10.1212/WNL.0000000000002550

Smith, M. T., McCrae, C. S., Cheung, J., et al. (2018). Use of actigraphy for the evaluation of sleep disorders and circadian rhythm sleep-wake disorders: An American Academy of Sleep

Medicine clinical practice guideline. *Journal of Clinical Sleep Medicine, 14*(7), 1231–1237. https://doi.org/10.5664/jcsm.7230

Stout, J. C., Andrews, S. C., & Glikmann-Johnston, Y. (2017). Cognitive assessment in Huntington disease clinical drug trials. *Handbook of Clinical Neurology, 144*, 227–244. https://doi.org/10.1016/B978-0-12-801893-4.00019-5

Tabrizi, S. J., Reilmann, R., Roos, R. A. C., et al. (2012). Potential endpoints for clinical trials in premanifest and early Huntington's disease in the TRACK-HD study: Analysis of 24 month observational data. *Lancet Neurology, 11*(1), 42–53. https://doi.org/10.1016/S1474-4422(11)70263-0

Tabrizi, S. J., Scahill, R. I., Owen, G., et al. (2013). Predictors of phenotypic progression and disease onset in premanifest and early-stage Huntington's disease in the TRACK-HD study: Analysis of 36-month observational data. *Lancet Neurology, 12*(7), 637–649. https://doi.org/10.1016/S1474-4422(13)70088-7

Terroba-Chambi, C., Bruno, V., Vigo, D. E., & Merello, M. (2021). Heart rate variability and falls in Huntington's disease. *Clinical Autonomic Research, 31*(2), 281–292. https://doi.org/10.1007/s10286-020-00669-2

Tortelli, R., Rodrigues, F. B., & Wild, E. J. (2021). The use of wearable/portable digital sensors in Huntington's disease: A systematic review. *Parkinsonism & Related Disorders, 83*, 93–104. https://doi.org/10.1016/j.parkreldis.2021.01.006

van Vugt, J. P. P., van Hilten, B. J., & Roos, R. A. C. (1996). Hypokinesia in Huntington's disease. *Movement Disorders, 11*(4), 384–388. https://doi.org/10.1002/mds.870110406

Vasudevan, S., Saha, A., Tarver, M. E., & Patel, B. (2022). Digital biomarkers: Convergence of digital health technologies and biomarkers. *NPJ Digital Medicine, 5*, 36. https://doi.org/10.1038/s41746-022-00583-z

Vogel, A. P., Shirbin, C., Churchyard, A. J., & Stout, J. C. (2012). Speech acoustic markers of early stage and prodromal Huntington's disease: A marker of disease onset? *Neuropsychologia, 50*(14), 3273–3278. https://doi.org/10.1016/j.neuropsychologia.2012.09.011

Waddell, E. M., Dinesh, K., Spear, K. L., et al. (2021). GEORGE®: A pilot study of a smartphone application for Huntington's disease. *Journal of Huntington's Disease, 10*(2), 293–301. https://doi.org/10.3233/JHD-200452

Warmerdam, E., Hausdorff, J. M., Atrsaei, A., et al. (2020). Long-term unsupervised mobility assessment in movement disorders. *Lancet Neurology, 19*(5), 462–470. https://doi.org/10.1016/S1474-4422(19)30397-7

Wiegand, M., Möller, A. A., Lauer, C. J., et al. (1991). Nocturnal sleep in Huntington's disease. *Journal of Neurology, 238*(4), 203–208. https://doi.org/10.1007/BF00314781

Zhang, Y., Ren, R., Yang, L., et al. (2019). Sleep in Huntington's disease: A systematic review and meta-analysis of polysomongraphic findings. *Sleep, 42*(10), 154. https://doi.org/10.1093/sleep/zsz154

Sex Differences in Huntington's Disease: Considerations for Clinical Care and Research Trials

Natalia P. Rocha, Antonio L. Teixeira, and Erin Furr Stimming

Abbreviations

ADHD	Attention deficit and hyperactivity disorder
CAP	CAG-Age-Product
CSF	Cerebrospinal fluid
HD	Huntington's disease
HDGECs	HD gene expansion carriers
HTT	Huntingtin
mHTT	Mutant huntingtin
NfL	Neurofilament light chain
tHTT	total huntingtin
UHDRS	Unified HD Rating Scale

1 Introduction

Sex is probably the most obvious distinguishing characteristic in humans. Sex differences in response to diseases have also been known for a very long time. More than 2000 years ago, Hippocrates reported that women were less affected than men by symptoms and mortality during a fever outbreak. Yet, sex differences have historically been neglected in biomedical research, and a strong rationale for studying males, but not females, was entrenched in both clinical and preclinical studies (Docherty et al., 2019). Preclinical studies have constantly omitted the sex of

N. P. Rocha (✉) · E. Furr Stimming
Department of Neurology, McGovern Medical School, The University of Texas Health Science Center, Houston, TX, USA
e-mail: npessoarocha@gmail.com

A. L. Teixeira
Department of Psychiatry and Behavioral Sciences, McGovern Medical School, University of Texas, Houston, TX, USA

E. A. Thomas, G. M. Parkin (eds.), *Biomarkers for Huntington's Disease*, Contemporary Clinical Neuroscience, https://doi.org/10.1007/978-3-031-32815-2_19

animals used or used only male animals blaming potential variabilities due to hormones and estrous cycle stages. Similarly, women have been excluded from clinical trials, with concerns that research may harm the female reproductive system or that the variability of menstrual cycle may impact the reliability of the results. The under-representation of women in clinical trials might have been accentuated by perceptions of the male body as the standard one (Merone et al., 2022).

The neglect of sex as a research variable has been addressed lately. There is the current consensus that sex differences in response to an experimental challenge can no longer be neglected. As a result, funding agencies, such as the USA National Institutes of Health and the Swedish Research Council, currently require consideration of sex and/or gender in all applications (Docherty et al., 2019). In addition, the FDA recommends that clinical studies include men and women "in numbers appropriate to allow the detection of clinically significant gender differences in drug responses" (US Food and Drugs Administration, 1993). These guidelines reversed the ban from 1977 when the recommendation was to exclude women of child-bearing potential from Phase I and early Phase II clinical research (US Food and Drugs Administration, 1977) in response to the Thalidomide disaster in the 1960s.

In neuroscience, accumulating evidence demonstrates significant sex differences in brain anatomy, physiology, and behavior from early neurodevelopmental stages to senescence. Earlier in life, females are more likely to suffer from depression and anxiety, while males are at higher risk for autism, attention deficit and hyperactivity disorder (ADHD), and schizophrenia. Sex differences are also observed in different phenotypes of the same disease, e.g., males with ADHD tend to exhibit the hyperactive-impulsive subtype while females tend to show the inattentive subtype. Sex-related differences are also important later in life, especially in the manifestation of neurodegenerative diseases. Male sex is a significant risk factor for Parkinson's disease, whereas females are more susceptible to Alzheimer's disease (Pinares-Garcia et al., 2018). The clinical environment replicates the research scenario: despite the well-established understanding that males and females differ in their propensity to develop neuropsychiatric disorders or on how they manifest such conditions, sex is rarely considered when making diagnostic or treatment decisions.

In this chapter, we will discuss sex-related differences in the presentation and progression of Huntington's disease (HD), their impact on biomarkers, and how this emerging knowledge can influence clinical practice and research. We used "sex" to refer to an individual's designation based on genital phenotype at birth, in which most individuals are categorized as males or females. While we recognize that sex is not strictly binary, we believe using the binary classification of sex to examine group-level differences is meaningful. An important limitation of the studies discussed in this chapter is that they rely on self-identification, and the typical anatomical and sex chromosome combinations used to define sexes are not usually confirmed. Lastly, biological sex differs from gender, a culturally defined and flexible concept. An individual's gender does not always align with biological sex (DeCasien et al., 2022). This is a complex matter, and it is worth investigating. Still, herein we will not discuss gender effects on HD but will focus on discussing biological sex as an important variable to be considered in HD research and clinical care.

2 Sex-Related Clinical Features in HD

Sex differences have not been well explored by studies describing the natural history of HD. This is partly because of the disease's autosomal dominant nature with equal penetrance and prevalence in both sexes (Hasnain et al., 2022; Meoni et al., 2020). However, a recent large American cohort study reported a slightly higher HD prevalence in females compared to male individuals (age-adjusted diagnostic frequency of 7.05 and 6.10 per 100,000 persons, respectively) (Bruzelius et al., 2019). Sex has also been known for its role in disease anticipation, i.e., expansion of CAG repeats that occur primarily in male intergenerational transmission due to instability in spermatogenesis. Accordingly, large expansions occur almost exclusively through paternal transmission in intergenerational CAG repeat length changes. At the same time, the offspring of affected mothers are more likely to show no change or even contractions in CAG length (Aziz et al., 2011).

One of the first studies (if not the first) highlighting phenotypic sex differences in HD gene expansion carriers (HDGECs) was published in 1969, describing more frequent aggressive behavior in males compared to females and more depression in females compared to males (Tamir et al., 1969). The next publication found on the topic is from 1999, when a study reported that female patients with HD have a longer duration of disease (based on age at onset) than males (Foroud et al., 1999). Later, a Serbian group found a longer survival (i.e., years of life after HD onset) in females compared to male patients with HD (Pekmezovic et al., 2007). Therefore, it is possible that HD progresses slower in female. Nevertheless, recent studies have not found significant sex differences in age at death in individuals with HD (Solberg et al., 2018). One may argue that the more prolonged disease duration/survival in females compared to male HDGECs is influenced by sex differences in HD clinical onset. Studies found no differences between male and female HDGECs regarding the time of HD clinical diagnosis when controlling for the CAG length or the CAG-Age-Product (CAP) score (Glidden et al., 2020; Hentosh et al., 2021).

It is well known that the CAG repeat length highly influences the age of HD clinical onset, but other genetic factors contribute to determining age at onset in HD. Sex-specific differences in genotypes and allele frequencies may influence age at onset differently in male and female patients. For example, a single nucleotide polymorphism in the *PPARGC1A* gene was associated with an earlier motor onset in male but not female HDGECs (Weydt et al., 2014). In addition, males with the apolipoprotein E genotype ε2ε3 have a significantly earlier onset than females of the same genotype than males with other genotypes. However, the average pathological CAG repeat length was slightly higher among ε2ε3 males than among females of the same genotype, which implies that their differences in age of onset may be explicable in terms of different CAG repeat length (Kehoe et al., 1999).

A series of studies by Zielonka et al. somehow contradicted the earlier findings by showing that females generally present with more severe symptoms and faster progression than male HDGECs. First, these authors described significant correlations between the number of CAG repeats in the expanded allele and Unified HD

Rating Scale (UHDRS) motor, cognitive, functional, and independence subscores in female (but not male) individuals. In addition, they reported a stronger impact of CAG repeats on the rate of disease progression in females compared to males (Zielonka et al., 2008). These results were confirmed by a subsequent study based on a much larger cohort from the REGISTRY project of the European Huntington's Disease Network. Female individuals presented with a more severe clinical phenotype and the longitudinal analyses showed a progression faster in females than males, especially in motor and functional domains (Zielonka et al., 2013). Next, they demonstrated that motor symptoms have the highest impact on function and independence in HD (followed by cognitive and behavioral symptoms) and that these symptoms affect females' function and independence more than males (Zielonka et al., 2018). In summary, not only the motor dysfunction was worse in women than in men, but also the motor domain affected function and independence in women more than in men, thus suggesting that the motor symptoms of HD have a significantly stronger impact on function in women. More recently, one study based on ENROLL-HD data confirmed that females have a more severe disease phenotype than male HDGECs. While both sexes showed gradual motor worsening over the course of four annual visits, female patients with HD consistently presented with more severe symptoms than males. Males also tended to perform better in the cognitive tests than female patients with HD (Hentosh et al., 2021). Importantly, males present higher health-related quality of life and lower societal costs than female HDGECs (Hawton et al., 2019; van Walsem et al., 2022). These differences may be explained by males being more and longer in the work force, in addition to living at home rather than at institutions, while females are more likely to be institutionalized due to less access to informal care (van Walsem et al., 2022).

Males and females may also manifest the disease differently, resulting in diverse clinical phenotypes. Analyses of the REGISTRY, PREDICT-HD, and ENROLL-HD data showed that while depression is more prevalent and severe in female HDGECs (Hentosh et al., 2021; Zielonka et al., 2013; Epping et al., 2013), history of alcohol abuse and smoking is more prevalent in males (Hentosh et al., 2021; Zielonka et al., 2013). In addition, female sex was described as a significant predictor of depression among HDGECs even when other confounders are considered (Rocha et al., 2022). Of note, sex does not seem to be associated with the risk of death by suicide among HDGECs (Alothman et al., 2022). On the other hand, higher rates of aggressive behavior were reported in males compared to female HDGECs (Tyler et al., 1983; Jensen et al., 1998). Accordingly, overall crime rates were significantly increased in male patients with HD compared to first-degree relatives and controls. Crime rates were not increased in female patients (Jensen et al., 1998). Figure 1 summarizes the main clinical differences between male and female HDGECs. Notably, there are no studies on cross-cultural differences in clinical presentations in HD, and the (scarce) findings above came from Westernized societies and may not be generalizable across the global HD population. The findings in HD align with sociocultural differences in mental health described in the general (Western) population. In this regard, men and women have different roles and societal expectations, with pronounced divisions in power, responsibilities, and autonomy. As a result, men tend to

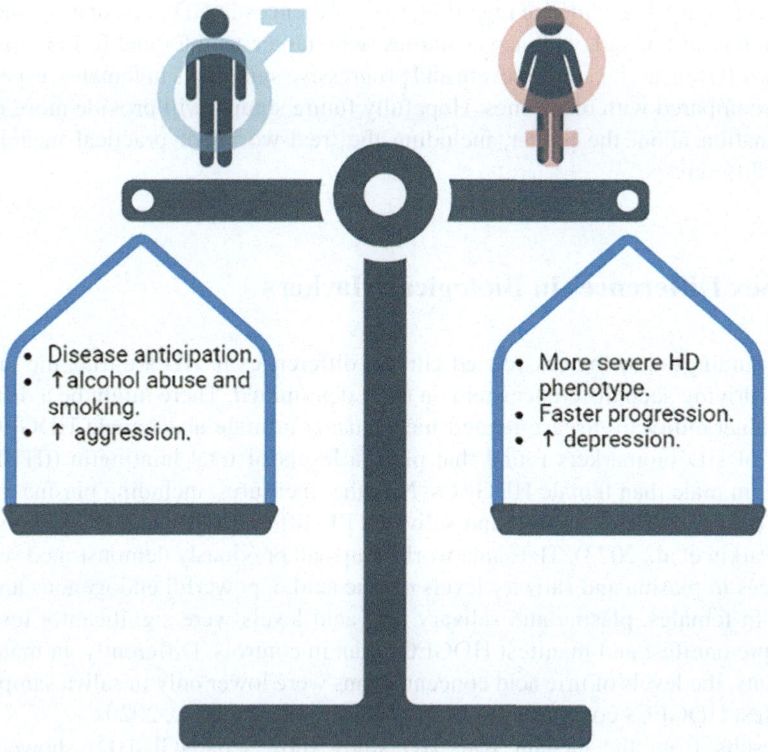

Fig. 1 Sex-related clinical features in HD

externalize stress, while women tend to internalize it, resulting in different clinical presentations of psychiatric diseases. While mental distress in females usually manifests as internalizing behaviors (e.g., depression, anxiety), males tend to exhibit more externalizing behaviors (e.g., substance abuse and aggressive behavior) (Rosenfield, 1999).

It is worth mentioning that, contradicting the reports above, several studies have failed to show any sex-related effect on HD progression or clinical phenotype. For instance, in an Italian cohort of 127 patients with HD, males and females presented a similar risk of progressing to a more severe stage in a 3-year follow-up (Volpi et al., 2021). In addition, a well-designed study conducted by the Huntington Study Group revealed that sex did not affect overall function, motor, cognitive, and behavior in a cohort of 960 patients with HD followed prospectively for a mean of 18.3 months at 43 sites (Marder et al., 2000). Also, a cross-sectional analysis of the REGISTRY data did not find sex as a significant predictor of depressive symptoms, anxiety, or general distress in the HD population (Dale et al., 2016). Finally, some older studies reported relatively similar rates of aggression in male and female HDGECs (Pflanz et al., 1991; Shiwach & Patel, 1993; Craufurd et al., 2001), reviewed in (Fisher et al., 2014).

In summary, the evidence regarding sex-differences in HD presentation and progression is still mixed, with newer studies (with larger samples and follow-up) indicating differences, i.e., more severe and progressive symptoms in females, especially when compared with older ones. Hopefully future studies will provide more robust information about the matter, including the 'real-world' or practical meaning of these differences.

3 Sex Differences in Biological Markers

If the findings showing sex-related clinical differences in HD are true, the mechanisms driving such differences remain to be determined. There might be a differential vulnerability to disease-related mechanisms in male and female HDGECs. A study of HD biomarkers found that plasma levels of total huntingtin (tHTT) are higher in male than female HDGECs. No other measures, including plasma mutant HTT (mHTT), saliva mHTT, and saliva tHTT differed significantly by individual sex (Parkin et al., 2023). The same workgroup had previously demonstrated sex differences in plasma and salivary levels of uric acid, a powerful endogenous antioxidant. In females, plasma and salivary uric acid levels were significantly lower in both premanifest and manifest HDGECs than in controls. Differently, in male participants, the levels of uric acid concentrations were lower only in saliva samples of manifest HDGECs compared to controls (Corey-Bloom et al., 2020).

Results from the elegant Kids-HD study (now ChANGE-HD) showed that increasing the length of CAG repeats within the huntingtin gene (*HTT*) below disease threshold (15–34 repeats) confers advantageous changes in both structure and function of the developing brain (children 6–18 years old) in a sex-specific manner. Among girls, higher repeats strongly predicted cerebral cortex thickness, and the repeat length also predicted better measures of global cognitive ability. Longer CAG repeats did not relate to cortical thickness in males as it did in the females. Whole-brain mapping correlations with repeat length showed fewer relationships in the males, and when present, they were most often in the opposite direction (higher CAG repeats associated with thinner cortex). With regard to the functional measures, repeat length did not significantly predict any cognitive measure in the male sample (Lee et al., 2017).

Regarding the pathological effects, one study investigated whether sex differences in the clinical worsening of HDGECs (i.e., a more severe phenotype and faster progression in females compared to male HDGECs) result from a differential impact of the underlying neurodegenerative processes on brain atrophy. The researchers measured the cerebrospinal fluid (CSF) levels of neurofilament light chain (NfL) and brain atrophy by magnetic resonance imaging as neurodegeneration markers. They found a significant NfL-by-sex interaction on brain atrophy and clinical severity. While the NfL CSF levels were not significantly increased in the female group compared to males, the correlation between increasing NfL levels, brain atrophy, and more severe clinical symptoms was more pronounced in females than in male HDGECs. These findings show that females are more vulnerable to neurodegeneration as measured by NfL than males. In other words, the same amount of measured NfL reflects more damage in females than males. Therefore, sex should be considered when interpretating markers of neurodegeneration in HDGECs. Importantly, clinical trials using NfL levels for disease monitoring should consider different reference limits for males and females in HD clinical trials (Sampedro et al., 2021).

The sex differences in HD-related markers are summarized in Box 1.

Box 1 Sex Differences in HD-Related Biomarkers

Biological marker	Methods	Main results	References
HTT levels. 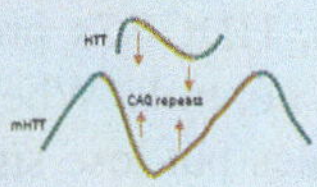	Quantification of total HTT (tHTT) and mutant HTT (mHTT) levels in plasma and saliva samples of HDGECs and controls.	Plasma levels of tHTT were significantly associated with sex, being higher in male than female HDGECs. No other measures (plasma tHTT and saliva/plasma mHTT) differed significantly by sex.	(Parkin, Corey-Bloom et al. 2023).
CAG repeat length below disease threshold (15-34 repeats) 	Evaluation of brain structure using MRI and brain function using cognitive tests in children aged 6 - 18.	Increasing the length of CAG repeats below disease threshold was associated with advantageous changes in both structure and function of the developing brain in a sex-specific manner (beneficial effects especially observed in females).	(Lee, Ding et al. 2017).
NfL and brain atrophy 	NfL levels in the CSF and brain atrophy by MRI as markers of neurodegeneration.	Females are more vulnerable to neurodegeneration as measured by NfL than males. While the NfL CSF levels were not significantly increased in the female group compared to males, the correlation between increasing NfL levels, brain atrophy, and more severe clinical symptoms was more pronounced in females than in male HDGECs.	(Sampedro, Martinez-Horta et al. 2021).

Male and female differential vulnerability to neurodegeneration is well-known in other neurodegenerative diseases with clear sex-related differences in prevalence. Alzheimer's disease is twice as frequent in females than in males. Conversely, Parkinson's disease and amyotrophic lateral sclerosis affect primarily male individuals (ratios 1.5:1 and 2:1, respectively); while spinal and bulbar muscular atrophy affect men only (Vegeto et al., 2020). In all these cases, sex hormones are thought to play an important role as either a risk or a protective factor depending on the disease. Studies on sex hormones in the HD population are very limited. There is evidence of differential effects of testosterone in male and female HDGECs. Male HDGECs present lower plasma testosterone levels than healthy controls. Decreased testosterone levels were associated with higher disease severity. In addition, low levels of testosterone were associated with cognitive impairment, but not with depression or psychotic features (Markianos et al., 2005). The testosterone protective effect was not observed in female HDGECs. There were no significant differences in hormone levels among patients with HD, HDGECs at risk, and controls, either in premenopausal or in postmenopausal states, and testosterone levels were not associated with disease severity or clinical symptoms. The subgroup of female patients with HD with depression had significantly lower testosterone levels compared to patients without depression, or to controls (Markianos et al., 2007). Studies about the association between HD clinical features and lifetime estrogen exposure/hormone replacement therapy in female HDGECs are lacking, mainly because of the young age of onset of the disease (Meoni et al., 2020).

As seen in other peripheral manifestations of HD, it is possible that impaired levels of testosterone might reflect direct dysfunction of testis and/or adrenal, an issue that deserves further investigation. Males and females also present differences in body composition that can affect drugs' pharmacokinetics and pharmacodynamics. Overall, females tend to be less affected than male HDGECs regarding body composition. For example, although females have lower bone mineral density than males, the differences against healthy controls are more pronounced in males than in female HDGECs (Goodman & Barker, 2011). In addition, HDGECs present lower lean body mass and fat mass (especially truncal fat) than controls. When the results are stratified by sex, all these differences remain in males but lose significance when female HDGECs were compared to female controls. Males also have lower body weight and body mass index than females with HD (Costa de Miranda et al., 2019). These results may be explained by evidence from preclinical studies showing that mHTT induces musculoskeletal and brain growth impairment, specifically in male BACHD rats, by disturbing the transcription of testosterone biosynthesis-related genes (Yu-Taeger et al., 2022). The reduced body weight and body mass index occur in HDGECs (Goodman & Barker, 2011) and animal models (Yu-Taeger et al., 2022) before the manifestation of neurological symptoms, suggesting that at least in the early disease stage, these changes are due to abnormal tissue growth rather than tissue atrophy. However, another publication from the ChANGE-HD study showed that children who carry the *HTT* expansion have lower intracranial, cerebrum, and cerebellum volumes than unaffected peers, but boys have higher volumes than girls (both HDGECs and controls). In contrast, in areas of primary pathologic importance—the striatum and the globus pallidus—there were no detectable sex-specific effects (van der Plas et al., 2019).

Sex differences in brain structure and brain structure–function relationships are robust and well-documented in the general population (Lee et al., 2017). Male individuals have larger brains than females (Williams et al., 2021a), which remains significant when adjusted for body size (Williams et al., 2021b). Two-thirds (409/620) of all brain measures show significant sex differences (36% larger in males, 29% larger in females) once the total cerebral measure is properly adjusted (Williams et al., 2021a, b). Because of that, sex has been considered an important confounder by studies measuring brain volume in HDGECs, such as IMAGE-HD (Dominguez et al., 2013, 2016), PREDICT-HD (Paulsen et al., 2010, 2014), and TRACK/TRACKON-HD (Tabrizi et al., 2009, 2011, 2012, 2013). Although these studies have properly controlled the results for sex, they did not report sex differences or how male and female HDGECs differ regarding brain atrophy in different brain regions. This is especially relevant when brain measures are used as biological markers. For example, putamen and caudate volumes are proposed as markers of stage progression (biomarkers of HD pathogenesis) in the Huntington's disease Integrated Staging System (Tabrizi et al., 2022). While the landmark cutoffs are described as age-dependent, sex differences should also be considered.

4 Sex-Related Differences in Response to Treatments

As well-noted in a systematic review by Meoni et al., no studies have specifically addressed sex differences in the pharmacological treatment for HD symptoms (Meoni et al., 2020). So far, the studies investigating the safety and efficacy of the dopamine receptor antagonist tetrabenazine to treat HD-related chorea showed similar results for males and females receiving the drug (Huntington Study Group, 2006; Frank et al., 2008; Frank, 2009). In addition, no sex differences were reported by a study analyzing the Enroll-HD to compare the efficacy of risperidone and olanzapine versus tetrabenazine for chorea management (Schultz et al., 2019). The study of deutetrabenazine (an analog of tetrabenazine with improved pharmacokinetic properties) did not report any analysis based on sex/gender (Huntington Study Group et al., 2016). Similarly, result reports of recently completed/terminated disease-modifying trials for HD, including the HTT lowering therapies (Ionis/Roche and Wave Life Sciences antisense oligonucleotides programs), did not mention any sex/gender difference analysis.

Given the sex differences in HD progression, phenotype, and biological markers noted above, the analysis of putative sex differences in clinical trials is of major relevance. Outcomes either related to disease signs/symptoms or biological markers (e.g. brain volume, CSF, or blood markers) can be strongly influenced by sex. In addition, the presence of comorbidities and the use of other medications that can influence a trial drug's safety/efficacy differ markedly between males and females. For instance, results from the PREDICT-HD study show that at early disease stages (prodromal HD), females are more likely to have anxiolytics and antidepressants prescribed than male HDGECs (Rowe et al., 2012). However, REGISTRY data show that although the use of anxiolytics is still more common in females later in the disease course (i.e., HDGECs with a clinical diagnosis of HD), the use of antidepressants is similar between male and female patients with HD (Dale et al., 2016). These differences suggest increased rates of antidepressant prescriptions for males as HD progresses (Meoni et al., 2020). Lastly, males and females may require different doses or treatment durations, as demonstrated by a preclinical study in which both male and female zQ175 HD mice had reduced HD neuropathology following the treatment with a mGluR5 negative allosteric modulator; however, sex had a clear impact on the efficacy of the treatment with females requiring longer treatment duration than males, and showing different results on the behavioral tests (Li et al., 2022).

Other considerations for clinical trials include sex differences in test performance and tolerability to trial procedures. In an analysis of the Q-Motor, a suite of motor tests originally designed to assess motor symptoms in HD, males' performance was associated with increased tapping speed and less irregularity, while females presented less involuntary movement (Teismann et al., 2022). Regarding tolerability, an analysis of 669 lumbar punctures from 497 HD Clarity participants revealed that younger age and female sex were associated with a higher risk of developing adverse events such as headache and back pain (Hassan et al., 2022). Another study, however, failed to show any association between sex and the rate of post-lumbar puncture headache (Rodrigues et al., 2022).

In summary, data on sex differences in response to treatments in HD are still scarce. Current and future trials should not only include equally males and females in the trials but also consider sex an important biological variable, thus reporting the differences between males and females regarding the response to treatments.

5 Final Remarks

Sex differences have been historically neglected by studies describing the natural history of HD. This picture is slowly changing by reports of relevant sex differences in HD progression and phenotype. These differences, however, are not well-understood; and further studies are needed as they could shed light on disease mechanisms and pathogenesis (Zielonka & Stawinska-Witoszynska, 2020). What we do know is that males and females have differential vulnerability to the disease pathological processes and well-established sex differences in HD should have important implications for the modeling, design, and interpretation of clinical trials (Sampedro et al., 2021). Researchers should pursue studies that consider sex as a biological variable and sex should be taken into consideration when interpreting trial outcomes and biomarkers of neurodegeneration in HD.

References

Alothman, D., Marshall, C. R., Tyrrell, E., Lewis, S., Card, T., & Fogarty, A. (2022). Risk of mortality from suicide in patients with Huntington's disease is increased compared to the general population in England. *Journal of Neurology, 269*(8), 4436–4439.

Aziz, N. A., van Belzen, M. J., Coops, I. D., Belfroid, R. D., & Roos, R. A. (2011). Parent-of-origin differences of mutant HTT CAG repeat instability in Huntington's disease. *European Journal of Medical Genetics, 54*(4), e413–e418.

Bruzelius, E., Scarpa, J., Zhao, Y., Basu, S., Faghmous, J. H., & Baum, A. (2019). Huntington's disease in the United States: Variation by demographic and socioeconomic factors. *Movement Disorders, 34*(6), 858–865.

Corey-Bloom, J., Haque, A., Aboufadel, S., Snell, C., Fischer, R. S., Granger, S. W., et al. (2020). Uric acid as a potential peripheral biomarker for disease features in Huntington's patients. *Frontiers in Neuroscience, 14*, 73.

Costa de Miranda, R., Di Lorenzo, N., Andreoli, A., Romano, L., De Santis, G. L., Gualtieri, P., et al. (2019). Body composition and bone mineral density in Huntington's disease. *Nutrition, 59*, 145–149.

Craufurd, D., Thompson, J. C., & Snowden, J. S. (2001). Behavioral changes in Huntington Disease. *Neuropsychiatry, Neuropsychology, and Behavioral Neurology, 14*(4), 219–226.

Dale, M., Maltby, J., Shimozaki, S., Cramp, R., Rickards, H., & Network RIotEHsD. (2016). Disease stage, but not sex, predicts depression and psychological distress in Huntington's disease: A European population study. *Journal of Psychosomatic Research, 80*, 17–22.

DeCasien, A. R., Guma, E., Liu, S., & Raznahan, A. (2022). Sex differences in the human brain: A roadmap for more careful analysis and interpretation of a biological reality. *Biology of Sex Differences, 13*(1), 43.

Docherty, J. R., Stanford, S. C., Panattieri, R. A., Alexander, S. P. H., Cirino, G., George, C. H., et al. (2019). Sex: A change in our guidelines to authors to ensure that this is no longer an ignored experimental variable. *British Journal of Pharmacology, 176*(21), 4081–4086.

Dominguez, D. J., Egan, G. F., Gray, M. A., Poudel, G. R., Churchyard, A., Chua, P., et al. (2013). Multi-modal neuroimaging in premanifest and early Huntington's disease: 18 month longitudinal data from the IMAGE-HD study. *PLoS One, 8*(9), e74131.

Domínguez, J. F., Stout, J. C., Poudel, G., Churchyard, A., Chua, P., Egan, G. F., et al. (2016). Multimodal imaging biomarkers in premanifest and early Huntington's disease: 30-month IMAGE-HD data. *The British Journal of Psychiatry, 208*(6), 571–578.

Epping, E. A., Mills, J. A., Beglinger, L. J., Fiedorowicz, J. G., Craufurd, D., Smith, M. M., et al. (2013). Characterization of depression in prodromal Huntington disease in the neurobiological predictors of HD (PREDICT-HD) study. *Journal of Psychiatric Research, 47*(10), 1423–1431.

Fisher, C. A., Sewell, K., Brown, A., & Churchyard, A. (2014). Aggression in Huntington's disease: A systematic review of rates of aggression and treatment methods. *Journal of Huntingtons Disease, 3*(4), 319–332.

Foroud, T., Gray, J., Ivashina, J., & Conneally, P. M. (1999). Differences in duration of Huntington's disease based on age at onset. *Journal of Neurology, Neurosurgery, and Psychiatry, 66*(1), 52–56.

Frank, S. (2009). Tetrabenazine as anti-chorea therapy in Huntington disease: An open-label continuation study. Huntington Study Group/TETRA-HD Investigators. *BMC Neurology, 9*, 62.

Frank, S., Ondo, W., Fahn, S., Hunter, C., Oakes, D., Plumb, S., et al. (2008). A study of chorea after tetrabenazine withdrawal in patients with Huntington disease. *Clinical Neuropharmacology, 31*(3), 127–133.

Glidden, A. M., Luebbe, E. A., Elson, M. J., Goldenthal, S. B., Snyder, C. W., Zizzi, C. E., et al. (2020). Patient-reported impact of symptoms in Huntington disease. *Neurology*. https://doi.org/10.1212/WNL.0000000000008906-10.1212/WNL

Goodman, A. O., & Barker, R. A. (2011). Body composition in premanifest Huntington's disease reveals lower bone density compared to controls. *PLoS Currents, 3*, RRN1214.

Hasnain, N., Arif, T. B., Shafaut, R., Zakaria, F., Fatima, S. Z., & Haque, I. U. (2022). Association between sex and Huntington's disease: An updated review on symptomatology and prognosis of neurodegenerative disorders. *Wiener Medizinische Wochenschrift (1946)*.

Hassan, Y. R., Brogueira Rodrigues, F., Zeun, P., Byrne, L. M., Estevez-Fraga, C., Tortelli, R., et al. (2022). Lumbar puncture safety and tolerability in premanifest and manifest Huntington's disease: A multi-analysis cross-sectional study. *Scientific Reports, 12*(1), 18377.

Hawton, A., Green, C., Goodwin, E., & Harrower, T. (2019). Health state utility values (QALY weights) for Huntington's disease: An analysis of data from the European Huntington's Disease Network (EHDN). *The European Journal of Health Economics, 20*(9), 1335–1347.

Hentosh, S., Zhu, L., Patino, J., Furr, J. W., Rocha, N. P., & Furr, S. E. (2021). Sex differences in Huntington's disease: Evaluating the enroll-HD database. *Movement Disorders Clinical Practice, 8*(3), 420–426.

Huntington Study Group. (2006). Tetrabenazine as antichorea therapy in Huntington disease: A randomized controlled trial. *Neurology, 66*(3), 366–372.

Huntington Study Group, Frank, S., Testa, C. M., Stamler, D., Kayson, E., Davis, C., et al. (2016). Effect of deutetrabenazine on chorea among patients with Huntington disease: A randomized clinical trial. *JAMA, 316*(1), 40–50.

Jensen, P., Fenger, K., Bolwig, T. G., & Sorensen, S. A. (1998). Crime in Huntington's disease: A study of registered offences among patients, relatives, and controls. *Journal of Neurology, Neurosurgery, and Psychiatry, 65*(4), 467–471.

Kehoe, P., Krawczak, M., Harper, P. S., Owen, M. J., & Jones, A. L. (1999). Age of onset in Huntington disease: Sex specific influence of apolipoprotein E genotype and normal CAG repeat length. *Journal of Medical Genetics, 36*(2), 108–111.

Lee, J. K., Ding, Y., Conrad, A. L., Cattaneo, E., Epping, E., Mathews, K., et al. (2017). Sex-specific effects of the Huntington gene on normal neurodevelopment. *Journal of Neuroscience Research, 95*(1–2), 398–408.

Li, S. H., Colson, T. L., Abd-Elrahman, K. S., & Ferguson, S. S. G. (2022). Metabotropic glutamate receptor 5 antagonism reduces pathology and differentially improves symptoms in male and female heterozygous zQ175 Huntington's mice. *Frontiers in Molecular Neuroscience, 15*, 801757.

Marder, K., Zhao, H., Myers, R. H., Cudkowicz, M., Kayson, E., Kieburtz, K., et al. (2000). Rate of functional decline in Huntington's disease. *Huntington Study Group Neurology, 54*(2), 452–458.

Markianos, M., Panas, M., Kalfakis, N., & Vassilopoulos, D. (2005). Plasma testosterone in male patients with Huntington's disease: Relations to severity of illness and dementia. *Annals of Neurology, 57*(4), 520–525.

Markianos, M., Panas, M., Kalfakis, N., & Vassilopoulos, D. (2007). Plasma testosterone, dehydroepiandrosterone sulfate, and cortisol in female patients with Huntington's disease. *Neuro Endocrinology Letters, 28*(2), 199–203.

Meoni, S., Macerollo, A., & Moro, E. (2020). Sex differences in movement disorders. *Nature Reviews Neurology, 16*(2), 84–96.

Merone, L., Tsey, K., Russell, D., & Nagle, C. (2022). Mind the gap: Reporting and analysis of sex and gender in health research in Australia, a cross-sectional study. *Women's Health Reports (New Rochelle), 3*(1), 759–767.

Parkin, G. M., Corey-Bloom, J., Snell, C., Smith, H., Laurenza, A., Daldin, M., et al. (2023). Salivary Huntingtin protein is uniquely associated with clinical features of Huntington's disease. *Scientific Reports, 13*(1), 1034.

Paulsen, J. S., Nopoulos, P. C., Aylward, E., Ross, C. A., Johnson, H., Magnotta, V. A., et al. (2010). Striatal and white matter predictors of estimated diagnosis for Huntington disease. *Brain Research Bulletin, 82*(3–4), 201–207.

Paulsen, J. S., Long, J. D., Ross, C. A., Harrington, D. L., Erwin, C. J., Williams, J. K., et al. (2014). Prediction of manifest Huntington's disease with clinical and imaging measures: A prospective observational study. *Lancet Neurology, 13*(12), 1193–1201.

Pekmezovic, T., Svetel, M., Maric, J., Dujmovic-Basuroski, I., Dragasevic, N., Keckarevic, M., et al. (2007). Survival of Huntington's disease patients in Serbia: Longer survival in female patients. *European Journal of Epidemiology, 22*(8), 523–526.

Pflanz, S., Besson, J. A., Ebmeier, K. P., & Simpson, S. (1991). The clinical manifestation of mental disorder in Huntington's disease: A retrospective case record study of disease progression. *Acta Psychiatrica Scandinavica, 83*(1), 53–60.

Pinares-Garcia, P., Stratikopoulos, M., Zagato, A., Loke, H., & Lee, J. (2018). Sex: A significant risk factor for neurodevelopmental and neurodegenerative disorders. *Brain Sciences, 8*(8).

Rocha, N. P., Tuazon, M. R., Patino, J., Furr Stimming, E., & Teixeira, A. L. (2022). Clinical correlates of depression and suicidality in Huntington disease: An analysis of the enroll-HD observational study. *Cognitive and Behavioral Neurology, 35*(2), 85–94.

Rodrigues, F. B., Owen, G., Sathe, S., Pak, E., Kaur, D., Ehrhardt, A. G., et al. (2022). Safety and feasibility of research lumbar puncture in Huntington's disease: The HDClarity cohort and bioresource. *Journal of Huntingtons Disease, 11*(1), 59–69.

Rosenfield, S. (1999). Gender and mental health: Do women have more psychopathology, men more, or both the same (and why)? In A. V. Horwitz & T. L. Scheid (Eds.), *A handbook for the study of mental health: Social contexts, theories, and systems* (pp. 348–360). Cambridge University Press.

Rowe, K. C., Paulsen, J. S., Langbehn, D. R., Wang, C., Mills, J., Beglinger, L. J., et al. (2012). Patterns of serotonergic antidepressant usage in prodromal Huntington disease. *Psychiatry Research, 196*(2–3), 309–314.

Sampedro, F., Martinez-Horta, S., Perez-Perez, J., Perez-Gonzalez, R., Horta-Barba, A., Campolongo, A., et al. (2021). Interaction between sex and neurofilament light chain on brain structure and clinical severity in Huntington's disease. *Annals of Clinical Translational Neurology, 8*(12), 2309–2313.

Schultz, J. L., Kamholz, J. A., Nopoulos, P. C., & Killoran, A. (2019). Comparing risperidone and olanzapine to tetrabenazine for the management of chorea in Huntington disease: An analysis from the enroll-HD database. *Movement Disorders Clinical Practice, 6*(2), 132–138.

Shiwach, R. S., & Patel, V. (1993). Aggressive behaviour in Huntington's disease: A cross-sectional study in a nursing home population. *Behavioural Neurology, 6*(1), 43–47.

Solberg, O. K., Filkukova, P., Frich, J. C., & Feragen, K. J. B. (2018). Age at death and causes of death in patients with Huntington disease in Norway in 1986-2015. *Journal of Huntingtons Disease, 7*(1), 77–86.

Tabrizi, S. J., Langbehn, D. R., Leavitt, B. R., Roos, R. A., Durr, A., Craufurd, D., et al. (2009). Biological and clinical manifestations of Huntington's disease in the longitudinal TRACK-HD study: Cross-sectional analysis of baseline data. *Lancet Neurology, 8*(9), 791–801.

Tabrizi, S. J., Scahill, R. I., Durr, A., Roos, R. A., Leavitt, B. R., Jones, R., et al. (2011). Biological and clinical changes in premanifest and early stage Huntington's disease in the TRACK-HD study: The 12-month longitudinal analysis. *Lancet Neurology, 10*(1), 31–42.

Tabrizi, S. J., Reilmann, R., Roos, R. A., Durr, A., Leavitt, B., Owen, G., et al. (2012). Potential endpoints for clinical trials in premanifest and early Huntington's disease in the TRACK-HD study: Analysis of 24 month observational data. *Lancet Neurology, 11*(1), 42–53.

Tabrizi, S. J., Scahill, R. I., Owen, G., Durr, A., Leavitt, B. R., Roos, R. A., et al. (2013). Predictors of phenotypic progression and disease onset in premanifest and early-stage Huntington's disease in the TRACK-HD study: Analysis of 36-month observational data. *Lancet Neurology, 12*(7), 637–649.

Tabrizi, S. J., Schobel, S., Gantman, E. C., Mansbach, A., Borowsky, B., Konstantinova, P., et al. (2022). A biological classification of Huntington's disease: The Integrated Staging System. *Lancet Neurology, 21*(7), 632–644.

Tamir, A., Whittier, J., & Korenyi, C. (1969). Huntington's chorea: A sex difference in psychopathological symptoms. *Diseases of the Nervous System, 30*(2), 103.

Teismann, H., Schubert, R., Reilmann, R., & Berger, K. (2022). Effects of age and sex on outcomes of the Q-Motor speeded finger tapping and grasping and lifting tests-findings from the population-based BiDirect Study. *Frontiers in Neurology, 13*, 965031.

Tyler, A., Harper, P. S., Davies, K., & Newcome, R. G. (1983). Family break-down and stress in Huntington's chorea. *Journal of Biosocial Science, 15*(2), 127–138.

US Food and Drugs Administration. (1977). *FDA guidance general considerations for the clinical evaluation of drugs*. Available from: https://www.fda.gov/media/71495/download

US Food and Drugs Administration. (1993). *FDA guidance study and evaluation of gender differences in the clinical evaluation of drugs*. Available from: https://www.fda.gov/media/75648/download

van der Plas, E., Langbehn, D. R., Conrad, A. L., Koscik, T. R., Tereshchenko, A., Epping, E. A., et al. (2019). Abnormal brain development in child and adolescent carriers of mutant huntingtin. *Neurology, 93*(10), e1021–e1e30.

van Walsem, M. R., Frich, J. C., Gomez Castaneda, M., Howe, E. I., Pihlstrom, L., Andelic, N., et al. (2022). Health related quality of life, service utilization and costs for patients with Huntington's disease in Norway. *BMC Health Services Research, 22*(1), 1527.

Vegeto, E., Villa, A., Della Torre, S., Crippa, V., Rusmini, P., Cristofani, R., et al. (2020). The role of sex and sex hormones in neurodegenerative diseases. *Endocrine Reviews, 41*(2), 273–319.

Volpi, E., Terenzi, F., Bagnoli, S., Latorraca, S., Nacmias, B., Sorbi, S., et al. (2021). Late-onset Huntington disease: An Italian cohort. *Journal of Clinical Neuroscience, 86*, 58–63.

Weydt, P., Soyal, S. M., Landwehrmeyer, G. B., Patsch, W., & European Huntington Disease N. (2014). A single nucleotide polymorphism in the coding region of PGC-1alpha is a male-specific modifier of Huntington disease age-at-onset in a large European cohort. *BMC Neurology, 14*, 1.

Williams, C. M., Peyre, H., Toro, R., & Ramus, F. (2021a). Neuroanatomical norms in the UK Biobank: The impact of allometric scaling, sex, and age. *Human Brain Mapping, 42*(14), 4623–4642.

Williams, C. M., Peyre, H., Toro, R., & Ramus, F. (2021b). Sex differences in the brain are not reduced to differences in body size. *Neuroscience and Biobehavioral Reviews, 130*, 509–511.

Yu-Taeger, L., Novati, A., Weber, J. J., Singer-Mikosch, E., Pabst, A. S., Cheng, F., et al. (2022). Evidences for mutant Huntingtin inducing musculoskeletal and brain growth impairments via disturbing testosterone biosynthesis in male Huntington disease animals. *Cell, 11*(23).

Zielonka, D., & Stawinska-Witoszynska, B. (2020). Gender differences in non-sex linked disorders: Insights from Huntington's disease. *Frontiers in Neurology, 11*, 571.

Zielonka, D., Niezgoda, A., Olejniczak, M., Krzyzosiak, W., Marcinkowski, J., & Kozubski, W. (2008). Gender differences in the CAG repeats and clinical picture correlations in Huntington's disease. *Ceskoslovenská Neurologie a Neurochirurgie, 71*, 688–694.

Zielonka, D., Marinus, J., Roos, R. A., De Michele, G., Di Donato, S., Putter, H., et al. (2013). The influence of gender on phenotype and disease progression in patients with Huntington's disease. *Parkinsonism & Related Disorders, 19*(2), 192–197.

Zielonka, D., Ren, M., De Michele, G., Roos, R. A. C., Squitieri, F., Bentivoglio, A. R., et al. (2018). The contribution of gender differences in motor, behavioral and cognitive features to functional capacity, independence and quality of life in patients with Huntington's disease. *Parkinsonism & Related Disorders, 49*, 42–47.

Index

E. A. Thomas, G. M. Parkin (eds.), *Biomarkers for Huntington's Disease*, Contemporary Clinical Neuroscience, https://doi.org/10.1007/978-3-031-32815-2

The manufacturer's authorised representative in the EU is Springer Nature Customer Service Centre GmbH, Europaplatz 3, 69115 Heidelberg, Germany. If you have any concerns regarding our products, please contact ProductSafety@springernature.com

Printed and bound by CPI Group (UK) Ltd, Croydon, CR0 4YY
29/09/2026
02224604-0001